Abstract Algebra

An Interactive Approach

Second Edition

TEXTBOOKS in MATHEMATICS

Series Editors: Al Boggess and Ken Rosen

PUBLISHED TITLES

ABSTRACT ALGEBRA: AN INQUIRY-BASED APPROACH
Jonathan K. Hodge, Steven Schlicker, and Ted Sundstrom

ADVANCED LINEAR ALGEBRA
Hugo Woerdeman

APPLIED ABSTRACT ALGEBRA WITH MAPLE™ AND MATLAB®, THIRD EDITION
Richard Klima, Neil Sigmon, and Ernest Stitzinger

APPLIED DIFFERENTIAL EQUATIONS: THE PRIMARY COURSE
Vladimir Dobrushkin

COMPUTATIONAL MATHEMATICS: MODELS, METHODS, AND ANALYSIS WITH MATLAB® AND MPI, SECOND EDITION
Robert E. White

DIFFERENTIAL EQUATIONS: THEORY, TECHNIQUE, AND PRACTICE, SECOND EDITION
Steven G. Krantz

DIFFERENTIAL EQUATIONS: THEORY, TECHNIQUE, AND PRACTICE WITH BOUNDARY VALUE PROBLEMS
Steven G. Krantz

DIFFERENTIAL EQUATIONS WITH MATLAB®: EXPLORATION, APPLICATIONS, AND THEORY
Mark A. McKibben and Micah D. Webster

ELEMENTARY NUMBER THEORY
James S. Kraft and Lawrence C. Washington

EXPLORING LINEAR ALGEBRA: LABS AND PROJECTS WITH MATHEMATICA®
Crista Arangala

GRAPHS & DIGRAPHS, SIXTH EDITION
Gary Chartrand, Linda Lesniak, and Ping Zhang

INTRODUCTION TO ABSTRACT ALGEBRA, SECOND EDITION
Jonathan D. H. Smith

INTRODUCTION TO MATHEMATICAL PROOFS: A TRANSITION TO ADVANCED MATHEMATICS, SECOND EDITION
Charles E. Roberts, Jr.

PUBLISHED TITLES CONTINUED

INTRODUCTION TO NUMBER THEORY, SECOND EDITION
Marty Erickson, Anthony Vazzana, and David Garth

LINEAR ALGEBRA, GEOMETRY AND TRANSFORMATION
Bruce Solomon

MATHEMATICAL MODELLING WITH CASE STUDIES: USING MAPLE™ AND MATLAB®, THIRD EDITION
B. Barnes and G. R. Fulford

MATHEMATICS IN GAMES, SPORTS, AND GAMBLING–THE GAMES PEOPLE PLAY, SECOND EDITION
Ronald J. Gould

THE MATHEMATICS OF GAMES: AN INTRODUCTION TO PROBABILITY
David G. Taylor

MEASURE THEORY AND FINE PROPERTIES OF FUNCTIONS, REVISED EDITION
Lawrence C. Evans and Ronald F. Gariepy

NUMERICAL ANALYSIS FOR ENGINEERS: METHODS AND APPLICATIONS, SECOND EDITION
Bilal Ayyub and Richard H. McCuen

ORDINARY DIFFERENTIAL EQUATIONS: AN INTRODUCTION TO THE FUNDAMENTALS
Kenneth B. Howell

RISK ANALYSIS IN ENGINEERING AND ECONOMICS, SECOND EDITION
Bilal M. Ayyub

TRANSFORMATIONAL PLANE GEOMETRY
Ronald N. Umble and Zhigang Han

TEXTBOOKS in MATHEMATICS

Abstract Algebra

An Interactive Approach

Second Edition

William Paulsen

Arkansas State University
Jonesboro, Arkansas, USA

CRC Press is an imprint of the
Taylor & Francis Group an **informa** business
A CHAPMAN & HALL BOOK

CRC Press
Taylor & Francis Group
6000 Broken Sound Parkway NW, Suite 300
Boca Raton, FL 33487-2742

International Standard Book Number-13: 978-1-4987-1976-6 (Hardback)

Library of Congress Cataloging-in-Publication Data

Names: Paulsen, William.
Title: Abstract algebra : an interactive approach / William Paulsen.
Description: 2nd edition. | Boca Raton : Taylor & Francis, 2016. | Series: Textbooks in mathematics ; 40 | "A CRC title." | Includes bibliographical references and index.
Identifiers: LCCN 2015037900 | ISBN 9781498719766 (alk. paper)
Subjects: LCSH: Algebra, Abstract--Textbooks.
Classification: LCC QA162 .P38 2016 | DDC 512/.02--dc23
LC record available at http://lccn.loc.gov/2015037900

Visit the Taylor & Francis Web site at
http://www.taylorandfrancis.com

and the CRC Press Web site at
http://www.crcpress.com

Contents

List of Figures

List of Tables

Preface

This textbook introduces a new approach to teaching an introductory course in abstract algebra. This text can be used for either an undergraduate-level course, or a graduate-level sequence. The undergraduate students would only cover the the basic material on groups and rings given in Chapters 0–4 and 9–12. A graduate-level sequence can be implemented by covering group theory in one semester (Chapters 1–8), and covering rings and fields the second semester (Chapters 9–15). (Graduate students should already know the contents of Chapter 0.) Alternatively, one semester could cover part of the group theory chapters and part of ring theory, while the second semester covers the remainder of the book.

This text covers many graduate-level topics that are not in most standard introductory abstract algebra courses. Some examples are semi-direct products (§6.4), polycyclic groups (§8.3), solving Rubik's Cube®-like puzzles (§8.4), and Wedderburn's theorem (§13.4). There are also some problem sequences that allow students to explore interesting topics in depth. For example, one sequence of problems outlines Fermat's two-square theorem, while another finds a principal ideal domain that is not a Euclidean domain. Hopefully, these extra titbits of information will satisfy the curiosity of the more advanced students.

What makes this book unique is the incorporation of technology into an abstract algebra course. Either *Mathematica*® or *Sage* can be used to give the students a hands-on experience with groups and rings. It is recommended that the instructor use at least one of these in the classroom to allow students to visualize the group and ring concepts. (*Sage* is totally free. See the section "*Sage* vs. *Mathematica*" for more information about both of these programs.) Every section includes many non-software exercises, so the students are not forced into using software. However, each section also has several interactive problems, so students can choose to use these programs to explore groups and rings. By doing these experiments, students can get a better grasp of the topic.

But in spite of the additional technology, this text is not short on rigor. There are still all of the classical proofs, although some of the harder proofs can be shortened with the added technology. For example, Abel's theorem is much easier to prove if we first assume that the groups A_5 and A_6 are simple, which *Mathematica* or *Sage* can verify in the classroom in a few seconds. In fact, the added technology allows students to study larger groups, such as some of the Chevalley groups.

This text has many tools that will aid the students. There is a symbols table, so if a student sees an unfamiliar symbol, he can look up the description in this table, and see where this symbol is first defined. The text is sprinkled with "Historical Diversions," one-page biographies of famous algebraic mathematicians and their contributions to abstract algebra. The answers to the odd-numbered problems are in the back, although the proofs are abbreviated. There is an extensive index that not only lists the relevant pages for a particular terminology, but also highlights the page where the term is first defined. A list of tables and figures allows students to find a multiplication table for a particular group or ring.

There have been many changes since the first edition. The biggest change is replacing GAP with *Sage*, which is very similar to *Mathematica*, so the text does not require as much software support. This allows for more non-computerized examples to be added. Also, there are more than twice the number of homework problems than in the first edition. The "Historical Diversions" have been added to reveal some of the tragic stories behind many of the mathematicians who contributed to abstract algebra. The preliminary chapter 0 has been added, along with discussion of new topics such as straight edge and compass constructions, and wreath products.

Acknowledgments

I am very grateful to Alexander Hulpke from Colorado State University for developing the GAP package "newrings.g" specifically for the first edition of my book. This package is currently incorporated into GAP, which in turn is included in *Sage*. Without this package, *Sage* would not be able to work with the examples that grace chapters 9–13. Other suggestions of his have proved to be invaluable.

I also must express my thanks to Shashi Kumar at the LaTeX help desk, who helped me with several different formatting issues throughout the text.

I also would like to express my appreciation to my wife Cynthia and my son Trevor for putting up with me during this past year, since this project ended up taking much more of my time than I first realized. They have been very patient with me and are looking forward to my finally being done.

About the Author

William Paulsen is a professor of mathematics at Arkansas State University. He has taught abstract algebra at both the undergraduate and graduate levels since 1997. He received his B.S. (summa cum laude), M.S., and Ph.D. degrees in mathematics at Washington University in St. Louis. He was on the winning team for the 45th William Lowell Putnam Mathematical Competition.

Dr. Paulsen has authored over 17 papers in abstract algebra and applied mathematics. Most of these papers make use of *Mathematica*®, including one which proves that Penrose tiles can be 3-colored, thus resolving a 30-year-old open problem posed by John H. Conway. He has also authored an applied mathematics textbook, "Asymptotic Analysis and Perturbation Theory," also published by CRC Press.

Dr. Paulsen has also programmed several new games and puzzles in Javascript and C++. One of these puzzles, Duelling Dimensions, has been syndicated through Knight Features. Other puzzles and games are available on the Internet.

Dr. Paulsen lives in Harrisburg, Arkansas with his wife Cynthia, his son Trevor, two pugs, and a dachshund.

Symbol Description

$\mathbb{Z}$	The set of integers	1
$\lfloor x \rfloor$	Greatest integer less than or equal to x	3
$\gcd(m, n)$	The greatest common divisor of m and n	7
$\operatorname{lcm}(m, n)$	The least common multiple of m and n	12
$f : A \to B$	The function f maps elements of A to elements of B.	13, 127
$f^{-1}(x)$	The inverse function of $f(x)$	17, 131
$x \textbf{ mod } n$	Modular arithmetic in base n	22
x^{-1}	The inverse of the element x	40
$(\text{mod } n)$	Modular equivalence in base n	43
$x \equiv y$	x and y are in the same equivalence class	43, 115
$x \cdot y$	Group multiplication	52
e	Identity element of a group	52
$x \in G$	x is a member of the set or group G	52
$\lvert G \rvert$	Number of elements in a set or group	15, 54
$A - \{a\}$	The set A with the element a removed	16
$f \circ g$	Composition of functions	16
$x * y$	Binary operation	18
Z_n	The group $\{0, 1, 2, \ldots, n-1\}$ using addition modulo n, or the ring of the same elements	52 319
Z_n^*	Numbers $< n$ coprime to n, with multiplication mod n	53
$\mathbb{Q}$	The group or field of rational numbers (fractions)	53
$\mathbb{Q}^*$	Non-zero rational numbers using multiplication	53
$\mathbb{R}$	The group or field of real numbers	53
$\mathbb{R}^*$	Non-zero real numbers using multiplication	128
x^n	x operated on itself n times	54
D_4	The group of symmetries of a square	42, 124
$\phi(n)$	Euler totient function	64
$\{\ldots \mid \ldots\}$	The set of elements ... such that ...	81
$H \cap K$	The intersection of H and K	76
$\bigcap_{H \in L} H$	The intersection of all sets in the collection L	77
$[S]$	Smallest subgroup containing the set S	77
$[x]$	Smallest subgroup containing the element x	78
$R_k(G)$	Number of solutions to $x^k = e$ in the group G	85
xH	A left coset of the subgroup H	91
Hx	A right coset of the subgroup H	91
$H \backslash G$	The collection of right cosets of H in the group G	91

G/H	The collection of left cosets of H in the group G, or the quotient group of G with respect to H	91 115
$G \approx M$	The group G is isomorphic to M	120
Q	The quaternion group	124
$\mathrm{Im}(f)$	The image (range) of the function f	131
$f^{-1}(H)$	The set of elements that map to an element of H	131
$\mathrm{Ker}(f)$	The kernel of the homomorphism f, which is $f^{-1}(e)$	132
$\binom{1\,2\,3\,4}{2\,3\,4\,1}$	Permutation notation	150
S_n	The symmetric group on n objects	151
$n!$	n factorial $= 1 \cdot 2 \cdot 3 \cdots n$	152
$(1\,2\,4\,6\,3)$	Cycle notation	157
$()$	The 0-cycle, the identity element of S_n	159
$\sigma(x)$	The signature function of the permutation x	161
A_n	The alternating group of permutations on n objects	162
$H \times K$	The direct product of the groups H and K	181
$P(n)$	The number of partitions of m	198
$\mathrm{Aut}(G)$	The group of automorphisms of the group G	203
$\mathrm{Inn}(G)$	The inner automorphisms of the group G	206
$\mathrm{Out}(G)$	The outer automorphisms of the group G	209
$N \rtimes_\phi H$	The semi-direct product of N with H through ϕ	213
D_n	The dihedral group with $2n$ elements	219
$G \mathrm{\ Wr\ } H$	Wreath product of G by H	223, 292
$Z(G)$	The center of the group G	226
$N_G(H)$	The normalizer of the subset H by the group G	231
$[H, K]$	The mutual commutator of the subgroups H and K	273
G'	The derived group of G, which is $[G, G]$	275
$\mathbb{H}$	The skew field of quaternions $a + bi + cj + dk$	303
$-x$	The additive inverse of x	304
nx	$x + x + x + \cdots + x$, n times	311
T_4	The smallest non-commutative ring	323
T_8	The smallest non-commutative unity ring	323
$\overline{x}$	The conjugate of x	309, 385
R/I	The quotient ring of the ring R by the ideal I	335
$X * Y$	The product of two cosets in R/I	335
$\langle S \rangle$	The smallest ideal containing the set S	339
$\langle a \rangle$	The smallest ideal containing the element a	339
PID	Principal ideal domain	340, 433
$n\mathbb{Z}$	Multiples of n (also written as $\langle n \rangle$)	340
kZ_{kn}	Multiples of k in the ring Z_{kn}	345
$\mathbb{C}$	The field of complex numbers	359, 380
$\mathbb{Z}[x]$	The polynomials with integer coefficients	362
$K[x]$	The polynomials with coefficients in the ring K	361
$\left(\frac{x}{y}\right)$	The equivalence class of ordered pairs containing (x, y)	373
$R[a]$	Smallest ring containing the ring R and the element a	398

Introduction

Most people use technology made possible by abstract algebra without realizing it. They go through checkout lines quickly via the UPC barcode, listen to music on a CD, and order items online through a secure website. Such actions are only possible due to error detection codes, error correction codes, and modern cryptography, which in turn rely on finite groups and finite fields.

Abstract algebra can also be used to prove that something is impossible. One of the classical problems from Greek geometry is to trisect an angle using only a straight edge and compass. For centuries, mathematicians have tried to produce such a construction to no avail. However, with a branch of abstract algebra called Galois theory, we can prove that such a trisection is impossible.

Another centuries-old problem proven impossible by Galois theory is solving a fifth-degree polynomial equation such as $x^5 + x - 1 = 0$ in terms of square roots, cube roots, and fifth roots. What makes the impossibility so surprising is that any fourth-degree polynomial equation *can* be solved in terms of roots. The change of behavior between the fourth-degree polynomials and the fifth-degree polynomials was proven by Galois in 1828, but he did not receive credit for his work until after his untimely death at the age of 20.

Because of the many applications to abstract algebra, there is an increased demand for students to learn this material. This book is devised to be a self-contained exposition of the structures of groups, rings, and fields that make up abstract algebra. It is designed to be used for a two-semester undergraduate course, but there is enough advanced material included to be used in a two-semester graduate sequence. It also is ideal for self-study, since it focuses on exploration of examples to find a general pattern, and then proves the patten persists. This is the interactive approach to learning.

Since undergraduate students are usually not accustomed to abstract thinking, there is a preliminary Chapter 0 that goes over the elementary properties of integers, functions, and real numbers. In the process, students are introduced to the technique of proofs, such as induction and *reductio ad absurdum*. Graduate students, on the other hand, would be able to begin with Chapter 1.

Although calculus is only needed for a handful of problems, it is recommended that students have had Calculus II, since a small amount of mathematical maturity is required. In fact, one of the goals of this textbook is to develop the mathematical maturity of the reader by introducing techniques for proofs, providing a bridge for higher-level mathematics courses.

One of the features unique to this textbook is the interactive approach. Often the book will focus on an example or two, and guide the reader into finding patterns in these examples. As the student looks into why these patterns appear, a proof is formulated. This interaction is made possible by the use of either *Sage* workbooks or *Mathematica* notebooks. There is both a workbook and a notebook corresponding to each chapter. These software packages allow students to experiment with different groups, rings, and fields, and allow the reader to visualize many of the important concepts. In order to use the bonus material, either *Mathematica* or *Sage* must be installed on a computer. There are several options for both of these programs, explained in detail in the appendix *Sage vs. Mathematica*. Since *Sage* is open source, and hence totally free, the examples in the text only refer to the *Sage* commands, but the corresponding *Mathematica* commands are usually similar, and are explained in the notebooks.

Another feature in this book are sequences of homework problems that together formulate new results not found in the text. For example, there is a sequence that outlines a proof of Fermat's two-square theorem, and another that finds an example of a PID that is not a Euclidean domain. These sequences are ideal for use as special projects for students taking the course with an honors option.

Dependency Diagram for the textbook

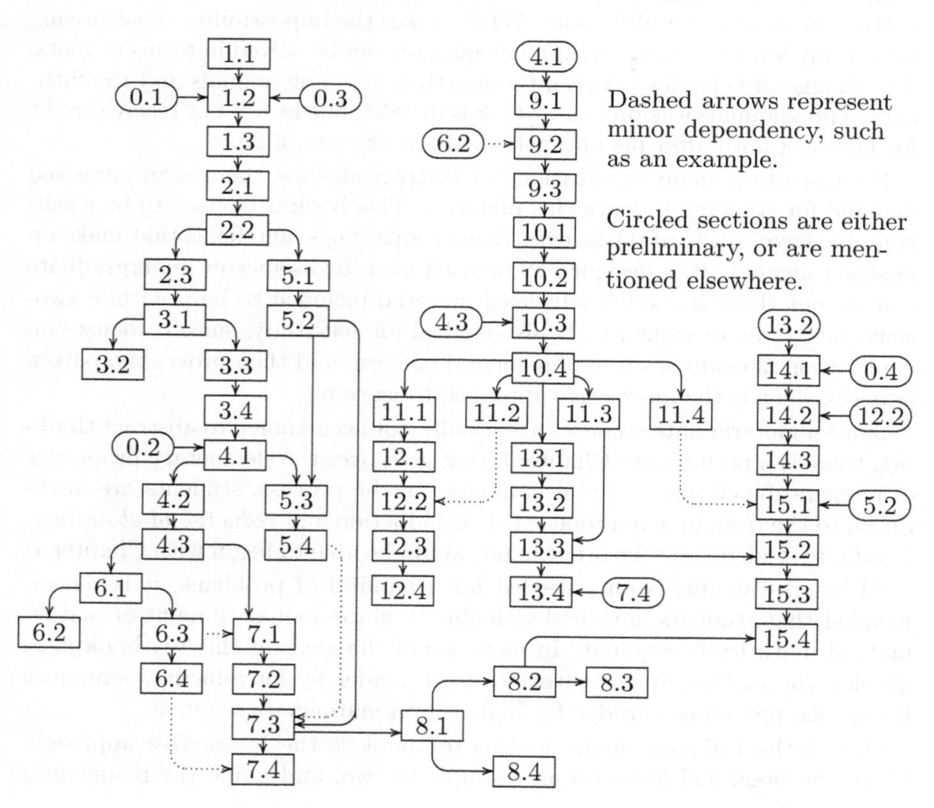

This textbook is also designed to work with a variety of different syllabi. A dependency diagram for the different sections is given above. To clarify this chart, here is a summary of each chapter with an explanation of some of the dependencies.

Chapter 0: Preliminaries. This chapter can be considered as a primer of the mathematics required to study abstract algebra. Undergraduate students should go over this material, although many sections will be familiar. The last section covers Cantor's diagonal theorem, which is actually not needed until an example in §14.1. Advanced students would have seen this material in other courses.

Chapter 1: Understanding the Group Concept. This chapter defines the group abstractly by first looking at several key examples, and observing the properties in common between the examples. The cyclic groups Z_n and the group of units Z_n^* are defined in terms of modular arithmetic. The non-abelian group D_3 is also introduced using the featured software. This chapter assumes the student is familar with integer factorization and modular arithmetic, which is covered in the preliminary chapter.

Chapter 2: The Structure within a Group. The basic properties of groups are developed in this chapter, including subgroups and generators. Also included in this section is a way to describe a group using generators and relations, giving us many more key examples of groups.

Chapter 3: Patterns within the Cosets of Groups. In this chapter, the notations of left and right cosets, normal groups, and quotient groups are developed. Section 3.2, which covers RSA encryption, is optional, but with the enhancement of the software packages it is a fun section to teach.

Chapter 4: Mappings between Groups. This chapter discusses group isomorphisms, and then generalizes the mappings to form group homomorphisms. This in turn leads to the three isomorphism theorems. Students are expected to understand abstract mappings, covered in the preliminary chapter.

Chapter 5: Permutation Groups. This chapter introduces another important class of groups, the symmetric groups S_n. The first two sections only require knowledge of §2.2, so these sections could in fact be taught earlier. But Cayley's theorem requires the concept of isomorphisms, requiring §4.1. The last section is optional, but introduces a notation for large subgroups of S_n, which comes in very handy for a number of examples.

Chapter 6: Building Larger Groups from Smaller Groups. As the name suggests, this chapter focuses on new ways to form groups, such as the direct product, the automorphism group, and the semi-direct product. Section 6.2, on the fundamental theorem of finite abelian groups, is not needed in the remaining sections on groups, but is referred to in a key exercise in §9.2 as we consider the additive group structure of a finite ring. The optional section on semi-direct products is more advanced, and would probably only be taught in a graduate-level course, even though it does provide some interesting examples.

Chapter 7: The Search for Normal Subgroups. This chapter explores the center of a group, the normalizer, and the conjugacy classes of a group. This leads to the class equation, which in turn leads to the three Sylow theorems. In this chapter we prove that the symmetric groups S_n from §5.2 are simple when $n \geq 5$, along with the group $L_2(3)$ with 168 elements, using the notation from §5.4. The last section on the Sylow theorems is optional, since it is only required for §13.4, which is also optional.

Chapter 8: Solvable and Insoluble Groups. This chapter looks at the subnormal series of a group, categorizing a group as either solvable or insoluble based on whether the composition factors are all cyclic. This is required for §15.4, which uses Galois theory to prove that fifth-degree polynomial equations cannot, in general, be solved in terms of radicals. The last two sections of Chapter 8 are optional, and rely heavily on *Sage*. Section 8.3 explains how *Sage* can do operations on a group much more efficiently if the group is entered into *Sage* using a polycylic subnormal series. The last section uses a special feature of *Sage*, which finds a way to express any element of a group in terms of a set of elements that generate the group. With this feature, we can solve Rubik's Cube^{TM}-like puzzles, giving an entertaining application of group theory.

Chapter 9: Introduction to Rings. This chapter introducing rings only requires §4.1, so one has the option of making a one-semester course covering half of the material on group theory, and half of the material on rings and fields. One exercise uses the fundamental theorem of finite abelian groups, but this can be avoided if that section was not covered.

Chapter 10: The Structure within Rings. This chapter focuses on the parallels between groups and rings, namely the similarities between normal groups and ideals. The chapter culminates with the first isomorphism theorem for rings, requiring only the counterpart in §4.3 from group theory.

Chapter 11: Integral Domains and Fields. This chapter appears in the dependency diagram horizontally instead of vertically, since each section is independent of the others. Nonetheless, these four sections are referred to in later chapters. The first section on polynomial rings is needed for Chapter 12, and the section on the field of quotients, §11.2, is also referred to in one of the corollaries of Chapter 12. Section 11.3 gives an overview of complex numbers, which is needed in §13.3 for cyclotomic polynomials. The last section on ordered commutative rings is optional, since this topic is not referred to elsewhere in the book. However, ring automorphisms are introduced in this section as a way to explain multiple ways of ordering certain rings, and these automorphisms are the key to Galois theory in Chapter 15.

Chapter 12: Unique Factorization. This chapter is dedicated to discovering which integral domains possess the unique factorization property.

The last two sections, on principal ideal domains and Euclidean domains, are not needed elsewhere in the book, but these topics are considered to be an important aspect of abstract algebra. On the other hand, §12.2 proves that a polynomial ring over a field is a unique factorization domain, and this result is needed to do any work in Galois theory.

Chapter 13: Finite Division Rings. Unlike most textbooks, this chapter covers finite fields before taking on infinite fields. Part of the reason is that finite fields are easy to visualize, but also finite fields can be completely classified. Section 13.3 takes a minor detour to discuss cyclotomic polynomials, a topic needed later in §15.2. The last section on Wedderburn's theorem is optional, but it gives a good example on how the class equation from §7.4 can be applied to finite fields.

Chapter 14: The Theory of Fields. The goal of this chapter is to explain the splitting fields of a polynomial, so it begins with a study of vector spaces, and then defines an extension field in terms of a vector space. A key example of an infinite dimensional vector space utilizes Cantor's diagonal theorem from §0.4. Other examples involve the finite fields of §13.2, but otherwise the chapter is self-contained.

Chapter 15: Galois Theory. The book comes to a climax with the discussion of Galois theory, along with its applications. For every polynomial there is a permutation group from §5.2 that describes the automorphism group hinted at in §11.4. By finding this permutation group for the cyclotomic polynomials of §13.3, we learn properties of the permutation group for the cases where the polynomial equation *is* solvable by radicals. Finally, using the composition series of §8.2, we prove that most fifth-degree polynomial equations cannot be solved in term of radicals. As a bonus, we also can prove the impossibility of two of the three famous construction problems of antiquity, trisecting an angle and duplicating the cube.

From the chapter summaries, it is clear that the textbook can be used to support a variety of different syllabi. For example, a junior-level one-semester course could consist of Chapters 0–5, only including the first isomorphism theorem in §4.3, then jumping to Chapters 9–10, finishing with selections from Chapter 11. On the other hand, there is enough material to cover a two-semester graduate-level sequence. Since there are both easy and challenging exercises, the textbook adapts well to both extremes, as well as a spectra of possible syllabi between these two.

Chapter 0

Preliminaries

This chapter gives the background material for studying abstract algebra. It introduces the concepts of sets and mappings, which are the foundations of all of modern mathematics. It also introduces some important strategies for writing proofs, such as induction and *reductio ad absurdum*. It is preferable to introduce this material here, since introducing this information at the point where it is needed interrupts the flow of the text.

Undergraduate students and those using the book for self-study are encouraged to go through this chapter, since it introduces concepts and notation that are used throughout the book. However, for most graduate students this material will be familiar, so such students could skip ahead to Chapter 1, referring back to this chapter whenever necessary.

0.1 Integer Factorization

Even in prehistoric times, there is evidence that societies developed a terminology for the counting numbers 1, 2, 3, etc. In fact, the Ishango bone suggests that prime numbers were contemplated as early as twenty thousand years ago. It is known that the early Egyptians understood prime numbers, but the Greeks of the fifth century B.C. get the credit for being the first to explore prime numbers for their own sake.

In this section we will explore the basic properties of integers stemming from the prime factorizations. We will prove an important theorem known to the Greeks, that all positive integers can be uniquely factored into prime numbers. In the process, we will learn some important techniques for proofs, which will be used throughout the book.

We begin by denoting the set of all integers,

$$\{\ldots, -3, -2, -1, 0, 1, 2, 3, \ldots\}$$

by the stylized letter $\mathbb{Z}$. This notation comes from the German word for number, *Zahl.* Many of the properties of factorizations refer only to positive integers, which are denoted $\mathbb{Z}^+$. Thus, we can write $n \in \mathbb{Z}^+$ to say that n is a positive integer.

We begin by defining a divisor of a number.

DEFINITION 0.1 We say that an integer a is a *divisor* of an integer b, denoted by $a|b$, if there is some integer c such that $b = ac$. Other ways of saying this is that a *divides* b, or a is a *factor* of b, or b is a *multiple* of a.

Example 0.1

Find the divisors of 30.

SOLUTION: Note that the definition allows for both negative and positive integers. Clearly if $30 = ac$ for integers a and c, $|a| \leq 30$. With a little trial and error, we find the divisors to be

$$\pm 1, \pm 2, \pm 3, \pm 5, \pm 6, \pm 10, \pm 15, \text{ and } \pm 30.$$

□

We can extend the idea of integer divisors to that of finding the quotient q and remainder r of integer division.

THEOREM 0.1: The Division Algorithm

Given any $x \in \mathbb{Z}$, and any $y \in \mathbb{Z}^+$, there are unique integers q and r such that

$$x = qy + r \qquad \textit{and} \qquad 0 \leq r < y.$$

PROOF: Since $y > 0$, we can consider the rational number x/y. Let q be the largest integer that is less than or equal to x/y. That is, we will pick the integer q so that

$$q \leq \frac{x}{y} < q + 1.$$

Multiplying by y, we have

$$yq \leq x < yq + y.$$

If we let $r = x - yq$, we have $0 \leq r < y$, and also $x = yq + r$, so we have found integers q and r that satisfy the required properties.

In order to show that q and r are unique, let us suppose that $\overline{q}$ and $\overline{r}$ are two other integers that satisfy the required conditions. Then $qy + r = \overline{q}y + \overline{r}$, so

$$(q - \overline{q})y = \overline{r} - r.$$

Since both r and $\overline{r}$ are between 0 and $y - 1$, the right-hand side is less than y in absolute value. But the left-hand side is at least y in absolute value unless $q = \overline{q}$. This in turn will force $r = \overline{r}$, so we see that the solution is unique. □

This is a constructive proof, since it gives an algorithm for finding q and r. This proof also demonstrates how to prove that a solution is *unique*. We assume there is another solution, and prove that the two solutions are in fact the same.

Example 0.2
Find integers q and r such that $849 = 31q + r$, with $0 \leq r < 31$.
SOLUTION: We can use *Sage* as a calculator. To find the numerical approximation of 849/31, enter

```
N(849/31)
  27.3870967741935
```

Note that the function **N( )** gives the numerical approximation. The largest integer less than this value is $q = 27$. Then we can compute r to be

```
849 - 27*31
  12
```

∎

The notation for finding the greatest integer function used in this algorithm is

$$\lfloor x \rfloor = \text{ the greatest integer less than or equal to } x.$$

Example 0.3
Find integers q and r such that $-925 = 28q + r$.
SOLUTION: Note that $-925/28 \approx -33.0357142857143$. But to find an integer less than this, we round *down*, so in the case of a negative number, it will increase in magnitude. Thus, $q = -34$, and $r = -925 - (-34)28 = 27$. ∎

We define a *prime* as an integer that has only two positive factors: 1 and itself. This definition actually allows negative numbers, such as -5, to be prime. Although this may seem to be a non-standard definition, it agrees with the generalized definition of primes defined in Chapters 10 and 12. The numbers 1 and -1 are not considered to be prime. The goal of this section is to prove that any integer greater than 1 can be uniquely factored into a product of positive primes.

We will begin by proving that every large number has at least one prime factor. This requires an assumption about the set of positive numbers, known as the *well-ordering axiom.*

The Well-Ordering Axiom:
Every non-empty subset of $\mathbb{Z}^+$ contains a smallest element.

The reason why this is considered to be an axiom is that it cannot be proven using only arithmetic operations. (Note that this statement is *not* true for rational numbers, which have the same arithmetic operations.) So this self-evident statement is assumed to be true, and is used to prove other properties of the integers.

LEMMA 0.1
Every number greater than 1 has a prime factor.

PROOF: Suppose that some number greater than 1 does not have a prime factor. Then we consider the set of all integers greater than 1 that do not have a prime factor, and using the well-ordering axiom, we find the smallest such number, called n. Then n is not prime, otherwise n would have a prime factor. Then by definition, n must have a positive divisor besides 1 and n, say m. Since $1 < m < n$, and n was the smallest number greater than 1 without a prime factor, m must have a prime factor, say p. Then p is also a prime factor of n, so we have a contradiction. Therefore, every number greater than 1 has a prime factor. □

Not only does the proof of Lemma 0.1 demonstrate how the well-ordering axiom is used, it also introduces an important strategy in proofs. Notice that to prove that every number greater than 1 had a prime factor, we assumed just the opposite. It was as if we admitted defeat from the very beginning! Yet from this we were able to reach a conclusion that was absurd—a number without a prime factor that did have a prime factor. This strategy is known as *reductio ad absurdum*, which is Latin for "reduce to the absurd." We assume what we are trying to prove is actually false, and proceed logically until we reach a contradiction. The only explanation would be that the assumption was wrong, which proves the original statement.

The prime factors lead to an important question. Is there a largest prime number? The Greek mathematician Euclid answered this question using *reductio ad absurdum* in the third century B.C. [11, p. 183]

THEOREM 0.2: Euclid's Prime Number Theorem
There are an infinite number of primes.

PROOF: Suppose there are only a finite number of prime numbers. Label these prime numbers

$$p_1 = 2, \quad p_2 = 3, \quad p_3 = 5, \quad p_4 = 7, \quad \ldots, \quad p_n.$$

Now consider the number

$$m = (2 \cdot 3 \cdot 5 \cdot 7 \cdot 11 \cdot 13 \cdots p_n) + 1.$$

This number is odd, so it cannot be divisible by 2. Likewise, m is one more than a multiple of 3, so it is not divisible by 3. In this way we see that m is not divisible by any of the prime numbers. But this is ridiculous, since m must have a prime factor by Lemma 0.1. Thus, the original assumption that there is a largest prime number is false, so there are an infinite number of prime numbers. □

Historical Diversion
Euclid of Alexandria (c. 300 BC)

Euclid of Alexandria is known as the "Father of Geometry," because of one great work that he wrote, *The Elements*. Euclid lived during the time of Ptolemy I (323–283 B.C.). Alexandria was the intellectual hub of its day, not only with the Great Library but also the *Museum* (meaning seat of the muses), which was their equivalent to a university. Although little is known about the life of Euclid, we can infer from his writings that he was a brilliant mathematician, being able to compile all known mathematical knowledge into a sequence of small steps, each proposition building on the previous in a well-defined order.

Although the *Elements* is primarily a treatise on geometry, books VII, VIII, and IX deal with number theory. Euclid was particularly interested in primes and divisibility. He proved that there were an infinite number of primes, and proved what is known as Euclid's lemma, that if a prime divides the product of two numbers, it must divide at least one of those numbers. This lemma then leads the the fundamental theorem of arithmetic, which says that any number greater than 1 can be uniquely factored into a product of primes. Euclid also considered the greatest common divisor of two numbers, and gave a constructive algorithm for finding the GCD of two numbers.

Euclid also defined a number as perfect if it equals the sum of its divisors other than itself. He proved that if $2^p - 1$ is prime, then $2^{p-1}(2^p - 1)$ is perfect.

In book X Euclid worked with irrational numbers, or *incommensurables* proving that $\sqrt{2}$ is irrational. This result was known to the school of Pythagoras, but was a closely guarded secret. The distinction between rational numbers and real numbers will play a vital role in future mathematics.

Euclid would have been aware of the three great construction problems of antiquity: trisecting an angle, duplicating the cube, and squaring the circle. The first problem is to divide any angle into 3 equal parts. The duplication of the cube involved constructing a line segment $\sqrt[3]{2}$ times another line segment. Finally, squaring the circle required construction of a square with the same area as a given circle. Euclid's *Elements* laid down the ground rules for a valid straight edge and compass constructions. Previous "solutions" done over a century earlier violated these rules. Although these seem like geometry problems, they were only proven to be impossible using algebraic methods in the nineteenth century. The first two were proven to be impossible using Galois theory. The last construction was proven impossible by Lindemann in 1882 when he proved π is transcendental.

Image source: Wikimedia Commons

In order to prove that every integer greater than 1 has a unique prime factorization, we must first prove that every such number can be expressed as a product of primes. This is easiest to do using the principle of *mathematical induction.* This principle stems from the well-ordering axiom, and is a powerful tool for proving statements about integers.

THEOREM 0.3: Principle of Mathematical Induction

Let S be a set of integers containing a starting value a. Suppose that S has the property that the integer n will be in S whenever all integers between a and n are in S. Then S contains all integers greater than or equal to a.

PROOF: Suppose there is some integer greater than a that is not in S. Let T be the set of integers greater than a but not in S. Since T is non-empty, by the well-ordering axiom we can let n be the smallest member of T. Note that $n \neq a$, since a is in S. Also, all integers between a and n would have to be in S, lest there be a smaller value in T. But by the property of S, n would have to be in S, hence not in T. This contradiction shows that there is no integer greater than a that is not in S, which is equivalent to saying all integers greater than or equal to a are in S. □

To use the principle of induction, we first prove a statement is true for a starting point a. Then we *assume* that the statement is true for all integers $a \leq k < n$. (Often we will be able to get by with just the previous case $k = n - 1$.) This gives us extra leverage to prove the statement is true for n. Here is an example of this principle in action.

LEMMA 0.2

Every integer $n \geq 2$ can be written as a product of one or more positive primes.

PROOF: In this case, our starting point is 2, so let us prove that statement is true for $n = 2$. Since 2 is prime, we can consider 2 to be the product of one prime, so we are done.

Let us now assume the statement is true for all integers $2 \leq k < n$, and work to prove the statement is true for the case n. If n is prime, we have n as the product of one prime. If n is not prime, then we can express $n = ab$, where both a and b are between 1 and n. By our assumption, a and b can both be expressed as a product of positive primes, and so n can also be expressed as a product of primes. Thus, by mathematical induction, the statement is true for all $n \geq 2$. □

In order to prove that the prime factorization is *unique*, we will first have to develop the concept of the greatest common divisor.

DEFINITION 0.2 We define the *greatest common divisor* (*GCD*) of two numbers to be the largest integer that divides both of the numbers. If the greatest common divisor is 1, this means that there are no prime factors in common. We say the numbers are *coprime* in this case. We denote the greatest common divisor of x and y by $\gcd(x, y)$.

We can use *Sage*'s **gcd** function to quickly test whether two numbers are coprime without having to factor them.

```
gcd(138153809229555633320199029, 14573040781012789119612213)
  1
```

There is an important property of the greatest common divisor, given in the following theorem.

THEOREM 0.4: The Greatest Common Divisor Theorem
Given two non-zero integers x and y, the greatest common divisor of x and y is the smallest positive integer that can be expressed in the form

$$ux + vy$$

with u and v being integers.

PROOF: Let A denote the set of all positive numbers that can be expressed in the form $ux + vy$. Note that both $|x|$ and $|y|$ can be written in the form $ux + vy$, so by the well-ordering axiom we can consider the smallest positive number n in A. Note that $\gcd(x, y)$ is a factor of both x and y, so $\gcd(x, y)$ must be a factor of n.

By the division algorithm (Theorem 0.1), we can find q and r, with $0 \le r < n$, such that $x = qn + r$. Then

$$r = x - qn = x - q(ux + vy) = (1 - qu)x + (-v)y,$$

which is in the set A. If $r \neq 0$, then r would be a smaller positive number in A than n, which contradicts the way we chose n. Thus, $r = 0$, and $n|x$. By a similar reasoning, n is also a divisor of y. Thus, n is a common divisor of x and y, and since the $\gcd(x, y)$ is in turn a divisor of n, n must be equal to $\gcd(x, y)$. ∎

Unfortunately, this is a *non-constructive proof.* Although this theorem proves the existence of the integers u and v, it does not explain how to compute them. Fortunately, there is an algorithm, known as the *Euclidean Algorithm*, which does compute u and v.

We start by assuming that $x > y > 0$, since we can consider absolute values if x or y are negative. We then repeatedly use the division algorithm to find

q_i and r_i such that

$$\begin{aligned} x &= q_1 y + r_1, & 0 \le r_1 < y, \\ y &= q_2 r_1 + r_2, & 0 \le r_2 < r_1, \\ r_1 &= q_3 r_2 + r_3, & 0 \le r_3 < r_2, \\ r_2 &= q_4 r_3 + r_4, & 0 \le r_4 < r_3, \ldots \end{aligned}$$

Because the integer sequence $\{r_1, r_2, r_3, \ldots\}$ is decreasing, this will reach 0 in a finite number of steps, say $r_m = 0$. Then r_{m-1} will be $\gcd(x, y)$. We can find the values for u and v by solving the second-to-the-last equation for r_{m-1} in terms of the previous two remainders r_{m-2} and r_{m-3}, and then using the previous equations recursively to express r_{m-1} in terms of the previous remainders. This will eventually lead to r_{m-1} expressed in terms of x and y, which is what we want. It helps to put the remainders r_i in parenthesis, as well as x and y, since these numbers are treated as variables.

Example 0.4

Find integers u and v such that $144u + 100v = \gcd(144, 100)$.

SOLUTION: Using the division algorithm repeatedly, we have

$$\begin{aligned} (144) &= 1 \cdot (100) + (44) \\ (100) &= 2 \cdot (44) + (12) \\ (44) &= 3 \cdot (12) + (8) \\ (12) &= 1 \cdot (8) + (4) \\ (8) &= 2 \cdot (4) + (0). \end{aligned}$$

Thus, we see that $\gcd(144, 100) = 4$. Starting from the second-to-the-last equation, we have

$$\begin{aligned} (4) &= (12) - (8) \\ &= (12) - [(44) - 3 \cdot (12)] = 4 \cdot (12) - (44) \\ &= 4 \cdot [(100) - 2 \cdot (44)] - (44) = 4 \cdot (100) - 9 \cdot (44) \\ &= 4 \cdot (100) - 9 \cdot [(144) - (100)] = 13 \cdot (100) - 9 \cdot (144). \end{aligned}$$

Thus, we have $u = -9$ and $v = 13$. □

Computational Example 0.5

Use *Sage* to find the numbers u and v such that

$$138153809229555633320199029\, u + 14573040781012789119612213\, v = 1.$$

SOLUTION: The command **xgcd** not only finds the gcd of the numbers, but also the values of u and v.

```
xgcd(138153809229555633320199029, 14573040781012789119612213)
   (1, -3653212340535639877557154, 34632799588881923050263359)
```

So the gcd is 1, and also

$$u = -3653212340535639877557154 \quad \text{and}$$

$$v = 34632799588881923050263359.$$

Note that these values were computed very quickly using the algorithm. □

We can now start to prove some familiar properties of prime numbers.

LEMMA 0.3: Euclid's Lemma

If a prime p divides a product ab, then either $p|a$ or $p|b$.

PROOF: Suppose that p does not divide a, so that p and a are coprime. By the greatest common divisor theorem (0.4), there exist integers u and v such that $ua + vp = 1$. Then

$$uab + vpb = b.$$

Since p divides both terms on the left-hand side, we see that $p|b$. Thus, p must divide either a or b. □

This lemma quickly generalizes using the principle of induction.

LEMMA 0.4

If a prime p divides a product $a_1a_2a_3 \cdots a_n$, then p divides a_i for some i.

PROOF: We will use induction on n. The starting case $n = 2$ is covered by Euclid's Lemma (0.3). Let us suppose the theorem is true for the case $n - 1$. That is, if p divides $a_1a_2a_3 \cdots a_{n-1}$, then p divides a_i for some i. If we let $b = a_1a_2a_3 \cdots a_{n-1}$, then $a_1a_2a_3 \cdots a_n = ba_n$. By Euclid's Lemma (0.3), if p divides ba_n, then p divides either b or a_n. But if p divides b, then by induction p divides a_i for some $1 \le i \le n - 1$. So in either case, p divides a_i for some $1 \le i \le n$. □

With this lemma, we can finally prove that integer factorization is unique.

THEOREM 0.5: The Fundamental Theorem of Arithmetic

Every integer greater than 1 can be factored into a product of one or more positive primes. Furthermore, this factorization is unique up to the rearrangement of the factors.

PROOF: Lemma 0.2 shows that all integers greater than 1 can be expressed as a product of positive primes. So we only have to show uniqueness. That is, we must show that if

$$p_1p_2p_3\cdots p_n = q_1q_2q_3\cdots q_m,$$

where $p_1, p_2, \ldots p_n, q_1, q_2, \ldots q_m$ are all positive primes, then $n = m$, and the q_i are a rearrangement of the p_i. We will use induction on n, the number of prime factors in the first factorization.

If $n = 1$, then $p_1 = q_1q_2q_3\cdots q_m$, and since p_1 is prime and cannot have more than one factor, we must have $m = 1$, and so $p_1 = q_1$.

By Lemma 0.4, since $p_n | q_1q_2q_3\cdots q_m$, p_n must divide one of the q_i's. Since p_n and q_i are both positive primes, we find that $p_n = q_i$. By rearranging the remaining q's, we can write

$$p_1p_2p_3\cdots p_n = q_1q_2q_3\cdots q_{m-1}p_n.$$

Thus,

$$p_1p_2p_3\cdots p_{n-1} = q_1q_2q_3\cdots q_{m-1}.$$

By induction we can assume that the statement is true for the case $n-1$, and so $n-1 = m-1$, hence $n = m$, and the q_i are a rearrangement of the p_i. □

The *Sage* command for finding the prime factorization of an integer is

```
factor(420)
    2^2 * 3 * 5 * 7
```

Note that *Sage* puts the primes in increasing order, and repeated prime factors are expressed using exponents. This is known as the *standard form* of the factorization. As long as the integers are less than about 60 digits long, *Sage* should not have any trouble factoring them. However, integer factorization is a difficult problem even with modern technology. The amount of time required is proportional to the square root of the second largest prime in the factorization. [14, p. 133]

On the other hand, determining whether or not a number is prime can be done quickly in *Sage*, even if the number has over 200 digits!

```
is_prime(10^200 + 357)
    True
```

How can *Sage* know for certain that this number is prime when it cannot begin to test for all possible factors? The answer lies in abstract algebra. Using the properties we will discover in this book, it is possible to prove whether or not a number is prime without knowing the factorization. This in turn will have many applications in Internet security and cryptology.

Problems for §0.1

For Problems **1** through **9**: Find integers q and r that satisfy $x = qy + r$ with $0 \le r < y$.

1 $x = 815,\ y = 32$
2 $x = 627,\ y = 41$
3 $x = -415,\ y = 23$
4 $x = -634,\ y = 31$
5 $x = 4827,\ y = 29$
6 $x = 9376,\ y = 107$
7 $x = 35,\ y = 215$
8 $x = -39,\ y = 254$
9 $x = 0,\ y = 7$

10 Use mathematical induction to show that $1 + n < n^2$ for all integers $n \ge 2$.

11 Use mathematical induction to show that $2^n < n!$ for all integers $n \ge 4$. (Recall that $n! = 1 \cdot 2 \cdot 3 \cdots n$.)

12 Use mathematical induction to show that $n^2 + 3n + 4$ is a multiple of 2 for all $n \ge 1$.

13 Use mathematical induction to show that $n^3 + 2n$ is a multiple of 3 for all $n \ge 1$.

14 Use mathematical induction to show that $4^n - 1$ is a multiple of 3 for all $n \ge 1$.

15 Use mathematical induction to show that $6^n + 4$ is a multiple of 20 for all $n \ge 2$.

16 Use mathematical induction to show that x is a positive real number, then $(1 + x)^n \ge 1 + xn$ for all positive integers n.

17 Use mathematical induction to prove that for all positive integers n,

$$1 + 2 + 3 + \cdots + n = \frac{n(n+1)}{2}.$$

18 Use mathematical induction to prove that for all positive integers n,

$$1 + 3 + 5 + \cdots + (2n - 1) = n^2.$$

19 Use mathematical induction to prove that for all positive integers n,

$$1^2 + 2^2 + 3^2 + \cdots + n^2 = \frac{n(n+1)(2n+1)}{6}.$$

20 Use mathematical induction to prove that for all positive integers n,

$$1^3 + 2^3 + 3^3 + \cdots + n^3 = \frac{n^2(n+1)^2}{4}.$$

21 Use mathematical induction to prove that for all positive integers n,

$$1 \cdot 2 + 2 \cdot 3 + 3 \cdot 4 + \cdots + n(n+1) = \frac{n(n+1)(n+2)}{3}.$$

22 Use mathematical induction to prove that for all positive integers n,

$$\frac{1}{1 \cdot 2} + \frac{1}{2 \cdot 3} + \frac{1}{3 \cdot 4} + \cdots + \frac{1}{n(n+1)} = \frac{n}{n+1}.$$

For Problems **23** through **31**: Find integers u and v that satisfy $ux + vy = \gcd(x, y)$. Note that there could be more than one solution.

23 $x = 24,\ y = 42$
24 $x = 100,\ y = 36$
25 $x = 102,\ y = 66$
26 $x = 464,\ y = 560$
27 $x = 1999,\ y = 29$
28 $x = 465,\ y = 105$
29 $x = -602,\ y = 252$
30 $x = 487,\ y = -119$
31 $x = 0,\ y = 7$

32 Show that if d is a positive integer, then $\gcd(da, db) = d \cdot \gcd(a, b)$.

33 Define the *least common multiple* of two positive integers x and y, denoted by $\mathrm{lcm}(x, y)$, to be the smallest integer that is a multiple of both x and y. Prove that the least common multiple will exist, and that $\mathrm{lcm}(x, y) | x \cdot y$

34 Prove that $\mathrm{lcm}(x, y) = (x \cdot y)/\gcd(x, y)$. See Problem 33.

For Problems **35** through **40**: Find the prime factorizations of the following numbers, and put the factorization into standard form.

35 32000
36 4002
37 5700
38 6293
39 26411
40 51207

Interactive Problems

41 Use *Sage* to find integers u and v such that

$$876543212345678\, u + 123456787654321\, v = 1.$$

42 Use *Sage* to find integers u and v such that

$$98765432123456789\, u + 12345678987654321\, v = 1.$$

43 Use *Sage* to find the factorization of 987654321.

44 Use *Sage* to find the factorization of 12345678987654321.

45 Use *Sage* to find the factorization of 98765432123456789.

0.2 Functions

The concept of a function is central to virtually every branch of mathematics. There are in fact various ways to define a function, but the concept remains the same. Standard functions in calculus map real numbers to real numbers, but we want to consider a more abstract definition for which the input and output can come from any set. We will then use this definition to introduce the concepts of a one-to-one and onto mapping. This in turn will lead to an important tool for proofs: the pigeonhole principle.

DEFINITION 0.3 Let A and B be two non-empty sets. A *function*, or *mapping*, from A to B is a rule that assigns to every element of A exactly one element of B. The set A is called the *domain* of the function, and the set B is called the *target.* If a function f assigns to a the element b, we write $f(a) = b$, and say that b is the *image* of a under f.

We will use the notation $f : A \to B$ to indicate that f is a function from the set A to the set B. The *range* of f, or the *image* of f, is the set of all y such that $y = f(x)$ for some x in A. This set is denoted by either $f(A)$ or $\text{Im}(f)$, and is a subset of B.

Example 0.6
Let A be the set of integers from 0 to 99, and let B be the set of English letters from a to z. Let ϕ map each integer to the first letter of the English word for that number. For example, $\phi(4) = f$. Then the range of ϕ is the set

$$\{e, f, n, o, s, t, z\}.$$

□

There are often different ways to denote the same element of the set A, so we must be careful that the rule for the function does not depend on the way the element is expressed. Had we extended the last example to include 100, this could be called either "a hundred" or "one hundred." Another example of an ambiguous definition is if we assign to each rational number a/b the value $1/b$. But by this rule, $f(1/2) \neq f(2/4)$, even though $1/2 = 2/4$. In order to show that a function is *well-defined*, we must show that if $x_1 = x_2$, then $f(x_1) = f(x_2)$.

Example 0.7
Consider the function from the set of rational functions (denoted by $\mathbb{Q}$) to itself, given by

$$f\left(\frac{a}{b}\right) = \frac{\gcd(a, b)}{b}.$$

Show that this function is well-defined.

SOLUTION: We need to show that if $x_1 = x_2$, then $f(x_1) = f(x_2)$. That is, if we have two ways of expressing the rational function $a/b = c/d$, then we must show that

$$\frac{\gcd(a,b)}{b} = \frac{\gcd(c,d)}{d}.$$

This is equivalent to showing $d \cdot \gcd(a,b) = b \cdot \gcd(c,d)$. Using the result of Problem 32 from §0.1, this is equivalent to $\gcd(ad, bd) = \gcd(bc, bd)$. But since $a/b = c/d$, we have $ad = bc$, so this function is well-defined. ∎

Many functions possess special properties that we want to explore.

DEFINITION 0.4 We say that a function $f : A \to B$ is *injective*, or *one-to-one*, if the only way in which $f(x) = f(y)$ is if $x = y$.

The function in Example 0.7 is not one-to-one, since $f(1/3) = f(2/3)$. In order to prove that a function is one-to-one, we assume that $f(x) = f(y)$, and work to prove that $x = y$.

Example 0.8

Consider the function $f : \mathbb{Z} \to \mathbb{Z}$ defined by

$$f(x) = \begin{cases} x+3 & \text{if } x \text{ is even,} \\ 2x & \text{if } x \text{ is odd.} \end{cases}$$

Show that $f(x)$ is one-to-one.

SOLUTION: We assume that $f(x) = f(y)$, and work to show that $x = y$. Because of the way that $f(x)$ is defined, there are several cases to consider.

Case 1) Both x and y are even. Then since $f(x) = f(y)$, $x + 3 = y + 3$, which implies that $x = y$.

Case 2) Both x and y are odd. Then since $f(x) = f(y)$, $2x = 2y$, so again $x = y$.

Case 3) x is even, and y is odd. Then $f(x) = f(y)$ implies that $x + 3 = 2y$, or $x = 2y - 3$. But this implies that x is odd, and we started out assuming that x is even. Thus, this case can never happen.

Case 4) x is odd, and y is even. This is a mirror image of case 3, so we find that this case also can never happen.

Thus, we have shown in all cases for which $f(x)$ *could* equal $f(y)$, then $x = y$. Hence f is one-to-one. ∎

We can also ask whether the range and the target of a given function are the same set.

DEFINITION 0.5 We say that a function $f : A \to B$ is *surjective*, or *onto*, if for every $b \in B$ there is at least one $a \in A$ such that $f(a) = b$. If a function is both one-to-one and onto, it is called a *bijection.*

Example 0.9
Determine whether the function in Example 0.8 is onto.
SOLUTION: Listing the first few values of $f(x)$,

$$f(0) = 3, \quad f(1) = 2, \quad f(2) = 5, \quad f(3) = 6, \quad f(4) = 7, \quad f(5) = 10, \ldots,$$

it seems that $f(x)$ is never 4. Let us suppose that $f(x) = 4$ and reach a contradiction.
Case 1) x is even. Then $x + 3 = 4$, so $x = 1$. But this contradicts that x is even.
Case 2) x is odd. Then $2x = 4$, so $x = 2$. But this contradicts that x is odd.
Since all cases reach a contradiction, we see that $f(x) \neq 4$, and so the function is not onto. ∎

Note that one counterexample is sufficient to prove that the function is not onto. The standard strategy for proving that a function $f : A \to B$ *is* onto is to show that for an arbitrary $y \in B$, there is some kind of formula for an element $x \in A$ such that $f(x) = y$.

Example 0.10
Let $f : \mathbb{Q} \to \mathbb{Q}$ be defined by $f(x) = 3x + 5$. Show that f is onto.
SOLUTION: If $f(x) = y$, we can solve for x to get $x = (y - 5)/3$. Note that this is defined for all rational numbers y, and produces a rational number. Then $f((y - 5)/3) = y$ for any $y \in \mathbb{Q}$, so f is onto. ∎

Often our functions will be defined on finite sets. In these cases, it is easy to determine whether or not a function is onto if we have already proven that it is one-to-one.

DEFINITION 0.6 For a finite set A, we denote the number of elements in the set by $|A|$. If A is infinite, we write $|A| = \infty$.

LEMMA 0.5
Let $f : A \to B$ be a function that is both one-to-one and onto, and suppose that A is a finite set. Then $|A| = |B|$.

PROOF: We will use induction on the size $n = |A|$. If A has only one element, a_1, then $f(a_1) = b_1$, and $B = \{b_1\}$. Let us suppose that the statement is true for $n - 1$.

If $A = \{a_1, a_2, a_3, \ldots a_n\}$, then $f(a_n) = b$ for some $b \in B$. If we let $\overline{A} = A - \{a_n\}$, that is, we remove the element a_n from the set A, and $\overline{B} = B - \{b\}$, then we can define the function $\overline{f} : \overline{A} \to \overline{B}$ by $\overline{f}(x) = f(x)$ for $x \in \overline{A}$. Since f is a bijection, so is $\overline{f}$, since no other element of A could map to b. By induction, we see that $|\overline{A}| = |\overline{B}|$, and so $|A| = |B|$. ∎

We can now prove an important principle that will help us to show whether a function is onto.

THEOREM 0.6: The Pigeonhole Principle
Let $f : A \to B$ be a function from a finite set A to a finite set B. If $|A| = |B|$ and f is one-to-one, then it is also onto.

PROOF: Let $\overline{B}$ be the range of f. Then the function $f : A \to \overline{B}$ would be both one-to-one and onto, so by Lemma 0.5 we have $|A| = |\overline{B}|$. Since we also have that $|A| = |B|$, then $B = \overline{B}$, so the function is onto. ∎

We will often need to apply two functions in succession, creating a new function.

DEFINITION 0.7 Let $f : B \to C$ and $g : A \to B$ be two functions. Then the mapping $(f \circ g) : A \to C$ is defined by

$$(f \circ g)(x) = f(g(x)) \qquad \text{for all } x \in A.$$

Note that in $f \circ g$, we apply the g function first, and then f. Some textbooks have $f \circ g = g(f(x))$, so care must be taken when referring to other texts.

Example 0.11
Let

$$f(x) = \begin{cases} x+3 & \text{if } x \text{ is even,} \\ 2x & \text{if } x \text{ is odd,} \end{cases} \qquad \text{and} \qquad g(x) = \begin{cases} 3x & \text{if } x \text{ is even,} \\ x-1 & \text{if } x \text{ is odd.} \end{cases}$$

Compute $f \circ g$ and $g \circ f$.
SOLUTION: For each computation, we need to consider the case x even and x odd separately. To find $(f \circ g)(x) = f(g(x))$:
Case 1) x is even. Then $g(x) = 3x$, which will also be even. Thus, $f(g(x)) = 3x + 3$.
Case 2) x is odd. Then $g(x) = x - 1$, which will be even, so $f(g(x)) = x + 2$. Thus,

$$f \circ g = \begin{cases} 3x+3 & \text{if } x \text{ is even,} \\ x+2 & \text{if } x \text{ is odd.} \end{cases}$$

To compute $(g \circ f)(x) = g(f(x))$, we also have to consider two cases.

Case 1) x is even. Then $f(x) = x+3$, which will be odd. So $g(f(x)) = x+2$.
Case 2) x is odd. Then $f(x) = 2x$, which will be even. So $g(f(x)) = 6x$.
Thus,

$$g \circ f = \begin{cases} x+2 & \text{if } x \text{ is even,} \\ 6x & \text{if } x \text{ is odd.} \end{cases}$$

□

Note that in this case, $f \circ g \neq g \circ f$. However, if we have three functions, with $f : C \to D$, $g : B \to C$, and $h : A \to B$, then $(f \circ g) \circ h = f \circ (g \circ h)$, since both of these expressions represent $f(g(h(x)))$.

If $f(x)$ is both one-to-one and onto, then we will be able to define the *inverse function* of f.

PROPOSITION 0.1
Let $f : A \to B$ be both one-to-one and onto. Then there exists a unique function $g : B \to A$ such that $g(f(x)) = x$ for all x in A, and $f(g(y)) = y$ for all $y \in B$.

PROOF: Because f is both one-to-one and onto, for every $y \in B$ there is a unique $x \in A$ such that $f(x) = y$. Thus, we can define $g(y)$ to be that value x such that $f(x) = y$. By the way $g(y)$ is defined, we see that $f(g(y)) = y$ for all $y \in B$. If we apply the function g to both sides of this equation, we have $g(f(g(y))) = g(y)$. Since every element $x \in A$ can be written as $g(y)$ for some $y \in B$, we can replace $g(y)$ with x to get $g(f(x)) = x$ for all $x \in A$.

To show that the function is unique, suppose there is another function $h(x) : B \to A$. Then

$$h(y) = h(f(g(y))) = (h \circ f)(g(y)) = g(y) \qquad \text{for all } y \in B.$$

Thus, $h = g$, showing that the function is unique. □

DEFINITION 0.8 The unique function in Proposition 0.1 is called the *inverse function* of $f(x)$, and is denoted by $f^{-1}(y)$.

Example 0.12
Consider the function $f : \mathbb{Z} \to \mathbb{Z}$ given by

$$f(x) = \begin{cases} x+3 & \text{if } x \text{ is even,} \\ x-1 & \text{if } x \text{ is odd.} \end{cases}$$

Show that this is both one-to-one and onto, and find $f^{-1}(y)$.
SOLUTION: If $f(x) = f(y)$, the only interesting case is if x is even, and y is odd. Then $x+3 = y-1$, or $y = x+4$, which would be even, not odd. Likewise, the case where x is odd and y is even leads to a contradiction. Thus, $x = y$, and f is one-to-one.

To show that f is onto, we must show that for every y, there is an x so that $f(x) = y$. We break this into two cases.
Case 1) y is even. Then $y + 1$ will be odd, so $f(y + 1) = (y + 1) - 1 = y$.
Case 2) y is odd. Then $y - 3$ is even, so $f(y - 3) = (y - 3) + 3 = y$.
In both cases, we found an x so that $f(x) = y$. In the process of determining that f is onto, we computed the inverse.

$$f^{-1}(y) = \begin{cases} y + 1 & \text{if } y \text{ is even}, \\ y - 3 & \text{if } y \text{ is odd}. \end{cases}$$

□

So far we have only considered functions with one input variable. But we could also consider functions with two input variables, $f(x, y)$. For simplicity we will only consider the cases where x and y come from the same set, which is also the target set.

DEFINITION 0.9 Let A be a non-empty set. A *binary operation* is a function that assigns to every x and y in A an element z in A.

Although we could denote a binary operation as $z = f(x, y)$, we will typically denote the operation by the infix notation $z = x * y$. The symbol $*$ does not always mean multiplication, but its definition depends on the binary operation. Often we will use a dot $(\cdot)$ instead of the asterisk, depending on the context.

Example 0.13
Define the binary operation $x * y$ defined on the set $\mathbb{Z}$ by

$$x * y = x + y + xy.$$

Show that $(x * y) * z = x * (y * z)$.
SOLUTION: Note that

$$\begin{aligned} (x * y) * z &= (x + y + xy) * z = x + y + xy + z + (x + y + xy)z \\ &= x + y + z + xy + xz + yz + xyz. \\ x * (y * z) &= x * (y + z + yz) = x + y + z + yz + x(y + z + yz) \\ &= x + y + z + xy + xz + yz + xyz. \end{aligned}$$

Thus, we see that $(x * y) * z = x * (y * z)$. □

DEFINITION 0.10 Let $*$ be a binary operation defined on a set A. We say that a subset B of A is *closed with respect to* $*$ if whenever both x and y are in B, then $x * y$ is in B.

Example 0.14
Let $*$ be the binary operation of Example 0.13. Show that the subset of odd integers is closed with respect to $*$.
SOLUTION: Let x and y be odd integers. Then we can express $x = 2m+1$ and $y = 2n+1$ for some integers m and n. Then

$$\begin{aligned} x * y &= (2m+1) * (2n+1) \\ &= (2m+1) + (2n+1) + (2m+1)(2n+1) \\ &= 2m+1+2n+1+4mn+2m+2n+1 \\ &= 4m+4n+4mn+3 \\ &= 2(2m+2n+2mn+1)+1. \end{aligned}$$

Thus, we see that $x * y$ is an odd integer, so the subset is closed. □

Problems for §0.2

1 Let ϕ be the mapping that sends every number from 0 to 99 to the *last* letter of the English word for that number. What would be the range of ϕ?

2 Show that the function $f : \mathbb{Q} \to \mathbb{Q}$ given by $f(a/b) = ab/(a^2 + b^2)$ is well-defined.

For Problems **3** through **8**: Part a) For the given $f : \mathbb{R} \to \mathbb{R}$, determine if the function is one-to-one. Part b) Determine if the function is onto. In both cases, prove your answer is correct.

3 $f(x) = |x|$

4 $f(x) = 3x + 5$

5 $f(x) = x^3$

6 $f(x) = x/3 - 2/5$

7 $x^2 - 4x$

8 $f(x) = 2x + |x|$

For Problems **9** through **14**: Part a) For the given $f : \mathbb{Z} \to \mathbb{Z}$, determine if the function is one-to-one. Part b) Determine if the function is onto. In both cases, prove your answer is correct.

9 $f(x) = \begin{cases} 2x+1 & \text{if } x \text{ is even,} \\ 2x & \text{if } x \text{ is odd.} \end{cases}$

10 $f(x) = \begin{cases} x-1 & \text{if } x \text{ is even,} \\ (x+1)/2 & \text{if } x \text{ is odd.} \end{cases}$

11 $f(x) = \begin{cases} x+1 & \text{if } x \text{ is even} \\ 2x & \text{if } x \text{ is odd.} \end{cases}$

12 $f(x) = \begin{cases} 2x+4 & \text{if } x \text{ is even,} \\ x-2 & \text{if } x \text{ is odd.} \end{cases}$

13 $f(x) = \begin{cases} (x+2)/2 & \text{if } x \text{ is even,} \\ (x-1)/2 & \text{if } x \text{ is odd.} \end{cases}$

14 $f(x) = \begin{cases} 3x & \text{if } x \text{ is even,} \\ 5x-1 & \text{if } x \text{ is odd.} \end{cases}$

15 Show that the function $f : \mathbb{Z} \to \mathbb{Z}$ given by $f(x) = 2x^2 + x$ is one-to-one.
Hint: Use the quadratic equation to solve $2x^2 + x = c$, and show that the two solutions cannot both be integers.

16 Let $f : A \to B$ be a function from a finite set A to a finite set B. If $|B| < |A|$, use Lemma 0.5 to show that f cannot be one-to-one.

17 Let $f : A \to B$ be a function from a finite set A to a finite set B. If $|B| > |A|$, show that f cannot be onto.

18 Let $f : A \to B$ be a function from a finite set A to a finite set B. If $|B| = |A|$, and f is onto, use Problem 17 to show that f is also one-to-one. Note that Problems 16 through 18 are three alternative ways to state the pigeonhole principle.

19 Use Problem 16 to show that there are two people in London with exactly the same number of hairs on their head. (Since the average number of hairs is about 150,000, assume no one can have more than 1,000,000 hairs.)

For Problems **20** through **25**: For the given $f : \mathbb{Z} \to \mathbb{Z}$ and $g : \mathbb{Z} \to \mathbb{Z}$, determine $(f \circ g)(x)$.

20 $f(x) = x^2 - 1$ $\qquad g(x) = x^2 + 1$

21 $f(x) = x^2$ $\qquad g(x) = x - |x|$

22 $f(x) = x^3 + 3x^2$ $\qquad g(x) = x - 1$

23 $f(x) = \begin{cases} 2x+5 & \text{if } x \text{ is even,} \\ x+2 & \text{if } x \text{ is odd.} \end{cases}$ $\qquad g(x) = \begin{cases} 2x+1 & \text{if } x \text{ is even,} \\ x-1 & \text{if } x \text{ is odd.} \end{cases}$

24 $f(x) = \begin{cases} 3x+2 & \text{if } x \text{ is even,} \\ x+|x| & \text{if } x \text{ is odd.} \end{cases}$ $\qquad g(x) = \begin{cases} x+4 & \text{if } x \text{ is even,} \\ 2x & \text{if } x \text{ is odd.} \end{cases}$

25 $f(x) = \begin{cases} x+3 & \text{if } x \text{ is even,} \\ (x-1)/2 & \text{if } x \text{ is odd.} \end{cases}$ $\qquad g(x) = \begin{cases} 2x-1 & \text{if } x \text{ is even,} \\ x+4 & \text{if } x \text{ is odd.} \end{cases}$

26 Let $f : B \to C$ and $g : A \to B$ both be one-to-one functions. Show that $f \circ g : A \to C$ is one-to-one.

27 Let $f : B \to C$ and $g : A \to B$ both be onto functions. Show that $f \circ g : A \to C$ is onto.

28 Let $f : B \to C$ and $g : A \to B$ be functions, and suppose that f is *not* onto. Show that $f \circ g : A \to C$ is not onto.

29 Let $f : B \to C$ and $g : A \to B$ be functions, and suppose that g is *not* one-to-one. Show that $f \circ g : A \to C$ is not one-to-one.

30 Show that the function $f : \mathbb{Z} \to \mathbb{Z}$, $f(x) = \begin{cases} x+5 & \text{if } x \text{ is even,} \\ x-3 & \text{if } x \text{ is odd} \end{cases}$ is a bijection, and find $f^{-1}(x)$.

31 Show that the function $f : \mathbb{R} \to \mathbb{R}$, $f(x) = \begin{cases} x^2 & \text{if } x \geq 0, \\ x & \text{if } x < 0 \end{cases}$ is a bijection, and find $f^{-1}(x)$.

For Problems **32** through **35**: Determine if the binary operation defined on the set $\mathbb{R}$ satisfies the condition $a * (b * c) = (a * b) * c$. If so the operation is called *associative*.

32 $x * y = x + y - 1$
33 $x * y = 2x + y$
34 $x * y = x - y$
35 $x * y = 2 - x - y + xy$

For Problems **36** through **41**: Determine if the subset S is closed with respect to the binary operation.

36 $x * y = x - y$ $S =$ set of even integers.
37 $x * y = x - y$ $S =$ set of odd integers.
38 $x * y = xy$ $S =$ set of even integers.
39 $x * y = xy$ $S =$ set of odd integers.
40 $f * g = f \circ g$ $S =$ set of all polynomial functions.
41 $x * y = x/y$ $S =$ non-zero integers.

Interactive Problems

42 Consider the function

$$f(x) = 2\lfloor x \rfloor - x.$$

Sage uses the function **floor** to denote $\lfloor x \rfloor$. Thus, we can have *Sage* plot this function with the commands

```
f(x) = 2*floor(x) - x
plot(f(x), [x, 0, 5])
```

Judging by the graph, is this function one-to-one? Is it onto? (Ignore the vertical lines in the graph.)

43 Define a function $g(x)$ in *Sage* such that $f(g(x)) = x$ for all x, using the function from Problem 42. Note that the formula must work for both integers and non-integers. Is $g(f(x))$ always equal to x?

0.3 Modular Arithmetic

There is an important operation on the set of integers $\mathbb{Z}$ that we will use throughout the book, based on the division algorithm. It is an abstraction of a counting method often used in every day life. For example, using standard 12-hour time, if it is 7:00 now, what time will it be 8 hours from now? The answer is not 15:00, since clock time "wraps around" every 12 hours, so the correct answer is 3:00. This type of arithmetic that "wraps around" is called modular arithmetic.

We formally define modular arithmetic as follows.

DEFINITION 0.11 Let $x, y \in \mathbb{Z}$, with $y > 0$. We define the operator

$$x \textbf{ mod } y,$$

pronounced "x modulo y," to be the unique value r from the division algorithm, which selects q and $0 \leq r < y$ such that $x = qy + r$. The number y is called the *modulus.*

The **mod** operation is almost, but not quite, a binary operation on $\mathbb{Z}$, since it is not defined if $y = 0$. Since there is a difference of opinion as to how the operator should be defined for $y < 0$, we will only use the operator for $y > 0$.

Example 0.15
Compute 8348 **mod** 43.
SOLUTION: Since $8342 = 194 \cdot 43 + 6$, we see that 8348 **mod** $43 = 6$. ☐

Computational Example 0.16
Compute 743532645703453453463 **mod** 257275073624623.
SOLUTION: For numbers this large, we will use *Sage* to help. We use the **%** symbol for the **mod** operator.

```
743532645703453453463 % 257275073624623
    221951157869396
```

☐

Sometimes the modulo operation is very easy to compute. For any positive x, x **mod** 10 will be the last digit in the decimal representation of the number. There are other tricks for small values of y. See Problem 13.

A familiar property of standard arithmetic is that the last digit of the sum and product of two positive numbers x and y can be computed using only the last digits of x and y. This can be generalized in the following proposition.

PROPOSITION 0.2
If x, y, and n are integers with $n > 0$, then

$$(x + y) \textbf{ mod } n = ((x \textbf{ mod } n) + (y \textbf{ mod } n)) \textbf{ mod } n, \quad (0.1)$$

and

$$(xy) \textbf{ mod } n = ((x \textbf{ mod } n) \cdot (y \textbf{ mod } n)) \textbf{ mod } n. \quad (0.2)$$

PROOF: In both equations, the two sides are between 0 and $n - 1$, so it is sufficient to show that the difference of the two sides is a multiple of n. Let

$a = x \textbf{ mod } n$, $b = y \textbf{ mod } n$, $c = (x + y) \textbf{ mod } n$, and $d = (xy) \textbf{ mod } n$. Then there are integers q_1, q_2, q_3, and q_4 such that

$$x = q_1 n + a, \quad y = q_2 n + b, \quad x + y = q_3 n + c, \quad xy = q_4 n + d.$$

For equation 0.1, we note that

$$\begin{aligned} c - (a + b) &= (x + y - q_3 n) - ((x - q_1 n) + (y - q_2 n)) \\ &= q_1 n + q_2 n - q_3 n = (q_1 + q_2 - q_3)n. \end{aligned}$$

Thus, the two sides of equation 0.1 differ by a multiple of n. Likewise, for equation 0.2, we see

$$\begin{aligned} d - ab &= (xy - q_4 n) - (x - q_1 n) \cdot (y - q_2 n) \\ &= q_1 q_2 n^2 - y q_1 n - x q_2 n - q_4 n = (q_1 q_2 n - y q_1 - x q_2 - q_4)n. \end{aligned}$$

So again, the two sides of equation 0.2 differ by a multiple of n. ∎

We can use Proposition 0.2 to compute powers modulo n. Since raising a number to an integer power is equivalent to repeated multiplication, we see that

$$(x^y) \textbf{ mod } n = (x \textbf{ mod } n)^y \textbf{ mod } n.$$

Example 0.17

Compute 234^5 **mod** 29.

SOLUTION: Since 234 **mod** 29 = 2, the answer is the same as 2^5 **mod** 29, and 32 **mod** 29 = 3. ∎

WARNING: It is not true that

$$(x^y) \textbf{ mod } n = (x \textbf{ mod } n)^{(y \textbf{ mod } n)} \textbf{ mod } n.$$

That is, we cannot apply the modulus to an exponent. However, there is a trick for simplifying a power in the case that the exponent is large—using the binary representation of the exponent y. The procedure is best explained by an example.

Example 0.18

Compute 25^{35} **mod** 29.

SOLUTION: The number 25^{35} is 49 digits long, and the base is already smaller than the modulus, so there is no obvious way of simplifying the expression. By looking at the binary representation of 35, we find that 35 = 32 + 2 + 1. Thus,

$$25^{35} = 25^{32} \cdot 25^2 \cdot 25.$$

In order to compute 25^{32} **mod** 29, we can *square* the number 5 times.

$$\begin{aligned}
25^2 \textbf{ mod } 29 &= 625 \textbf{ mod } 29 = 16, \\
25^4 \textbf{ mod } 29 &= 16^2 \textbf{ mod } 29 = 256 \textbf{ mod } 29 = 24, \\
25^8 \textbf{ mod } 29 &= 24^2 \textbf{ mod } 29 = 576 \textbf{ mod } 29 = 25, \\
25^{16} \textbf{ mod } 29 &= 25^2 \textbf{ mod } 29 = 625 \textbf{ mod } 29 = 16, \\
25^{32} \textbf{ mod } 29 &= 16^2 \textbf{ mod } 29 = 256 \textbf{ mod } 29 = 24.
\end{aligned}$$

Finally, we see that

$$25^{35} \textbf{ mod } 29 = 25^{32} \cdot 25^2 \cdot 25^1 \textbf{ mod } 29 = 24 \cdot 16 \cdot 25 \textbf{ mod } 29 = 9600 \textbf{ mod } 29 = 1.$$

Note that we never had to deal with numbers more than 4 digits long. ∎

The *Sage* command **PowerMod(x, y, n)** uses this algorithm to compute x^y **mod** n.

Computational Example 0.19

Use *Sage* to find

$$743532645703453453463^{42364872163462467234} \textbf{ mod } 257275073624623326487 2.$$

SOLUTION:

```
PowerMod(743532645703453453463, 42364872163462467234,
2572750736246233264872)
    1270976212484154802393
```

Note that *Sage* was able to do this computation fast. We will see that the ability for computers to quickly compute large powers modulo n has applications in Internet security. ∎

There is another property of modular arithmetic involving coprime numbers that will be used often throughout the book, known to the ancient Chinese since before 240 C.E.

THEOREM 0.7: The Chinese Remainder Theorem

If x and y in $\mathbb{Z}^+$ are coprime, then given any a and b in $\mathbb{Z}$, there is a unique k in $\mathbb{Z}$ such that

$$0 \le k < xy,$$

$$k \textbf{ mod } x = a \textbf{ mod } x,$$

and

$$k \textbf{ mod } y = b \textbf{ mod } y.$$

PROOF: We will begin by showing that there cannot be more than one such number. Suppose we have two different numbers, k and m, which satisfy the above conditions. Then

$$(k - m) \textbf{ mod } x = 0 \qquad \text{and} \qquad (k - m) \textbf{ mod } y = 0.$$

Thus, $k - m$ must be a multiple of both x and y. But since x and y are coprime, the least common multiple of x and y is xy. (See Problem 34 from §0.1.) Thus, $k - m$ is a multiple of xy.

However, both k and m are less then xy. So the only way this is possible is for $k - m = 0$, which contradicts our assumption that k and m were distinct solutions.

To show that there is a solution, we first note that since x and y are coprime, by the greatest common divisor theorem (0.4), there are integers u and v such that $ux + vy = 1$. Then we can consider the number

$$k = (avy + bux) \textbf{ mod } (xy).$$

Clearly $0 \le k < xy$, so we only have to show that $k \textbf{ mod } x = a \textbf{ mod } x$ and $k \textbf{ mod } y = b \textbf{ mod } y$. Since $vy = 1 - ux$,

$$\begin{aligned} k \textbf{ mod } x &= (avy + bux) \textbf{ mod } x \\ &= (a(1 - ux) + bux) \textbf{ mod } x \\ &= (a + ux(b - a)) \textbf{ mod } x = a \textbf{ mod } x. \end{aligned}$$

Likewise, since $ux = 1 - vy$,

$$\begin{aligned} k \textbf{ mod } y &= (avy + bux) \textbf{ mod } y \\ &= (avy + b(1 - vy)) \textbf{ mod } y \\ &= (b + vy(a - b)) \textbf{ mod } y = b \textbf{ mod } y. \end{aligned}$$

□

This is a constructive proof, since it gives us a formula for finding the value of k.

Example 0.20

Find a non-negative number k less than 210 such that

$$k \textbf{ mod } 14 = 3, \qquad \text{and}$$
$$k \textbf{ mod } 15 = 7.$$

SOLUTION: Since 14 and 15 are coprime, we begin by finding u and v such that $14u + 15v = 1$. But there is the obvious solution

$$14(-1) + 15(1) = 1.$$

Then we compute k to be $avy + bux = 3 \cdot 15 + 7 \cdot (-14) = -53$. But since this is negative, we can add $14 \cdot 15$ to get another solution, 157. □

There is a *Sage* command **crt(a, b, x, y)** that finds k given the 2 sets $\{a, b\}$ and $\{x, y\}$.

Computational Example 0.21

Use *Sage* to find a number k such that

$$k \textbf{ mod } 771234712398742343 = 573457203572345239 \quad \text{and}$$
$$k \textbf{ mod } 642374682348623642 = 568134658235924534.$$

SOLUTION:

```
crt(573457203572345239, 568134658235924534,
771234712398742343, 642374682348623642)
    155720011750587503187230769057470234
```

We can verify that this solution is correct.

```
155720011750587503187230769057470234 %  7712347123987423437
    573457203572345239
155720011750587503187230769057470234 %  642374682348623642
    568134658235924534
```

□

The Chinese remainder theorem has many applications. One of these is in the distribution of classified information among two or more people in such a way that no one person can see the information. Each would receive one of the two (or more) modulus conditions, which is not enough information to determine the number k. Only when all of the pieces of the problem are assembled can k be determined, which can be decoded.

Another application is in solving linear congruence equations of the form

$$(ax) \textbf{ mod } n = b.$$

This can be solved by letting $k = ax$. Then

$$k \textbf{ mod } a = 0, \qquad \text{and}$$
$$k \textbf{ mod } n = b.$$

Since k is known, we can find x.

Example 0.22
Solve the linear congruence equation

$$12x \textbf{ mod } 19 = 3.$$

SOLUTION: We need to solve k **mod** $12 = 0$ and k **mod** $19 = 3$. Thus, we must first find u and v such that $12u + 19v = 1$. Using the Euclidean algorithm, we find that

$$8 \cdot 12 + (-5) \cdot 19 = 1.$$

Using these values of u and v, we have that

$$k = avy + bux = 0 \cdot (-5) \cdot 19 + 3 \cdot 8 \cdot 12 = 192.$$

Finally, $x = 12k$, so $x = 24$. Note that we can add or subtract multiples of 19 to get other solutions, so $x = 5$ also works. ∎

Problems for §0.3

For Problems **1** through **12**: Evaluate the following modular arithmetic problems.

1 $297 \textbf{ mod } 31$
2 $5643 \textbf{ mod } 127$
3 $953 \cdot 823 \textbf{ mod } 38$
4 $1432 \cdot 234 \textbf{ mod } 47$
5 $279^7 \textbf{ mod } 23$
6 $302^6 \textbf{ mod } 37$
7 $21^{49} \textbf{ mod } 31$
8 $33^{43} \textbf{ mod } 37$
9 $893^{57} \textbf{ mod } 23$
10 $1045^{29} \textbf{ mod } 47$
11 $8923^{31} \textbf{ mod } 103$
12 $5927^{61} \textbf{ mod } 113$

13 A trick for computing x **mod** 9 for any positive x is to add the digits of the number x. If this number is greater than 9, add the digits of the new number. Eventually the number will be between 1 and 9. If the result is 9, x **mod** $9 = 0$, otherwise x **mod** 9 is the final number produced. Prove that this method will always work.

For Problems **14** through **25**: Use the Chinese remainder theorem to find the smallest non-negative number that satisfies the system of modular equations.

14 $\begin{cases} k \textbf{ mod } 12 = 7, \\ k \textbf{ mod } 13 = 4. \end{cases}$

15 $\begin{cases} k \textbf{ mod } 17 = 4, \\ k \textbf{ mod } 11 = 8. \end{cases}$

16 $\begin{cases} k \textbf{ mod } 18 = 7, \\ k \textbf{ mod } 13 = 2. \end{cases}$

17 $\begin{cases} k \textbf{ mod } 23 = 5, \\ k \textbf{ mod } 12 = 7. \end{cases}$

18 $\begin{cases} k \textbf{ mod } 21 = 10, \\ k \textbf{ mod } 16 = 9. \end{cases}$

19 $\begin{cases} k \textbf{ mod } 34 = 13, \\ k \textbf{ mod } 27 = 10. \end{cases}$

20 $\begin{cases} k \textbf{ mod } 51 = 19, \\ k \textbf{ mod } 49 = 28. \end{cases}$

21 $\begin{cases} k \textbf{ mod } 61 = 37, \\ k \textbf{ mod } 73 = 58. \end{cases}$

22 $\begin{cases} k \textbf{ mod } 83 = 48, \\ k \textbf{ mod } 79 = 62. \end{cases}$

23 $\begin{cases} k \textbf{ mod } 103 = 78, \\ k \textbf{ mod } 97 = 48. \end{cases}$

24 $\begin{cases} k \textbf{ mod } 107 = 23, \\ k \textbf{ mod } 128 = 35. \end{cases}$

25 $\begin{cases} k \textbf{ mod } 113 = 47, \\ k \textbf{ mod } 142 = 84. \end{cases}$

26 Let u, v, and w be three positive integers that are *mutually* coprime. That is, each is coprime to the other two. Given any x, y, and z in $\mathbb{Z}$, prove that there is a unique number k such that

$$0 \le k < u \cdot v \cdot w,$$

$$k \equiv x \pmod{u},$$

$$k \equiv y \pmod{v},$$

and

$$k \equiv z \pmod{w}.$$

Hint: Use the Chinese remainder theorem (0.7).

For Problems **27** through **38**: Solve the following linear congruence equations.

27	$8x$ **mod** $11 = 7$	**31**	$7x$ **mod** $31 = 10$	**35**	$32x + 20$ **mod** $51 = 17$
28	$4x$ **mod** $13 = 9$	**32**	$12x$ **mod** $37 = 13$	**36**	$16x + 37$ **mod** $61 = 29$
29	$7x$ **mod** $18 = 11$	**33**	$18x$ **mod** $29 = 7$	**37**	$14x + 71$ **mod** $83 = 48$
30	$9x$ **mod** $23 = 13$	**34**	$27x$ **mod** $41 = 8$	**38**	$23x + 49$ **mod** $91 = 39$

Interactive Problems

39 Use *Sage*'s **PowerMod** function to compute

$$2351557923579239475 29^{7528934797293539 0234} \ \textbf{mod}\ 4623452735792375925234.$$

40 Use *Sage*'s **PowerMod** function to compute

$$938457289347272352 3452^{2452352345234521 6644} \ \textbf{mod}\ 8376258362352836587697.$$

41 Use *Sage*'s **crt** function to find the solution to the system

$$k \ \textbf{mod}\ 9243798374502516137 = 2375216462433536 26 \quad \text{and}$$
$$k \ \textbf{mod}\ 1978654573572351516 = 2632567324568422 3.$$

42 Use *Sage*'s **crt** function to find the solution to the system

$$k \ \textbf{mod}\ 8675612376265160933543 = 1523523523462547535 48, \quad \text{and}$$
$$k \ \textbf{mod}\ 6226345262345235236201 = 5263523462345735234 64.$$

43 Use *Sage* to solve the linear congruence equation

$$7289475362034522153x \ \textbf{mod}\ 915156238625161124 = 210982524590982446.$$

44 Use *Sage* to solve the linear congruence equation

$$9357298518686215025x \ \textbf{mod}\ 19651562734986125 12 = 1871551633523628256.$$

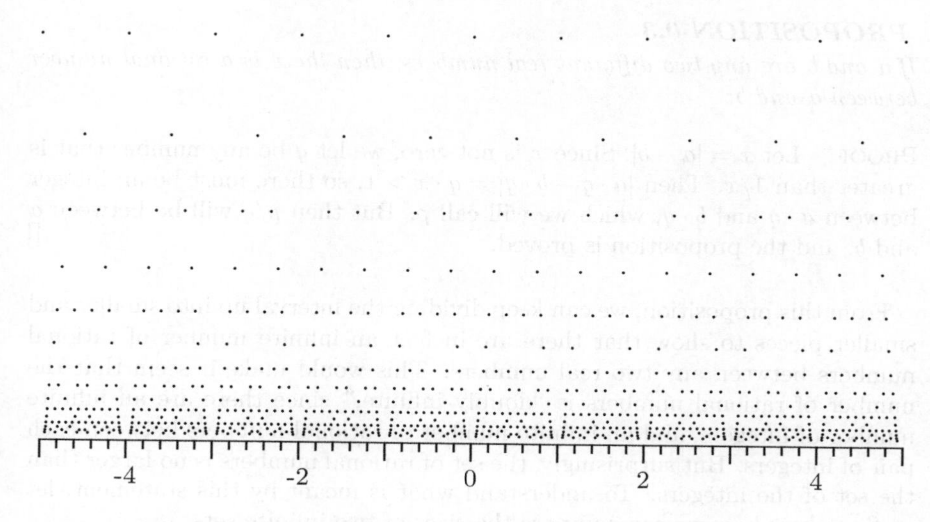

FIGURE 0.1: Plot depicting the rational numbers

0.4 Rational and Real Numbers

In this section, we will explore some properties of rational numbers and real numbers. In the process we will find an Earth-shattering result: The set of real numbers is "more infinite" than the set of rational numbers. When Georg Cantor first proved this theorem, it was met with fierce opposition (see the Historical Diversion on page 33.) We will later utilize Cantor's result to create some counter-examples in Chapter 14. Because of the importance of Cantor's theorem to almost every field of modern mathematics, it is included in this preliminary chapter.

The set of rational numbers $\mathbb{Q}$ can be described as the numbers of the form p/q, where p is an integer and q is a positive integer.

Although the set of rationals $\mathbb{Q}$ is easy to define, it is often hard to visualize. One way to illustrate the rationals graphically can be seen by the *Sage* command

```
ShowRationals(-5, 5)
```

which draws Figure 0.1. This figure helps to visualize the rational numbers from -5 to 5 using a sequence of rows. The nth row represents the rational numbers with denominator n when expressed in simplest form. In principle there would be an infinite number of rows, getting closer and closer to each other as they get close to the axis.

Figure 0.1 suggests the following.

PROPOSITION 0.3

If a and b are any two different real numbers, then there is a rational number between a and b.

PROOF: Let $x = |a - b|$. Since x is not zero, we let q be any number that is greater than $1/x$. Then $|a \cdot q - b \cdot q| = q \cdot x > 1$, so there must be an integer between $a \cdot q$ and $b \cdot q$, which we will call p. But then p/q will be between a and b, and the proposition is proved. □

From this proposition, we can keep dividing the interval up into smaller and smaller pieces to show that there are in fact an infinite number of rational numbers between any two real numbers. This would make it seem that the number of rational numbers is "doubly infinite," since there are an infinite number of integers, and an infinite number of rational numbers between each pair of integers. But surprisingly, the set of rational numbers is no larger than the set of the integers. To understand what is meant by this statement, let us first show how we can compare the sizes of two infinite sets.

DEFINITION 0.12 A set S is called *countable* if there is an infinite sequence of elements from the set that includes every member of the set.

What do sequences have to do with comparing the sizes of two sets? A sequence can be considered as a function between the set of positive integers and the set S. If a sequence manages to include every member of the set S, then it stands to reason that there are at least as "many" positive integers as there are elements of S. The shocking fact is that even though it would first appear that there must be infinitely many more rational numbers than integers, in fact the two sets have the same size.

PROPOSITION 0.4

The set of rationals forms a countable set.

PROOF: In order to show that the rationals are countable, we need a sequence that will eventually contain every rational somewhere in the sequence. Equivalently, we can connect the dots of Figure 0.1 using a pattern that would, in principle, reach every dot of Figure 0.1 extended to infinity. There are of course many ways to do this, but one way is given in Figure 0.2. This path starts at 0, and swings back and forth, each time hitting the rationals on the next row. Since there are an infinite number of rows, we can extend this pattern indefinitely, and every rational number will eventually be hit by this path. This path gives rise to the sequence

$$\{0, 1, \frac{1}{2}, \frac{-1}{2}, -1, -2, \frac{-3}{2}, \frac{-2}{3}, \frac{-1}{3}, \frac{1}{3}, \frac{2}{3}, \frac{3}{2}, 2, 3, \ldots\},$$

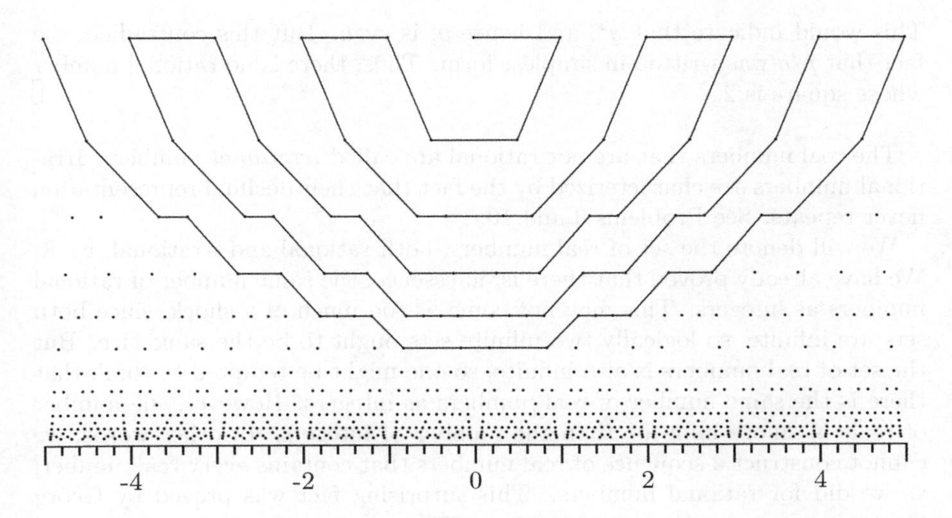

FIGURE 0.2: Sample path going through every rational

which contains every rational number, so we have shown that the rationals form a countable set. ∎

There of course are many other ways of creating a sequence of rational numbers that includes every rational. Problems 1 through 8 explore a recursively defined sequence that contains all of the positive rational numbers.

Even though we have shown that there are an infinite number of rational numbers between any two numbers, the natural question to ask is whether there are numbers that are not rational. The first discovery of a number that was not rational was $\sqrt{2}$, proven by the Greeks [12, p. 82].

PROPOSITION 0.5

There is no rational number p/q such that $(p/q)^2 = 2$.

PROOF: Suppose that there was such a rational number, p/q. Let us further suppose that p/q is in simplest form, so that p and q are integers with no common factors. We could rewrite the equation $(p/q)^2 = 2$ as

$$p^2 = 2q^2.$$

This would indicate that p^2 is an even number, which implies that p is even.

Next, we make the substitution $p = 2r$, where r is an integer. This produces the equation

$$(2r)^2 = 2q^2, \quad \text{or} \quad 2r^2 = q^2.$$

This would indicate that q^2, and hence q, is even. But this contradicts the fact that p/q was written in simplest form. Thus, there is no rational number whose square is 2. ∎

The real numbers that are not rational are called *irrational* numbers. Irrational numbers are characterized by the fact that their decimal representation never repeats. See Problems 9 and 10.

We will denote the set of real numbers, both rational and irrational, by $\mathbb{R}$. We have already proven that there is, in essence, the same number of rational numbers as integers. This may not come as too much of a shock, since both sets are infinite, so logically two infinite sets ought to be the same size. But the set of real numbers is also infinite, so one might be tempted to think that there is the same number of real numbers as integers. However, the number of reals is "more infinite" than the number of integers. In other words, we cannot construct a sequence of real numbers that contains every real number, as we did for rational numbers. This surprising fact was proved by Georg Cantor using a classic argument [11, p. 670].

THEOREM 0.8: Cantor's Diagonalization Theorem
The set of all real numbers between 0 and 1 is uncountable. *That is, there cannot be a sequence of numbers that contains every real number between 0 and 1.*

PROOF: We begin by assuming that we can form such a sequence

$$\{a_1, a_2, a_3, \ldots\}$$

and work to find a contradiction. The plan is to find a number b that cannot be in this list. We can do this by forcing b to have a different first digit than a_1, a different second digit than a_2, a different third digit than a_3, and so on. The only technical problem with this is that some numbers have two decimal representations, such as

$$0.348600000000000000\ldots = 0.348599999999999999\ldots.$$

For these numbers, all we need to do is require that *both* representations are in the list. (That is, some rational numbers will appear twice on the list with different decimal representations.)

We now can find a number b using any number of procedures, such as letting the nth digit of b be one more than the nth digit of a_n, **mod** 10. For example, if the list of numbers is

$$a_1 = 0.94837490123798570\ldots$$
$$a_2 = 0.83840000000000000\ldots$$
$$a_3 = 0.83839999999999999\ldots$$
$$a_4 = 0.34281655343424444\ldots$$

Historical Diversion
Georg Cantor (1845–1918)

Georg Cantor was born in St. Petersburg, Russia. When he was eleven, his father became ill, so his family moved to Germany to escape the cold climate. He graduated with distinction from the Realschule in Darmstadt. In 1862, he entered the University of Zürich, but shifted his studies to the University of Berlin after the death of his father. Cantor attended lectures by Leopold Kronecker and Ernst Kummer.

Cantor completed his dissertation on number theory in 1867, and took up a position at the University of Halle. He began his work on set theory in 1874, being the first mathematician to consider infinite sets. He was able to prove that the set of real numbers is "more numerous" than the set of integers. He was also the first mathematician to appreciate the importance of a one-to-one mapping.

However, his work was met with opposition, particularly from Kronecker. Cantor often proved the existence of sets which had certain properties, without giving any examples of such sets. He assumed that one is allowed to make an infinite number of decisions in the construction of a set, an assumption we currently call the Axiom of Choice. Kronecker, a well-established mathematician, had a constructive viewpoint of mathematics, and called Cantor a "scientific charlatan," and a "renegade." While Cantor tried to publish one of his papers in *Acta Mathematica* , the publisher Mittag-Leffler asked Cantor to withdraw the paper, since it was "about one hundred years too soon."

In 1884, Cantor suffered his first bout with depression, and spent some time in a sanitarium. Cantor soon recovered, and returned to his research, producing his famous diagonal argument and Cantor's theorem. Cantor also tried to prove, in vain, the Continuum Hypothesis, which states that there is no set that is both strictly larger than the set of integers, but strictly smaller than the set of reals. Today we know that the Continuum Hypothesis, like the Axiom of Choice, is undecidable, that is, it can be neither proven or disproven.

In 1899, Cantor returned to the sanatorium. Soon afterwards, Cantor's youngest son died suddenly. Cantor's passion for mathematics was completely drained, and he suffered from chronic depression for the rest of his life, going in and out of sanatoriums. Although he still made mathematical lectures, he retired in 1913, and died in poverty on January 6, 1918 in a sanatorium.

Image source: Wikimedia Commons

then $b = 0.0499\ldots$. Certainly b is missing from the list, since it differs from each member of the list by at least one digit. This contradiction proves the theorem. $\square$

We will use the sets $\mathbb{Q}$ and $\mathbb{R}$ throughout this book, so knowing the properties of these two sets will be important in many of the examples.

Problems for §0.4

1 Although we exhibited a sequence that contains every element of $\mathbb{Q}$, there are other ways to accomplish this. One way is to consider the sequence defined recursively by

$$a_0 = 0, \qquad \text{and} \qquad a_{n+1} = \frac{1}{1 + 2\lfloor a_n \rfloor - a_n} \qquad \text{for } n \geq 0.$$

(Recall $\lfloor a_n \rfloor$ means the largest integer that is less than or equal to a_n.) Write out the first 17 terms of this sequence. (Problems 2 through 7 show this sequence contains all of the non-negative elements of $\mathbb{Q}$.)

2 Show that in the sequence defined by Problem 1, the numerator of a_{n+1} is the denominator of a_n, when the fractions are expressed in lowest terms. (Assume integers have a denominator of 1.)

3 Define the integer sequence b_n to be the numerator of a_n in Problem 1. Show that this sequence satisfies

$$b_0 = 0, \quad b_1 = 1, \quad \text{and} \quad b_{n+2} = b_n + b_{n+1} - 2(b_n \ \textbf{mod}\ b_{n+1}) \quad \text{for } n \geq 0.$$

This sequence is known as *Stern's diatomic sequence.* (Hint: by Problem 2, $a_n = b_n/b_{n+1}$.)

4 Use induction to show that the sequence in Problem 3 satisfies

$$b_{2n} = b_n, \qquad \text{and} \qquad b_{2n+1} = b_n + b_{n+1}$$

for all integers $n > 0$.

5 Use Problem 4 to show that the sequence in Problem 1 satisfies

$$a_{2n} = \frac{a_n}{1 + a_n}$$

for integers $n > 0$. Note that $a_n = b_n/b_{n+1}$.

6 Use Problem 4 to show that the sequence in Problem 1 satisfies

$$a_{2n+1} = a_n + 1$$

for integers $n > 0$. Note that $a_n = b_n/b_{n+1}$.

7 Use Problems 5 and 6 to show that the sequence in Problem 1 contains every non-negative rational number.

Hint: If $x = p/q$, let $n = p + q$, and assume true for previous n. Either $x - 1$ or $x/(1 - x)$ will have a smaller n.

8 Use Problem 7 to show that no rational number is mentioned twice in the sequence given by Problem 1.

Hint: if $a_i = a_j$ for $i > j$, what is a_{2i-j}?

9 For a given rational number p/q, consider the sequence that begins $a_0 = p$, and

$$a_{n+1} = (10a_n) \textbf{ mod } q.$$

Show that this sequence will eventually repeat. See the hint for Problem 8.

10 Use Problem 9 to show that the decimal expansion of a rational number p/q will eventually repeat. (1/2 can be considered as $.500000000000\cdots$)

11 Show that if the decimal expansion of a number eventually repeats,

$$x = n.d_1 d_2 d_3 \ldots d_i \overline{d_{i+1} d_{i+2} \ldots d_{i+j}},$$

the number is rational. Here, $d_1, d_2, \ldots$ are the digits, and the overlined digits will repeat.

Hint: Sum a geometric series.

For Problems **12** through **17**: Prove that the following numbers are irrational.

12 $\sqrt[3]{2}$
13 $\sqrt{3}$
14 $\sqrt{5}$
15 $\sqrt{6}$
16 $\sqrt[3]{3}$
17 $\sqrt[3]{4}$

18 Prove that if a is irrational, then $1/a$ is irrational.

19 Prove that if a is rational and b is irrational, then $a + b$ is irrational.

20 Prove that between any two distinct real numbers, there is an irrational number.

Hint: Use Problem 19 along with Proposition 0.3.

21 Prove that if a is rational and nonzero, and b is irrational, then $a \cdot b$ is irrational.

22 Prove that $y = \sqrt{2} + \sqrt{3}$ is irrational.

Hint: First show that y^2 is irrational.

23 The number $e \approx 2.718281828\ldots$ can be expressed by the series

$$e = \sum_{n=0}^{\infty} \frac{1}{n!} = 1 + 1 + \frac{1}{2} + \frac{1}{6} + \frac{1}{24} + \frac{1}{120} + \cdots.$$

Show that e is irrational.

Hint: If $e = p/q$, put an upper bound on the sum of the non-integral terms of $q! \cdot e$.

24 Is the sum of two irrational numbers always irrational? If not, find a counter-example.

Interactive Problems

25 Notice that in *Sage*, the plot of rational numbers between 0.03 and 0.1,

```
S = ShowRationals(0.03, 0.1); S
```

shows most of the points lying on a curve. Try to find the equation of this curve, using the fact that each dot is three fourths closer to the x-axis than the previous dot. Verify your answer by plotting the curve with the points, using the following command:

```
var("x")
P = plot(function goes in here, [x, 0.03, 0.1]); P + S
```

Hint: Scale the function so that $f(0.1) = 1$.

26 If we begin the sequence in Problem 1 with an irrational number, all terms of the sequence will be irrational. Explore what happens if we consider the same formula, but start with $a_0 = \sqrt{2}$.

```
a = sqrt(2); a
    sqrt(2)
a = Together(1/(1 + 2*floor(a) - a)); a
    1/7*sqrt(2) + 3/7
```

Here, **floor(a)** calculates $\lfloor a \rfloor$, and **Together** rationalizes the denominator. By repeatedly evaluating the last statement, we can compute the sequence $\{a_0, a_1, a_2, a_3, \ldots\}$. Note that a_6 is $\sqrt{2}$ plus an integer. When is the next time in the sequence that a_n is an integer plus $\sqrt{2}$?

Chapter 1

Understanding the Group Concept

The goal of this chapter is to formulate the definition of a group. This is done by first exploring many different examples for which there is a binary operator defined on a set, for which some interesting patterns seem to persist. By observing the minimum requirements for these patterns to appear, we can create the simplest definition of a group that will apply to all of the examples we encountered, plus many other new examples. This will produce an abstract definition of a group.

1.1 Introduction to Groups

This section focuses on one particular group, and then explores this group to find different patterns within the structure of the group. As we strive to determine why these patterns exist, we begin to find proofs that will later be valid for all groups. This example also introduces the concept of *non-commutativity*, since $x \cdot y$ is not always equal to $y \cdot x$. For students not exposed to linear algebra, non-commutativity takes some time to get used to, hence it is important to introduce it early.

To help introduce us to the concept of groups, let us meet a triangle whose dance steps give us an unusual kind of arithmetic. Terry the triangle is a simple looking three-colored triangle that appears by the *Sage* command

```
ShowTerry()
```

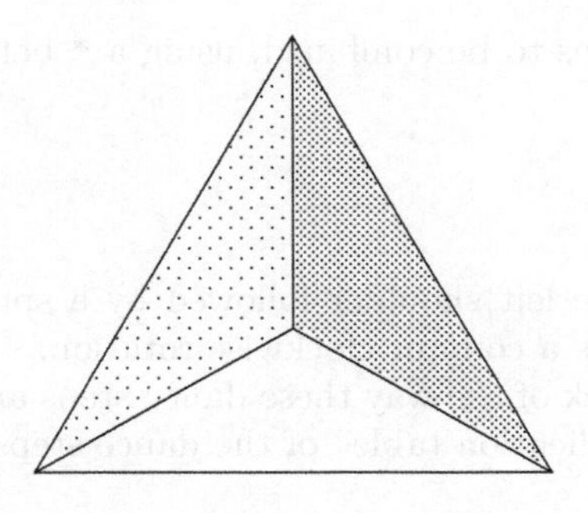

TABLE 1.1: Terry the triangle's dance steps

RotRt	rotate clockwise 120 degrees.
RotLft	rotate counterclockwise 120 degrees.
Spin	spins in three dimensions, keeping the top fixed.
FlipRt	flips over the right shoulder.
FlipLft	flips over the left shoulder.
Stay	does nothing.

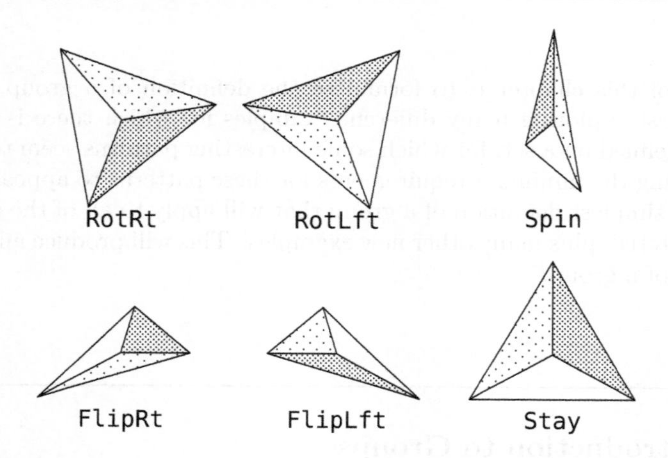

FIGURE 1.1: Scenes from Terry's animated dance steps

Terry can perform the dance steps listed in Table 1.1. Although *Sage* animates these dance steps, one can understand the six steps without *Sage* by observing scenes in Figure 1.1, taken from the animation close to the completion of each step.

Terry can combine these dance steps to form a dance routine. But in any routine, the ending position of the triangle is the same as that of performing just one dance step. Thus, when the triangle gets "lazy," it can perform just one dance step instead of several. The *Sage* command

```
InitTerry()
    {Stay, FlipRt, RotRt, FlipLft, RotLft, Spin}
```

allows these dance steps to be combined, using a * between the dance steps. So we find that:

```
FlipRt * Spin
    RotLft
```

That is, a flip over the left shoulder followed by a spin puts the triangle in the same orientation as a counter-clockwise rotation.

In order to keep track of the way these dance steps are multiplied together, we can form a "multiplication table" of the dance steps. The *Sage* command

TABLE 1.2: Multiplication table for Terry's dance steps

	Stay	FlipRt	RotRt	FlipLft	RotLft	Spin
Stay	Stay	FlipRt	RotRt	FlipLft	RotLft	Spin
FlipRt	FlipRt	Stay	FlipLft	RotRt	Spin	RotLft
RotRt	RotRt	Spin	RotLft	FlipRt	Stay	FlipLft
FlipLft	FlipLft	RotLft	Spin	Stay	FlipRt	RotRt
RotLft	RotLft	FlipLft	Stay	Spin	RotRt	FlipRt
Spin	Spin	RotRt	FlipRt	RotLft	FlipLft	Stay

```
MultTable([Stay, FlipRt, RotRt, FlipLft, RotLft, Spin])
```

forms the table shown in Table 1.2.

To read this table, the first of the dance steps is located on the left side of the table, and the second dance step is found on the top. This table is called the *Cayley table* of the dance steps. Thus, one can use the Cayley table to see that **FlipRt** · **Spin** = **RotLft**. This table allows us to combine dance steps without the help of *Sage*.

We can notice several things from the multiplication table of the dance steps:

1. The *order* in which the dance steps are performed are important. For example, **Spin** · **FlipRt** ≠ **FlipRt** · **Spin**.
2. The combination of any two dance steps is equivalent to one of the six dance steps. In other words, there are no "holes" in Table 1.2.
3. The order in which a dance routine is simplified does not matter. That is,

$$x \cdot (y \cdot z) = (x \cdot y) \cdot z$$

where x, y, and z represent three dance steps.

4. Any dance step combined with **Stay** yields the same dance step. This is apparent by looking at the row and column corresponding to **Stay** in Table 1.2.
5. Every dance step has another dance step that "undoes" it. That is, for every x there is a y such that $x \cdot y =$ **Stay**. For example, the step that undoes **RotRt** is **RotLft**.

We will introduce the following mathematical terminology to express each of these properties:

1. The dance steps are not *commutative.*
2. The dance steps are *closed* under multiplication.

3. The dance steps are *associative.*
4. There is an *identity* dance step.
5. Every dance step has an *inverse.*

With just these properties, we are able to prove the following.

PROPOSITION 1.1

If y is an inverse of x, then x is an inverse of y. Furthermore, x will be the only *inverse of y.*

PROOF: Let z be any inverse of y. Our job is to show that z is in fact equal to x. Consider the product $x \cdot y \cdot z$. According to the associative property,

$$x \cdot (y \cdot z) = (x \cdot y) \cdot z.$$

On the left side, we see that $y \cdot z$ is an identity element, so $x \cdot (y \cdot z) = x$. But on the right side, we find that $x \cdot y$ is an identity element, so $(x \cdot y) \cdot z = z$. Thus, $x = z$, and so x is an inverse of y. Therefore, the inverse of an inverse gives us back the original element.

But as a bonus, we see that inverses are unique! We let z be any inverse of y, and found that it had to equal x. Thus, y has only one inverse, namely x. But if we apply the argument again, reversing the roles of x and y, we see that x has only one inverse, namely y. Thus, all inverses are unique. ∎

Notice that we did not yet assume that there is only one identity element. However, this fact immediately follows from Proposition 1.1. (See Problems 3 and 4.)

DEFINITION 1.1 We use the notation x^{-1} for the unique inverse of the element x.

Proposition 1.1 can now be expressed simply as $(x^{-1})^{-1} = x$. This raises the question as to whether other familiar exponential properties hold. For example, does $(x \cdot y)^{-1}$ always equal $x^{-1} \cdot y^{-1}$?

```
(FlipRt * Spin)^-1
   RotRt
FlipRt^-1 * Spin^-1
   RotLft
```

These results can be verified by looking at Table 1.2. Apparently $(x \cdot y)^{-1}$ is not always equal to $x^{-1} \cdot y^{-1}$. Yet it is not hard to determine the correct way to simplify $(x \cdot y)^{-1}$.

PROPOSITION 1.2

$$(x \cdot y)^{-1} = y^{-1} \cdot x^{-1}.$$

PROOF: Since the inverse $(x \cdot y)^{-1}$ is the unique dance step z such that

$$(x \cdot y) \cdot z = \texttt{Stay},$$

it suffices to show that $y^{-1} \cdot x^{-1}$ has this property. We see that

$$(x \cdot y) \cdot (y^{-1} \cdot x^{-1}) = x \cdot (y \cdot y^{-1}) \cdot x^{-1} = x \cdot \texttt{Stay} \cdot x^{-1} = x \cdot x^{-1} = \texttt{Stay}.$$

So $(x \cdot y)^{-1} = y^{-1} \cdot x^{-1}$. ∎

Another pattern of the multiplication table of the dance steps is that each row and each column in the interior part of the table contain all six dance steps. For example, **RotRt** appears only once in the row beginning with **Spin**. That is, there is only one solution to $\texttt{Spin} \cdot x = \texttt{RotRt}$. We can show why this pattern holds in general using inverses.

PROPOSITION 1.3

If a and b are given, then there exists a unique x such that

$$a \cdot x = b.$$

PROOF: Suppose that there is an x such that $a \cdot x = b$. We can multiply both sides of the equation on the *left* by a^{-1} to give us

$$a^{-1} \cdot (a \cdot x) = a^{-1} \cdot b.$$

Then

$$(a^{-1} \cdot a) \cdot x = a^{-1} \cdot b.$$

$$\texttt{Stay} \cdot x = a^{-1} \cdot b.$$

So

$$x = a^{-1} \cdot b.$$

Thus, if there is a solution, this must be the unique solution $x = a^{-1} \cdot b$. Let us check that this is indeed a solution.

$$a \cdot (a^{-1} \cdot b) = (a \cdot a^{-1}) \cdot b = \texttt{Stay} \cdot b = b.$$

Thus, there is only one solution to the equation, namely $a^{-1} \cdot b$. ∎

This last proposition, when combined with Problem 6, shows that the interior of the multiplication table forms a *Latin square*. A Latin square is a formation in which every row and every column contain each item once and only once. The Latin square property is easy to check visually.

Even though there are very few of Terry's dance steps, we already can see some of the patterns that can appear when we consider the multiplication of these dance steps. In the next section, we will consider another operation that has many of the same patterns.

Problems for §1.1

1 Suppose that Terry the Triangle has a friend who is a square. (Most of us have had such a friend from time to time.) How many dance steps would the square have? Construct a multiplication table of all of the square's dance steps. This set of dance steps is referred to as D_4.

2 Suppose that Terry has a friend who is a regular tetrahedron. (A tetrahedron is a triangular pyramid.) How many dance steps would this tetrahedron have?

3 Using only the four basic properties of Terry's dance steps, prove that there can be only one identity element. That is, there cannot be two elements e and e' for which $x \cdot e = e \cdot x = x$ and $x \cdot e' = e' \cdot x = x$ for all $x \in G$.

4 Using only the four basic properties of Terry's dance steps, prove that an element cannot have two different inverses. That is, show that there cannot be two elements y and y' such that both $x \cdot y = e$ and $x \cdot y' = e$.

5 Prove the cancellation law holds for Terry's dance steps. That is, if $a \cdot b = a \cdot c$ for dance steps a, b, and c, then $b = c$.

6 Prove that if a and b are two of Terry's dance steps, then there is a unique dance step x such that

$$x \cdot a = b.$$

This shows that every column in the multiplication table contains one and only one of each element.

7 If two of Terry's dance steps are chosen at random, what are the chances that these two dance steps will commute?

Hint: There are 36 ways of choosing two dance steps. Count the number of combinations that satisfy the equation $x \cdot y = y \cdot x$.

8 Three of Terry's dance steps are types of flips, **`FlipRt`**, **`FlipLft`**, and **`Spin`**. Does the product of two different flips always produce a rotation? Explain why this is so.

9 Is the product of a flip and a rotation always a flip? Explain why this is so. See Problem 8.

10 Find dance steps a, b, and c such that $a \cdot b = b \cdot c$, but $a \neq c$.

11 Find dance steps a, b such that $(a \cdot b)^{-1} \neq a^{-1} \cdot b^{-1}$.

12 Find dance steps a, b such that $(a \cdot b)^2 \neq a^2 \cdot b^2$.

Interactive Problems

13 If Terry was only allowed to do the dance steps **`FlipRt`** or **`FlipLft`**, could it get itself into all six possible positions? If possible, express the other four dance steps in terms of these two. The command

```
InitTerry()
```

reloads Terry's group.

14 Repeat Problem 13, only allow Terry to do only the steps **`RotRt`** and **`RotLft`**.

15 Can you find a dance routine that includes each of Terry's 6 dance steps once, and only once, and that puts Terry back into the initial position?

1.2 Modular Congruence

We have already seen that one operation, namely the combination of Terry's dance steps, produces some interesting properties such as the Latin square property. In this section we will find some other operations that have this same property, using ordinary integers and modulo arithmetic.

We have already introduced modular arithmetic in §0.3. We defined x **mod** n as the remainder r when x is divided by n, using the division algorithm. But we can also say that two integers x and y are *equivalent* if

$$x \textbf{ mod } n = y \textbf{ mod } n.$$

We will introduce another notation for this relation.

DEFINITION 1.2 Let x, y, and n be integers. We say x and y are *equivalent modulo* n, written

$$x \equiv y \pmod{n}$$

if, and only if, there is an integer k such that

$$(x - y) = k\,n.$$

Note the slight difference in notation between the operator **mod** (expressed in boldface) and the above notation (where mod is not in boldface). The two notations are clearly related, since $x \equiv y \pmod{n}$ means that $x \mathbf{~mod~} n = y \mathbf{~mod~} n$.

The new notation also satisfies three very important properties for equivalence (mod n).

1. (Reflexive) Every integer x is equivalent to itself.
2. (Symmetric) If x is equivalent to y, then y is equivalent to x.
3. (Transitive) If x is equivalent to y, and y in turn is equivalent to z, then x is equivalent to z.

DEFINITION 1.3 Any relation that satisfies these three properties is called an *equivalence relation.* We will use the notation $x \sim y$ to say that x is equivalent to y for a generic equivalence relation.

Let us prove that equivalence (mod n) forms an equivalence relation.

PROPOSITION 1.4
Let n be a positive integer. Then the definition of

$$x \equiv y \pmod{n}$$

forms an equivalence relation on the set of integers.

PROOF: To show that this definition is reflexive, we need to show that $x \equiv x \pmod{n}$, which is clear since $x - x = 0 \cdot n$.

To show that this definition is symmetric, suppose that $x \equiv y \pmod{n}$. Then $x - y = kn$ for some integer k, hence $y - x = -kn$ for the integer $-k$. Thus, $y \equiv x \pmod{n}$.

Finally, to show this definition is transitive, suppose both $x \equiv y \pmod{n}$ and $y \equiv z \pmod{n}$. Then $x - y = k_1 n$ and $y - z = k_2 n$, so

$$x - z = (x - y) + (y - z) = k_1 n + k_2 n = (k_1 + k_2)n.$$

Hence, we find that $x \equiv z \pmod{n}$. □

Whenever an equivalence relation is defined on a set, the set can be broken up into disjoint *equivalence classes*, where each equivalence class is the set of elements related to one element in the class.

DEFINITION 1.4 *Let $x \sim y$ be an equivalence relation defined on a set S. Then the* equivalence class $[a]$ *is the set of elements of S related to a. That is,*

$$[a] = \{s \in S \mid s \sim a\}.$$

Example 1.1
In the relation $x \equiv y \pmod{10}$, the set [3] will be the set of integers equivalent to 3 (mod 10), giving the set

$$[3] = \{\ldots - 37, -27, -17, -7, 3, 13, 23, 33, 43, \ldots\}$$

Other equivalence classes in this relation are similar. □

It is not hard to show that the set of integers can be broken up into disjoint sets using the equivalence classes.

PROPOSITION 1.5
If $x \sim y$ is an equivalence relation on a set S, then S is the disjoint union of equivalence classes.

PROOF: For any $a \in S$, we have by the reflexive property that $a \in [a]$, so $[a]$ is non-empty, and the union of all equivalence classes will be all of S. Next, let us show that if there is an element c in common with two equivalence classes $[a]$ and $[b]$, then these classes are the same. Since $c \sim a$ and $c \sim b$, we have by the symmetric and transitive properties that $a \sim b$. Hence, for every $x \in [a]$, $x \sim a$, so $x \in [b]$ as well, indicating $[a] \subseteq [b]$. By similar logic, $[b] \subseteq [a]$, so $[a] = [b]$. □

Many of the properties of modular arithmetic found in §0.3 can be translated in terms of equivalence relations. For example, Proposition 0.2 can be restated by saying that if

$$x \equiv a \pmod{n} \qquad \text{and} \qquad y \equiv b \pmod{n},$$

then $x + y \equiv a + b \pmod{n}$ and $xy \equiv ab \pmod{n}$.

These statements make it clear that to add or multiply two numbers modulo n, we can choose any representative element from the equivalence class.

Computational Example 1.2
Consider the set of numbers from 0 to 9, with the binary operation being $x * y = (x + y)$ **mod** 10. We can have *Sage* define this binary operation with the command

```
Z = AddMod(10); Z
    {0, 1, 2, 3, 4, 5, 6, 7, 8, 9}
```

Although the elements of **Z** are displayed as integers, we will soon see that they have different properties than ordinary integers. We will continue to use the star to indicate the operation, as we did for Terry's dance steps. In order to access the elements in the set **Z**, we will put a number in brackets to indicate the location of the element in the set. So here is how we can combine the fourth and seventh elements in **Z**:

TABLE 1.3: Addition (mod 10)

	0	1	2	3	4	5	6	7	8	9
0	0	1	2	3	4	5	6	7	8	9
1	1	2	3	4	5	6	7	8	9	0
2	2	3	4	5	6	7	8	9	0	1
3	3	4	5	6	7	8	9	0	1	2
4	4	5	6	7	8	9	0	1	2	3
5	5	6	7	8	9	0	1	2	3	4
6	6	7	8	9	0	1	2	3	4	5
7	7	8	9	0	1	2	3	4	5	6
8	8	9	0	1	2	3	4	5	6	7
9	9	0	1	2	3	4	5	6	7	8

```
Z[4] * Z[7]
    1
```

So with the dot meaning "addition modulo 10", we find that $4 \cdot 7 = 1$. Although it seems strange to use the dot instead of the plus sign, for consistency we always uses the dot for the binary operation, whatever that operator is. So the one thing we must remember is that *the dot does not always mean multiplication.* Rather, the dot represents the operation in the current context. For Terry's group, the dot represented combining two dance steps. Here, it represents addition modulo 10.

We will still use the command **MultTable** to give the Cayley table of the set, even though the operation is more like addition. Thus the command

```
MultTable(Z)
```

produces Table 1.3. □

By looking at the table for addition modulo 10, we are able to establish the following properties:

1. For any two numbers x and y in $\{0, 1, 2, 3, 4, 5, 6, 7, 8, 9\}$, $x \cdot y$ is in the set. (Recall that we are using the dot to indicate the operation, regardless of what that operation is. In this example, the operation is addition modulo 10.)

2. $(x \cdot y) \cdot z = x \cdot (y \cdot z)$ for any x, y, and z.

3. $x \cdot 0 = x$ and $0 \cdot x = x$ for all x.

4. For any x, there is a y such that $x \cdot y = 0$.

5. For any x and y, $x \cdot y = y \cdot x$.

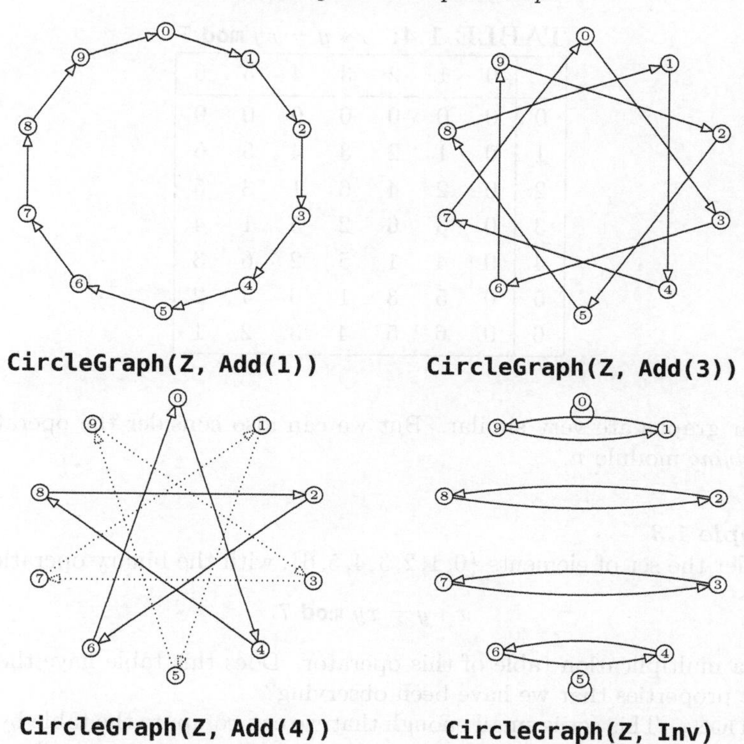

FIGURE 1.2: Circle graphs for modulo 10 operations

This operation can also be pictured by means of circular graphs. The *Sage* command

```
CircleGraph(Z, Add(1))
```

gives us the first picture in Figure 1.2, which draws an arrow from each point to the point given by "adding 1 modulo 10." Figure 1.2 also shows what happens if we replace the 1 with 3 or 4. We get different-looking graphs, but all with the same amount of symmetry. The *Sage* command

```
CircleGraph(Z, Add(1), Add(2), Add(3), Add(4), Add(5))
```

combines several of these circular graphs together, each drawn in a different color. The last picture in Figure 1.2 shows the additive inverse of each digit. This graph was created with the command

```
CircleGraph(Z, Inv)
```

Of course, we could do these same experiments by considering addition modulo n with any other base as well as $n = 10$. The patterns formed by the

TABLE 1.4: $x * y = xy$ **mod** 7

	0	1	2	3	4	5	6
0	0	0	0	0	0	0	0
1	0	1	2	3	4	5	6
2	0	2	4	6	1	3	5
3	0	3	6	2	5	1	4
4	0	4	1	5	2	6	3
5	0	5	3	1	6	4	2
6	0	6	5	4	3	2	1

circular graphs are very similar. But we can also consider the operation of *multiplying* modulo n.

Example 1.3

Consider the set of elements $\{0, 1, 2, 3, 4, 5, 6\}$, with the binary operation

$$x * y = xy \textbf{ mod } 7.$$

Form a multiplication table of this operator. Does this table have the Latin square properties that we have been observing?

SOLUTION: This set is small enough that we can compute the table by hand, producing Table 1.4. Although the first row and first column are all zeros, we notice that if we removed the 0 and only considered the digits $\{1, 2, 3, 4, 5, 6\}$, we would get a Latin square. The identity element is 1, and each of the numbers has an inverse. ▯

If we try Example 1.3 with a different base, we get a surprise. To display the multiplication table for (mod 10) arithmetic, we can use the *Sage* command

```
Z = MultMod(10)
MultTable(Z)
```

to produce a table similar to Table 1.5. We find several rows that do not contain any 1's. These rows indicate the numbers without inverses modulo 10. Only $1, 3, 7$, and 9 have inverses. If we try this using 15 instead of 10, we find only $1, 2, 4, 7, 8, 11, 13$, and 14 have inverses.

Computational Example 1.4

What if we consider the multiplication table of just those numbers that have inverses modulo 15? We can use the *Sage* commands

```
Z = MultMod(15)
L = [Z[1], Z[2], Z[4], Z[7], Z[8], Z[11], Z[13]]
MultTable(L)
```

TABLE 1.5: Multiplication (mod 10)

	0	1	2	3	4	5	6	7	8	9
0	0	0	0	0	0	0	0	0	0	0
1	0	1	2	3	4	5	6	7	8	9
2	0	2	4	6	8	0	2	4	6	8
3	0	3	6	9	2	5	8	1	4	7
4	0	4	8	2	6	0	4	8	2	6
5	0	5	0	5	0	5	0	5	0	5
6	0	6	2	8	4	0	6	2	8	4
7	0	7	4	1	8	5	2	9	6	3
8	0	8	6	4	2	0	8	6	4	2
9	0	9	8	7	6	5	4	3	2	1

to produce Table 1.6. Once again, many of the same patterns are found that were in Terry's multiplication, namely:

1. For any two numbers x and y in $\{1, 2, 4, 7, 8, 11, 13, 14\}$, $x \cdot y$ is in that set.
2. $(x \cdot y) \cdot z = x \cdot (y \cdot z)$ for any x, y, and z.
3. $x \cdot 1 = x$ and $1 \cdot x = x$ for all x.
4. For any x, there is a y such that $x \cdot y = 1$.
5. For any x and y, $x \cdot y = y \cdot x$. ∎

We can generalize these patterns to multiplication modulo n for any n.

TABLE 1.6: Invertible elements (mod 15)

	1	2	4	7	8	11	13	14
1	1	2	4	7	8	11	13	14
2	2	4	8	14	1	7	11	13
4	4	8	1	13	2	14	7	11
7	7	14	13	4	11	2	1	8
8	8	1	2	11	4	13	14	7
11	11	7	14	2	13	1	8	4
13	13	11	7	1	14	8	4	2
14	14	13	11	8	7	4	2	1

PROPOSITION 1.6

For n a positive integer greater than 1, let the dot $(\cdot)$ denote multiplication modulo n. Let G be the set of all non-negative numbers less than n that have inverses modulo n. Then the set G has the following properties:

1. *For any two numbers x and y in G, $x \cdot y$ is in G.*
2. *$(x \cdot y) \cdot z = x \cdot (y \cdot z)$ for any x, y, and z.*
3. *$x \cdot 1 = 1 \cdot x = x$ for all x.*
4. *For any x that is in G, there is a y in G such that $x \cdot y = 1$.*
5. *For any x and y, $x \cdot y = y \cdot x$.*

PROOF: Properties 2, 3, and 5 come from the properties of standard multiplication.

Property 1 comes from Proposition 1.2. If x and y are both invertible, then $y^{-1} \cdot x^{-1}$ is an inverse of $x \cdot y$, and so $x \cdot y$ is invertible modulo n.

Property 4 seems obvious, since if x is invertible modulo n, we let $y = x^{-1}$ making $x \cdot y = 1$. But we must check that y is also invertible, which it is since $y^{-1} = x$. ∎

Of course, this does not tell us *which* of the numbers less than n have inverses modulo n. The following proposition will help us out.

PROPOSITION 1.7

Let n be in $\mathbb{Z}^+$. Then for x between 0 and $n-1$, x has a multiplicative inverse modulo n if, and only if, x is coprime to n.

PROOF: If x and n are not coprime, then there is a common prime factor p. In order for x to have a multiplicative inverse, there must be a y such that

$$x \cdot y \equiv 1 \pmod{n}.$$

But this means that $xy = 1 + wn$ for some w. This is impossible, since xy is a multiple of p, but $1 + wn$ is one more than a multiple of p.

Now suppose that x and n are coprime. By the greatest common divisor theorem (0.4), there are u and v in $\mathbb{Z}$ such that

$$ux + vn = \gcd(x, n) = 1.$$

But then

$$ux = 1 + (-v)n,$$

and so $u \cdot x \equiv 1 \pmod{n}$. Hence, u is a multiplicative inverse of x. ∎

We now have seen several binary operations, such as Terry's dance steps, addition modulo n, and multiplication modulo n, which have many properties in common. In the next section we will generalize these examples to produce many more interesting examples, but in such a way that they will all have the important properties that we have seen.

Problems for §1.2

For Problems **1** through **6**: Construct a Cayley table for the set of numbers using addition modulo n.

1 $\{0, 1, 2, 3, 4\}$, $n = 5$

2 $\{0, 1, 2, 3, 4, 5\}$, $n = 6$

3 $\{0, 1, 2, 3, 4, 5, 6, 7\}$, $n = 8$

4 $\{0, 2, 4, 6\}$, $n = 8$

5 $\{0, 2, 4, 6, 8, 10\}$, $n = 12$

6 $\{0, 3, 6, 9, 12, 15, 18, 21\}$, $n = 24$

For Problems **7** through **12**: Construct a Cayley table for the set of numbers using multiplication modulo n.

Hint: Since these are the numbers that have multiplicative inverses modulo n, Proposition 1.6 shows that the multiplication table has the same properties as Terry's dance steps, in particular, the Latin square property.

7 $\{1, 3, 5, 7\}$, $n = 8$

8 $\{1, 2, 4, 5, 7, 8\}$, $n = 9$

9 $\{1, 5, 7, 11\}$, $n = 12$

10 $\{1, 3, 5, 9, 11, 13\}$, $n = 14$

11 $\{1, 5, 7, 11, 13, 17\}$, $n = 18$

12 $\{1, 5, 7, 11, 13, 17, 19, 23\}$, $n = 24$

13 Let S be a set, and suppose S can be described as the union of a collection of non-empty, disjoint subsets. Show that there is an equivalence relation such that the equivalence classes are precisely the given collection of disjoint subsets.

14 Let $f : S \to T$ be a function defined on a set S. Define $x \sim y$ if $f(x) = f(y)$. Show that this defines an equivalence relation on S.

For Problems **15** through **20**: Find the multiplicative inverse for the element in the following group.

15 $7 \in Z_{16}^*$

16 $8 \in Z_{17}^*$

17 $10 \in Z_{21}^*$

18 $5 \in Z_{18}^*$

19 $7 \in Z_{20}^*$

20 $9 \in Z_{22}^*$

Interactive Problems

For Problems **21** through **26**: Proposition 1.7 explains how to use **xgcd** to find the multiplicative inverse modulo n. Use *Sage* to find the multiplicative inverse of a modulo n.

21 $a = 3$, $n = 100$

22 $a = 5$, $n = 121$

23 $a = 7$, $n = 360$

24 $a = 11$, $n = 900$

25 $a = 13$, $n = 1200$

26 $a = 17$, $n = 1500$

27 We saw that there were exactly four numbers less than 10 that were invertible modulo 10. For what other values of n are there exactly four numbers less than n that are invertible modulo n? Use *Sage*'s circle graph to graph the inverse functions.

1.3 The Definition of a Group

In this chapter, we have seen several different ways of combining numbers or dance steps. Yet, all of the different "products" had many properties in common. We are now ready to try to generalize these examples. Our strategy is to define a *group* abstractly by requiring the same patterns we observed to continue. Thus, we make the following definition based upon the first four properties we saw in all of our examples.

DEFINITION 1.5 A *group* is a set G together with a binary operation $(\cdot)$ such that the following four properties hold:

1. (closure) For any x and y in G, $x \cdot y$ is in G.
2. (identity) There exists a member e in G which has the property that, for all x in G, $e \cdot x = x \cdot e = x$.
3. (inverse) For every x in G, there exists a y in G, called the *inverse* of x, such that $x \cdot y = e$.
4. (associative law) For any x, y, and z in G, then $(x \cdot y) \cdot z = x \cdot (y \cdot z)$.

Terry's dance steps give us the first example of a group, more commonly referred to as the group of symmetries of a triangle, D_3.

The members of the group, whether they are numbers, dance steps, or even ordered pairs, are called the *elements* of the group. The element e that satisfies property 2 is called the *identity element* of the group.

The mathematical notation for an element x to be in a group G is

$$x \in G.$$

Since Propositions 1.1, 1.2, and 1.3 used only these four properties, the proofs are valid for all groups, using the identity element e in place of the dance step `Stay`.

Other examples of groups come from modular arithmetic. For n in $\mathbb{Z}^+$, we considered the elements

$$\{0, 1, 2, ..., n-1\},$$

with the operator $(\cdot)$ being the sum modulo n. This group will be denoted Z_n. In fact, the *Sage* command **ZGroup** will load the group Z_n.

```
G = ZGroup(10); G
    {0, 1, 2, 3, 4, 5, 6, 7, 8, 9}
```

We also considered having the operator $(\cdot)$ denote the product modulo n, and considered only the set of numbers less than n that are coprime to n. Proposition 1.6 shows that this set also has the four properties of groups. We will refer to this group by Z_n^*. This group can be loaded in *Sage* by the command **ZStar**.

```
G = ZStar(15); G
    {1, 2, 4, 7, 8, 11, 13, 14}
```

The groups Z_n and Z_n^* had a fifth property—the multiplication tables were symmetric about the northwest-to-southeast diagonal. Not all groups have this property, but those that do are important enough to give this property a special name.

DEFINITION 1.6 A group G is *abelian* (or *commutative*) if $x \cdot y = y \cdot x$ for all $x, y \in G$.

Although these definitions appear to be ad hoc, in fact the four properties of groups have been carefully chosen so that they will apply to many different aspects of mathematics. Here are some important examples of groups that appear on other contexts besides group theory:

Example 1.5
The set of integers $\mathbb{Z}$, with the binary operation being the sum of two numbers. The identity element is 0, and $-x$ is the inverse of x. This forms an abelian group. □

Example 1.6
Consider the set of rational numbers, denoted by $\mathbb{Q}$. We will still use addition for our binary operation. This is also an abelian group. □

Example 1.7
Consider the set of all rational numbers except for 0. This time we will use multiplication instead of addition for our group operation. The identity element is now 1, and the inverse of an element is the reciprocal. This abelian group will be denoted by $\mathbb{Q}^*$. □

Example 1.8
Consider the set of all *linear* functions of the form $f(x) = mx + b$, with $m, b \in \mathbb{R}$, $m \neq 0$. (The $\mathbb{R}$ represents the real numbers.) We multiply two linear functions together by function composition. That is, if $f(x) = mx + b$ and $g(x) = nx + c$, then

$$f \cdot g = f(g(x)) = m(nx + c) + b = (mn)x + (mc + b).$$

Note that in $f \cdot g$, we do g first, then f, so we apply the functions from right to left. We can find the inverse of $f(x)$ as well:

$$f^{-1}(x) = \frac{1}{m}x - \frac{b}{m},$$

which is also a linear function. This group satisfies all of the group properties, but is not abelian. For example, if $f(x) = 2x + 3$ and $g(x) = 3x + 2$, then $f \cdot g = f(g(x)) = 6x + 7$, whereas $g \cdot f = g(f(x)) = 6x + 11$. □

DEFINITION 1.7 The number of elements in a group G is called the *order* of the group, and is denoted $|G|$. If G is has an infinite number of elements, we say that $|G| = \infty$.

Examples 1.5 though 1.8 have infinite order, and hence we cannot form Cayley tables for these groups. On the other hand, the *smallest* possible group is given by the following example.

Example 1.9
Consider the group containing just the identity element, $\{e\}$. We can have *Sage* give a Cayley table of this group by the following commands:

```
InitGroup("e")
MultTable([e])
```

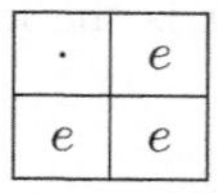

We call this group the *trivial group.* The first of these *Sage* commands introduces a new command—**InitGroup**. This command designates the new identity element, and sets the stage for entering a new group. □

Note that sometimes the operator $(\cdot)$ means addition, sometimes it means multiplication, and sometimes it means neither. Nonetheless, we can define x^n to mean x operated on itself n times. Thus,

$$x = x^1,$$

$$x \cdot x = x^2,$$

$$x \cdot x \cdot x = x^3,$$

etc.

We want to formally define x^n for any integer n. We let $x^0 = e$, the identity element. We then define, for $n > 0$,

$$x^n = x^{n-1} \cdot x.$$

By defining the nth power in terms of the previous power, we have defined x^n whenever n is a positive integer.

Finally, we can define negative powers by letting

$$x^{-n} = (x^n)^{-1} \quad \text{if} \quad n > 0.$$

This is an *inductive* definition, since it defines each power in terms of a previous power. This type of definition works well for proving simple propositions about x^n.

PROPOSITION 1.8

If x is an element in a group G, and m and n are integers, then

$$x^{m+n} = x^m \cdot x^n.$$

PROOF: If m or n are 0, this proposition is very easy to verify:

$$x^{m+0} = x^m = x^m \cdot e = x^m \cdot x^0, \qquad x^{0+n} = x^n = e \cdot x^n = x^0 \cdot x^n.$$

We will now prove the statement when m and n are positive integers. If n is 1, then we have

$$x^{m+1} = x^{(m+1)-1} \cdot x = x^m \cdot x^1,$$

using the inductive definition of the power of x.

We will now proceed by means of *induction.* That is, we will assume that the statement is true for $n = k - 1$, and then prove that it is then true for $n = k$. Then we will have that, since the statement is true for $n = 1$, and it is true for each number that follows, it must be true for all positive n.

Thus, we will assume that

$$x^{m+(k-1)} = x^m \cdot x^{k-1}.$$

But then

$$x^{m+k} = x^{m+k-1} \cdot x = x^m \cdot x^{k-1} \cdot x = x^m \cdot x^k.$$

Thus, by assuming the statement is true for $n = k - 1$, we found that it was also true for $n = k$. By induction, this proves that $x^{m+n} = x^m \cdot x^n$ for all positive n.

Once we have the statement true for positive m and n, we can take the inverse of both sides to give us

$$(x^{m+n})^{-1} = (x^n)^{-1} \cdot (x^m)^{-1}.$$

But by the definition of negative exponents, this is

$$x^{(-n)+(-m)} = x^{-n} \cdot x^{-m}$$

which, by letting $M = -n$ and $N = -m$, proves the proposition for the case of both exponents being negative.

Finally, if m and n have different signs, then $(m+n)$ will either have the same sign as $-n$, or the same sign as $-m$. If $(m+n)$ has the same sign as $-n$, then we have already shown that

$$x^m = x^{(m+n)+(-n)} = x^{m+n} \cdot x^{-n}.$$

So we have $x^m \cdot (x^{-n})^{-1} = x^{m+n} \cdot x^{-n} \cdot (x^{-n})^{-1}$, and hence $x^{m+n} = x^m \cdot x^n$.

If $(m+n)$ has the same sign as $-m$, then we have already shown that

$$x^n = x^{(-m)+(m+n)} = x^{-m} \cdot x^{m+n}.$$

So we have $(x^{-m})^{-1} \cdot x^n = (x^{-m}) \cdot x^{-m} \cdot x^{m+n}$, and hence $x^{m+n} = x^m \cdot x^n$.

Thus we have proven the proposition for all integers m and n. □

This last proof utilizes an important method of proving theorems called *induction*, which was introduced in §0.1. Induction is based on the well-ordering axiom, which states that any non-empty subset of positive integers contains a smallest element.

Although this last proof introduced the variable k, this really was not necessary. To prove a statement for all positive integers n, we can first prove the statement is true for $n = 1$, and then we can assume that the statement is true for the previous case $n - 1$. This extra information often gives us the leverage we need to be able to prove the statement is true for n. Here is another example of the use of induction.

PROPOSITION 1.9

If x is an element in a group G, and m and n are in $\mathbb{Z}$, then

$$(x^m)^n = x^{(mn)}.$$

PROOF: Notice that this statement is trivial if $n = 0$ and $n = 1$:

$$(x^m)^0 = e = x^{m\cdot 0}, \qquad (x^m)^1 = x^m = x^{(m\cdot 1)}.$$

We will again proceed by means of induction, which means we can assume that the statement is true for the previous case, with n replaced by $n - 1$. That is, we can assume that

$$(x^m)^{n-1} = x^{m\cdot(n-1)}.$$

Note that

$$(x^m)^n = (x^m)^{n-1} \cdot x^m = x^{m\cdot(n-1)} \cdot x^m.$$

By Proposition 1.8, this is equal to $x^{m\cdot(n-1)+m} = x^{mn}$.

So by induction, the proposition holds for positive n. To see that it holds for negative n as well, simply note that

$$(x^m)^n = ((x^m)^{-n})^{-1} = (x^{-mn})^{-1} = x^{mn}.$$

If n is negative, then $-n$ is positive, so the second step is valid. $\square$

Propositions 1.8 and 1.9 show that the common laws of exponents hold for elements of a group. In the next section, we will use the powers of elements to explore the properties of a group.

Problems for §1.3

1 Consider the following multiplication table:

$\cdot$	e	a	b	c	d
e	e	a	b	c	d
a	a	e	c	d	b
b	b	d	e	a	c
c	c	b	d	e	a
d	d	c	a	b	e

Notice that this multiplication table satisfies the "Latin square" property, hence this multiplication satisfies Proposition 1.3. Does this set form a group? Why or why not?

2 Consider the following multiplication table:

$\cdot$	e	a	b	c
e	e	a	b	c
a	a	e	c	b
b	b	c	e	a
c	c	b	a	e

Notice that this multiplication table satisfies the "Latin square" property, hence this multiplication satisfies Proposition 1.3. Does this set form a group? Why or why not?

For Problems **3** through **14**: Decide whether each set forms a group using the given binary operation. If it is not a group, state which parts of Definition 1.5 fails to hold.

3 $G =$ rational numbers, $x * y = x + y$.
4 $G =$ irrational numbers, $x * y = x + y$.
5 $G =$ non-negative real numbers, $x * y = xy$.
6 $G =$ positive rational numbers, $x * y = xy$.
7 $G =$ positive irrational numbers, $x * y = xy$.
8 $G =$ non-negative integers, $x * y = x + y$.
9 $G =$ even integers, $x * y = x + y$.
10 $G =$ odd integers, $x * y = x + y$.
11 $G =$ odd integers, $x * y = xy$.
12 $G =$ all integers, $x * y = xy$.
13 $G = \{1, -1\}$, $x * y = xy$.
14 $G =$ all integers, $x * y = x + y + 3$.

15 Note that in Definition 1.5, we only required the inverse of x to have the property that $x \cdot y = e$. Show that this element will also satisfy $y \cdot x = e$.

16 Show that a group can have at most one identity element.

17 Show that the inverse of an element must be unique.

18 Show that in any group, $(x \cdot y)^{-1} = y^{-1} \cdot x^{-1}$.

19 Show that if $a \cdot x = a \cdot y$ in a group, then $x = y$.

20 Suppose that S is a finite set (not necessarily a group) that is closed under the operator $(\cdot)$. Suppose also that the equation

$$a \cdot x = a \cdot y$$

holds if, and only if, $x = y$. Prove Proposition 1.3 holds for the set S, even if S is not a group.

Hint: Use the pigeonhole principle.

21 Let G be a group. Show that G is commutative if, and only if, $(a \cdot b)^2 = a^2 \cdot b^2$ for all a and b in G.

22 If G is a group such that $x^2 = e$ for all elements x in G, prove that G is commutative.

23 Let G be a finite group that contains an even number of elements. Show that there is at least one element besides the identity such that $a^2 = e$.

Hint: Show that there are an even number of elements for which $a^2 \neq e$.

24 Let G be a finite group. Show that there are an odd number of elements that satisfy the equation $a^3 = e$.

For Problems **25** through **27**: Fill in the remaining spaces in this multiplication table so that the resulting set forms a group.

Hint: Determine what the identity element must be. Once the row and column of the identity element are filled in, the remaining table can be finished using only the Latin square property.

Problem 25

$\cdot$	a	b	c	d
a	b			
b				
c		b		
d				

Problem 26

$\cdot$	a	b	c	d
a				
b				
c	d			
d			b	

Problem 27

	a	b	c	d	e	f	g	h
a	b		d					c
b	g	e		h				
c						e	d	g
d		h		b			f	
e			c					
f			e			b		a
g	e	a			g		b	
h			a				c	

Interactive Problems

28 Use *Sage*'s **`ZStar`** command to find the size of Z_n^* for $n = 9$, 27, 81, 243, 5, 25, 125. Make a conjecture about the size of Z_n^* when n is a power of an odd prime. Note that you can use the **`len(_)`** command to have *Sage* count the elements for you.

29 Use *Sage*'s **`ZStar`** command to find the size of Z_n^* for $n = 18$, 54, 162, 486, 50, 250, 98, 686. Make a conjecture about the size of Z_n^* when n is twice the power of an odd prime.

30 Use *Sage*'s **`ZStar`** command to make a conjecture about the size of Z_{mn}^*, where m and n are coprime, in terms of the sizes of Z_m^* and Z_n^*.

Chapter 2

The Structure within a Group

We have already seen some patterns within a group, such as the Latin square property. However, in order to determine more patterns, we need to consider the possibility of a smaller group sitting inside of a larger group. For example, the group of integers is inside of the group of real numbers. Whenever this happens, we say the smaller group is a *subgroup* of the larger group. Subgroups will lead to even more important properties of groups. But before we determine the subgroups of a given group, we need to understand the generators of a group.

2.1 Generators of Groups

In this section we will explore the set of elements within the group. We will find that some elements may possess an important property, allowing every element to be expressible in terms of that one element. We can then define a group as *cyclic* if it possesses such an element.

Cyclic groups turn out to be very important in the study of groups. In fact, we will discover that every finite abelian group can be expressed using the cyclic groups as building blocks.

Knowing about cyclic groups will also help us to define other groups in programs such as *Sage*. Many of these groups will be fairly large, and so rather than giving *Sage* the entire group, we will define a group using a very small number of elements. From these few elements, *Sage* will be able to reconstruct the entire group.

We begin by studying *finite* groups, such as Terry's group, Z_n, and Z_n^*. By observing the properties of a single element within such a group, we gain insight on how to program *Sage* to work with finite groups.

Computational Example 2.1

Study the powers of the elements 3 and 4 in the group Z_{10}.

This group is loaded into *Sage* with the command

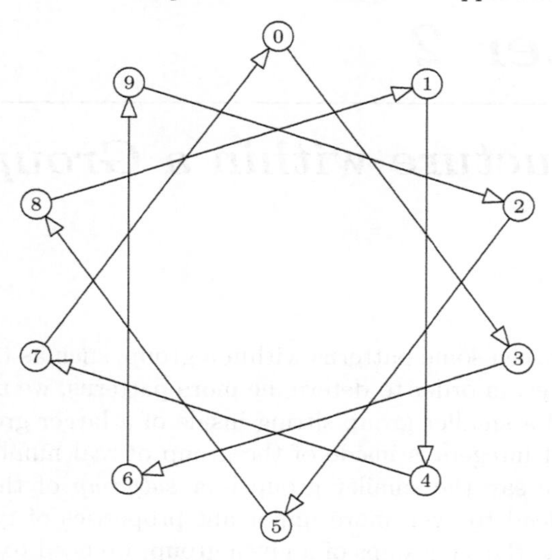

FIGURE 2.1: Circle graph of adding 3 **mod** 10

```
G = ZGroup(10); G
    {0, 1, 2, 3, 4, 5, 6, 7, 8, 9}
```

We can map each element x to the element $x \cdot 3$ with a circle graph

```
CircleGraph(G, Add(3) )
```

which produces Figure 2.1

This graph allows us to visualize powers of 3 in the group Z_{10}. If we follow the arrows starting with 0, we have the sequence $\{0, 3, 6, 9, 2, 5, 8, 1, 4, 7, 0 \ldots\}$. This tells us that

$$3^0 = 0, \qquad 3^1 = 3, \qquad 3^2 = 6, \qquad 3^3 = 9, \qquad 3^4 = 2, \quad \text{etc.}$$

Recall that for this group the dot represents addition modulo 10, so an exponent would represent repeated addition. Note that every element in the group can be expressed as a power of 3.

This property does not hold for all elements, since the powers of 4 are seen to be $\{0, 4, 8, 2, 6, 0, 4, 8, \ldots\}$, which does *not* include all of the elements. ∎

DEFINITION 2.1 We'll say that the element $g \in G$ is a *generator* of the group G if every element of G can be expressed as a power of g.

We can have *Sage* list all of the generators of a group for us. In the case of $G = Z_{10}$, the generators are:

```
Generators(G)
    [1, 3, 7, 9]
```

So there are 4 generators to the group Z_{10}.

Example 2.2
Find all of the generators of the group Z_7^*.
SOLUTION: This group is small enough to do by hand. For each of the elements in $Z_7^* = \{1, 2, 3, 4, 5, 6\}$, we raise the element to different powers until we reach the identity.

$$\begin{aligned}
&1^2 = 1. \\
&2^2 = 4, \quad 2^3 = 1. \\
&3^2 = 2, \quad 3^3 = 6, \quad 3^4 = 4, \quad 3^5 = 5, \quad 3^6 = 1. \\
&4^2 = 2, \quad 4^3 = 1. \\
&5^2 = 4, \quad 5^3 = 6, \quad 5^4 = 2, \quad 5^5 = 3, \quad 5^6 = 1. \\
&6^2 = 1.
\end{aligned}$$

Thus we see that 3 and 5 are generators. ∎

The natural question that arises is whether a given element is a generator of a group. There isn't an obvious pattern for the group Z_7^*, but is not difficult for the group Z_n.

PROPOSITION 2.1
The generators of Z_n are precisely the integers between 0 and n that are coprime to n.

PROOF: Suppose that g is a generator of Z_n. Then 1 is able to be expressed as a power of g, so we have that

$$g^v = 1 \text{ in } Z_n$$

for some v. Since the group action of Z_n is addition, raising to a power is equivalent to repeated addition, or standard multiplication. Thus, we have that

$$gv \equiv 1 \pmod{n}.$$

By Proposition 1.7, there is such a v if, and only if, g is coprime to n.

Now suppose that g is coprime to n. By Proposition 1.7, there is a v such that

$$gv \equiv 1 \pmod{n}, \text{ hence } g^v = 1 \text{ in } Z_n.$$

So 1 can be expressed as a power of g. But 1 is a generator of Z_n, and so every element of Z_n can be expressed as a power of 1, say 1^w. Then that element can be written as $g^{(vw)} = (g^v)^w = 1^w$. So every element can be expressed as a power of g, hence g is a generator of Z_n. ∎

TABLE 2.1: Table of Euler's totient function $\phi(n)$

n	$\phi(n)$	n	$\phi(n)$	n	$\phi(n)$	n	$\phi(n)$
1	1	10	4	19	18	28	12
2	1	11	10	20	8	29	28
3	2	12	4	21	12	30	8
4	2	13	12	22	10	31	30
5	4	14	6	23	22	32	16
6	2	15	8	24	8	33	20
7	6	16	8	25	20	34	16
8	4	17	16	26	12	35	24
9	6	18	6	27	18	36	12

The count of numbers less than n that are coprime to n is called the *Euler totient function* of n, and is denoted $\phi(n)$. Thus, the number of generators of Z_n is precisely $\phi(n)$. A small table of this function up to $n = 36$ is given in Table 2.1.

For larger values of n, we can use the *Sage* command **EulerPhi**.

```
EulerPhi(60)
    16
```

Hence, there are 16 generators of Z_{60}. *Sage* uses the following formula for the totient function based on the prime factorization of the number.

THEOREM 2.1: The Totient Function Theorem
If the prime factorization of n is given by

$$n = p_1^{r_1} \cdot p_2^{r_2} \cdots p_k^{r_k},$$

where $p_1, p_2, p_3, \ldots, p_k$ are distinct primes, and $r_1, r_2, r_3, \ldots, r_k$ are positive integers, then the count of numbers less than n that are coprime to n is

$$\phi(n) = (p_1 - 1) \cdot p_1^{(r_1-1)} \cdot (p_2 - 1) \cdot p_2^{(r_2-1)} \cdot \cdots \cdot (p_k - 1) \cdot p_k^{(r_k-1)}.$$

PROOF: To begin, let us show that if p is a prime, then $\phi(p^r) = (p-1)p^{r-1}$.

Note that the only numbers that are *not* coprime to p^r will be multiples of p. So of the numbers between 1 and p^r, exactly $1/p$ of them will be multiples of p. The remaining $(1 - 1/p) \cdot p^r$ will be coprime, and this can be simplified to $(p-1)p^{r-1}$.

Next we want to show that if n and m are coprime, then $\phi(nm) = \phi(n)\phi(m)$. Let A denote the set of numbers that are less than n, but coprime to n. Let B denote the set of numbers that are less than m, but coprime to m.

Then for any number x coprime to nm, x **mod** n must be in the set A, while x **mod** m must be in B. Yet for every a in A and b in B, there is, by the Chinese remainder theorem (0.7), a unique number less than nm that is equivalent to $a \pmod n$ and $b \pmod m$. This number will be coprime to both n and m, and hence will be coprime to nm.

Therefore, we have a one-to-one correspondence between ordered pairs (a, b), where a is in A, and b is in B, and numbers coprime to nm. Thus, we have

$$\phi(n \cdot m) = \phi(n) \cdot \phi(m).$$

Finally, we can combine these results together. By simply noting that if

$$n = p_1^{r_1} \cdot p_2^{r_2} \cdots p_k^{r_k},$$

then $p_1^{r_1}, p_2^{r_2}, p_3^{r_3}, \ldots, p_k^{r_k}$ will all be coprime. Hence, we can find ϕ for each of these terms, and multiply them together, giving us our formula. ▯

We can also consider finding generators for the groups of the form Z_n^*.

Example 2.3

The group Z_{10}^* has four elements, $\{1, 3, 7, 9\}$, and looking at the powers of the elements, we see that

$$\begin{aligned} 1^2 &= 1. \\ 3^2 &= 9, \quad 3^3 = 7, \quad 3^4 = 1. \\ 7^2 &= 9, \quad 7^3 = 3, \quad 7^4 = 1. \\ 9^2 &= 1. \end{aligned}$$

so 3 and 7 are generators. ▯

Example 2.4

Z_8^* also has four elements, $\{1, 3, 5, 7\}$, but

$$\begin{aligned} 1^2 &= 1. \\ 3^2 &= 1. \\ 5^2 &= 1. \\ 7^2 &= 1. \end{aligned}$$

so *none* of these elements are generators of the group! This becomes apparent as we look at the multiplication table for Z_8^*.

```
G = ZStar(8)
MultTable(G)
```

·	1	3	5	7
1	1	3	5	7
3	3	1	7	5
5	5	7	1	3
7	7	5	3	1

Notice that the square of every element is equal to 1. Hence no element of Z_8^* can generate the whole group. We can see this by asking *Sage* to list all of the generators.

```
Generators(G)
    [ ]
```

□

From these examples, we see that some groups have generators, while others do not. This leads us to the following definition.

DEFINITION 2.2 We say a group is *cyclic* if there is one element that can generate the entire group.

Although we have seen an example of a finite group that is not cyclic, we will later see that the structure of *any* finite abelian group can be expressed in terms of the cyclic groups.

Even when a group is not cyclic, we sometimes can find *two* elements by which every element of the group can be expressed. For example, consider the two elements 3 and 5 from the group Z_8^*. Since $1 = 3 \cdot 3$ and $7 = 3 \cdot 5$, we find that all four elements of the group can be written as some *combination* of 3 and 5. We say that the *set* $\{3, 5\}$ generates the group.

Finally, consider the group of the dancing triangle, whose multiplication table is given in Table 1.2. By experimenting, we find that no single element can generate the entire group. However, there are many ways in which we can have *two* elements generating the entire group. For example, if we pick the two elements **RotRt** and **Spin**, we find that the other four elements can be expressed in terms of these two: **Stay** = **Spin** · **Spin**, **FlipRt** = **Spin** · **RotRt** **FlipLft** = **RotRt** · **Spin**, and **RotLft** = **RotRt** · **RotRt**.

One of the keys for entering a group into *Sage* is finding one or two elements (or sometimes even three are needed) that will generate the entire group. This information begins to reveal the structure of the group itself.

Problems for §2.1

For Problems **1** through **12**: Find all of the generators of the following groups. How many generators are there? (Note some groups will not have generators.)

1 Z_{12}	**4** Z_{24}	**7** Z_{12}^*	**10** Z_{16}^*
2 Z_{14}	**5** Z_9^*	**8** Z_{14}^*	**11** Z_{18}^*
3 Z_{16}	**6** Z_{11}^*	**9** Z_{15}^*	**12** Z_{20}^*

For Problems **13** through **20**: Use the totient function theorem (2.1) to find the size of the following groups:

13 Z_{100}^*	**15** Z_{490}^*	**17** Z_{1260}^*	**19** Z_{2100}^*
14 Z_{360}^*	**16** Z_{1200}^*	**18** Z_{1331}^*	**20** Z_{3675}^*

21 Prove that $\phi(n)$ is even for $n > 2$.

22 Using the totient function theorem (2.1), prove that there is no value of n for which $\phi(n) = 14$.

Interactive Problems

23 Use *Sages*'s circle graph to find all of the generators of the group Z_{21}.

24 Use *Sage*'s circle graph to see if there is an element of Z_{25}^* that generates Z_{25}^*. If so, how many such elements are there?

25 By using *Sage*'s `Generators()` command, determine whether Z_n^* is cyclic for $n = 9, 27, 81, 243, 5, 25, 125$. Make a conjecture about when Z_n^* is cyclic if n is a power of an odd prime.

26 By using *Sage*'s `Generators()` command, determine whether Z_n^* is cyclic for $n = 18, 54, 162, 486, 50, 250, 98, 686$. Make a conjecture about when Z_n^* is cyclic if n is twice the power of an odd prime.

27 By using *Sage*'s `Generators()` command, see if you can find an n for which Z_n^* is cyclic, and n doesn't fit into the categories of Problems 25 or 26.

2.2 Defining Finite Groups in *Sage*

For some groups there is a single element that generates the entire group, whereas in other groups two or more elements are required. In this section we will show how a finite group can be entered into *Sage* using a set of elements that generates the group. This in turn will give us a host of new groups to study, some of which will be very important as we explore the properties of groups.

We will begin with a cyclic group Z_n, which has a single generator that we will call x. From the circle graphs of Z_n, we could see that the sequence of n

elements

$$\begin{aligned} e &= x^0, \\ x &= x^1, \\ x \cdot x &= x^2, \\ x \cdot x \cdot x &= x^3, \\ \cdots &\quad \cdots \\ x \cdot x \cdot x \cdot \cdots \cdot x &= x^{(n-1)}, \end{aligned}$$

must mention every element of Z_n exactly once. This gives us a way to label the elements of Z_n in terms of the generator x. We also find that $x^n = e$. Thus, we can define the group Z_n merely by saying "x is a generator of the group, and n is the smallest number such that x^n is the identity."

Computational Example 2.5
Define the group Z_5 in *Sage*.

This group is cyclic, so we can use a single generator **x** to describe the group. First we define **e** to be the identity element with the command

```
InitGroup("e")
```

Next, we define the symbol **x** to be the group variable.

```
AddGroupVar("x")
```

Finally, we define x^5 to be e.

```
Define(x^5, e)
```

This is all we need to define the group Z_5. □

To view this group, we use the command

```
Z5 = ListGroup(); Z5
    {e, x, x^2, x^3, x^4}
```

which gives a list of all of the elements in the group, and assigns this list to the identifier **Z5**. The multiplication table for this group produced by the **MultTable** command is shown in Table 2.2.

Although the notation $\{0, 1, 2, 3, 4\}$ is more concise for this particular example, the use of generators is more versatile, since almost all finite groups can be expressed easily using generators.

Computational Example 2.6
Define the group Z_8^* in *Sage*.

This is not cyclic, but the group can be generated by $a = 3$ and $b = 5$. This group can be entered into *Sage* with the commands:

TABLE 2.2: Table of Z_5

·	e	x	x^2	x^3	x^4
e	e	x	x^2	x^3	x^4
x	x	x^2	x^3	x^4	e
x^2	x^2	x^3	x^4	e	x
x^3	x^3	x^4	e	x	x^2
x^4	x^4	e	x	x^2	x^3

```
InitGroup("e")
AddGroupVar("a", "b")
Define(a^2, e)
Define(b^2, e)
Define(b*a, a*b)
```

Note that we needed an extra **Define** statement to let *Sage* know that a and b commute with each other. To list the elements of the group, we could either use the **ListGroup** command as we did for Z_5, or we can find the group generated by the elements a and b with the **Group** command.

```
G = Group(a, b); G
    {e, a, b, a*b}
```

∎

We can define several groups in *Sage* at the same time, and by listing the generators with the **Group** command, *Sage* will know which group we are refering to. In contrast, **ListGroup()** will only list the most recently defined group.

We can now display the multiplication table for this group.

```
MultTable(G)
```

·	e	a	b	a*b
e	e	a	b	a*b
a	a	e	a*b	b
b	b	a*b	e	a
a*b	a*b	b	a	e

Computational Example 2.7

Suppose we have three different books on a shelf, and we consider rearrangements of the books. Enter this group into *Sage.*

Such a group of arrangements can be illustrated with the command

```
InitBooks(3)
```

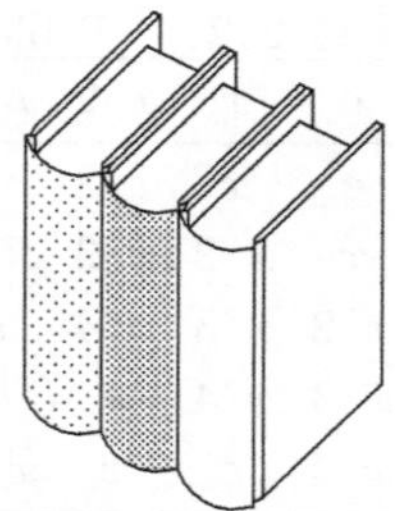

FIGURE 2.2: Visualizing arrangements of three books

which begins by showing three differently colored books, as in Figure 2.2. Two ways we could rearrange the books are to swap the first two books, or move the first book to the other end, sliding the other two books to the left. These two operations can be animated in *Sage* by

MoveBooks(First)
MoveBooks(Left)

By letting e be the identity element, a be the first rearrangement, and b be the rearrangement moving the books to the left, we find that all possible permutations of the books are generated by a and b. Since we clearly have $a^2 = b^3 = e$, we can use this to help define the group. As in Z_8^*, the plan is to express $b \cdot a$ in terms of a and b in alphabetical order. Since the combination $b \cdot a$ essentially switches the first and last books, we see that $(b \cdot a)^2 = e$, or

$$b \cdot a = (b \cdot a)^{-1} = a^{-1} \cdot b^{-1} = a \cdot b^2.$$

Thus, we can define this group by

```
InitGroup("e")
AddGroupVar("a", "b")
Define(a^2, e)
Define(b^3, e)
Define(b*a, a*b^2)
```

□

If we use the **Group** command to find the list of elements,

```
Group(a, b)
    {e, a, a*b, b, a*b*a, b*a}
```

we find that the last two elements are not written in standard order. In fact, if we compare this list to the **ListGroup** output,

```
G = ListGroup(); G
    {e, a, b, a*b, b^2, a*b^2}
```

TABLE 2.3: Multiplication table for S_3

·	e	a	b	a*b	b^2	a*b^2
e	e	a	b	a*b	b^2	a*b^2
a	a	e	a*b	b	a*b^2	b^2
b	b	a*b^2	b^2	a	e	a*b
a*b	a*b	b^2	a*b^2	e	a	b
b^2	b^2	a*b	e	a*b^2	b	a
a*b^2	a*b^2	b	a	b^2	a*b	e

we find that $a \cdot b \cdot a$ is really b^2. *Sage* is able to tell that these are the same element,

```
a*b*a == b^2
    True
```

but will not immediately simplify an expression involving group elements.

```
b^7
    b^7
```

Sage will, however, simplify expressions when putting them in a multiplication table. The output of

```
MultTable(G)
```

is shown in Table 2.3.

Is this really a group? We can tell from the multiplication table that G is closed with respect to multiplication, and that there is an identity element, e. We also recognize the familiar Latin square property that we have seen in all of the other multiplication tables. Since every row and every column contains exactly one e, every element has a unique inverse. The only property that we cannot check directly using the multiplication table is the associativity property. But this property is guaranteed by the way *Sage* defines groups. This group is called S_3, the permutation group on three objects. (Obviously it makes no difference what the three objects are. Books are just one possibility.)

Can *Sage* determine the inverse of an element?

```
(a*b)^-1
    b^-1*a^-1
```

Apparently, *Sage* is using Proposition 1.2, $(u \cdot v)^{-1} = v^{-1} \cdot u^{-1}$, but is not reducing it any further. However, there is a command, **SetReducedMult**, which will force all group operations to simplify.

```
SetReducedMult()
(a*b)^-1
    b^-1*a
```

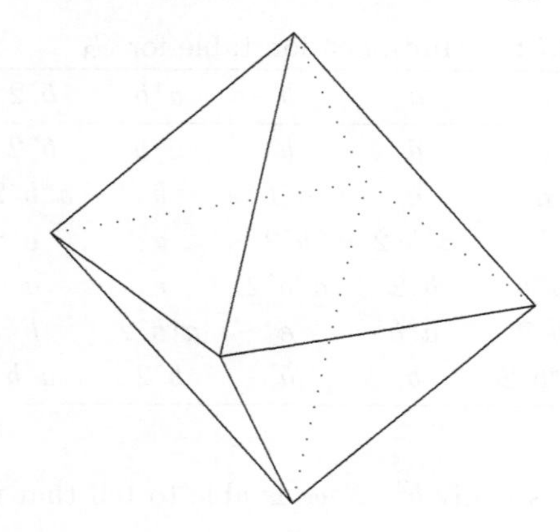

FIGURE 2.3: Octahedron with eight equilateral triangles

This may not seem like much of a simplification, but let us look at the whole group.

```
Group(a, b)
    {e, a, b^-1*a, b, b^-1, b*a}
```

It is clear that *Sage* is preferring to express b^2 as b^{-1}, and seems to prefer having the b's before the a's. The advantage of using **SetReducedMult** is that all group operations will be reduced to one of the six forms given above.

```
b^7
    b
b*a*b
    a
```

The multiplication tables for Terry's group and S_3 are very similar. By color coding the elements in the table, we see that the color patterns of the two multiplication tables are identical. Thus, these two groups behave in exactly the same way, even though the elements have different names. We say that these groups are *isomorphic*. We will cover isomorphic groups in Chapter 4.

Group have many applications. For example, the shape of an uncut diamond, as well as many other gemstones, is shown in Figure 2.3. This figure is reproduced by the *Sage* command

```
InitOctahedron()
```

One problem a gem cutter often faces is determining the orientation into which he should put the gemstone before he starts to cut. In such a case, he needs to know all of the possible ways the octahedron can be rotated. The set of rotations would form a group, similar to Terry's dance steps.

Computational Example 2.8
Consider the group of rotations on the octahedron, and enter this group into *Sage*.

There are eight triangles forming this solid. Three ways of rotating this figure are given by

```
RotateOctahedron(a)
RotateOctahedron(b)
RotateOctahedron(c)
```

The first of these flips the front horizontal edge, turning it upside down. The second rotates the closest face counter-clockwise, while the third rotates the closest vertex clockwise. If we let e be the identity element of this group, it is easy to see that

$$a^2 = e, \qquad b^3 = e, \qquad c^4 = e.$$

After some experimenting, we find that $b \cdot a \cdot b \cdot a = e$, $c \cdot b \cdot c \cdot c \cdot a = e$, and $c \cdot a \cdot c^3 \cdot a \cdot b = e$. From these identities, we can come up with the identities

$$b \cdot a = (b \cdot a)^{-1} = a^{-1} \cdot b^{-1} = a \cdot b^2.$$

$$c \cdot b = (c \cdot c \cdot a)^{-1} = a^{-1} \cdot c^{-1} \cdot c^{-1} = a\, c^3 \cdot c^3 = a \cdot c^2 \cdot c^4 = a \cdot c^2.$$

$$c \cdot a = (c^{-1} \cdot a \cdot b)^{-1} = b^{-1} \cdot a^{-1} \cdot c = b^2 \cdot a \cdot c = b \cdot a \cdot b^2 \cdot c = a \cdot b^4 \cdot c = a \cdot b \cdot c.$$

This allows us to define $b \cdot a$, $c \cdot a$, and $c \cdot b$ in terms of operations that are performed in *alphabetical order*. Although this is not mandatory, it is a good strategy to ensure each element will have a natural representation.

```
InitGroup("e")
AddGroupVar("a", "b", "c")
Define(a^2, e)
Define(b^3, e)
Define(c^4, e)
Define(b*a, a*b^2)
Define(c*a, a*b*c)
Define(c*b, a*c^2)
G = ListGroup(); G
    {e, a, b, a*b, b^2, a*b^2, c, a*c, b*c, a*b*c, b^2*c,
    a*b^2*c, c^2, a*c^2, b*c^2, a*b*c^2, b^2*c^2, a*b^2*c^2,
    c^3, a*c^3, b*c^3, a*b*c^3, b^2*c^3, a*b^2*c^3}
```

Since we told *Sage* how to express each combination of two generators out of order in terms of a combination in alphabetical order, *Sage* can express every element as a combination of generators in alphabetical order. ∎

We call this group the *octahedral* group, which will be an important example later on. The command

```
len(G)
  24
```

shows this group has 24 elements. (We could also have found this number using geometry on the octahedron. See Problem 7 of §2.3.) This group is too large to print a complete multiplication table, but *Sage* is able to produce a color-coded table for groups of up to 27 elements.

With *Sage*, we are able to create new groups to study. These examples help us to find pattens in the structure of all groups. In the next section we will study the substructure of a group, by finding smaller groups within a group.

Problems for §2.2

1 Show that if $a^2 = b^2 = e$, then saying that $b \cdot a = a \cdot b$ is equivalent to saying that $a \cdot b \cdot a \cdot b = e$.

2 In defining S_3, we used three facts about the group: $a^2 = e$, $b^3 = e$, and $b \cdot a = a \cdot b^2$. Using just these facts without *Sage*, prove that $b^2 \cdot a = a \cdot b$.

3 The group defined in Problem 18 has elements a and b such that $a^5 = e$, $b^4 = e$, and $b \cdot a = a^2 \cdot b$. Using just these facts without *Sage*, prove that $b^3 \cdot a = a^3 \cdot b^3$.

4 Write down the multiplication table for the group of rotations of a regular tetrahedron.

Hint: Consider an octahedron with four of the faces (colored by **sage** as red, yellow, orange, and cyan) extended so as to cover the other four faces. This gives us a tetrahedron, so the symmetries of a tetrahedron must be a subgroup of the octahedral group. Number the elements $1, 2, 3, \ldots, 9, T, E, W$, with 1 as the identity element. Then fill in the rest of the table. Once several elements are put in, use the Latin square property to speed up the process.

For Problems **5** through **16**: Recall the octahedral group was defined with 3 generators such that $a^2 = b^3 = c^4 = e$, $b \cdot a = a \cdot b^2$, $c \cdot a = a \cdot b \cdot c$, and $c \cdot b = a \cdot c^2$. Using just these facts without *Sage*, simplify the following expressions into a product that is in the form $a^i \cdot b^j \cdot c^k$, with $i < 2$, $j < 3$, and $k < 4$.

5 $b^2 \cdot a$ **8** $c \cdot b \cdot a$ **11** $c^2 \cdot b \cdot a$ **14** $c \cdot b \cdot a \cdot b$
6 $c^2 \cdot b$ **9** $c \cdot b^2$ **12** $c^2 \cdot b^2$ **15** $b \cdot c^2 \cdot a$
7 $c^2 \cdot a$ **10** $b \cdot c \cdot b$ **13** $c \cdot b^2 \cdot a$ **16** $b \cdot c^2 \cdot b$

17 Suppose we considered rearranging four books on a shelf instead of three. How many ways could we rearrange the books?

Interactive Problems

18 Use *Sage* to define a group that has two elements, a and b, such that $a^5 = b^4 = e$, and $b \cdot a = a^2 \cdot b$. How many elements does this group have?

19 Since the elements b and c could generate the octahedral group, define this group in *Sage* using only b and c. (Note: This will not work in *Mathematica.*)
Hint: Besides $b^3 = e$ and $c^4 = e$, *Sage* will need one more equation. What is the order of $b^2 \cdot c$?

20 Define a group in *Sage* that is generated by two elements a and b, with $a^3 = b^5 = (a \cdot b)^2 = e$. How big is the group? (Note: This will not work in *Mathematica.*)

2.3 Subgroups

A natural question to ask is whether we can have a smaller group inside of a particular group. If so, then this smaller group will yield additional structure to the overall group. In fact, we will learn how to find *all* smaller groups within a larger group, provided that the size of the group is not too large.

We begin by saying that H is a *subset* of a group G, denoted $H \subseteq G$, if H consists only of the elements of G. The empty set $\{\ \}$ is always considered to be a subset, but we will restrict our attention to non-empty subsets.

DEFINITION 2.3 We say that H is a *subgroup* of G if H is a non-empty subset of G and H is a group with respect to the operation $(\cdot)$ of G.

It should be noted that all non-trivial groups have at least two subgroups. One subgroup contains just the identity element $\{e\}$, while another contains all of the elements of G. These two subgroups are called the *trivial subgroups.*

To see if a subset H is a group, we must test all four of the group properties. But the associative property of H is guaranteed because the original group G is associative. The remaining three properties,

1. H is closed under multiplication. That is, $x \cdot y \in H$ whenever x and $y \in H$.
2. The identity element of G is in H.
3. Every element of H has its inverse in H. That is, $x^{-1} \in H$ whenever $x \in H$.

can be combined into one simple test.

PROPOSITION 2.2
Let $H \subseteq G$ and $H \neq \{\ \}$. Then H is a subgroup of G if, and only if, we have

$$x \cdot y^{-1} \in H \qquad \textit{for all} \qquad x, y \in H.$$

PROOF: First of all, we need to see that if H is a subgroup, then $x \cdot y^{-1}$ is in H whenever x and y are in H. By property (3), y^{-1} is in H, and so by property (1), $x \cdot y^{-1}$ is in H.

Conversely, let us suppose that $H \subseteq G$, $H \neq \{\ \}$, and whenever $x, y \in H$, then $x \cdot y^{-1} \in H$. We need to see that properties (1) through (3) are satisfied.

Since H is not the empty set, there is an element x in H, and so $x \cdot x^{-1} = e$ is in H. Thus, property (2) holds.

Next, we have that if y is in H, then $e \cdot y^{-1} = y^{-1}$ is in H, and so property (3) holds.

Finally, if x and y are in H, then y^{-1} is in H, and so $x \cdot (y^{-1})^{-1} = x \cdot y$ is in H. Thus, property (1) also holds. ◻

Example 2.9
Let us find a subgroup of S_3, defined in *Sage* by the commands:

```
InitGroup("e")
AddGroupVar("a", "b")
Define(a^2, e)
Define(b^3, e)
Define(b*a, a*b^2)
G = ListGroup(); G
    {e, a, b, a*b, b^2, a*b^2}
```

We can find smaller groups within this one, such as

$$H = \{e, b, b^2\}.$$

It is easy to see that if x and y are in H, then $x \cdot y^{-1}$ is in H. Therefore, this is a subgroup. There are other subgroups within this group, such as $\{e, a\}$. ◻

One of the main tools we will use to find subgroups of a group is the *intersection.* Given two subsets H and K of G, the *Sage* command **Intersection** finds the set of elements that are in both subsets, denoted $H \cap K$.

```
H = [e, b, b^2]
K = [e, a]
Intersection(H, K)
    [e]
```

Note that sets are *entered* in *Sage* using square brackets, even though they are often displayed using curly braces. (Technically, using square brackets produce a list of elements, which acts similar to a set. But the *Sage* routines know to treat a list as if it were a set.) Moreover, we can consider taking the intersection of a collection of many sets. If we let

```
L = [[e, a, b], [e, a*b, b], [e, a, b, b^2]]
```

then L represents a "set of sets." We can take the intersection of all of the sets in this collection with the command

```
Intersection(L)
    [e, b]
```

The mathematical notation for this intersection is

$$\bigcap_{H \in L} H.$$

We could ask whether the intersection of two *subgroups* of G forms a subgroup of G. The next proposition shows us that indeed, the intersection of subgroups forms a new subgroup.

PROPOSITION 2.3
Given a group G and a non-empty collection of subgroups, donated by L, then the intersection of all of the subgroups in the collection

$$H^* = \bigcap_{H \in L} H$$

is a subgroup of G.

PROOF: First of all, note that H^* is not the empty set, since the identity element is in each H in the collection. We now can apply Proposition 2.2. Let x and y be two elements in H^*. Then, for every $H \in L$ we have $x, y \in H$. Since each H is a subgroup of G, we have

$$x \cdot y^{-1} \in H.$$

Therefore, $x \cdot y^{-1}$ is in H^*, and so H^* is a subgroup of G. □

This proposition allows us to generate a subgroup of G from any subset of G.

DEFINITION 2.4 Given a subset S of a group G, we define the *subgroup generated by S* to be

$$[S] = \bigcap_{H \in L} H$$

where L denotes the collection of subgroups of G that contain the set S.

Actually, $[S]$ is the *smallest* subgroup of G that contains S. For if H is a subgroup of G containing S, then $H \in L$, so that $[S] \subseteq H$.

We can determine $[S]$ another way. It is clear that $[S]$ contains all of the products of the form

$$x_1 \cdot x_2 \cdot x_3 \cdot \cdots \cdot x_n,$$

where either

$$x_k \in S \quad \text{or} \quad x_k^{-1} \in S \qquad (1 \leq k \leq n).$$

But the set of all such products forms a subgroup H of G that contains S. Thus, $H = [S]$.

The command **Group** finds $[S]$ for any set S. Thus, we can find the subgroup of S_3 generated by the element b by the *Sage* command

```
SetReducedMult()
Group(b)
    {e, b, b^-1}
```

Note that we use the **SetReducedMult** command, so that the elements will appear in a consistant, albeit non-standard, format. Notice that this produces the same subgroup $\{e, b, b^2\}$ we observed before.

The subgroup generated by the set $\{b, a \cdot b\}$ is

```
Group(b, a*b)
    {e, a, b^-1, b, b^-1*a, b*a}
```

which produces the entire group. Had we not used the **SetReducedMult** command, the elements would have appeared in unusual combinations, yet we could still see that we had all 6 elements.

In order to find all of the subgroups of a given group G, we will begin by finding all of the *cyclic* subgroups. Notice that if we pick any element x of G, then $[\{x\}]$ will always be a cyclic subgroup of G, since x is the generator. This subgroup is usually denoted by $[x]$.

Example 2.10

Find all of the cyclic subgroups of S_3.

SOLUTION: The process of finding all of the cyclic subgroups is similar to finding the generators of a group. For each element, we consider raising that element to higher and higher powers until we produce the identity element. By referring to Table 2.3, we see that:

$$(e)^2 = e.$$
$$(a)^2 = e.$$
$$(b)^2 = b^2, \qquad (b)^3 = e.$$

$$(a \cdot b)^2 = e.$$
$$(b^2)^2 = b, \qquad (b^2)^3 = e.$$
$$(a \cdot b^2)^2 = e.$$

Thus, there are 5 cyclic subgroups, $\{e\}$, $\{e, a\}$, $\{e, b, b^2\}$, $\{e, a \cdot b\}$, and $\{e, a \cdot b^2\}$. Notice that none of the elements were generators, so the group itself is not cyclic. ∎

The easiest way to keep track of the cyclic subgroups is to note the size of the subgroup generated by each element.

DEFINITION 2.5 Let G be a group and let x be an element in G. We define the *order* of x to be $|[x]|$. That is, if $[x]$ is finite, the order of x is the number of elements in $[x]$. If $[x]$ is an infinite group we define the order of x to be infinity.

For each element in Example 2.10, the power of the element eventually reached the identity element, indicating that we have finished finding the cyclic subgroup. Here is a proof that shows this will always happen for a finite subgroup.

PROPOSITION 2.4
Suppose that the element x has finite order n. Then n is the smallest positive integer such that $x^n = e$. Furthermore,

$$[x] = \{e, x, x^2, x^3, \ldots, x^{n-1}\}.$$

PROOF: Since $[x]$ is finite, not all of the elements $\{x^0, x^1, x^2, x^3, x^4, \ldots\}$ can be distinct. Suppose that $x^a = x^b$ for two integers a and b, with $b > a$. Then $x^{b-a} = e$ and $b - a > 0$. So there exists a positive integer c such that $x^c = e$. We can let n be the smallest such integer. We want to prove that

$$[x] = \{e = x^0, x^1, x^2, x^3, \ldots, x^{n-1}\}$$

with these elements distinct. Indeed, if $x^a = x^b$ with $0 \le a < b \le n - 1$, then $x^{b-a} = e$ and $0 < b - a < n$, which contradicts the definition of n. Therefore, the elements in

$$\{e = x^0, x^1, x^2, x^3, \ldots, x^{n-1}\}$$

are all distinct.

Finally, we need to show that if y is in $[x]$, then there exists an r such that $x^r = y$, with $0 \le r \le n - 1$. But $y = x^k$ for some $k \in \mathbb{Z}$. We can define $r = k \textbf{ mod } n$. Then $0 \le r \le n - 1$ and furthermore, there is an integer q such that $k - r = nq$. Thus,

$$y = x^k = x^{(nq+r)} = (x^n)^q \cdot x^r = e^q \cdot x^r = x^r.$$

So every element of $[x]$ is of the form x^r, with $0 \le r \le n-1$. ∎

Example 2.11

Find the cyclic subgroups of the group $Z_{15}^* = \{1, 2, 4, 7, 8, 11, 13, 14\}$, showing the orders of the elements.

SOLUTION: We compute powers of each element until we reach the identity.

$$\begin{array}{lll} 1^2 = 1. & & \\ 2^2 = 4, & 2^3 = 8, & 2^4 = 1. \\ 4^2 = 1. & & \\ 7^2 = 4, & 7^3 = 13, & 7^4 = 1. \\ 8^2 = 4, & 8^3 = 2, & 8^4 = 1. \\ 11^2 = 1. & & \\ 13^2 = 4, & 13^3 = 7, & 13^4 = 1. \\ 14^2 = 1. & & \end{array}$$

Thus, we see that the cyclic subgroups are $[1] = \{1\}$, $[2] = [8] = \{1, 2, 4, 8\}$, $[4] = \{1, 4\}$, $[7] = [13] = \{1, 4, 7, 13\}$, $[11] = \{1, 11\}$, $[14] = \{1, 14\}$. We also see that 1 has order 1, the elements 4, 11, and 14 have order 2, and the elements 2, 7, 8, and 13 have order 4. Note this group lacks a generator. ∎

We can make a similar observation if we have an *infinite* cyclic subgroup.

PROPOSITION 2.5

Suppose that x has infinite order. Then x^n is not the identity element for all nonzero integers n. Furthermore,

$$[x] = \{\ldots, x^{-3}, x^{-2}, x^{-1}, x^0 = e, x^1, x^2, x^3, \ldots\},$$

where the powers of x are all distinct.

PROOF: Suppose that $x^n = e$ for some nonzero n. It suffices to consider the case $n > 0$, for if $x^n = e$, then $x^{-n} = e$.

By exactly the same reasoning as was used to prove Proposition 2.4, we see that

$$[x] = \{e = x^0, x^1, x^2, x^3, \ldots, x^{n-1}\}.$$

But this contradicts the fact that $[x]$ was infinite. Therefore, $x^n = e$ only if $n = 0$.

Moreover, if $x^a = x^b$, then $x^{b-a} = e$ and so $b - a = 0$ by what we have just proved. Thus, the powers of x are all distinct. ∎

Even though the group in Proposition 2.5 is infinite, we can still define it in *Sage*. In fact, we defined an infinite group in the process of defining all of the other groups. If we have x as the generator of an infinite group, then the group is defined by the following:

```
InitGroup("e")
AddGroupVar("x")
```

At this point, we have an infinite group defined.

```
x^4 * x^-7
x^-3
Order(x)
    +Infinity
```

Granted, we cannot display all of the elements as we did for the other groups (**Group(x)** would require interrupting *Sage*), but we can still multiply elements of this group.

Because of Propositions 2.4 and 2.5, we know that any cyclic group G is either a finite group

$$G = \{e, x, x^2, x^3, \ldots, x^{n-1}\}$$

that resembles the group Z_n, or is an infinite group

$$G = \{\ldots, x^{-3}, x^{-2}, x^{-1}, x^0 = e, x^1, x^2, x^3, \ldots\},$$

which resembles the group $\mathbb{Z}$. From this, we can quickly determine the nature of a subgroup of a cyclic group.

PROPOSITION 2.6

A subgroup of a cyclic group must be cyclic.

PROOF: Let g be a generator of the cyclic group G. The trivial subgroup $\{e\}$ is considered cyclic, so let S be a non-trivial subgroup. Then every element of S can be written as g^i for some i. Since both g^i and g^{-i} would then be in S, we see that g^i is in S for some positive i. Let k be the smallest positive integer such that g^k is in S. Then g^{mk} is in S for all integers m.

If there were some other element in S not in $[g^k]$, then this element is g^y for some integer y. Then $y = qk+r$ for some $0 < r < k$. Then $g^r = g^y \cdot (g^k)^{-q} \in S$, but we chose k to be the smallest positive integer for which $g^k \in S$. Thus, $S = [g^k]$, and so S is cyclic. □

Example 2.12

Find *all* the subgroups of the group $\mathbb{Z}$.

SOLUTION: Since $\mathbb{Z}$ is cyclic, we know that all subgroups are cyclic, hence can be expressed as $[k]$ for some integer k. But $[k]$ would be the multiples of k,

$$[k] = \{k \cdot x \mid x \in \mathbb{Z}\}.$$

We will denote the subgroup of the multiples of k by $k\mathbb{Z}$. Note that $0\mathbb{Z} = \{0\}$, and $1\mathbb{Z} = \mathbb{Z}$, so even the trivial subgroups are of this form. □

Finding all of the subgroups of a *non-cyclic* group is trickier, since we have to consider subgroups generated by two or more elements. *Sage* can speed up the process.

Computational Example 2.13
Find all of the subgroups of the group S_3.
SOLUTION: We found all of the cyclic subgroups in Example 2.10: $\{e\}$, $\{e, a\}$, $\{e, b, b^2\}$, $\{e, a \cdot b\}$, and $\{e, a \cdot b^2\}$. Note that any subgroup containing b must also contain b^2, and vice-versa. Also all subgroups will contain e. So to find subgroups that require two elements, we have 6 combinations to try:

```
InitGroup("e")
AddGroupVar("a", "b")
Define(a^2, e)
Define(b^3, e)
Define(b*a, a*b^2)
SetReducedMult()
Group(a, b)
    {e, a, b^-1, b, b^-1*a, b*a}
Group(a, a*b)
    {e, a, b^-1, b, b^-1*a, b*a}
Group(a, a*b^2)
    {e, a, b^-1, b, b^-1*a, b*a}
Group(b, a*b)
    {e, a, b^-1, b, b^-1*a, b*a}
Group(b, a*b^2)
    {e, a, b^-1, b, b^-1*a, b*a}
Group(a*b, a*b^2)
    {e, a, b^-1, b, b^-1*a, b*a}
```

In each case, we produced the entire group. This shows that the only non-cyclic subgroup of S_3 is S_3 itself. Thus, we have found a total of 6 subgroups of S_3. □

Computational Example 2.14
Find the orders of the elements of the octahedral group.
SOLUTION: If we reload the octahedral group,

```
InitGroup("e")
AddGroupVar("a", "b", "c")
Define(a^2, e); Define(b^3, e); Define(c^4, e)
Define(b*a, a*b^2); Define(c*a, a*b*c); Define(c*b, a*c^2)
```

```
G = ListGroup(); G
    {e, a, b, a*b, b^2, a*b^2, c, a*c, b*c, a*b*c, b^2*c,
    a*b^2*c, c^2, a*c^2, b*c^2, a*b*c^2, b^2*c^2, a*b^2*c^2,
    c^3, a*c^3, b*c^3, a*b*c^3, b^2*c^3, a*b^2*c^3}
```

we can find the order of the element $a \cdot c$ by typing

```
Order(a*c)
    3
```

to see that the order of this element is 3. There is a trick for finding the orders of all of the elements of the group at the same time.

```
[ Order(x) for x in G ]
    [1, 2, 3, 2, 3, 2, 4, 3, 4, 3, 2, 2, 2, 4, 3, 4, 3, 2, 4, 3, 2,
    3, 4, 2]
```

We find that every element of this group besides the identity has order 2, 3, or 4. In fact, there are 9 elements of order 2, 8 elements of order 3, and 6 elements of order 4. In Problem 7, you are asked to derive these values purely by considering the geometry of the octahedron. ∎

Let us now consider the orders of the elements of a cyclic group, such as Z_{12}.

```
G = ZGroup(12); G
    {0, 1, 2, 3, 4, 5, 6, 7, 8, 9, 10, 11}
[ Order(x) for x in G ]
    [1, 12, 6, 4, 3, 12, 2, 12, 3, 4, 6, 12]
```

We see that there is only one element of order 2, two elements each of order 3, 4, and 6, and four elements of order 12.

It is apparent that finding the number of elements of order k involves finding the number of solutions to the equation $x^k = e$. To help us find the number of solutions for a cyclic group, let us first prove the following proposition about modular multiplication.

PROPOSITION 2.7

Let n and k be two positive integers. Then

$$x \cdot k \equiv 0 \pmod{n}$$

if, and only if,

$$x = \frac{a \cdot n}{\gcd(n, k)}$$

for some integer a.

PROOF: First of all, notice that if

$$x = \frac{a \cdot n}{\gcd(n,k)},$$

then

$$x \cdot k = \frac{a \cdot n \cdot k}{\gcd(n,k)} = a \cdot n \cdot \frac{k}{\gcd(n,k)}.$$

and since $\gcd(n,k)$ is a divisor of k, we see that $x \cdot k$ is a multiple of n. Thus,

$$x \cdot k \equiv 0 \pmod{n}.$$

Now suppose that $x \cdot k$ is a multiple of n. We want to show that

$$a = \frac{x \cdot \gcd(n,k)}{n}$$

is in fact an integer. By the greatest common divisor theorem (0.4), there exist integers u and v such that $\gcd(n,k) = u \cdot n + v \cdot k$. Then

$$a = \frac{x \cdot (u \cdot n + v \cdot k)}{n} = x \cdot u + \frac{x \cdot k \cdot v}{n}.$$

Since $x \cdot k$ is a multiple of n, we see that a is an integer. Thus,

$$x = \frac{a \cdot n}{\gcd(n,k)}$$

for some integer a. □

We can now find the number of elements in a cyclic group that satisfies the equation $x^k = e$.

COROLLARY 2.1

Let G be a cyclic group of order n. Then there are precisely $\gcd(n,k)$ elements of G such that $x^k = e$.

PROOF: Let g be a generator of G, and let $x = g^y$ be an element of G. Then $x^k = (g^y)^k = g^{y \cdot k}$, which is equal to the identity if and only if

$$y \cdot k \equiv 0 \pmod{n}.$$

By Proposition 2.7, this is true if and only if

$$y = \frac{a \cdot n}{\gcd(n,k)}$$

for some integer a. Hence, the number of possible values of y between 0 and $n-1$ for which $z^{y \cdot k} = e$ is

$$\frac{n}{n/\gcd(n,k)} = \gcd(n,k).$$

Each such value of y between 0 and $n-1$ produces a different solution $x = z^y$, so there are exactly $\gcd(n, k)$ solutions. ∎

Finding the number of solutions to the equation $x^k = e$ in a group will become important as we classify the different groups. We will give a notation to this count.

DEFINITION 2.6 *Let G be a group, and k a positive integer. Then the number of elements of G for which $x^k = e$ is called the kth* root count *of G, and is denoted by*

$$R_k(G) = \big|\{x \in G \mid x^k = e\}\big|.$$

Corollary 2.1 can now be expressed in the new notation. If G is a cyclic group, then

$$R_k(G) = \gcd(|G|, k).$$

Sage has a command **RootCount(G, k)** that will compute $R_k(G)$. For example, to find the number of solutions to the equation $x^8 = e$ in Z_{12}, we can enter:

```
G = ZGroup(12)
RootCount(G, 8)
  4
```

We are now ready to consider a more complicated group. One of the puzzles that is related to the Rubik's Cube® is called the Pyraminx™. The Pyraminx™ consists of a triangular pyramid, with each of the four triangular sides partitioned into nine smaller triangles. The four "tips" can rotate, but this does not affect the puzzle. The command

```
InitPuzzle()
```

shows a simplified puzzle with the four tips chopped off, as in Figure 2.4. In fact, removing the four tips gives us the advantage of being able to see the faces on the back side of the puzzle through the hole created. Now the four corners of this puzzle can rotate clockwise, using the commands

```
RotatePuzzle(f)
RotatePuzzle(b)
RotatePuzzle(l)
RotatePuzzle(r)
```

We can always put the puzzle back into its original form with the command

```
InitPuzzle()
```

The set of all actions on the puzzle forms a group, called the Pyraminx™ group. This group is generated by the elements $\{t, b, r, l\}$, and has over 900,000 elements! We can animate a sequence of moves as we did for the octahedron:

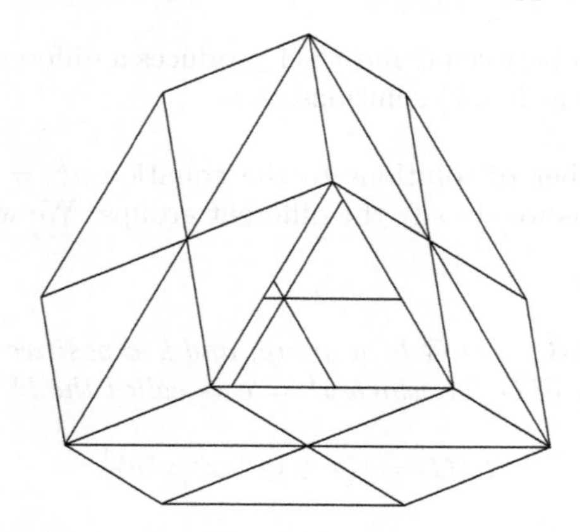

FIGURE 2.4: The Pyraminx^{TM} puzzle without tips

```
RotatePuzzle(b, f)
```

We can find the order of this element by repeatedly executing this command until the puzzle is back in order. In this particular case, the order of the element $b \cdot f$ is 15, meaning that we have to execute this procedure 15 times before we are back where we started.

Throughout this course, we will develop tools to work with groups that will help us to solve this puzzle, and others like it. The solution to the Pyraminx^{TM}, for example, is covered in §8.4.

Problems for §2.3

For Problems **1** through **6**: Find all of the subgroups of the following groups.

1 Z_{12}	**3** Z_{21}	**5** Z_8^*
2 Z_{20}	**4** Z_9^*	**6** Z_{15}^* (see Table 1.6 on page 49)

7 Use geometry to figure out how many elements of the octahedral group are of order 4 (rotations by 90 degrees). How many elements are of order 3? Of order 2? Check these figures by adding up these numbers, and adding one for the identity element, and show that this gives 24.

8 Using either the result of Problem 7 or the results of Example 2.14, find $R_2(G)$, $R_3(G)$, $R_4(G)$, and $R_6(G)$ for the octahedral group. Is $R_k(G)$ always a multiple of k?

9 Prove that no element of the Pyraminx^{TM} group can have order greater than 30.

Hint: Consider corners and edges separately. See the hint for Problem 25.

10 Use Corollary 2.1 to find the number of solutions to the equation $x^9 = e$ in the group Z_{18}. How many solutions are there to the equation $x^3 = e$ in this group? How many elements of order 9 are in this group?

Hint: For an element to be of order 9, it must solve $x^9 = e$, and *not* solve $x^n = e$ for any lower value of n.

11 Using only Corollary 2.1, determine the number of elements of Z_{42} that are of order 6. (See the hint for Problem 10.)

12 Prove that if k is a divisor of n, then there are exactly $\phi(k)$ elements of the group Z_n that are of order k.

Hint: First do the case when $n = k$. Then use Corollary 2.1 to show that the number of elements of order k for the groups Z_n and Z_k is the same.

13 Use Problem 12 to show that

$$n = \sum_{k|n} \phi(k)$$

where the sum has one term for each positive divisor k of n.

14 If a cyclic group has an element of infinite order, how many elements of finite order does it have? Prove your answer.

15 Let p be a prime number. If a group G has more than $p - 1$ elements of order p, prove that G cannot be a cyclic group.

16 Let G be an abelian group. Show that the set of elements of G that has finite order forms a subgroup of G. This subgroup is called the *torsion subgroup* of G.

17 Let G be an arbitrary group, with a and b two elements of G. Show that $a \cdot b$ and $b \cdot a$ have the same order.

Hint: First show by induction that $(a \cdot b)^n = a \cdot (b \cdot a)^{(n-1)} \cdot b$.

18 Suppose that G is a group with exactly one element of order 2, say x. Prove that $x \cdot y = y \cdot x$ for all y in G.

19 Let p be an odd prime number, and let $G = Z_p^*$. Show that the set

$$H = \{x^2 \mid x \in Z_p^*\}$$

forms a subgroup of G of order $(p-1)/2$. This subgroup H is called the group of *quadratic residues modulo p*.

Hint: Once you have shown that H is a subgroup, show that

$$x^2 \equiv 1 \pmod{p}$$

has exactly two solutions. Finally show that every element of H is derived from exactly two elements of Z_p^*.

20 Let G be a group with an even number of elements. Prove that $R_2(G)$ is even. See the hint for Problem 23 in §1.3.

Interactive Problems

21 Use Problem 18 from §2.2 to find the subgroup generated by the set $\{a, b^2\}$. How many elements does this subgroup have?

22 Use *Sage* to investigate the relationship between the order of an element, and the order of its inverse. First we pick a large group:

```
G = ZStar(360)
len(G)
    96
```

The following command selects a random element from the group.

```
a = G[randint(1, len(G)) - 1]; a
```

Then compare the results of the following operations.

```
Order(a)
Order(a^-1)
```

What do you observe? Try this with several random elements. Can you make a conjecture?

23 Repeat Problem 22, only using the octahedral group.

```
InitGroup("e")
AddGroupVar("a", "b", "c")
Define(a^2, e); Define(b^3, e); Define(c^4, e)
Define(b*a, a*b^2); Define(c*a, a*b*c); Define(c*b, a*c^2)
G = ListGroup()
```

24 Use *Sage* to find the order of the elements $b \cdot f$, $b \cdot f \cdot r \cdot f \cdot f$, and $f \cdot b \cdot r$ in the Pyraminx™ group.

25 Can you use *Sage* to find an element of the Pyraminx™ group that has order 30?

Hint: Exactly five of the six edges must be moved out of place. The sixth edge must flip as well.

Chapter 3

Patterns within the Cosets of Groups

We introduced subgroups in the last chapter, but left many questions unanswered. For example, is there any relationship between the size of the group and the size of one of its subgroups?

In this chapter we will introduce the tool of *cosets* to determine many of the properties of subgroups, including what possible sizes the subgroups could be. This in turn will allow us to create an encryption scheme that is virtually impossible to crack. The cosets will also reveal that some subgroups have a special property, which we will call *normal subgroups*. Normal subgroups will become an important tool for many important theorems, such as proving that a fifth-degree polynomial cannot be solved in terms of radicals.

3.1 Left and Right Cosets

In this section we will use cosets to prove Lagrange's theorem, which states that the order of the subgroup must divide the order of the group. This has some important ramifications in many fields such as Internet security.

To understand cosets, let us begin by looking at some cases where an element does *not* generate the group, in hopes of finding some patterns in the circle graphs. For example, consider the element 4 from the group Z_{10}. This element does not generate the entire group, as evident from the two types of arrows in the circle graph. The commands

```
ZGroup(10)
CircleGraph(Z, Add(4))
```

generate the graph in Figure 3.1. The solid arrows connect the set of points $\{0, 2, 4, 6, 8\}$, while the dotted arrows connect the points $\{1, 3, 5, 7, 9\}$. Thus, the group is partitioned into two sets, and no arrow connects these two.

One of the two sets is actually a subgroup of Z_{10}, the subgroup generated by the element 4. The other set is obtained by adding 1 to each element of the subgroup. Similar patterns arise when we use different elements of Z_{10} instead of 4.

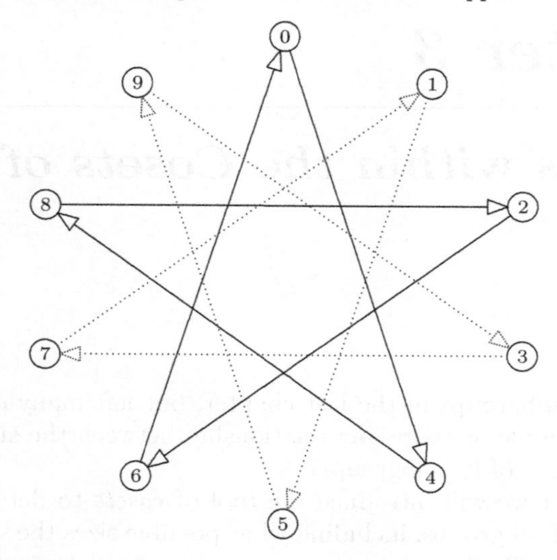

FIGURE 3.1: Circle graph of adding 4 (mod 10)

We can try a similar partitioning on non-abelian groups, such as Terry's group. If we consider forming a circle graph that sends each element to that element multiplied by **Spin**, we find we have a choice as to whether we have x map to $x \cdot$ **Spin** or to **Spin** $\cdot\, x$. The circle graph for the first option is shown in the left half of Figure 3.2. This leads to a partition of the group into the sets {**Stay**, **Spin**}, {**RotRt**, **FlipLft**}, and {**RotLft**, **FlipRt**}. The latter option, shown on the right side of Figure 3.2, is to multiply on the right instead of the left, giving the partition {**Stay**, **Spin**}, {**RotRt**, **FlipRt**}, and {**RotLft**, **FlipLft**}. In both cases, one of the sets in the partition is the subgroup

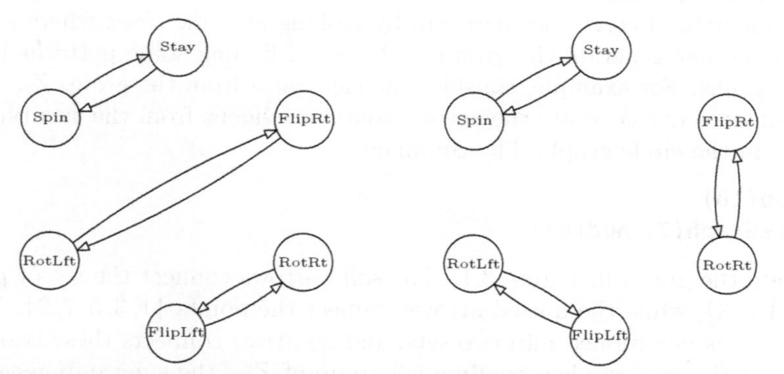

FIGURE 3.2: Circle graphs revealing cosets of Terry's group

```
G = InitTerry()
H = Group(Spin); H
    {Stay, Spin}
```

but the other sets are different.

DEFINITION 3.1 Let G be a group, and let H be a subgroup of G. If x is an element of G, we define the set

$$xH = \{x \cdot y \mid y \in H\}.$$

The set xH is called a *left coset of H*. Likewise,

$$Hx = \{y \cdot x \mid y \in H\}$$

is a *right coset of H*.

Sage mimics this notation. Thus,

```
H * RotRt
    {RotRt, FlipRt}
```

forms a right coset by multiplying every element in H by RotRt. Likewise

```
RotRt * H
    {RotRt, FlipLft}
```

forms a left coset.

We will denote the set of all *left* cosets of the subgroup H of G by G/H, and will denote the set of all *right* cosets of this subgroup by $H\backslash G$. Notice that the notation for right cosets uses a backward slash. In both cases, the subgroup can be considered to be on the "bottom," but since a right coset Hx has the subgroup on the left, we use $H\backslash G$, which also has H on the left, to list all such right cosets.

Sage finds all left and right cosets of G with H with the commands

```
LftCoset(G, H)
    {{Stay, Spin}, {RotLft, FlipRt}, {RotRt, FlipLft}}
RtCoset(G, H)
    {{Stay, Spin}, {RotRt, FlipRt}, {RotLft, FlipLft}}
```

Each coset is displayed as a list of elements, so we end up with a "list of lists," giving all of the cosets. These are exactly the partitions we observed in the circle graphs of **LeftMult(Spin)** and **RightMult(Spin)**. In fact, we begin to see some patterns in the cosets. First of all, all of the cosets are the same size. Also, every element of the group appears once, and only once, in each of the two coset lists. We will prove that these patterns are true in general with two lemmas.

LEMMA 3.1
Let G be a group and H be a finite subgroup of G. Then all left and right cosets of G with respect to H contain $|H|$ elements.

PROOF: It is clear from the definitions that Hu and uH each contains at most $|H|$ elements. In order to prove that the number is exactly $|H|$ we need to show that two distinct elements of H produce two different elements in the cosets. Suppose that this were not the case in a right coset. We would have two different elements x and y in H, for which

$$x \cdot u = y \cdot u,$$

but multiplying on the right by u^{-1} gives $x = y$, a contradiction. Similar reasoning works for left cosets. If

$$u \cdot x = u \cdot y,$$

multiplying on the left by u^{-1} shows that $x = y$. □

Next we must show that every element of G is in exactly one left coset and one right coset. This can be worded as follows:

LEMMA 3.2
If two left or two right cosets have an element in common, they are in fact the same coset. That is,

$$Hx \cap Hy \neq \{\ \} \quad \textit{implies that} \quad Hx = Hy,$$

and

$$xH \cap yH \neq \{\ \} \quad \textit{implies that} \quad xH = yH.$$

PROOF: We begin with right cosets. Suppose there is an element

$$g \in Hx \cap Hy.$$

Then there are elements h and k in H such that

$$g = h \cdot x = k \cdot y.$$

Therefore,

$$x = h^{-1} \cdot k \cdot y,$$

and so

$$(*) \qquad Hx = Hh^{-1} \cdot k \cdot y.$$

Since H is a subgroup, $h^{-1} \cdot k \in H$, so that $Hh^{-1} \cdot k \subseteq H$. Moreover, if u is in H, then

$$u = (u \cdot k^{-1} \cdot h)(h^{-1} \cdot k) \in Hh^{-1} \cdot k.$$

Therefore

$$H \subseteq Hh^{-1} \cdot k,$$

and we have shown that $H = Hh^{-1} \cdot k$. Combining this with (*) gives us $Hx = Hy$.

We can do left cosets in the same way. If there is an element $g \in xH \cap yH$, then there are elements h and k in H such that

$$g = x \cdot h = y \cdot k.$$

Therefore,

$$x = y \cdot k \cdot h^{-1},$$

and so

$$xH = y \cdot k \cdot h^{-1}H = yH. \qquad \square$$

Example 3.1

Find all of the left and right cosets of the subgroup $\{1, 11\}$ of the group Z_{15}^*.

SOLUTION: Since Z_{15}^* is abelian, the left and right cosets are the same. By Lemmas 3.1 and 3.2, the cosets will be disjoint, and all have 2 elements. One of the cosets will be the subgroup $\{1, 11\}$. We pick an element not in the subgroup, say 2, and multiply each element of the subgroup by 2, producing the coset $\{2, 7\}$. We pick another element not yet in a coset, and repeat the process. To find the coset containing 4, we multiply the subgroup by 4, to produce the coset $\{4, 14\}$. At this point, only 2 elements are unaccounted for, so they must be in their own coset, $\{8, 13\}$. So the list of cosets are

$$\{\{1, 11\}, \{2, 7\}, \{4, 14\}, \{8, 13\}\}. \qquad \square$$

With these two lemmas, we can show that the size of any subgroup is related to the size of the original group.

THEOREM 3.1: Lagrange's Theorem

Let G be a finite group, and H a subgroup of G. Then the order of H divides the order of G. That is, $|G| = k \cdot |H|$ for some positive integer k.

PROOF: We can use either left cosets or right cosets to prove this, so let us use right cosets. Every element of x in G is contained in at least one right coset. For example, x is contained in Hx. Let k be the number of distinct right cosets. Then, if the right cosets are

$$Hx_1, Hx_2, Hx_3, \ldots, Hx_k,$$

we can write

$$G = Hx_1 \cup Hx_2 \cup Hx_3 \cup \cdots \cup Hx_k.$$

The ∪'s represent the union of the cosets. But by Lemma 3.2, there are no elements in common among these sets, and so this union defines a partition of G. By Lemma 3.1, each coset contains $|H|$ elements. So $|G| = k \cdot |H|$. ∎

Lagrange's theorem (3.1), which seems apparent when looking at the cosets of a subgroup, turns out to have some far-reaching consequences. Let us look at some of the results that can be obtained using Lagrange's theorem.

COROLLARY 3.1

Let G be a finite group, and let x be an element of G. Then the order of x divides $|G|$.

PROOF: The order of x equals the order of the subgroup $[x]$ of G. Therefore, by Lagrange's theorem (3.1), the assertion follows. ∎

COROLLARY 3.2

Let G be a finite group of order n and let x be an element of G. Then

$$x^n = e.$$

PROOF: Let m denote the order of x. By Corollary 3.1, $n = mk$ for some integer k. Then we have $x^n = x^{mk} = (x^m)^k = e^k = e$. ∎

COROLLARY 3.3

A group of prime order is cyclic.

PROOF: Suppose G is of order p, which is prime. Then the only positive divisors of p are 1 and p, so by Lagrange's theorem (3.1) any subgroup must be of order 1 or p. If x is any element of G besides the identity, then $[x]$ contains x as well as the identity. Thus, $G = [x]$ so G is cyclic. ∎

COROLLARY 3.4

Let n be a positive integer, and x a number coprime to n. Then

$$x^{\phi(n)} \equiv 1 \pmod{n},$$

where $\phi(n)$ is Euler's totient function.

PROOF: We simply apply Corollary 3.2 to the group Z_n^*. This group has $\phi(n)$ elements, and if x is coprime to n then x is a generator of Z_n, so x is in Z_n^*. ∎

In particular, when $n = p$ is prime, we have

$$x^{p-1} \equiv 1 \pmod{p}.$$

This result is known as Fermat's little theorem. (See Historical Diversion on page 96.)

DEFINITION 3.2 If H is a subgroup of G, we define the *index* of H in G, denoted $[G\colon H]$, to be the number of right cosets in $H\backslash G$. Of course this is the same as the number of left cosets in G/H.

Notice that when G is a finite group we have by the argument in Lagrange's theorem (3.1) that $|G| = |H| \cdot [G\colon H]$.

Problems for §3.1

For Problems **1** through **8**: Find all of the cosets of the subgroup H of the group G. Since these groups are abelian, the left and right cosets are the same.

1 $G = Z_{10}$, $H = \{0,5\}$.
2 $G = Z_{12}$, $H = \{0,4,8\}$.
3 $G = Z_{15}$, $H = \{0,5,10\}$.
4 $G = Z_{15}^*$, $H = \{1,4\}$
5 $G = Z_{15}^*$, $H = \{1,14\}$
6 $G = Z_{16}^*$, $H = \{1,7\}$
7 $G = Z_{16}^*$, $H = \{1,9\}$
8 $G = Z_{24}^*$, $H = \{1,5\}$

9 List all of the left and right cosets of the subgroup { `Stay`, `FlipRt` } of Terry's group. Are the left and right cosets the same?

10 List all of the left and right cosets of the subgroup $\{e, a \cdot b\}$ of S_3. Are the left and right cosets the same? See Table 2.3 for the Cayley table of S_3.

For Problems **11** through **22**: Without using *Sage*, but rather by taking advantage of Corollary 3.4, compute the following modular powers.

11 5^{157} **mod** 7.
12 7^{185} **mod** 13.
13 13^{247} **mod** 15.
14 177^{203} **mod** 14
15 213^{317} **mod** 16
16 249^{343} **mod** 20.
17 323^{407} **mod** 21.
18 483^{479} **mod** 24
19 527^{429} **mod** 29
20 617^{579} **mod** 31
21 739^{629} **mod** 37
22 823^{729} **mod** 41

23 Prove that the order of Z_n^* is even whenever $n > 2$.
Hint: Find a subgroup of order 2.

24 Show that if H is a subgroup of G, and the left coset xH is also a subgroup of G, then x is in H.

25 Show that if an element y of a group G is in the right coset Hx, where H is a subgroup of G, then $Hy = Hx$.

26 Let $|G| = 33$. What are the possible orders for the elements of G? Show that G must have an element of order 3.

Historical Diversion

Pierre de Fermat (1601–1665)

Pierre de Fermat was a French lawyer and amateur mathematician. Although mathematics was only a hobby for him, he made several important contributions to the field. He came up with a method, which he called *adequality*, to find the maxima and minima of functions, and then adapted this method to find the tangent lines to curves. This would later be developed into differentiable calculus. He also made notable contributions in analytic geometry, probability and optics.

Fermat also did significant research in number theory. He studied perfect numbers (numbers equal to the sum of their positive divisors excluding the number itself), and amicable numbers, which would later be called Fermat numbers. While researching perfect numbers, he discovered Fermat's little theorem, which states that if p is a prime number, then $a^p - a$ is a multiple of p for all integers a.

But perhaps his greatest contribution to mathematics was accidental. He had a translation of *Arithmetica*, written by the Greek Diophantus, which in one section explained how to find solutions to the equation $x^2 + y^2 = z^2$ where x, y, and z are integers. Fermat wrote in the margin of his book, in Latin,

> It is impossible to write a cube as a sum of two cubes, a fourth power as a sum of two fourth powers, and, in general, any power beyond the second as a sum of two similar powers. For this, I have discovered a truly remarkable proof, but this margin is too small to contain it.

This note, discovered 30 years after Fermat's death by his son, claims that there is no positive integer solution to the equation $x^n + y^n = z^n$ for $n > 2$. Historians figure that his proof of "Fermat's last theorem" was probably flawed, as was the proof of countless mathematicians after him who tried to prove the statement. Yet, because of Fermat's "mistake," several new developments in mathematics occurred in attempt to find a proof. Countless advances in number theory were found in order to prove the theorem for small values of n. Ernst Kummer discovered rings and ideals in an attempt to correct a proof using unique factorization. (See Historical Diversion on page 432.) Finally, in 1994, Andrew Wiles produced the first successful proof, using the concepts of elliptic curves and modular forms, both of which would have been unknown to Fermat.

Image source: Wikimedia Commons

27 Suppose G is a group of order pq, where p and q are prime. Show that every non-trivial subgroup is cyclic.

28 Suppose G is a group of order pq, where p and q are prime. Suppose there is only one subgroup of order p, and one subgroup of order q. Prove that G is cyclic.

29 Find all subgroups of the group Z_{16}^*.
Hint: What does Lagrange's theorem say about a non-trivial, non-cyclic subgroup?

30 If G is a finite group, and p is prime, show that the number of elements of G of order p is a multiple of $p - 1$.

Interactive Problems

31 Find the left and right cosets of the subgroup $\{e, c, c^2, c^3\}$ of the octahedral group, given by:

```
InitGroup("e"); AddGroupVar("a", "b", "c")
Define(a^2, e); Define(b^3, e); Define(c^4, e)
Define(b*a, a*b^2); Define(c*a, a*b*c); Define(c*b, a*c^2)
G = ListGroup()
```

Are the left and right cosets the same?

32 Find the left and right cosets of the subgroup $\{e, c^2, a \cdot b^2 \cdot c, a \cdot b^2 \cdot c^3\}$ of the octahedral group, given by:

```
InitGroup("e"); AddGroupVar("a", "b", "c")
Define(a^2, e); Define(b^3, e); Define(c^4, e)
Define(b*a, a*b^2); Define(c*a, a*b*c); Define(c*b, a*c^2)
G = ListGroup()
```

Are the left and right cosets the same?

3.2 Writing Secret Messages

It was mentioned in the last section that Lagrange's theorem (3.1) has some far-reaching implications. One of these implications is the ability to write a message that no one can read except for the person to whom the message is sent, *even if the whole world knows the code!* This code has applications in Internet security and secure data transmissions.

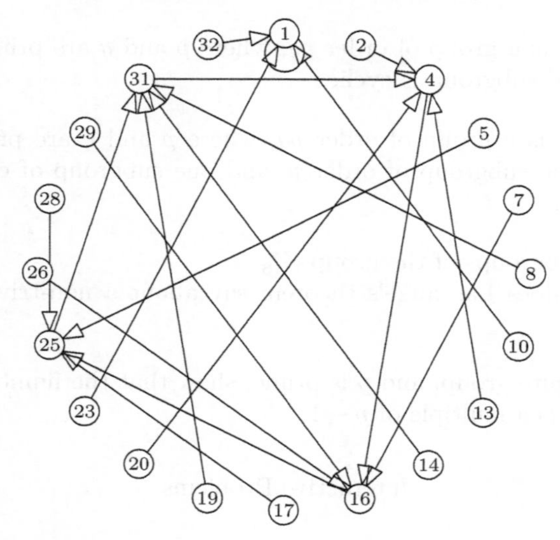

FIGURE 3.3: Circle graph for squaring in Z_{33}^*

Motavational Example 3.2
To introduce this code, we begin by considering the group Z_{33}^*, whose order is $\phi(33) = 20$. The elements of Z_{33}^* are

$$\{1, 2, 4, 5, 7, 8, 10, 13, 14, 16, 17, 19, 20, 23, 25, 26, 28, 29, 31, 32\}.$$

Consider the mapping that sends every element to its square. In essence we are defining a function $f(x) = x^2$ on this group. We can make a circle graph in *Sage* that maps each element to its square by the command

```
G = ZStar(33)
CircleGraph(G, Pow(2))
```

which produces Figure 3.3.

This graph is rather perplexing. The squares of 2, 13, 20, and 31 are all 4. The elements having "square roots" have four of them, while the majority of the elements do not have square roots.

If we try cubing each element instead, using the command

```
CircleGraph(G, Pow(3))
```

we get Figure 3.4. This graph has a very different behavior: no two elements have the same cube. We see from Figure 3.4 that the cube function is both one-to-one and onto. Thus, every element has a unique cube root. □

To understand this example, we notice that the cube root of any element in this group can be found by taking the seventh power of the element! This

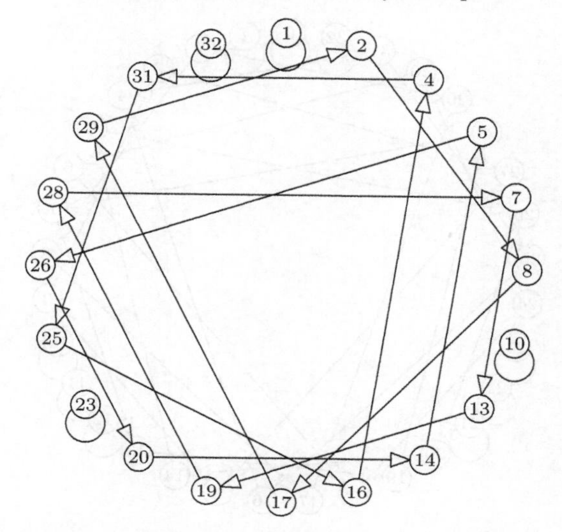

FIGURE 3.4: Circle graph for cubing in Z_{33}^*

is because $\phi(33) = 20$, so using Corollary 3.4,

$$(x^3)^7 = x^{21} = x^{20} \cdot x = e \cdot x = x.$$

The key difference between the squaring function and the cubing function stems from the fact that 3 is coprime to $\phi(33) = 20$, whereas 2 is not.

PROPOSITION 3.1

Suppose G is a finite group of order m, and that r is some integer that is coprime to m. Then the function $f(x) = x^r$ is one-to-one and onto. In other words, we can always find the unique r^{th} root of any element in G.

PROOF: Since G is of order m, we have by Corollary 3.2 that $x^m = e$ for all x in G. If r and m are coprime, then r is a generator in the additive group Z_m. But this means that r is an element of the group Z_m^*, and so there is an inverse element $s = r^{-1}$. Thus, $s \cdot r = 1$ in Z_m^*. Another way we could say this is

$$sr = km + 1$$

for some integer k.

Now we are ready to take the rth root of an element. If y is an element of G, then the rth root of y in G is merely y^s. To see this, note that

$$(y^s)^r = y^{sr} = y^{(km+1)} = (y^m)^k \cdot y = e^k \cdot y = y.$$

So y^s is one r^{th} root of y. But y^s must be a different element for every y in G, since the rth power of y^s is different. Since the rth root of every element of

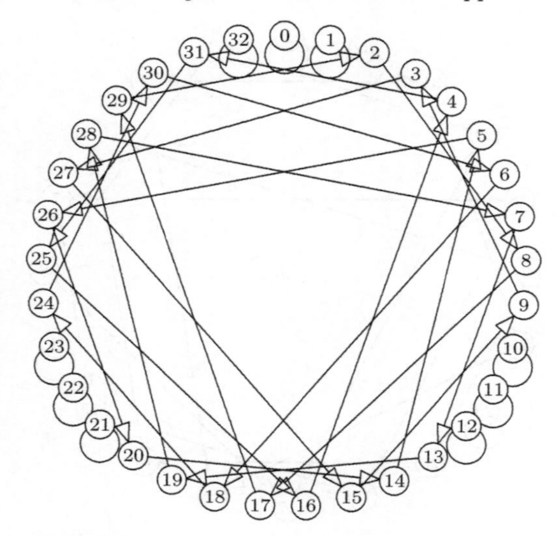

FIGURE 3.5: Circle graph for cubing modulo 33

G is accounted for, by the pigeonhole principle there cannot be *two* r^{th} roots to any element. Thus, y^s gives the unique r^{th} root of y in G. ∎

Motivational Example 3.3

Let us now consider the cubes of all numbers from 0 to 32. This will no longer be a group, since we have included non-invertible elements. But with the circle graph shown in Figure 3.5, we find that the mapping $x \to x^3$ is still one-to-one and onto. Thus, we can still find the cube root of a number modulo 33 by taking the seventh power modulo 33. ∎

The reason is given in the next proposition.

PROPOSITION 3.2

Suppose n is a product of two distinct primes and

$$r \cdot s \equiv 1 \pmod{\phi(n)}.$$

Then for all values of x less than n,

$$(x^r)^s \equiv x \pmod{n}.$$

PROOF: The proposition is trivial if $x = 0$, so we will assume that $x > 0$.

If x is coprime to n, then the proposition is true by Proposition 3.1. Suppose x is not coprime to $n = p \cdot q$, where p and q are the two distinct primes. By the totient function theorem (2.1), $\phi(n) = (p-1) \cdot (q-1)$. The number x

TABLE 3.1: Standard code sending letters to numbers

A	=	1	J	=	10	S	=	19
B	=	2	K	=	11	T	=	20
C	=	3	L	=	12	U	=	21
D	=	4	M	=	13	V	=	22
E	=	5	N	=	14	W	=	23
F	=	6	O	=	15	X	=	24
G	=	7	P	=	16	Y	=	25
H	=	8	Q	=	17	Z	=	26
I	=	9	R	=	18	Space	=	0

would be a multiple of either p or q, say p. Then

$$x^{r \cdot s} = (p \cdot a)^{r \cdot s} = p^{r \cdot s} \cdot a^{r \cdot s}$$

will be a multiple of p. Also, x is *not* a multiple of q since x is less than n. Since $r \cdot s \equiv 1 \pmod{(p-1)(q-1)}$, $r \cdot s \equiv 1 \pmod{(q-1)}$. Thus, by Proposition 3.1 again, we have

$$x^{rs} \equiv x \pmod{q}.$$

Since we also have $x^{rs} \equiv x \pmod{p}$, by the Chinese remainder theorem (0.7), we have, since p and q are coprime,

$$x^{rs} \equiv x \text{ (mod } pq = n). \qquad \square$$

Example 3.4

The function $x \to x^3$ is not only one-to-one and onto, but also mixes up the numbers 0 through 32 fairly well. This suggests an encryption scheme. We can first convert a message to a sequence of numbers using Table 3.1. For example,

CAN YOU READ THIS

becomes

3, 1, 14, 0, 25, 15, 21, 0, 18, 5, 1, 4, 0, 20, 8, 9, 19.

The encryption scheme is to replace each number with its cube, modulo 33. This gives us

27, 1, 5, 0, 16, 9, 21, 0, 24, 26, 1, 31, 0, 14, 17, 3, 28.

To decipher this, one would take the seventh power of each number in the sequence modulo 33, and convert back to letters in the alphabet. $\square$

The main drawback with this code is that, for longer messages, the letter E which encodes to 26 would appear most frequently in the encoded string.

Someone who didn't know the code might deduce that 26 stands for E without knowing anything about algebra. But also anyone who knew how to encrypt the message could use Proposition 3.2 to decipher the message, for they could deduce that 7 is the inverse of 3 modulo 20. What we need is a code in which everyone would know how to encrypt a message, but only the person who originated the code could decipher.

We can solve both of these problems just by picking n to be the product of two huge prime numbers p and q, say 80 digits each. Then $\phi(n) = (p-1) \cdot (q-1)$. We then pick r to be a number of at least four digits that is coprime to $\phi(n)$. The encryption scheme is then

$$x \to y = x^r \textbf{ mod } n.$$

We decode this by finding $s = r^{-1}$ in the group $Z^*_{\phi(n)}$. By Proposition 3.2, the operation

$$y \to x = y^s \textbf{ mod } n$$

"undoes" the encryption, since

$$(x^r)^s \equiv x \pmod{n}.$$

One big advantage of using huge numbers for the code is that we can encrypt an entire *line* at a time. For example,

CAN YOU READ THIS

can be encrypted by the single number

0301140025152100180501040020080919

by having every two digits represent one letter (still using Table 3.1). This prevents cracking the code using the frequencies of the letters. But the unusual advantage of this code is that only the originator of the code can decipher a message, even if the encryption scheme and the values of n and r were made public.

In order to decode a message, one must know the value of s, which is given by the inverse of $r \pmod{\phi(n)}$. This is easy to do with *Sage* once $\phi(n)$ is known, but how difficult it is to find $\phi(n)$! One needs to know the prime factorization of n, which would be about 160 digits long. Even *Sage* could not factor this in a reasonable amount of time. In fact, adding two digits to p and q makes the factorization 10 times harder. So by making the prime numbers larger, we can be assured that the factorization cannot be done within one's lifetime [6, p. 21]. Thus, without knowing the original primes p and q that were multiplied together, it is virtually impossible to determine s.

This encryption scheme is called the Rivest-Shamir-Adleman encryption [6, p. 374]. *Sage* has built in routines that allow us to experiment with RSA encryption.

Computational Example 3.5
The function

```
p = NextPrime(12345678901234567890123456789012345678901234567890123456789012345678
  90123456789012345678901234567890); p
   1234567890123456789012345678901234567890123456789012345678901234567\
   8901234567890123456789012345678901234567997
```

finds the next prime number larger than that 80 digit number. Since we want n to be the product of two large primes, we will find another large prime q, and multiply these primes together.

```
q = NextPrime(98765432109876543210987654321098765432109876543210987654321098765432
  10987654321098765432109876543210); q
   9876543210987654321098765432109876543210987654321098765432109876543\
   2109876543210987654321098765432109876543391
```

Although the input lines shown here are broken up to allow it to be printed, in *Sage* the input would be all on one line. The output uses a backslash to show that the output continues to the next line.

Sage uses a variation of the Agrawal, Kayal, and Saxena primality test to find the next prime number. This test can definitely determine whether a number is prime, in a time that is a polynomial function of the number of digits in p and q. Hence, we can quickly know for certain that the numbers p and q are prime.

Next, we multiply the two numbers together, and broadcast this number, n.

```
n = p*q; n
   12193263113702179522618503273386678859451150739156363359\
   36761164455788599298917889041106664075578553924704644144\
   851432895898221647614501039932917991510457827
```

The number n can be made public, along with any four-digit number r that is coprime to both $p-1$ and $q-1$. For simplicity, we will use a four-digit prime number.

```
r = NextPrime(1234); r
   1237
```

We can verify that this is coprime to $(p-1)(q-1)$ by computing

```
gcd((p-1)*(q-1), r)
   1
```

which returns 1.

To encrypt a message, the command

```
x = MessageToNumber("HERE IS A MESSAGE"); x
   805180500091900010013051919010705
```

converts any sentence into a number. Note we put the message in quotation marks. This number can now be encrypted by the command

```
y = PowerMod(x, r, n); y
    14724730500997597506102032344396082021733211823548530l293\
    32813791066600978417459038796026101371461452068807307578l\
    5860390004768255761553771456042827540589693 44
```

∎

Deciphering a message is very similar, only we will use the secret number s instead of r.

Computational Example 3.6

Suppose a friend, knowing the above values of n and r, gives the message

```
y = 6955740514702440687061142665742560438277560654407470323877
0078844683078352538833128853882711316059576508050596669314319 9
918635215093570816224139063616551830794
```

Use *Sage* to decipher the message.

Solution: To decode the message, we first need to know the value of s, which is the inverse of r modulo $(p-1)(q-1)$. Thus, the command to find s is given by

```
s = PowerMod(r, -1, (p-1)*(q-1) ); s
    116609783860223754044120366793989014476400253228956975375\
    724239753849619344952453906961891044114511747360397424479\
    600495150691225836271908768698156641698660 2133
```

Next, compute $y^s \pmod{n}$ by the command

```
x = PowerMod(y, s, n); x
    135555700063550051700037403330006693639300525558596454007\
    05855006958555493
```

Finally, the command

```
NumberToMessage(x)
    'Meet me at 7:30 p.m. behind the shed.'
```

puts the message into readable form. ∎

You may notice that the encryption in Table 3.1 has been expanded to allow lower case letters and punctuation. There are many other applications to this code besides sending secret messages. For example, suppose to get an account at the Electronic Bank, you pick two large random prime numbers, p and q. The bank then gives you the account number $n = p \cdot q$, and a number r, and makes these public. The bank also gives you the secret number

$$s = r^{-1} \pmod{(p-1)(q-1)}.$$

You use the number s to *decode* messages such as

```
y = MessageToNumber(
"Check 1034: Pay to the order of John Brown $43.50"); y
    358555361003130333440001651750070650070585500656854556800\
    65560010655864000268657364008334339335 30
x = PowerMod(y, s, n); x
    582856389557555731159430339514715251572029961076124346556\
    829718227800157027543664564994630786322333669442864481876\
    98381380453782773148309350448224286100193 3825
```

This number, along with your account number and the number r, is sent to John Brown. His bank can verify that this number is in fact a check as follows:

```
y = PowerMod(x, r, n)
NumberToMessage(y)
    'Check 1034: Pay to the order of John Brown $43.50'
```

This proves that the only person knowing s sent this message. Hence, the encryption acts as a *signature* to the check. Using this method, one can send an "electronic check" (even through e-mail) that is virtually impossible to forge.

Problems for §3.2

For Problems **1** through **4**: Use the code in Example 3.4 to encript the following messages.

1 RSA WORKS
2 TRUST NO ONE
3 NO PROBLEM
4 DONT PANIC

For Problems **5** through **8**: Use the code in Example 3.4 to decipher the following messages.

5 14, 17, 3, 28, 0, 3, 28, 26, 1, 20, 16.
6 1, 12, 12, 0, 28, 16, 28, 14, 26, 19, 28, 0, 13, 9.
7 19, 1, 11, 26, 0, 3, 14, 0, 28, 9.
8 24, 26, 22, 26, 24, 28, 26, 0, 4, 9, 12, 1, 24, 3, 14, 16.

9 Show that Proposition 3.2 is still true if n is the product of *three* distinct primes. In fact, many applications of the RSA code use three large primes instead of two.

10 Show that Proposition 3.2 is no longer true if we let $n = p^2$ for some prime p.

For Problems **11** through **18**: Find the inverse of the following functions. Note that some of these require the result of Problem 9.

11 $f(x) = x^3$ **mod** 51

12 $f(x) = x^7$ **mod** 55

13 $f(x) = x^5$ **mod** 91

14 $f(x) = x^7$ **mod** 143

15 $f(x) = x^{11}$ **mod** 221

16 $f(x) = x^{13}$ **mod** 437

17 $f(x) = x^7$ **mod** 1001

18 $f(x) = x^{11}$ **mod** 2717

19 Use the public key $n = 2773$ and $r = 17$ to encript "PASCAL" two letters at a time, using Table 3.1. How would you decipher this message?

20 Figure 3.3 shows that whenever an element of Z_{33}^* has a square root, it has 4 of them. Generalize this to any abelian group. If $R_k(G) = n$ for an abelian group G, and $y^k = b$ for some element b, then there are precisely n solutions to the equation $x^k = b$.

Interactive Problems

21 This exercise is required in order to do the RSA encryption Problems 22 or 23. Using *Sage*'s **NextPrime** command, find two large prime numbers p and q, at least 80 digits each. The digits should be random enough so that no one can spot a pattern. This is done by the two commands

```
p = NextPrime( large number goes here ); p
q = NextPrime( another large number goes here ); q
```

We will use the value $r = 10007$. Verify that this number is coprime to $p - 1$ and $q - 1$ by executing the following:

```
gcd( (p - 1)*(q - 1), 10007)
```

If this yields 10007 instead of 1, go back and find new values for p and q. Once the GCD is 1, compute $n = p \cdot q$, and save this to a file. To do this, enter

```
n = p*q;
print 'n =', n
```

If the output is continued over several lines using backslashes, clicking on the left side of the output will convert it to a single line output. This line can then be copied and pasted into a text file, using a text editor such as Notepad or TextEditor. If you are using VirtualBox, make sure that the shared clipboard is set to "Bidirectional" in the Settings → General → Advanced tab. Note: In *Mathematica*, the commands are different. See the *Mathematica* notebooks for the correct commands.

Next, find the secret number s, which deciphers a message:

```
s = PowerMod(10007, -1, (p - 1)*(q - 1))
```

You will need to save this number for future reference. Enter

```
print 's =', s
```

and copy and paste the single line version output to the same text file. Save this file with a name of your choice.

Finally, copy and paste just the n number into the body of an e-mail message, sent to the professor. Do not send the value of the secret number s, but save it for a future exercise.

22 Using the values of n and s from Problem 21, send an "electronic check" to your favorite professor for $100.00. This check will be in the form of a huge number, x. Once this number is found, enter

```
print 'x =', x
```

then copy and paste the single line version of the output into the body of a letter.

23 After doing Problem 21, your instructor will send you a response with a value of y. Copy and paste this line into an input cell of *Sage* and evaluate it. Also copy and paste the n and s lines from the text file you created in Problem 21, and execute these as well.

Using these values of n and s, decode the message y and hand in (on paper) what it says.

24 B. L. User tried creating his encryption number with the two primes

```
p = NextPrime(715870273457197548734156715678567821637415615197
37155752525673649286739584756092); p
q = NextPrime( p + 1 ); q
```

When he publicized the product $n = pq$, along with the value $r = 6367$, he received a message from a friend:

```
y = 3092722521993064335403878476414515883199432204869058005976
1407250735465231068482494915312824566404543856784721076165212
42043590910817888839981759972041752306977
```

What did this message say?

3.3 Normal Subgroups

In this section we will discover that some subgroups are special, for they have a property that other subgroups do not have. Such subgroups will be called *normal subgroups*. We will find that general subgroups do not always behave like we expect them to for certain operations, but normal subgroups will allow more operations to be done on the subgroups. As a result, normal subgroups become the cornerstone for most group theory results.

When we defined left cosets and right cosets, we were in essence defining how we could take an element of a group G and multiply it with a subgroup of G. But this definition can apply to any sub*set* of G. We can define a product of any *subset* of a group G by an element of G in the same way that we defined a product of a subgroup and an element. That is, if X is any subset of G, we can define

$$Xu = \{x \cdot u \mid x \in X\}, \quad \text{and}$$
$$uX = \{u \cdot x \mid x \in X\}.$$

We can also, using the same idea, multiply two subsets of G together.

DEFINITION 3.3 *If X and Y are two subsets of a group G, we can define*

$$X \cdot Y = \{x \cdot y \mid x \in X \text{ and } y \in Y\}.$$

By defining the product of subsets in this way, we find that $\{u\} \cdot X = uX$. We also discover that

$$X \cdot (Y \cdot Z) = (X \cdot Y) \cdot Z.$$

This raises some interesting questions. If X and Y are subgroups of G, will $X \cdot Y$ be a subgroup? Suppose X and Y are cosets of G with respect to a subgroup H. Will $X \cdot Y$ be a coset of G?

Exploratory Example 3.7

We will use the octahedral group of order 24 to experiment. In *Sage*, this can be reloaded with the commands

```
InitGroup("e")
AddGroupVar("a", "b", "c")
Define(a^2, e)
Define(b^3, e)
Define(c^4, e)
Define(b*a, a*b^2)
Define(c*a, a*b*c)
Define(c*b, a*c^2)
G = ListGroup(); G
    {e, a, b, a*b, b^2, a*b^2, c, a*c, b*c, a*b*c, b^2*c,
    a*b^2*c, c^2, a*c^2, b*c^2, a*b*c^2, b^2*c^2, a*b^2*c^2,
    c^3, a*c^3, b*c^3, a*b*c^3, b^2*c^3, a*b^2*c^3}
```

Two sample subgroups of order 4 are given by

```
H = Group(c); H
    {e, c^2, c, c^3}
```

```
K = Group(b*c); K
    {e, (b*c)^2, b*c, (b*c)^3}
```

which, unfortunately, are not displayed in their standard form because the command **SetReducedMult** was not enabled. Alternatively, we can conform the list of elements to appear as they do in the group G by using the **Conform** command.

```
Conform(K, G)
    {e, a*b^2*c^3, b*c, a*b*c^2}
```

We can now explore what happens when we multiply two subgroups together.

```
Conform(H*K, G)
    {e, a*b^2, a*b^2*c, c^2, c, a*b^2*c^2, b^2, a*b^2*c^3, a*b,
    c^3, a*c, b^2*c^2, b*c, a*b*c^2, a*c^3, b*c^3}
```

We can count the number of elements in the set by the command:

```
len(_)
    16
```

So $H \cdot K$ has 16 elements. Apparently, each element of H, when multiplied by an element in K, produces a unique element. This cannot be a subgroup by Lagrange's theorem (3.1), since 16 is not a factor of 24. □

Let us try again using the cosets of a subgroup instead of two subgroups.

Exploratory Example 3.8

The right cosets of H are given by

```
RtCoset(G, H)
    {{e, c^2, c, c^3}, {a, a*b*c, b*c^2, b^2*c^3}, {b, b^2*c,
    a*c^2, a*b*c^3}, {a*b, b^2*c^2, b*c, a*c^3}, {b^2, a*c,
    a*b*c^2, b*c^3}, {a*b^2, a*b^2*c, a*b^2*c^2, a*b^2*c^3}}
```

Let us try multiplying two right cosets of H, say the third and the fifth.

```
X = Conform(H*b, G); X
    {b, b^2*c, a*c^2, a*b*c^3}
Y = Conform(H*a*c, G); Y
    {b^2, a*c, a*b*c^2, b*c^3}
Conform(X * Y, G)
    {e, a*b^2, b, a*b^2*c, c^2, a, c, a*b^2*c^2, a*b*c, b*c^2,
    a*b^2*c^3, b^2*c, a*c^2, c^3, a*b*c^3, b^2*c^3}
```

This also produces something with 16 elements, so this cannot be a subgroup. However, a left coset multiplied by a right coset produces a glimmer of hope:

```
W = Conform(b*H, G); W
    {b, b*c^2, b*c, b*c^3}
Conform(W * Y, G)
    {e, a*b^2*c, a*c^2, b^2*c^3}
```

This, in fact, turns out to be a subgroup! In fact, any left coset times a right coset will produce a set with 4 elements. □

So what happens if we find a subgroup for which the right cosets and the left cosets are the same? Then the product of a left coset and a right coset would merely be the product of two cosets.

Motivational Example 3.9

An example of a subgroup for which the left and right cosets are the same is

```
M = Group(a*b*c^2, c^2); M
    {e, a*b^2*c, c^2, a*b^2*c^3}
```

which we can verify in *Sage* by the commands

```
RtCoset(G, M)
    {{e, a*b^2*c, c^2, a*b^2*c^3}, {a, b^2*c, a*c^2, b^2*c^3},
    {b, a*b*c, b*c^2, a*b*c^3}, {a*b, b*c, a*b*c^2, b*c^3},
    {b^2, a*c, b^2*c^2, a*c^3}, {a*b^2, c, a*b^2*c^2, c^3}}
LftCoset(G, M)
    {{e, a*b^2*c, c^2, a*b^2*c^3}, {a, b^2*c, a*c^2, b^2*c^3},
    {b, a*b*c, b*c^2, a*b*c^3}, {a*b, b*c, a*b*c^2, b*c^3},
    {b^2, a*c, b^2*c^2, a*c^3}, {a*b^2, c, a*b^2*c^2, c^3}}
```

Two of these cosets are

```
H = a*M; H
    {a, a^2*b^2*c, a*c^2, a^2*b^2*c^3}
K = Conform(b*M, G); K
    {b, a*b*c, b*c^2, a*b*c^3}
```

and the product

```
Conform(H * K, G)
    {a*b, b*c, a*b*c^2, b*c^3}
```

turns out to be another coset. In fact, the product of any two cosets of the subgroup M will yield a coset of M. □

First, let us give some terminology for this special type of subgroup.

DEFINITION 3.4 A subgroup H of the group G is said to be *normal* if all left cosets are also right cosets, and conversely, all right cosets are also left cosets. That is, H is normal if $G/H = H\backslash G$.

Next, we need a way to test whether a subgroup is normal.

PROPOSITION 3.3

A subgroup H is a normal subgroup of G if, and only if, $gHg^{-1} = H$ for all elements g in G.

PROOF: First of all, suppose H is normal, and let g be an element of G. Then gH and Hg both contain the element g. Since the left and right cosets are the same, we have

$$gH = Hg.$$

Multiplying both sides on the right by g^{-1} gives

$$gHg^{-1} = Hg \cdot g^{-1} = H.$$

Now, suppose that $gHg^{-1} = H$ for all elements g in G. Then

$$Hg = (gHg^{-1}) \cdot g = gHe = gH.$$

Thus, every left coset is also a right coset, and vice versa. ∎

This gives us a way to determine if a subgroup is normal, but we can improve on this test.

PROPOSITION 3.4

Let H be a subgroup of G. Then H is normal if, and only if,

$$g \cdot h \cdot g^{-1} \in H$$

for all elements $g \in G$, and $h \in H$.

PROOF: If H is a normal subgroup of G, then $g \cdot h \cdot g^{-1} \in gHg^{-1}$, which is H by Proposition 3.3.

Let us suppose that for all g in G and h in H, $g \cdot h \cdot g^{-1} \in H$. Then

$$gHg^{-1} \subseteq H.$$

In particular, if we replace every g with g^{-1}, we get

$$g^{-1}H(g^{-1})^{-1} \subseteq H.$$

Multiplying both sides of the equation by g on the left gives us $Hg \subseteq gH$, and multiplying on the right by g^{-1} gives us $H \subseteq gHg^{-1}$. Since we also have that $gHg^{-1} \subseteq H$, we can conclude that $gHg^{-1} = H$. Then from Proposition 3.3, H is normal. ∎

There are many other examples of normal subgroups. For example, if G is any group, then the subgroups $\{e\}$ and G are automatically normal. These normal subgroups are said to be *trivial*. If G is commutative, then any subgroup will be a normal subgroup. Here is another way to tell a subgroup is normal.

PROPOSITION 3.5

If H is a subgroup of G with index 2, then H is a normal subgroup.

PROOF: Since H is a subgroup of G with index 2, there are two left cosets and two right cosets. One of the left cosets is eH, which is the set of elements in H. The other left coset must then be the set of elements not in H. But the same thing is true for the right cosets, so the left and right cosets are the same. Thus, H is normal. ∎

When we have a normal subgroup, the set of cosets will possess more properties than for standard subgroups. We will explore these in the next section.

Problems for §3.3

1 Show that if H is a subgroup of a group G, then $H \cdot H = H$, where the product of two sets is defined in Definition 3.3.

2 Find all of the normal subgroups of D_3. (This is Terry's group.)

3 Let H be a subgroup of G such that every left coset $a \cdot H$ is also a right coset $H \cdot b$. Prove that H is a normal subgroup of G.

4 Prove that the intersection of two normal subgroups of G is a normal subgroup of G.

5 Let N be a normal subgroup of G, and let H be a subgroup of G that contains the subgroup N. Prove that N is a normal subgroup of H.

6 Show that if G is an abelian group, and X and Y are two subgroups of G, then $X \cdot Y$ is a subgroup of G.

7 We saw in Example 3.9 that M was a normal subgroup of the octahedral group. Find a normal subgroup of M that is *not* a normal subgroup of the octahedral group.

8 Let G be a group, and let Z be the set of elements in G that commute with *all* the elements of G. That is,

$$Z = \{x \in G \mid g \cdot x = x \cdot g \text{ for all } g \in G\}.$$

Show that Z is a subgroup of G.

9 Let Z be the subgroup of Problem 8. Show that Z is in fact a normal subgroup of G.

10 Suppose a group G has a normal subgroup H with only two elements. Show that H is contained in the subgroup Z from Problem 8.

11 Let H be a normal *cyclic* subgroup of a finite group G, and let K be a subgroup of H. Show that K is a normal subgroup of G. (This would not be true if the word *cyclic* was left out. See Problem 7.)

12 Let G be the group from Example 1.8 in §1.3, the group of *linear functions* of the form $f(x) = mx + b$, with $m, b \in \mathbb{R}$, $m \neq 0$. Let N be the subset of G for which $m = 1$, that is,

$$N = \{\phi(x) = x + b \mid b \in \mathbb{R}\}.$$

Show that N is a normal subgroup of G.

13 Let G be the group of *linear functions* as in Problem 12. Let T be the subset of G for which $b = 0$, that is,

$$T = \{\phi(x) = mx \mid m \in \mathbb{R}, m \neq 0\}.$$

Show that T is a subgroup of G, but not a normal subgroup. If $f(x) = 2x+3$, describe both the left and right cosets $f \cdot T$ and $T \cdot f$.

14 If H is a subgroup of G, and K is a normal subgroup of G, show that $H \cdot K = K \cdot H$.

15 Use Problem 14 to show that $H \cdot K$ is a subgroup of G.

16 Let H be a subgroup of G, and K a normal subgroup of G. Show that $H \cap K$ is a normal subgroup of H.

17 Use Problem 15 to show that if *both* H and K are normal subgroups of G, then $H \cdot K$ is a normal subgroup of G.

18 Let G be a group of order $2p$, where p is prime. Show that if H is a subgroup that is *not* normal, then H has precisely two elements.

Interactive Problems

19 Show that there is a group Q which is generated by two elements a and b, for which

$$a^4 = e, \qquad b^2 = a^2, \qquad b \cdot a = a^3 \cdot b, \qquad a^2 \neq e.$$

This can be entered into *Sage* with the command

```
InitGroup("e")
AddGroupVar("a", "b")
Define(a^4, e)
Define(b^2, a^2)
Define(b*a, a^3*b)
Q = Group(a, b); Q
```

Find all subgroups of this group, and show that all subgroups are normal, even though the group is non-abelian. (Write down the list of left cosets and right cosets for each subgroup found.)

20 Use *Sage*, along with a bit of trial and error, to find a subgroup of order 12 of the octahedral group. Show that this subgroup is a normal subgroup. The following reloads the octahedral group:

```
InitGroup("e"); AddGroupVar("a", "b", "c")
Define(a^2, e); Define(b^3, e); Define(c^4, e)
Define(b*a, a*b^2); Define(c*a, a*b*c); Define(c*b, a*c^2)
G = ListGroup()
```

3.4 Quotient Groups

In this section, we will take advantage of the special properties that normal subgroups have. In fact, we will be able to create a new group from the normal subgroup, which in many aspects acts like a division of two groups. Hence we will call these new groups quotient groups.

In the last section we observed a case where H was a normal subgroup of G, and the product of two cosets yielded another coset. Let us begin by proving that this will always happen for normal subgroups.

LEMMA 3.3

If N is a normal subgroup of G, then the product of two cosets of N produces a coset of N. In fact,

$$aN \cdot bN = (a \cdot b)N.$$

PROOF: We simply observe that

$$aN \cdot bN = a \cdot (Nb) \cdot N = a \cdot (bN) \cdot N = (a \cdot b) \cdot (N \cdot N) = (a \cdot b)N.$$

Note that $Nb = bN$ because N is a normal subgroup. ∎

This proposition is very suggestive. Since we can multiply two cosets together, can the set of all cosets form another group? This is, in fact, exactly what happens.

THEOREM 3.2: The Quotient Group Theorem
Let N be a normal subgroup of G. Then the set of all cosets is a group, which is denoted by G/N, called the quotient group *of G with respect to N.*

PROOF: We simply have to check that G/N satisfies the four requirements in Definition 1.5. The closure property is given by Lemma 3.3. To check associativity,

$$\begin{aligned} aN \cdot (bN \cdot cN) &= aN \cdot (b \cdot c)N = (a \cdot (b \cdot c))N \\ &= ((a \cdot b) \cdot c)N = (a \cdot b)N \cdot cN = (aN \cdot bN) \cdot cN. \end{aligned}$$

The identity element is $eN = N$, and we can check that

$$eN \cdot aN = (e \cdot a)N = aN, \quad \text{and}$$
$$aN \cdot eN = (a \cdot e)N = aN.$$

Finally, the inverse of aN is $a^{-1}N$, since

$$aN \cdot a^{-1}N = (a \cdot a^{-1})N = eN = N, \quad \text{and}$$
$$a^{-1}N \cdot aN = (a^{-1} \cdot a)N = eN = N.$$

Thus, the set of all cosets forms a group. ∎

Example 3.10
One of the easiest groups to consider is the group of integers $\mathbb{Z}$ under addition. A subgroup of $\mathbb{Z}$ would consist of all multiples of k, with $k \geq 0$. ($k = 0$ and $k = 1$ produce the two trivial subgroups.) We denoted this normal subgroup of $\mathbb{Z}$ by $k\mathbb{Z}$. All elements in each coset would be equivalent modulo k. Thus, there would be k cosets of $k\mathbb{Z}$ (except when $k = 0$). Hence, $\mathbb{Z}/k\mathbb{Z}$ is essentially the same group as Z_k. That is, x and y will be in the same coset if, and only if,

$$x \equiv y \pmod{k}.$$

∎

We can extend this notation to any normal subgroup. We say that

$$x \equiv y \pmod{N}$$

to indicate x and y belong in the same coset of G with respect to N. In fact, if $x \equiv y \pmod{N}$, then $N \cdot x = N \cdot y$, so $N \cdot x \cdot y^{-1} = N$, giving us $x \cdot y^{-1} \in N$. Thus, we have

$$x \equiv y \pmod{N} \quad \text{if, and only if,} \quad x \cdot y^{-1} \in N.$$

In §1.2, we defined an equivalence relation as a relation satisfying the three properties

1. (Reflexive) Every element x is equivalent to itself.
2. (Symmetric) If x is equivalent to y, then y is equivalent to x.
3. (Transitive) If x is equivalent to y, and y in turn is equivalent to z, then x is equivalent to z.

Because of the fact that the two elements are equivalent if they are in the same coset, it is clear that $x \equiv y \pmod{N}$ is an equivalence relation. The equivalence classes would be the cosets of N for which the relation is defined.

Computational Example 3.11

In the last section we found a normal subgroup of the octahedral group, namely

```
M = Group(a*b*c^2, c^2); M
    {e, a*b^2*c, c^2, a*b^2*c^3}
```

The cosets, or equivalence classes, with respect to this subgroup are given by the command

```
Q = LftCoset(G, M); Q
    {{e, a*b^2*c, c^2, a*b^2*c^3}, {a, b^2*c, a*c^2, b^2*c^3},
    {b, a*b*c, b*c^2, a*b*c^3}, {a*b, b*c, a*b*c^2, b*c^3},
    {b^2, a*c, b^2*c^2, a*c^3}, {a*b^2, c, a*b^2*c^2, c^3}}
```

We can use the *Sage* command **MultTable(Q)** to give us a multiplication table of the quotient group Q, shown in Figure 3.6. Since the names of the elements are so long, *Sage* uses a color code for the elements, which is shown here as shading. □

Notice that this table is very similar to the table for the group S_3. This group is already defined in as a subset of the octahedral group, so we can look at its multiplication table.

```
H = Conform(Group(a, b), G); H
    {e, a*b^2, b, a, b^2, a*b}
MultTable(H)
```

This produces the table in Table 3.2. With this particular arrangement of the elements, we see that the color patterns for Q and H match. In Chapter 4, we will define two groups that have the same color pattern as being *isomorphic*.

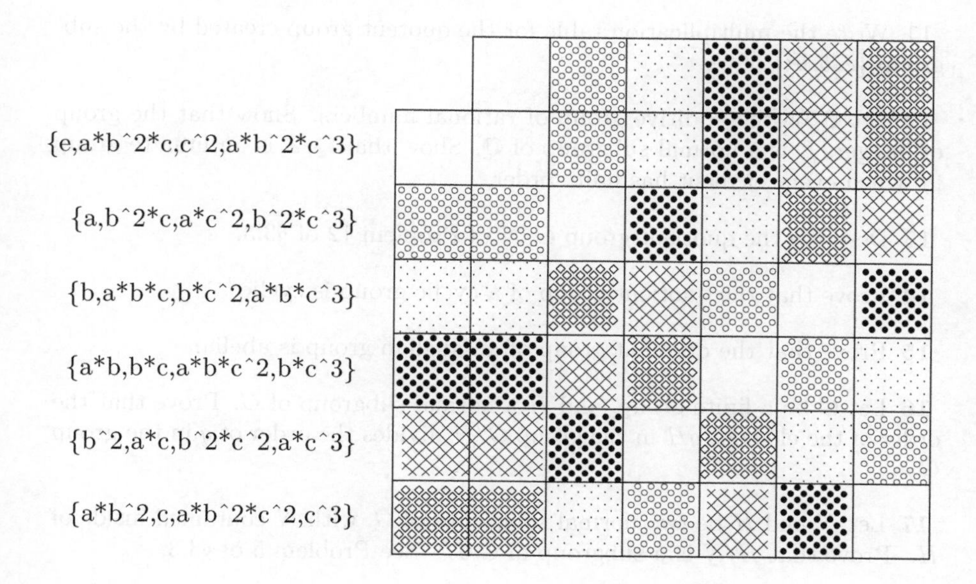

FIGURE 3.6: Multiplication table for the quotient group

Problems for §3.4

For Problems **1** through **9**, write the multiplication table for the following quotient groups:

1 $Z_{10}/\{0,5\}$
2 $Z_{12}/\{0,4,8\}$
3 $Z_{15}/\{0,5,10\}$
4 $Z_{12}/\{0,6\}$
5 $Z_{15}^*/\{1,4\}$
6 $Z_{15}^*/\{1,14\}$
7 $Z_{16}^*/\{1,7\}$
8 $Z_{13}^*/\{1,3,9\}$
9 $Z_{24}^*/\{1,5\}$

10 Write the multiplication table for the quotent group created by the subgroup {**Stay**, **RotRt**, **RotLft**} of Terry's group.

TABLE 3.2: Another multiplication table for S_3

$\cdot$	e	a*b^2	b	a	b^2	a*b
e	e	a*b^2	b	a	b^2	a*b
a*b^2	a*b^2	e	a	b	a*b	b^2
b	b	a*b	b^2	a*b^2	e	a
a	a	b^2	a*b	e	a*b^2	b
b^2	b^2	a	e	a*b	b	a*b^2
a*b	a*b	b	a*b^2	b^2	a	e

11 Write the multiplication table for the quotent group created by the subgroup $\{e, b, b^2\}$ of S_3.

12 Let $\mathbb{Q}$ be the additive group of rational numbers. Show that the group of integers $\mathbb{Z}$ is a normal subgroup of $\mathbb{Q}$. Show that $\mathbb{Q}/\mathbb{Z}$ is an infinite group in which every element has finite order.

13 Describe the quotient group G/N of Problem 12 of §3.3.

14 Prove that the quotient group of a cyclic group is cyclic.

15 Prove that the quotient group of an abelian group is abelian.

16 Let G be a finite group, and H a normal subgroup of G. Prove that the order of the element gH in the group G/H divides the order of g in the group G.

17 Let N and H be two normal subgroups of G, with N contained inside of H. Prove that H/N is a subgroup of G/N. See Problem 5 of §3.3.

18 Let N and H be two normal subgroups of G, with N contained inside of H. Show that H/N is a *normal* subgroup of G/N. See Problem 17.

Interactive Problems

19 Define in *Sage* the group $G = Z^*_{105}$. How many elements does this group have? Consider the subgroup H generated by the element 11. A circle graph demonstrating the cosets G/H can be obtained by the command

```
CircleGraph(G, Mult(11))
```

By looking at the circle graph, determine the cosets of G with respect to H. What is the order of the element $2 \cdot H$ in the quotient group G/H?

20 Here is a group of order 20 from Problem 18 of §2.2:

```
InitGroup("e")
AddGroupVar("a", "b")
Define(a^5, e); Define(b^4, e); Define(b*a, a^2*b)
G = ListGroup()
```

Find a normal subgroup H of order 5, and form the quotient group G/H.

Chapter 4

Mappings between Groups

So far we have not considered the possibility of a *function* defined on a group. This chapter explores the idea of a function, or mapping, which sends elements of one group to another. With such mappings, we will find a way to determine whether two groups are essentially the same. We also will find a connection between group functions and normal subgroups. Finally, we will use function composition to prove three very important theorems in group theory.

4.1 Isomorphisms

The quotient group G/M we saw at the end of the last chapter turned out to be very similar to the group S_3. They are technically distinct, since the names for their elements are totally different. In this section we will demonstrate the relationship between these two groups, using the concept of a mapping from one group to the other.

We begin by finding a correlation between the elements of the two groups so that the corresponding multiplication tables would have identical color patterns.

Motivational Example 4.1

Here is one such possible correlation between the two groups:

$$\begin{aligned} e &\leftrightarrow \{e,\, a\cdot b^2\cdot c,\, c^2,\, a\cdot b^2\cdot c^3\} \\ a\cdot b^2 &\leftrightarrow \{a,\, b^2\cdot c,\, a\cdot c^2,\, b^2\cdot c^3\} \\ b &\leftrightarrow \{b,\, a\cdot b\cdot c,\, b\cdot c^2,\, a\cdot b\cdot c^3\} \\ a &\leftrightarrow \{a\cdot b,\, b\cdot c,\, a\cdot b\cdot c^2,\, b\cdot c^3\} \\ b^2 &\leftrightarrow \{b^2,\, a\cdot c,\, b^2\cdot c^2,\, a\cdot c^3\} \\ a\cdot b &\leftrightarrow \{a\cdot b^2,\, c,\, a\cdot b^2\cdot c^2,\, c^3\} \end{aligned}$$

Suppose we use this correlation to define a *function* $f(x)$ sending each element of S_3 to an element of G/M. Thus,

$$\begin{aligned} f(e) &= \{e,\ a\cdot b^2\cdot c,\ c^2,\ a\cdot b^2\cdot c^3\} \\ f(a\cdot b^2) &= \{a,\ b^2\cdot c,\ a\cdot c^2,\ b^2\cdot c^3\} \\ f(b) &= \{b,\ a\cdot b\cdot c,\ b\cdot c^2,\ a\cdot b\cdot c^3\} \\ f(a) &= \{a\cdot b,\ b\cdot c,\ a\cdot b\cdot c^2,\ b\cdot c^3\} \\ f(b^2) &= \{b^2,\ a\cdot c,\ b^2\cdot c^2,\ a\cdot c^3\} \\ f(a\cdot b) &= \{a\cdot b^2,\ c,\ a\cdot b^2\cdot c^2,\ c^3\} \end{aligned}$$

The fact that the corresponding multiplication tables have the same color patterns can now be expressed simply by

$$f(x\cdot y) = f(x)\cdot f(y).$$

Also, the function $f(x)$ maps different elements of S_3 to different elements of G/M. That is, $f(x)$ is one-to-one, or *injective.* Finally, every element of G/M appears as $f(x)$ for some element x. This is expressed by saying that $f(x)$ is onto, or *surjective.* □

DEFINITION 4.1 Let G_1 and G_2 be two groups. An *isomorphism* from G_1 to G_2 is a one-to-one function sending elements of G_1 to elements of G_2 such that

$$f(x\cdot y) = f(x)\cdot f(y) \quad \text{for all } x, y \in G_1.$$

If there exists an isomorphism from G_1 to G_2 that is also onto, then we say that G_1 and G_2 are *isomorphic*, denoted by

$$G_1 \approx G_2.$$

For example,

$$S_3 \approx G/M$$

because of the existence of the function $f(x)$, which we saw was both one-to-one and onto.

It should be noted that $\approx$ is an equivalence relation on groups. (Reflexive property is obvious, symmetric and transitive properties are covered in Problems 1 and 2.) One of the important yet extremely hard problems in group theory is to find all of the non-isomorphic groups of a given order. Although this is still an unsolved problem, we have the following upper bound for the number of groups.

PROPOSITION 4.1
There are at most $n^{(n^2)}$ non-isomorphic groups of order n.

PROOF: If two groups have the same multiplication table, they are isomorphic, so a group is completely determined by its multiplication table. Notice that each element of this table must be one of n elements, and there are n^2 entries in the table. So there are $n^{(n^2)}$ ways of creating such a table. ∎

Of course, not very many of these tables will actually form a group. In fact, in some cases we can show that there is only one non-isomorphic group of order n.

PROPOSITION 4.2

For n a positive integer, every cyclic group of order n is isomorphic to Z_n.

PROOF: Let G be a group of order n, and let g be a generator of G. For clarity, we will let $\cdot$ denote the group operation of G, and $*$ denote the group operation of Z_n. Since $g^n = e$, we have

$$G = \{e = g^0, g^1, g^2, g^3, \ldots, g^{n-1}\}.$$

Define $f : Z_n \to G$ by

$$f(x) = g^x \qquad (0 \le x \le n-1).$$

That is, f will map the elements of Z_n to elements of G. Clearly f is one-to-one and onto, and we would like to show that it is an isomorphism. Suppose x and y satisfy

$$0 \le x, y \le n-1.$$

We let $z = x * y = (x + y)$ **mod** n. Then we can find an m such that $x + y = mn + z$. Now, $f(x * y) = f(z) = g^z$ by the definition of f. Thus,

$$f(x * y) = g^z = g^{(x+y-mn)} = g^x \cdot g^y \cdot (g^n)^{-m} = g^x \cdot g^y = f(x) \cdot f(y).$$

Since f is an isomorphism of Z_n onto G, we have $Z_n \approx G$. ∎

In particular if p is prime, Corollary 3.3 indicates all groups of order p are cyclic. Thus all groups of order p are isomorphic to Z_p.

For example, there is only one group each, up to isomorphism, of sizes 2, 3, 5, and 7, namely Z_2, Z_3, Z_5, and Z_7. Our goal for this section is to find all of the possible groups, up to isomorphism, up to order 8. To help us in this endeavor we have the following lemma.

LEMMA 4.1

Suppose a group G whose order is greater than 2 has all non-identity elements being of order 2. Then G has a subgroup isomorphic to Z_8^*.

PROOF: Since the order of G is greater than 2, there are two elements a

and b besides the identity element e. Since these will have order 2, we have $a^2 = b^2 = e$. Consider the product $a \cdot b$. It can be neither a nor b since this would imply the other was the identity. On the other hand, $a \cdot b = e$ implies

$$a = a \cdot e = a \cdot (b \cdot b) = (a \cdot b) \cdot b = e \cdot b = b.$$

So $a \cdot b$ is not the identity either. So there must be a fourth element in G, which we will call c, such that $a \cdot b = c$. This element will also be of order 2, so we have $c^2 = e$.

Finally, note that

$$b \cdot a = e \cdot b \cdot a \cdot e = a \cdot a \cdot b \cdot a \cdot b \cdot b = a \cdot (a \cdot b)^2 \cdot b = a \cdot c^2 \cdot b = a \cdot e \cdot b = a \cdot b = c.$$

With this we can quickly find the remaining products involving a, b, and c.

$$c{\cdot}a = b{\cdot}a{\cdot}a = b, \qquad c{\cdot}b = a{\cdot}b{\cdot}b = a, \qquad a{\cdot}c = a{\cdot}a{\cdot}b = b, \qquad b{\cdot}c = b{\cdot}b{\cdot}a = a.$$

Hence, the set $H = \{e, a, b, c\}$ is closed under multiplication, contains the identity, and also contains the inverses of every element in the set. Hence, H is a subgroup of G. The multiplication table for H

·	e	a	b	c
e	e	a	b	c
a	a	e	c	b
b	b	c	e	a
c	c	b	a	e

shows that this is isomorphic to Z_8^* using the mapping

$$f(e) = 1,$$
$$f(a) = 3,$$
$$f(b) = 5,$$
$$f(c) = 7.$$

□

We can now find all non-isomorphic groups of order up to 8. For example, if we have a group of order 6, any element of order 6 would imply that it is isomorphic to Z_6. We cannot have all non-identity elements be of order 2, or else Lemma 4.1 would give a subgroup of order 4, violating Lagrange's theorem (3.1). Thus, there must be an element a of order 3. Then $H = \{e, a, a^2\}$ is a normal subgroup of order 3 by Proposition 3.5. If b be any element not in H, then the two cosets of H are

$$\{\{e, a, a^2\}, \{b, a \cdot b, a^2 \cdot b\}\}.$$

TABLE 4.1: Multiplication table for Z_{24}^*

$\cdot$	1	5	7	11	13	17	19	23
1	1	5	7	11	13	17	19	23
5	5	1	11	7	17	13	23	19
7	7	11	1	5	19	23	13	17
11	11	7	5	1	23	19	17	13
13	13	17	19	23	1	5	7	11
17	17	13	23	19	5	1	11	7
19	19	23	13	17	7	11	1	5
23	23	19	17	13	11	7	5	1

We see that b^2 is in H, and if b^2 is a or a^2, then b is of order 6, so to get something different b^2 must be e. Then since H is normal $b \cdot a$ is either b, $a \cdot b$, or $a^2 \cdot b$.

If $b \cdot a = b$, then clearly we have the contradiction $a = e$. If $b \cdot a = a \cdot b$, then we find that $a \cdot b$ has order 6. Only the final possibility $b \cdot a = a^2 \cdot b$ gives a non-cyclic group. Since we know of a non-cyclic group of order 6, namely S_3, this must be it. Hence, there are two non-isomorphic groups of order 6, Z_6 and S_3.

A similar exhaustive search can be used to find all groups of order 8. If such a group has all non-identity elements of order 2, then by Lemma 4.1 there is a subgroup $\{e, a, b, a \cdot b\}$. By Problem 22, the group is commutative, so if we pick c to be any other element, then $c^2 = e$, $c \cdot a = a \cdot c$, and $c \cdot b = b \cdot c$.

```
InitGroup("e")
AddGroupVar("a", "b", "c")
Define(a^2, e)
Define(b^2, e)
Define(c^2, e)
Define(b*a, a*b)
Define(c*a, a*c)
Define(c*b, b*c)
G = ListGroup(); G
    {e, a, b, a*b, c, a*c, b*c, a*b*c}
```

So there is only one group of order 8 for which all non-identity elements are of order 2. But we can find such a group—Z_{24}^*, whose table is given in Table 4.1.

If $|G| = 8$ and G is not isomorphic to either Z_8 or Z_{24}^*, then there must be an element a of order 4. Then $H = \{e, a, a^2, a^3\}$ is a normal subgroup, and we can let b be any element not in H. Since G/H has order 2, b^2 must be in H, but if either $b^2 = a$ or $b^2 = a^3$, then b will have order 8. Hence, b^2 is either e or a^2. Also, $b \cdot a \notin S$, but $b \cdot a \neq b$, since this would force $a = e$. So $b \cdot a$ is either $a \cdot b$, $a^2 \cdot b$, or $a^3 \cdot b$. These six possibilities can be tried out in *Sage*.

TABLE 4.2: Multiplication table for D_4

$\cdot$	e	a	a^2	a^3	b	$a \cdot b$	$a^2 \cdot b$	$a^3 \cdot b$
e	e	a	a^2	a^3	b	$a \cdot b$	$a^2 \cdot b$	$a^3 \cdot b$
a	a	a^2	a^3	e	$a \cdot b$	$a^2 \cdot b$	$a^3 \cdot b$	b
a^2	a^2	a^3	e	a	$a^2 \cdot b$	$a^3 \cdot b$	b	$a \cdot b$
a^3	a^3	e	a	a^2	$a^3 \cdot b$	b	$a \cdot b$	$a^2 \cdot b$
b	b	$a^3 \cdot b$	$a^2 \cdot b$	$a \cdot b$	e	a^3	a^2	a
$a \cdot b$	$a \cdot b$	b	$a^3 \cdot b$	$a^2 \cdot b$	a	e	a^3	a^2
$a^2 \cdot b$	$a^2 \cdot b$	$a \cdot b$	b	$a^3 \cdot b$	a^2	a	e	a^3
$a^3 \cdot b$	$a^3 \cdot b$	$a^2 \cdot b$	$a \cdot b$	b	a^3	a^2	a	e

If $b \cdot a = a \cdot b$, and b^2 is either e or a^2, the group become isomorphic to Z_{15}^*, which we have seen before. Also, both combinations for which $b \cdot a = a^2 \cdot b$ fail to produce a group. If $b \cdot a = a^3 \cdot b$ and $b^2 = e$, we get the group

```
InitGroup("e")
AddGroupVar("a", "b")
Define(a^4, e)
Define(b^2, e)
Define(b*a, a^3*b)
G = ListGroup()
     {e, a, a^2, a^3, b, a*b, a^2*b, a^3*b}
```

This gives rise to the group D_4, the symmetry group of the square studied in Problem 1 of §1.1. The multiplication table is shown in Table 4.2.

The final possibility is that $b \cdot a = a^3 \cdot b$, and $b^2 = a^2$. This produces a new group called the quaternion group Q, described by the following:

```
InitGroup("e")
AddGroupVar("a", "b")
Define(a^4, e)
Define(b^2, a^2)
Define(b*a, a^3*b)
Q = ListGroup(); Q
     {e, a, a^2, a^3, b, a*b, a^2*b, a^3*b}
```

Although the group can be defined in terms of only two generators, it is more natural to use the notation that appears in Table 4.3. Note that i, j, and k sometimes behave like the vector cross product:

$$i \cdot j = k, \quad j \cdot k = i, \quad \text{and} \quad k \cdot i = j,$$

and sometimes act like complex numbers:

$$i^2 = -1, \quad j^2 = -1, \quad \text{and} \quad k^2 = -1.$$

In summary, we have the following groups up to order 8:

TABLE 4.3: Multiplication table for Q

$\cdot$	1	i	j	k	-1	$-i$	$-j$	$-k$
1	1	i	j	k	-1	$-i$	$-j$	$-k$
i	i	-1	k	$-j$	$-i$	1	$-k$	j
j	j	$-k$	-1	i	$-j$	k	1	$-i$
k	k	j	$-i$	-1	$-k$	$-j$	i	1
-1	-1	$-i$	$-j$	$-k$	1	i	j	k
$-i$	$-i$	1	$-k$	j	i	-1	k	$-j$
$-j$	$-j$	k	1	$-i$	j	$-k$	-1	i
$-k$	$-k$	$-j$	i	1	k	j	$-i$	-1

$n = 1$: The one element must be the identity, so we have just the trivial group, $\{e\}$.

$n = 2$: Since 2 is prime, all groups are isomorphic to Z_2.

$n = 3$: Since 3 is prime, all groups are isomorphic to Z_3.

$n = 4$: By Lemma 4.1, the only two non-isomorphic groups are Z_4 and Z_8^*.

$n = 5$: Since 5 is prime, all groups are isomorphic to Z_5.

$n = 6$: There are only two non-isomorphic groups: Z_6 and the non-abelian group S_3.

$n = 7$: Since 7 is prime, all groups are isomorphic to Z_7.

$n = 8$: There are three abelian groups, Z_8, Z_{15}^*, and Z_{24}^* and two non-abelian groups, D_4 and Q.

Finally, Table 4.4 gives the number of non-isomorphic groups of order n, when n is not prime.

Problems for §4.1

1 Prove that if f is a surjective isomorphism from a group G to a group M, then f^{-1} is a surjective isomorphism from M to G.

2 If G_1, G_2, and G_3 are three groups, and f is an isomorphism from G_1 to G_2, and ϕ is an isomorphism from G_2 to G_3, prove that $\phi(f(x))$ is an isomorphism from G_1 to G_3.

3 Find an isomorphism between D_3 (Terry's group) and S_3.

TABLE 4.4: Number of groups of order n for composite n

n	groups	n	groups	n	groups	n	groups	n	groups
4	2	26	2	46	2	65	1	85	1
6	2	27	5	48	52	66	4	86	2
8	5	28	4	49	2	68	5	87	1
9	2	30	4	50	5	69	1	88	12
10	2	32	51	51	1	70	4	90	10
12	5	33	1	52	5	72	50	91	1
14	2	34	2	54	15	74	2	92	4
15	1	35	1	55	2	75	3	93	2
16	14	36	14	56	13	76	4	94	2
18	5	38	2	57	2	77	1	95	1
20	5	39	2	58	2	78	6	96	230
21	2	40	14	60	13	80	52	98	5
22	2	42	6	62	2	81	15	99	2
24	15	44	4	63	4	82	2	100	16
25	2	45	2	64	267	84	15	102	4

4 Find an isomorphism between the group consisting of the four complex numbers

$$\{1, -1, i, -i\}$$

and the group Z_4.

For Problems **5** through **13**: Find an isomorphism between the two groups.

5 Z_6 and Z_7^*
6 Z_6 and Z_9^*
7 Z_6 and Z_{14}^*
8 Z_6 and Z_{18}^*
9 Z_{10} and Z_{11}^*
10 Z_{10} and Z_{22}^*
11 Z_{12} and Z_{13}^*
12 Z_{12} and Z_{26}^*
13 Z_8^* and Z_{12}^*

14 Let G be an arbitrary group. Prove or disprove that $f(x) = x^{-1}$ is an isomorphism from G to G.

15 Prove that any infinite cyclic group is isomorphic to $\mathbb{Z}$.

16 Let $\mathbb{R}$ be the group of real numbers under addition, and let G be the group of positive real numbers under multiplication. Prove that $\mathbb{R} \approx G$, with $\phi(x) = e^x$.

17 Let ϕ be an isomorphism from a group G to a group M. Prove that a and $\phi(a)$ have the same order.

Interactive Problems

18 Prove that there are exactly two non-isomorphic groups of order 10. Find these two groups, and have *Sage* produce the multiplication tables.
Hint: Follow the logic for $n = 6$.

For Problems **19** through **21**: Each of the following groups is of order 8. Which of the known five groups (Z_8, Z_{24}^*, Z_{15}^*, D_4, or Q) is each of these isomorphic to? First have *Sage* display a table of the new group, and then rearrange the elements of one of the five known groups so that the color patterns in the two tables are identical.

19 Z_{16}^* **20** Z_{20}^* **21** Z_{30}^*

4.2 Homomorphisms

It is easy to see the application of isomorphisms, since these functions show how two groups are essentially the same. But suppose we have a function between two groups for which $f(x \cdot y) = f(x) \cdot f(y)$, but this function may not be one-to-one or onto. Can we still glean some information about the groups from this function? In this section we will find that functions with this property are very special indeed.

DEFINITION 4.2 Let G and M be two groups. A function

$$f : G \to M$$

mapping elements of G to elements of M is called a *homomorphism* if it satisfies

$$f(x \cdot y) = f(x) \cdot f(y) \quad \text{for all } x, y \in G.$$

The group G is called the *domain* of the homomorphism, and the group M is called the *target* of the homomorphism. Note that a homomorphism need not be either one-to-one or onto.

Of course, all isomorphisms are also homomorphisms. But we can have many other homomorphisms, as the following examples show.

Example 4.2
Let G be any group, and let M be a group with identity e. If we let

$$f(x) = e \quad \text{for all } x \in G$$

then f will obviously be a homomorphism, since

$$f(x \cdot y) = e = e \cdot e = f(x) \cdot f(y).$$

This is called the *trivial homomorphism.* □

Example 4.3
Let $\mathbb{R}^* = \mathbb{R} - \{0\}$ be the group of nonzero real numbers under multiplication, and let $f(x) = x^2$. This forms a homomorphism

$$f : \mathbb{R}^* \to \mathbb{R}^*,$$

so this gives an example of a homomorphism that maps a group onto itself. Note that this homomorphism is neither one-to-one nor onto since $f(-2) = f(2) = 4$, yet there is no real number such that $f(x) = -1$. □

Example 4.4
We can generalize Example 4.3 as follows: Let G be any commutative group, and let n be any integer. We can define $f(x) = x^n$. Then $f(x)$ is a homomorphism from G to itself, since

$$f(x \cdot y) = (x \cdot y)^n = x^n \cdot y^n = f(x) \cdot f(y).$$ □

We can prove a few properties that must be true of all homomorphisms.

PROPOSITION 4.3
Let $f : G \to M$ be a homomorphism. Let e denote the identity of G. Then $f(e)$ is the identity element of M.

PROOF: Since $e \cdot e = e$ in the group G, we have

$$f(e) = f(e \cdot e) = f(e) \cdot f(e).$$

Multiplying both sides by $[f(e)]^{-1}$ gives us that $f(e)$ is the identity element of M. □

PROPOSITION 4.4
If $f : G \to M$ is a homomorphism, then $f(a^{-1}) = [f(a)]^{-1}$.

PROOF: We merely need to show that $f(a) \cdot f(a^{-1})$ is the identity element of M. If e represents the identity element of G, then

$$f(a) \cdot f(a^{-1}) = f(a \cdot a^{-1}) = f(e).$$

By Proposition 4.3 this is the identity element of M. So

$$f(a^{-1}) = [f(a)]^{-1}.$$ □

Example 4.5
Find a homomorphism from Z_{15}^* to Z_4 such that $f(2) = f(7) = 1$.

SOLUTION: We know from Proposition 4.3 that the identity must map to the identity, so $f(1) = 0$. Also, $f(4) = f(2)^2 = 1^2 = 2$. (Recall the operation of Z_4 is *addition* **mod** 4.) Likewise, $f(8) = f(2)^3 = 3$, $f(13) = f(7)^3 = 3$, $f(14) = f(7) \cdot f(2) = 2$, and $f(11) = f(13) \cdot f(2) = 0$. □

To define homomorphisms using *Sage*, we must first define the two groups G and M simultaneously, using different sets of letters for the generators.

Computational Example 4.6

Let us create a homomorphism from the octahedral group to the quaternion group.

We first load the octahedral group with the following commands:

```
InitGroup("e")
AddGroupVar("a", "b", "c")
Define(a^2, e); Define(b^3, e); Define(c^2, e)
Define(b*a, a*b^2); Define(c*a, a*b*c); Define(c*b, a*c^2)
Oct = ListGroup(); Oct
    {e, a, b, a*b, b^2, a*b^2, c, a*c, b*c, a*b*c, b^2*c,
    a*b^2*c, c^2, a*c^2, b*c^2, a*b*c^2, b^2*c^2, a*b^2*c^2,
    c^3, a*c^3, b*c^3, a*b*c^3, b^2*c^3, a*b^2*c^3}
```

Next let us define the quaternion group Q from the last section. The easiest way to load this group is with the command

```
Q = InitQuaternions(); Q
    {1, i, j, k, -1, -i, -j, -k}
```

Let us define a homomorphism F from Q to `Oct`. First we tell *Sage* that F will be a homomorphism.

```
F = Homomorph(Q, Oct)
```

We need only define the homomorphism on the *generators* of where the generators are sent, since *Sage* would then be able to use the properties of the homomorphism to determine where the other elements map to. Thus, to define the mapping

$$\begin{aligned} 1 &\to e, \\ i &\to c^2, \\ -1 &\to e, \\ -i &\to c^2, \\ j &\to a \cdot b^2 \cdot c, \\ k &\to a \cdot b^2 \cdot c^3, \\ -j &\to a \cdot b^2 \cdot c, \\ -k &\to a \cdot b^2 \cdot c^3; \end{aligned}$$

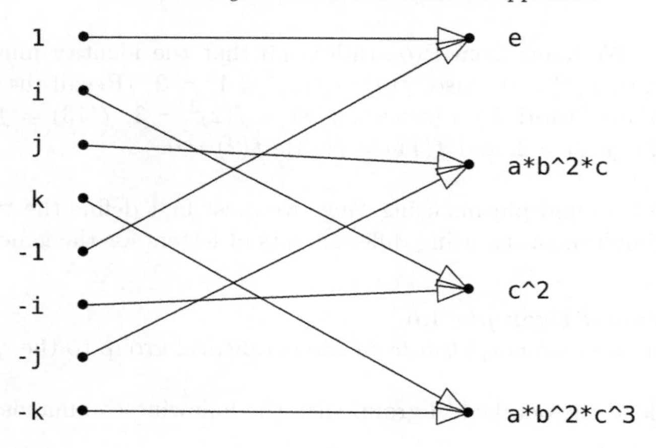

FIGURE 4.1: Diagram of a typical homomorphism

we have only to define $F(i)$ and $F(j)$. This is done with the `HomoDef` command.

```
HomoDef(F, i, c^2)
HomoDef(F, j, a*b^2*c)
```

Sage can check whether this function can be expanded to form a homomorphism by the command

```
FinishHomo(F)
    'Homomorphism defined'
```

This shows that the function F is indeed a homomorphism. The command

```
GraphHomo(F)
```

will draw a picture of this homomorphism as shown in Figure 4.1. □

We can apply a homomorphism f to a *set* of elements by applying the homomorphism to each element in the set, and consider the set of all possible results. For example, consider the set of real numbers $S = \{-2, -1, 1, 2, 3, 4\}$. Let $f(x)$ be the homomorphism in Example 4.3 above, $f(x) = x^2$. Then

$$f(S) = \{1, 4, 9, 16\}.$$

The set $f(S)$ is smaller than the set S, since the homomorphism mapped two elements to both 1 and 4.

To apply the homomorphism to a set of elements in *Sage*, we can use the **Image** command using a list for the second argument.

```
Image(F, [i, k, -i, -k])
    {c^2, a*b^2*c^3}
Image(F, [1, i, -1, -i])
    {e, c^2}
```

In the last example, we see the image of a subgroup of Q being a subgroup of the octahedral group. It is not hard to prove that this will be the case in general.

PROPOSITION 4.5
If $f : G \to M$ is a homomorphism and H is a subgroup of G, then $f(H)$ is a subgroup of M.

PROOF: We want to show that $f(H)$ is a subgroup using Proposition 2.2. If u and v are elements in $f(H)$, there must be elements x and y in H such that $f(x) = u$, and $f(y) = v$.

Then $x \cdot y^{-1}$ is in H, and so

$$f(x \cdot y^{-1}) = f(x) \cdot f(y^{-1}) = f(x) \cdot [f(y)]^{-1} = u \cdot v^{-1}$$

is in $f(H)$. So by Proposition 2.2, $f(H)$ is a subgroup of M. ∎

DEFINITION 4.3 If

$$f : G \to M$$

is a homomorphism, then the group $f(G)$ is called the *range*, or *image* of the homomorphism f. We denote this set by

$$\mathrm{Im}(f).$$

We can also consider taking the inverse homomorphism f^{-1} of an element or a set of elements. Because homomorphisms are not always one-to-one, $f^{-1}(x)$ may not represent a single element. Thus, we will define $f^{-1}(x)$ to be the *set* of numbers such that $f(y) = x$. Likewise, we define

$$f^{-1}(S) = \{y \mid f(y) \in S\}.$$

We can use *Sage*'s **HomoInv** command to take the inverse homomorphism of an element or set of elements.

```
HomoInv(F, c^2)
    {-i, i}
```

finds $F^{-1}(c^2)$, whereas

```
HomoInv(F, [a, b, a*b^2*c])
    {-j, j}
```

finds the inverse of a set of elements. Note that not all of the elements in the set have to be in the image of F. There is one inverse image that is very important.

DEFINITION 4.4 If f is a homomorphism from G to M and e is the identity element of M, then we define the *kernel* of f to be the set

$$\text{Ker}(f) = f^{-1}(e).$$

The command

```
Kernel(F)
    {-1, 1}
```

can be used to find the kernel of a homomorphism.

PROPOSITION 4.6

If f is a homomorphism from G to M, then the kernel of f is a normal subgroup of the domain G.

PROOF: First we need to show that the kernel of f is a subgroup of G. If e is the identity element of M, and if a and b are two elements of $\text{Ker}(f)$, then

$$f(a \cdot b^{-1}) = f(a) \cdot f(b)^{-1} = e \cdot e^{-1} = e,$$

so $a \cdot b^{-1}$ is also in the kernel of f. Thus, by Proposition 2.2, $\text{Ker}(f)$ is a subgroup.

Now let us show that $\text{Ker}(f)$ is a normal subgroup of G. Let a be an element in $\text{Ker}(f)$, and g be any element in G. Then by Proposition 3.4, since

$$f(g \cdot a \cdot g^{-1}) = f(g) \cdot f(a) \cdot f(g^{-1}) = f(g) \cdot e \cdot [f(g)]^{-1} = e,$$

$g \cdot a \cdot g^{-1}$ is in $\text{Ker}(f)$, and so $\text{Ker}(f)$ is a normal subgroup. □

Figure 4.1 is very suggestive. The inverse image of any element is a coset of $\{-1, 1\}$. The next proposition explains why this is so.

PROPOSITION 4.7

Let f be a homomorphism from the group G to the group M. Suppose that y is in the image of f, and that $f(x) = y$. Then

$$f^{-1}(y) = x \cdot \text{Ker}(f).$$

PROOF: First let us consider an element $z \in x \cdot \text{Ker}(f)$. Then $z = x \cdot k$ for some element k in the kernel of f. Therefore,

$$f(z) = f(x \cdot k) = f(x) \cdot f(k) = f(x) \cdot e = f(x)$$

since k is in $\text{Ker}(f)$. Here, e is the identity element of M. But $f(x) = y$, and so $z \in f^{-1}(y)$. Thus we have proved that

$$f^{-1}(y) \subseteq x \cdot \text{Ker}(f).$$

To prove the inclusion the other way, note that if $z \in f^{-1}(y)$, then $f(z) = y$, and so we have

$$f(x^{-1} \cdot z) = f(x)^{-1} \cdot f(z) = y^{-1} \cdot y = e.$$

Thus, $x^{-1} \cdot z$ is in the kernel of f, and since $z = x \cdot (x^{-1} \cdot z) \in x \cdot \text{Ker}(f)$, we have

$$x \cdot \text{Ker}(f) \subseteq f^{-1}(y).$$ ∎

We now have a quick way to determine if a homomorphism is an isomorphism.

COROLLARY 4.1

Let $f : G \to M$ be a homomorphism. Then f is an injection (one-to-one) if, and only if, the kernel of f is the identity element of G.

PROOF: If f is an injection, clearly the kernel would just be the identity element. Suppose that the kernel is just the identity. Then Proposition 4.7 states that if h is in the image of f, then $f^{-1}(h)$ consists of exactly one element. Therefore, f is one-to-one. ∎

In particular, if the image of a homomorphism $f : G \to M$ is all of M, and the kernel is $\{e\}$, then $G \approx M$.

We can also consider what happens if we take the inverse image of a subgroup.

COROLLARY 4.2

Let $f : G \to M$ be a homomorphism. Let H be a subgroup of M. Then $f^{-1}(H)$ is a subgroup of G. Furthermore, if H is a normal subgroup of M, then $f^{-1}(H)$ is a normal subgroup of G.

PROOF: Let x and y be in $f^{-1}(H)$. Then since $f(x \cdot y^{-1}) = f(x) \cdot f(y)^{-1}$, which is in H, we have that $x \cdot y^{-1}$ is in $f^{-1}(H)$. Thus, by Proposition 2.2, $f^{-1}(H)$ is a subgroup of G.

Now suppose that H is a normal subgroup of M. Then if y is in $f^{-1}(H)$, and g is in G, then $f(g \cdot y \cdot g^{-1}) = f(g) \cdot f(y) \cdot f(g)^{-1}$. Since $f(y)$ is in H, which is normal in M, we have that $f(g) \cdot f(y) \cdot f(g)^{-1}$ is in H. Thus, $g \cdot y \cdot g^{-1}$ is in $f^{-1}(H)$, and so by Proposition 3.4, $f^{-1}(H)$ is normal in G. ∎

We are now in a position to show how homomorphisms can be used to reveal relationships between different groups. There are three such relationships to be revealed, and these are covered in the next section.

Problems for §4.2

1 If ϕ is a homomorphism from an abelian group G to a group M, show that $\text{Im}(\phi)$ is abelian.

2 If ϕ is a homomorphism from a cyclic group G to a group M, show that $\text{Im}(\phi)$ is a cyclic group.

3 Let $\mathbb{Z}$ be the group of integers using addition. Show that the function $\phi(x) = 2x$ is a homomorphism from $\mathbb{Z}$ to itself. What is the image of this homomorphism? What is the kernel?

4 Let $\mathbb{Z}$ be the group of integers using addition. Show that the function $\phi(x) = -x$ is a homomorphism from $\mathbb{Z}$ to itself. Show that this mapping is in fact one-to-one and onto.

5 Let $\mathbb{Z}$ be the group of integers using addition. Show that the function $\phi(x) = x + 3$ is *not* a homomorphism from $\mathbb{Z}$ to itself.

6 Let $\mathbb{R}^*$ denote the group of nonzero real numbers, using multiplication as the operation. Let $\phi(x) = x^6$. Show that ϕ is a homomorphism from $\mathbb{R}^*$ to $\mathbb{R}^*$. What is the kernel of this homomorphism? What is the image of the homomorphism?

7 Let $\mathbb{R}^*$ denote the group of nonzero real numbers, using multiplication as the operation. Let $\phi(x) = 2x$. Show that ϕ is *not* a homomorphism from $\mathbb{R}^*$ to $\mathbb{R}^*$.

8 Let $\mathbb{R}^*$ denote the group of nonzero real numbers, using multiplication as the operation. Recall that $\mathbb{R}$ is the group of real numbers using addition for the operation. Let $\phi(x) = \ln|x|$. Show that ϕ is a homomorphism from $\mathbb{R}^*$ to $\mathbb{R}$. What is the kernel of this homomorphism?

9 Let $\mathbb{R}^*$ denote the group of nonzero real numbers, using multiplication as the operation. Recall that $\mathbb{R}$ is the group of real numbers using addition for the operation. Let $\phi(x) = e^x$. Show that ϕ is a homomorphism from $\mathbb{R}$ to $\mathbb{R}^*$. What is the image of this homomorphism?

10 Let $\mathbb{R}[t]$ denote the group of all polynomials in t with real coefficients under addition, and let ϕ denote the mapping $\phi(f) = f'$, which sends each polynomial to its derivative. Show that ϕ is a homomorphism from $\mathbb{R}[t]$ to $\mathbb{R}[t]$. What is the kernel of ϕ?

11 Let $\mathbb{R}[t]$ denote the group of all polynomials in t with real coefficients under addition. Prove that the mapping from $\mathbb{R}[t]$ into $\mathbb{R}$ given by $f(t) \to f(3)$ is a homomorphism. Give a description of the kernel of this homomorphism.

12 Find a homomorphism ϕ from Z_{15}^* to Z_{15}^* with kernel $\{1, 11\}$ and with $\phi(2) = 7$.

13 Find a homomorphism ϕ from Z_{30}^* to Z_{30}^* with kernel $\{1, 11\}$ and with $\phi(7) = 13$.

14 Find a homomorphism ϕ from Z_{32}^* to Z_{32}^* with $\phi(7) = 1$ and $\phi(11) = 9$.

15 Find a homomorphism from the quaternion group Q onto Z_8^*.
Hint: The kernel must be a normal subgroup of order 2. See Table 4.3 for a multiplication table of Q.

16 Let k be a divisor of n. Show that the mapping $\phi(x) = x \pmod{k}$ is a homomorphism from Z_n^* to Z_k^*. Find a formula for the number of elements in the kernel.

17 Let $f : G \to M$ be a homomorphism from a finite group G onto M, and let $f^{-1}(H)$ be the subgroup from Corollary 4.2. Show that the size of this subgroup is $|H| \cdot |\text{Ker} f|$.

18 Let $f : G \to M$ be a homomorphism from G onto M, and let H be a normal subgroup of G. Prove that $f(H)$ is a normal subgroup of M.

Interactive Problems

19 Define Terry's group in *Sage* with the command

```
Terry = InitTerry()
```

and then define the group S_3.

```
InitGroup("e")
AddGroupVar("a", "b")
Define(a^2, e)
Define(b^3, e)
Define(b*a, a*b^2)
S3 = ListGroup()
```

Now define an isomorphism F from S_3 to Terry's group. Use *Sage*'s **FinishHomo** to verify that your function is a homomorphism. Finally, find the kernel of F to prove that F is an isomorphism.

20 Use *Sage* to find all of the homomorphisms from S_3 to itself. Label these homomorphisms $F1$, $F2$, $F3$, etc. How many of these are isomorphisms? The following reloads S_3 into *Sage*:

```
InitGroup("e")
AddGroupVar("a", "b")
Define(a^2, e)
Define(b^3, e)
Define(b*a, a*b^2)
S3 = ListGroup()
```

4.3 The Three Isomorphism Theorems

We have seen in the last section that the kernel K of a homomorphism is always a normal subgroup of the domain G. Furthermore, Proposition 4.7 proves what is suggested by Figure 4.1, that the inverse image of any element is essentially a coset of K. Hence, the inverse image $f^{-1}(y)$ can be considered as an element of the quotient group G/K. This leads us to the first of three very useful theorems for finding isomorphisms between groups.

THEOREM 4.1: The First Isomorphism Theorem
Let $f : G \to M$ be a homomorphism with $\mathrm{Ker}(f) = K$, *and* $\mathrm{Im}(f) = I$. *Then there is a natural isomorphism*

$$\phi : I \to G/K$$

which is onto. Thus, $I \approx G/K$.

PROOF: Note that this theorem states more than just $I \approx G/K$, but also that there is a *natural* isomorphism between these two groups. This isomorphism is given by

$$\phi(h) = f^{-1}(h).$$

Proposition 4.7 states that whenever h is in the image of f, $f^{-1}(h)$ is a member of the quotient group $G/\mathrm{Ker}(f)$. Thus, $\phi : I \to G/K$ is properly defined.

Let us show that the mapping ϕ is one-to-one. Suppose $\phi(x) = \phi(y)$ for two different elements of I. Then $f(\phi(x)) = f(\phi(y))$. But $f(\phi(x)) = f(f^{-1}(x))$ is the set containing just the element x, and also $f(\phi(y))$ is the set containing just the element y. Thus, $x = y$, and we have shown that ϕ is one-to-one.

Now let us show that ϕ is onto. If xK is an element of G/K, then $f(x) \in I$. Thus,

$$x \in f^{-1}(f(x)) = \phi(f(x)) \in G/K.$$

So we have that x is an element of both cosets xK and $\phi(f(x))$. Since two different cosets have no elements in common, we must have $\phi(f(x)) = xK$. We therefore have that any coset in G/K is mapped by ϕ from an element in I, so ϕ is onto.

Finally, we want to show that ϕ is a homomorphism. That is, we wish to show that

$$f^{-1}(v) \cdot f^{-1}(w) = f^{-1}(v \cdot w).$$

Let $x \in f^{-1}(v)$ and $y \in f^{-1}(w)$. Then $f(x) = v$ and $f(y) = w$, so we have

$$f(x \cdot y) = f(x) \cdot f(y) = v \cdot w.$$

Hence,

$$x \cdot y \in f^{-1}(v \cdot w).$$

Since $f^{-1}(v) \cdot f^{-1}(w)$ and $f^{-1}(v \cdot w)$ are two cosets in G/K, and both contain the element $x \cdot y$, they must be the same coset. So we have that

$$\phi(v) \cdot \phi(w) = \phi(v \cdot w).$$

∎

This theorem says that whenever we have a homomorphism f from G to M with an image I, then we get a natural isomorphism ϕ from I to $G/\mathrm{Ker}(f)$.

This suggests that there ought to be a mapping that goes directly from G to $G/\mathrm{Ker}(f)$ without involving the homomorphism f. The next proposition shows how this can be done.

PROPOSITION 4.8

Let G be a group, and N be a normal subgroup of G. Then there is a natural homomorphism

$$i_N : G \to G/N$$

given by $i_N(a) = a \cdot N$. This homomorphism is surjective, and $\mathrm{Ker}(i_N) = N$.

PROOF: To show that i_N is a homomorphism, we note that if a and b are elements of G, then

$$i_N(a \cdot b) = a \cdot b \cdot N = a \cdot N \cdot b \cdot N = i_N(a) \cdot i_N(b).$$

Also, i_N is clearly surjective. To find the kernel of i_N, we note that the identity element of G/N is $eN = N$, and so x is in the kernel if, and only if,

$$i_N(x) = N \Longleftrightarrow x \cdot N = N \Longleftrightarrow x \in N.$$

Therefore, the kernel of i_N is N. ∎

We call the homomorphism i_N the *canonical homomorphism associated with* N. We can make a diagram of this homomorphism, along with the homomorphisms f and ϕ, to produce Figure 4.2.

Notice that we now have two ways of getting from G to $G/\mathrm{Ker}(f)$, one route though the canonical homomorphism, and the other route through f and ϕ. Yet we have drawn this diagram to indicate that $\phi(f(x)) = i_N(x)$ for all elements in G. Thus, the two routes from G to $G/\mathrm{Ker}(f)$ produce the same function. We express this fact by saying that the *diagram is commutative*. In other words, for a commuting diagram, the functions defined by two paths with the same beginning and ending points produce the same composition function. In this diagram there are arrows going in both directions for the function ϕ to indicate that this is an isomorphism, hence invertible. Hence, by the commuting diagram, we also have the result $\phi^{-1}(i_N(x)) = f(x)$. We will

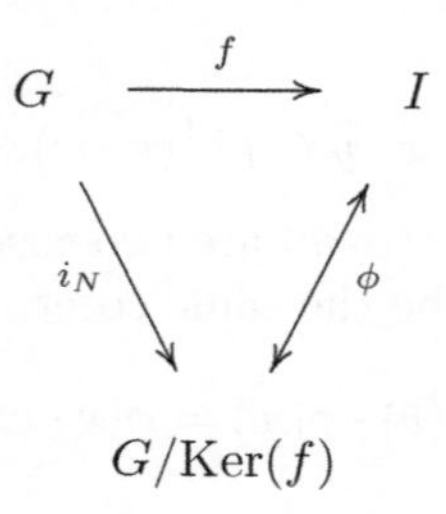

FIGURE 4.2: Commuting diagram for first isomorphism theorem

later be able to visualize many theorems about homomorphisms by means of commuting diagrams.

We observed in §3.3 that the product of two subgroups H and K was not necessarily a subgroup. However, it is possible that if one of the groups is normal, then indeed the product $H \cdot K$ would be a subgroup. (In fact, this was proven in Problem 15 of §3.3.) Let us try it on the octahedral group.

Motivational Example 4.7

Explore the product of two subgroups of the octahedral group, one of which is normal.

```
InitGroup("e")
AddGroupVar("a", "b", "c")
Define(a^2, e); Define(b^3, e); Define(c^2, e)
Define(b*a, a*b^2); Define(c*a, a*b*c); Define(c*b, a*c^2)
G = ListGroup()
M = Group(a*b^2*c, c^2)
    {e, a*b^2*c, c^2, a*b^2*c^3}
H = Group(c); H
    {e, c^2, c, c^3}
L = Conform(H * M, G); L
    {e, a*b^2, a*b^2*c, c^2, c, a*b^2*c^2, a*b^2*c^3, c^3}
```

The purpose of the **Conform** in the last statement is to put all of the elements of $H \cdot M$ in the form in which they appear in the group G. *Sage* can verify that these 8 elements form a subgroup. What happens if we try multiplying H and M in the other order?

```
Conform(M * H, G)
    {e, a*b^2, a*b^2*c, c^2, c, a*b^2*c^2, a*b^2*c^3, c^3}
```

We discovered that not only is $H \cdot M$ a subgroup, but also $M \cdot H$ is exactly the same as $H \cdot M$. ∎

It is not hard to see the connection between these two facts.

LEMMA 4.2

Suppose H and K are two subgroups of G. Then $H \cdot K$ is a subgroup if, and only if,

$$H \cdot K = K \cdot H.$$

PROOF: First suppose that $H \cdot K$ is a subgroup. Let $h \in H$ and $k \in K$. We wish to show that the element $h \cdot k$ in $H \cdot K$ is also in $K \cdot H$. Since $H \cdot K$ is a subgroup, $(h \cdot k)^{-1}$ is in $H \cdot K$. Thus, $(h \cdot k)^{-1} = x \cdot y$ for some $x \in H$ and $y \in K$. But then, $h \cdot k = (x \cdot y)^{-1} = y^{-1} \cdot x^{-1}$, and $y^{-1} \cdot x^{-1}$ is in $K \cdot H$. Thus,

$$H \cdot K \subseteq K \cdot H.$$

By a similar argument, the inverse of any element in $K \cdot H$ must be in $H \cdot K$, and so $K \cdot H \subseteq H \cdot K$. Therefore, we have $H \cdot K = K \cdot H$.

Now, let us suppose that $H \cdot K = K \cdot H$. We want to show that $H \cdot K$ is a subgroup. Let $h_1, h_2 \in H$ and $k_1, k_2 \in K$ so both $h_1 \cdot k_1$ and $h_2 \cdot k_2$ are elements of $H \cdot K$. By Proposition 2.2, it is enough to show that $(h_1 \cdot k_1) \cdot (h_2 \cdot k_2)^{-1}$ is in $H \cdot K$. But $(k_1 \cdot k_2^{-1}) \cdot h_2^{-1}$ is in $K \cdot H = H \cdot K$, and so there must be two elements $h_3 \in H$ and $k_3 \in K$ such that $(k_1 \cdot k_2^{-1}) \cdot h_2^{-1} = h_3 \cdot k_3$. Then we have

$$(h_1 \cdot k_1) \cdot (h_2 \cdot k_2)^{-1} = h_1 \cdot k_1 \cdot k_2^{-1} \cdot h_2^{-1} = (h_1 \cdot h_3) \cdot k_3$$

which is in $H \cdot K$. Thus, $H \cdot K$ is a subgroup if, and only if, $H \cdot K = K \cdot H$. □

We are now in a position to show that $H \cdot K$ is a subgroup if one of the subgroups H or K is normal.

LEMMA 4.3

If H is a subgroup of G, and N is a normal subgroup of G, then $H \cdot N$ is a subgroup of G.

PROOF: If $h \in H$ and $n \in N$, then $h \cdot n \cdot h^{-1}$ is in N, since N is normal. Then

$$h \cdot n = (h \cdot n \cdot h^{-1}) \cdot h$$

is in $N \cdot H$. Thus, $H \cdot N \subseteq N \cdot H$.

By a similar argument $N \cdot H \subseteq H \cdot N$, so $H \cdot N = N \cdot H$. Therefore, $H \cdot N$ is a group by Lemma 4.2. □

Since we have found a new subgroup of G that contains the normal subgroup M, the natural question is whether it is a normal subgroup. We can try to find the left and right cosets of $H \cdot M$ from Example 4.7.

```
LftCoset(G, H * M)
    {{e, a*b^2, a*b^2*c, c^2, c, a*b^2*c^2, a*b^2*c^3, c^3},
    {a, b^2, b^2*c, a*c^2, a*c, b^2*c^2, b^2*c^3, a*c^3},
    {b, a*b*c, b*c^2, a*b, a*b*c^3, b*c, a*b*c^2, b*c^3}}
RtCoset(G, H * M)
    {{e, a*b^2, a*b^2*c, c^2, c, a*b^2*c^2, a*b^2*c^3, c^3},
    {b, a, a*b*c, b*c^2, b^2*c, a*c^2, a*b*c^3, b^2*c^3},
    {b^2, a*b, a*c, b^2*c^2, b*c, a*b*c^2, a*c^3, b*c^3}}
```

We see that these are not the same, so in general, $H \cdot N$ is not a normal subgroup if only N is normal. (Note that if both H and N were normal, then Problem 17 of §3.3 shows that $H \cdot N$ is normal.)

But would M be a normal subgroup of $H \cdot M$?

```
LftCoset(L, M)
    {{e, a*b^2*c, c^2, a*b^2*c^3}, {a*b^2, c, a*b^2*c^2, c^3}}
RtCoset(L, M)
    {{e, a*b^2*c, c^2, a*b^2*c^3}, {a*b^2, c, a*b^2*c^2, c^3}}
```

We can quickly see in this case it is normal, since M contains half of the elements of $H \cdot M$. But we can prove that this will happen in general, using the fact that $H \cdot M$ is a subgroup of G.

LEMMA 4.4

Let N be a normal subgroup of G, and suppose that H is a subgroup of G which contains N. Then N is a normal subgroup of H.

PROOF: Since N is a group, and is contained in H, N is a subgroup of H. For any x in H, we have that

$$x \cdot N \cdot x^{-1} = N$$

since x is also in G. Therefore, by Proposition 3.4, N is a normal subgroup of H. ∎

Given two subgroups of a group G, there is another way of forming a new subgroup. Proposition 2.3 tells us that the intersection of two subgroups will again be a subgroup. Recall that the *Sage* command

```
R = Intersection(H, M); R
    {e, c^2}
```

finds the intersection of two subgroups. If, as in Lemma 4.3, one of the two subgroups is normal, we have the following.

LEMMA 4.5

If N is a normal subgroup of G, and H is a subgroup of G, then

$$H \cap N$$

is a normal subgroup of H.

PROOF: Given elements $h \in H$ and $x \in H \cap N$, we note that since x is in N, which is a normal subgroup of G, $h \cdot x \cdot h^{-1}$ is in N. Also, x is in H, so $h \cdot x \cdot h^{-1}$ is in H. Thus,

$$h \cdot x \cdot h^{-1} \in H \cap N,$$

and so by Proposition 3.4, the intersection is a normal subgroup of H. □

We can ask whether there is a relationship between the two quotient groups $H/(H \cap N)$ and $(H \cdot N)/N$. We can calculate both quotient groups in *Sage*.

```
LftCoset(H, R)
    {{e, c^2}, {c, c^3}}
LftCoset(L, M)
    {{e, a*b^2*c, c^2, a*b^2*c^3}, {a*b^2, c, a*b^2*c^2, c^3}}
```

Notice that each coset in $(H \cdot M)/M$ contains one of the cosets from H/R. In fact, if we threw out all elements in a coset of $(H \cdot M)/M$ that were not an element of H, we would get a coset of H/R. This provides us the mechanism to prove the isomorphism.

THEOREM 4.2: The Second Isomorphism Theorem

Suppose that N is a normal subgroup of G, and that H is a subgroup of G. Then

$$H/(H \cap N) \approx (H \cdot N)/N.$$

PROOF: By Lemma 4.3, $H \cdot N$ is a subgroup, and by Lemma 4.4, N is a normal subgroup of $H \cdot N$. Also, by Lemma 4.5, $H \cap N$ is a normal subgroup of H, and so both of the quotient groups are defined.

We will use the two homomorphisms

$$i : H \to H \cdot N$$

$$f : H \cdot N \to (H \cdot N)/N$$

where i is the identity mapping $i(h) = h$, and f is the canonical homomorphism.

We can now consider the combination of the two,

$$f(i(h)) : H \to (H \cdot N)/N.$$

Let us call the composition function $\psi(h) = f(i(h))$. We want to find the kernel of ψ, for then we can use the first isomorphism theorem (4.1). If we let

$$\begin{array}{ccc} H & \xrightarrow{i} & H \cdot N \\ \phi \downarrow & & \downarrow f \\ H/(H \cap N) & \longleftrightarrow & (H \cdot N)/N \end{array}$$

FIGURE 4.3: Commuting diagram for second isomorphism theorem

e denote the identity element of $(H \cdot N)/N$, then

$$\begin{aligned} h \in \text{Ker}(\psi) &\Longleftrightarrow f(i(h)) = e \\ &\Longleftrightarrow i(h) \in \text{Ker}(f) = N \\ &\Longleftrightarrow h \in N \quad \text{and} \quad h \in H \\ &\Longleftrightarrow h \in H \cap N. \end{aligned}$$

So by the first isomorphism theorem (4.1), we have

$$(H \cdot N)/N \approx H/(H \cap N).$$

∎

We can describe the second isomorphism theorem (4.2) pictorially through the diagram in Figure 4.3, which is commutative according to the first isomorphism theorem (4.1): Note that this diagram demonstrates that

$$|H|/|H \cap N| = |H \cdot N|/|N|.$$

In fact, we can show that $|H|/|H \cap N| = |H \cdot N|/|N|$ even when neither of the groups H nor N is a normal subgroup.

PROPOSITION 4.9

Let H and K be two subgroups of a finite group G. Then the number of elements in the product $H \cdot K$ is given by

$$|H \cdot K| = \frac{|H|\,|K|}{|H \cap K|}.$$

PROOF: Even though $H \cdot K$ may not be a group, it still makes sense to consider the set of left cosets $(H \cdot K)/K$. A typical left coset belonging to $(H \cdot K)/K$ would be $h \cdot k \cdot K$, where h is an element of H, and k is an element of K. By Lemma 3.1, all cosets contain $|K|$ elements, and by Lemma 3.2 two cosets would intersect if, and only if, they are equal. Thus the elements of

$H \cdot K$ are distributed into non-overlapping cosets, each having $|K|$ elements. Thus, the number of cosets in $(H \cdot K)/K$ is

$$|(H \cdot K)/K| = \frac{|H \cdot K|}{|K|}.$$

Likewise, we have

$$|H/(H \cap K)| = \frac{|H|}{|H \cap K|}.$$

Thus, if we can show that $|H/(H \cap K)| = |(H \cdot K)/K|$, we will have proven the proposition. Let us define a mapping (not a homomorphism) that will relate the elements of these two sets. Let

$$\phi : (H \cdot K)/K \rightarrow H/(H \cap K)$$

be defined by

$$\phi(h \cdot K) = h \cdot (H \cap K).$$

To see that this is well defined, note that if $h_1 \cdot K = h_2 \cdot K$ for two elements h_1 and h_2 in H, then $h_2^{-1} \cdot h_1 \cdot K = K$, so $h_2^{-1} \cdot h_1$ must be in K. But $h_2^{-1} \cdot h_1$ is also in H, hence in the intersection. Thus,

$$h_2 \cdot (H \cap K) = h_2 \cdot (h_2^{-1} \cdot h_1) \cdot (H \cap K) = h_1 \cdot (H \cap K).$$

So we see that if $h_1 \cdot K = h_2 \cdot K$, then $\phi(h_1 \cdot K) = \phi(h_2 \cdot K)$, and the function ϕ is well defined.

On the other hand, if $h_1 \cdot (H \cap K) = h_2 \cdot (H \cap K)$, then $h_2^{-1} \cdot h_1$ would have to be in the intersection of H and K. So then, $h_1 \cdot K = h_2 \cdot K$. Hence the mapping is one-to-one. It is clear that the mapping is also surjective, so ϕ is a bijection, and the proposition is proved. □

If we consider a group with two normal subgroups, one of which is a subgroup of the other, we begin to see more patterns. Let us reload the octahedral group, and look at two normal subgroups.

```
InitGroup("e")
AddGroupVar("a", "b", "c")
Define(a^2, e); Define(b^3, e); Define(c^2, e)
Define(b*a, a*b^2); Define(c*a, a*b*c); Define(c*b, a*c^2)
G = ListGroup()
```

Motivational Example 4.8

The octahedral group has two non-trivial normal subgroups, one being the subgroup of the other. Explore the possible quotient groups.

The two normal subgroups this is referring to are

```
M = Group(a*b^2*c, c^2); M
    {e, a*b^2*c, c^2, a*b^2*c^3}
H = Group(b, c^2)
H = Conform(H, G); H
    {e, b, a*b^2*c, c^2, b^2, a*b*c, b*c^2, a*b^2*c^3, a*c,
    b^2*c^2, a*b*c^3, a*c^3}
```

The first normal subgroup we have seen before. The latter subgroup H has 12 elements, so by Proposition 3.5, H is a normal subgroup.

Since both H and M are normal subgroups, we can consider two different quotient groups.

```
Q1 = RtCoset(G, H); Q1
    {{e, b, a*b^2*c, c^2, b^2, a*b*c, b*c^2, a*b^2*c^3, a*c,
    b^2*c^2, a*b*c^3, a*c^3}, {a*b^2, a, c, a*b^2*c^2, a*b,
    b^2*c, a*c^2, c^3, b*c, a*b*c^2, b^2*c^3, b*c^3}}
Q2 = RtCoset(G, M); Q2
    {{e, a*b^2*c, c^2, a*b^2*c^3}, {a, b^2*c, a*c^2, b^2*c^3},
    {b, a*b*c, b*c^2, a*b*c^3}, {a*b, b*c, a*b*c^2, b*c^3},
    {b^2, a*c, b^2*c^2, a*c^3}, {a*b^2, c, a*b^2*c^2, c^3}}
```

At this point there doesn't seem to be much connection between these. But notice that M is also a subgroup of H. By Lemma 4.4, M will be a normal subgroup of H. This gives us a third quotient group to consider:

```
Q3 = RtCoset(H, M); Q3
    {{e, a*b^2*c, c^2, a*b^2*c^3}, {b, a*b*c, b*c^2, a*b*c^3},
    {b^2, a*c, b^2*c^2, a*c^3}}
```

Note that H/M will be a subgroup of G/M. Could this be a normal subgroup? In the case we are looking at, **Q3** $= H/M$ contains half of the elements of **Q2** $= G/M$, so it is normal, giving us a *fourth* quotient group:

```
Q4 = LftCoset(Q2, Q3)
Q4 = Conform(Q4, G); Q4
    {{{e, a*b^2*c, c^2, a*b^2*c^3}, {b, a*b*c, b*c^2, a*b*c^3},
    {b^2, a*c, b^2*c^2, a*c^3}}, {{a*b^2, c, a*b^2*c^2, c^3},
    {a, b^2*c, a*c^2, b^2*c^3}, {a*b, b*c, a*b*c^2, b*c^3}}}
```

□

Before we try to interpret this mess, let us first see why H/N will be a normal subgroup of G/N in general.

LEMMA 4.6

If H and N are normal subgroups of G, and if N is a subgroup of H, then H/N is a normal subgroup of G/N.

PROOF: From Lemma 4.4, N is a normal subgroup of H. A typical element of G/N is gN, where g is an element of G. A typical element of H/N is hN, where h is an element of H. Thus, H/N is contained in G/N, and so H/N is a subgroup of G/N.

To show that H/N is in fact a normal subgroup of G/N, we will use Proposition 3.4. That is, we will see if

$$(gN) \cdot (hN) \cdot (gN)^{-1}$$

will always be in H/N. But this simplifies to $(g \cdot h \cdot g^{-1}) \cdot N$, and $g \cdot h \cdot g^{-1}$ is in H since H is a normal subgroup of G. Therefore, $(g \cdot h \cdot g^{-1}) \cdot N$ is in H/N, and hence H/N is a normal subgroup of G/N. ∎

The "quotient group of quotient groups" **Q4** $= (G/N)/(H/N)$ is a list containing two lists, each of which contains several lists of elements. If this is too many nested lists for you to handle, imagine what would happen if we removed the innermost curly brackets. This would simplify the output to just a list of two lists, each of which contains 12 elements. But by looking carefully, we can see that we would get *exactly* **Q1**. We can use the canonical homomorphisms as a tool to strip away these inside-level brackets.

THEOREM 4.3: The Third Isomorphism Theorem

Let H and N be normal subgroups of G, and let N be a subgroup of H. Then

$$(G/N)/(H/N) \approx G/H.$$

PROOF: We will use the example to guide us in finding a mapping from $(G/N)/(H/N)$ to a set of elements in G. We have a canonical mapping from G to G/N, and another canonical mapping from G/N to $(G/N)/(H/N)$. Let us call these mappings ϕ and f, respectively.

For an element x in G, the composition homomorphism $f(\phi(x))$ gives the element of $(G/N)/(H/N)$ that contains x somewhere inside of it. Let us call this composition homomorphism ψ. Since f and ϕ are both surjective, the composition $\psi(x) = f(\phi(x))$ is surjective. Thus, the inverse of this homomorphism, $\psi^{-1}(y)$, gives a list of elements of G that are somewhere inside of the element y. This inverse is the mapping that removes the interior curly brackets. We only need to check that this is in fact a coset of G/H. Let us determine the kernel of the composition homomorphism $\psi(x)$.

Note that if x is in G, and e is the identity element of $(G/N)/(H/N)$, then

$$\begin{aligned} x \in \mathrm{Ker}(\psi) &\Longleftrightarrow f(\phi(x)) = e \\ &\Longleftrightarrow \phi(x) \in \mathrm{Ker}(f) = H/N \\ &\Longleftrightarrow x \in \phi^{-1}(H/N) = H. \end{aligned}$$

Therefore, the kernel of the composition ψ is H, and so from the first isomorphism theorem (4.1),

$$(G/N)/(H/N) \approx G/H.$$

∎

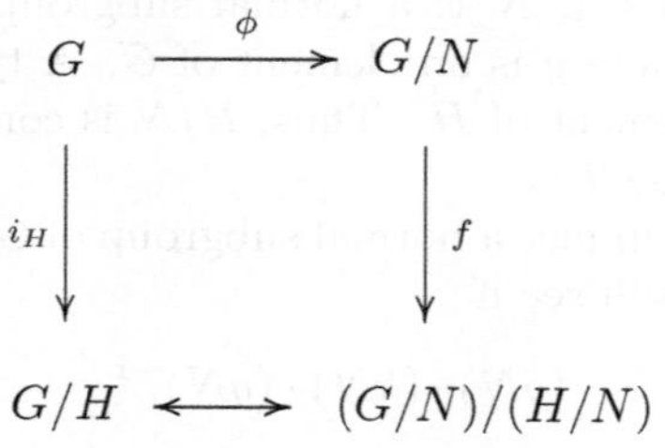

$$\begin{array}{ccc} G & \xrightarrow{\phi} & G/N \\ \downarrow i_H & & \downarrow f \\ G/H & \longleftrightarrow & (G/N)/(H/N) \end{array}$$

FIGURE 4.4: Commuting diagram for third isomorphism theorem

We can describe the third isomorphism theorem visually by the diagram in Figure 4.4. Since H is the kernel of the composition homomorphism

$$f(\phi) : G \rightarrow (G/N)/(H/N)$$

we have by the first isomorphism theorem that this diagram commutes.

The three isomorphism theorems work not only for groups, but many other objects as well, such as the rings we will study in Chapter 9. Because the same theorems apply to many different types of objects, an abstraction of these theorems can be made that would apply to any object for which there are natural mappings defined, called *morphisms.* This introduces a rich field called *category theory.* Although category theory is another level of abstraction beyond group theory, there are applications in physics and computer languages.

Problems for §4.3

For Problems **1** through **8**: Find, up to isomorphism, the possible homomorphic images of the following groups. That is, for all possible homomorphisms from G to G', what possible images could the homomorphism have?

1 Z_{10}.	**5** Q
2 Z_{12}.	**6** S_3
3 Z_{15}^*.	**7** Z_{24}^*
4 D_4	**8** The octahedral group (See Example 4.8.)

9 Prove that the homomorphic image of a cyclic group is cyclic.

10 Find all of the homomorphisms from Z_4 to Z_8^*.

11 Find all of the homomorphisms from Z_8^* to S_3.

12 Prove that there can be no nontrivial homomorphisms from S_3 to Z_3. Hint: What are the normal subgroups of S_3?

13 Suppose that there is a homomorphism from a finite group G onto Z_6. Prove that there are normal subgroups of G with index 2 and 3.

14 Let X, Y, and Z be three subgroups of a finite group G, with Y normal. Use Proposition 4.9 to find a formula for the number of elements in $X \cdot Y \cdot Z$.

15 Suppose that H and K are distinct subgroups of G of index 2. Prove that $H \cap K$ is a normal subgroup of G of index 4 and that $G/(H \cap K) \approx Z_8^*$.
Hint: Use the second isomorphism theorem.

16 Demonstrate the third isomorphism theorem using the subgroups $\{e, a^2\}$ and $\{e, a, a^2, a^3\}$ from D_4.

17 Demonstrate the third isomorphism theorem using the subgroups $\{1, 4\}$ and $\{1, 2, 4, 8\}$ from Z_{15}^*.

18 Prove or disprove: If H and N are two normal subgroups of G, with N a subgroup of H, then

$$(G/N)/(G/H) \approx H/N.$$

Interactive Problems

19 Use *Sage* to find a non-trivial homomorphism from the octahedral group to Q. (Hint: According to the first isomorphism theorem, what could the image be?)

20 Use *Sage* to find a homomorphism from the octahedral group onto S_3. (Hint: Use the first isomorphism theorem to determine what the kernel must be.)

Chapter 5

Permutation Groups

Although we have defined a group abstractly, they were not always defined in this way. When Galois introduced the term *groupe*, he only referred to a subset of permutations that was closed under multiplication. Hence, he only was considering the subgroups of a special type of group, known as *permutation groups*. However, with these permutation groups, he was able to prove that most fifth-degree polynomials cannot be solved in terms of roots. Hence, permutation groups have historically been at the core of abstract algebra.

5.1 Symmetric Groups

This section will introduce the notation for an important class of groups, known as the *permutation groups* or the *symmetric groups*,. Although at first they seem like a small number of examples of groups, in fact we will see that every finite group is isomorphic to a subgroup of these symmetric groups. So by studying these groups, by proxy we are studying all finite groups.

We have already seen one example of a symmetric group, S_3. We can easily generalize this group, and consider the group of all permutations of n objects. For example, with four books the beginning position would be

```
InitBooks(4)
```

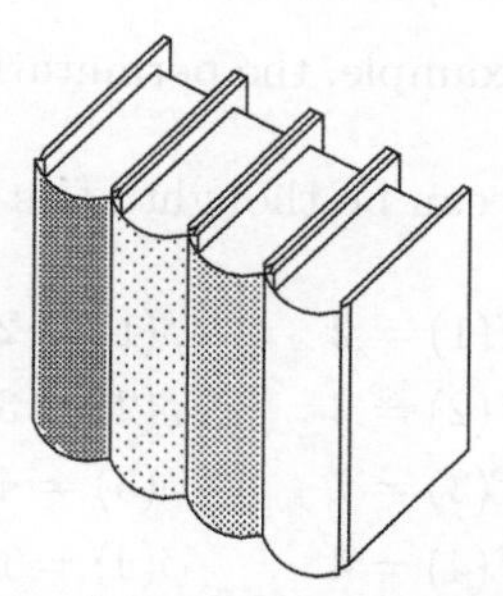

There are six *Sage* operations that rearrange these books.

`MoveBooks(First)`	swap the first two books.
`MoveBooks(Last)`	swap the last two books.

MoveBooks(Left)	move the first book to the end, sliding the other books to the left.
MoveBooks(Right)	move the last book to the beginning, sliding the other books to the right.
MoveBooks(Rev)	reverse the order of the books.
MoveBooks(Stay)	leave the books as they are.

For three books, any permutation can be obtained by just one of these six commands. But with four books it is a bit tricky to arrange the books in a particular order. With even more books, it becomes very cumbersome. Let us introduce a notation for a permutation of books that explicitly states where each book ends up.

One natural way to do this is to number the books in consecutive order, and determine the numbers in the final position. For example, if we put the books in their original order, and then shift the books to the left with **MoveBooks(Left)**, we find that if the books started in 1, 2, 3, 4 order, the final position will be 2, 3, 4, 1. We write the ending position below the starting position, as follows.

$$\begin{pmatrix} 1 & 2 & 3 & 4 \\ 2 & 3 & 4 & 1 \end{pmatrix}.$$

We can multiply the permutations using the new notation. For example, to calculate **Left·Last**, we have

$$\begin{pmatrix} 1 & 2 & 3 & 4 \\ 2 & 3 & 4 & 1 \end{pmatrix} \cdot \begin{pmatrix} 1 & 2 & 3 & 4 \\ 1 & 2 & 4 & 3 \end{pmatrix} = \begin{pmatrix} 1 & 2 & 3 & 4 \\ 2 & 3 & 1 & 4 \end{pmatrix}.$$

On the other hand, **Last·Left** is given by

$$\begin{pmatrix} 1 & 2 & 3 & 4 \\ 1 & 2 & 4 & 3 \end{pmatrix} \cdot \begin{pmatrix} 1 & 2 & 3 & 4 \\ 2 & 3 & 4 & 1 \end{pmatrix} = \begin{pmatrix} 1 & 2 & 3 & 4 \\ 2 & 4 & 3 & 1 \end{pmatrix}.$$

Obviously, **Left·Last** does not equal **Last·Left**.

We can also interpret each permutation as a *function* whose domain is a subset of the integers. For example, the permutations $f(x) = \begin{pmatrix} 1 & 2 & 3 & 4 \\ 2 & 3 & 1 & 4 \end{pmatrix}$ and $\phi(x) = \begin{pmatrix} 1 & 2 & 3 & 4 \\ 2 & 3 & 4 & 1 \end{pmatrix}$ can be thought of as two functions for which

$$\begin{aligned} f(1) &= 2 & \phi(1) &= 2 \\ f(2) &= 3 & \phi(2) &= 3 \\ f(3) &= 1 & \phi(3) &= 4 \\ f(4) &= 4 & \phi(4) &= 1. \end{aligned}$$

Note that $f(x)$ appears directly below x in the permutation $\begin{pmatrix} 1 & 2 & 3 & 4 \\ 2 & 3 & 1 & 4 \end{pmatrix}$. The product of the permutations is the same as the composition of the two functions. Thus, $f \cdot \phi$ would be

$$f(\phi(1)) = f(2) = 3$$
$$f(\phi(2)) = f(3) = 1$$
$$f(\phi(3)) = f(4) = 4$$
$$f(\phi(4)) = f(1) = 2.$$

Thus, the composition function $f(\phi(x))$, that is, of doing ϕ first, and then f, is $f \cdot \phi = f(\phi(x)) = \begin{pmatrix} 1 & 2 & 3 & 4 \\ 3 & 1 & 4 & 2 \end{pmatrix}$.

There is something curious here. When we view permutations as ways to rearrange a set of objects, such as books, the permutations are multiplied from left to right, which is the natural order. But when we view permutations as functions, the permutations are multiplied from right to left, which is again the natural order for function composition.

DEFINITION 5.1 *For the set $\{1, 2, 3, \ldots n\}$, we define the group of permutations on the set by S_n. That is, S_n is the set of functions that are one-to-one and onto on the set $\{1, 2, 3, \ldots n\}$. The group operation is function composition.*

To enter a permutation into *Sage*, only the bottom line is needed. A permutation in S_n can be entered as

$$P(x_1, x_2, x_3, \ldots, x_n),$$

where $x_1, x_2, x_3, \ldots x_n$ are distinct integers ranging from 1 to n. This permutation corresponds to the function

$$\begin{aligned} f(1) &= x_1 \\ f(2) &= x_2 \\ f(3) &= x_3 \\ &\cdots \\ f(n) &= x_n. \end{aligned}$$

Thus the product

```
P(5,4,1,2,3) * P(4,3,5,1,2)
    P(2, 1, 3, 5, 4)
```

yields $P(2, 1, 3, 5, 4)$. On the other hand, multiplying these permutations in the other order

```
P(4,3,5,1,2) * P(5,4,1,2,3)
    P(2, 1, 4, 3)
```

yields a different result.

Since the composition function maps 5 to itself, *Sage* drops the 5, treating this as a permutation on four objects instead. Since all permutations in S_4 can be expressed in terms of some combinations of the **Left** and **Last** book rearrangements, we can find all of the elements of S_4.

```
S4 = Group(P(2, 3, 4, 1), P(1, 2, 4, 3)); S4
    {P(), P(2, 1), P(1, 3, 2), P(3, 1, 2), P(2, 3, 1), P(3, 2, 1),
    P(1, 2, 4, 3), P(2, 1, 4, 3), P(1, 4, 2, 3), P(4, 1, 2, 3),
    P(2, 4, 1, 3), P(4, 2, 1, 3), P(1, 3, 4, 2), P(3, 1, 4, 2),
    P(1, 4, 3, 2), P(4, 1, 3, 2), P(3, 4, 1, 2), P(4, 3, 1, 2),
    P(2, 3, 4, 1), P(3, 2, 4, 1), P(2, 4, 3, 1), P(4, 2, 3, 1),
    P(3, 4, 2, 1), P(4, 3, 2, 1)}
len(S4)
    24
```

Note that the identity element of S_4 is denoted by **P()**, since the corresponding function leaves all objects fixed. We can determine the size of the group S_n in general, by counting the number of one-to-one and onto functions from the set $\{1, 2, 3, \ldots n\}$ to itself. We have n choices for $f(1)$, but then there will be only $n-1$ choices for $f(2)$, $n-2$ choices for $f(3)$, and so on. Thus, the size of the group S_n is given by

$$n \cdot (n-1) \cdot (n-2) \cdot (n-3) \cdots 2 \cdot 1.$$

This product is denoted by $n!$, read as "n factorial." Table 5.1 gives a short table for $n!$.

Both S_4 and the octahedral group have 24 elements, so we could ask if these two groups are isomorphic. The octahedral group can be reloaded by the commands

```
InitGroup("e")
AddGroupVar("a", "b", "c")
Define(a^2, e); Define(b^3, e); Define(c^2, e)
Define(b*a, a*b^2); Define(c*a, a*b*c); Define(c*b, a*c^2)
Oct = ListGroup(); Oct
    {e, a, b, a*b, b^2, a*b^2, c, a*c, b*c, a*b*c, b^2*c,
    a*b^2*c, c^2, a*c^2, b*c^2, a*b*c^2, b^2*c^2, a*b^2*c^2,
    c^3, a*c^3, b*c^3, a*b*c^3, b^2*c^3, a*b^2*c^3}
```

TABLE 5.1: $n!$ for $n \leq 10$

1!	= 1	6!	= 720
2!	= 2	7!	= 5040
3!	= 6	8!	= 40320
4!	= 24	9!	= 362880
5!	= 120	10!	= 3628800

Let us begin by defining a homomorphism from the subgroup generated by a and b to S_3, since we know that this is an isomorphism.

```
F = Homomorph(Oct, S4)
HomoDef(F, a, P(2,1) )
HomoDef(F, b, P(2,3,1) )
FinishHomo(F)
    'Homomorphism consistent, but not defined for the whole
    domain.'
```

This shows that so far, the homomorphism is consistent. To finish this homomorphism we only need to define $F(c)$. Since c must map to an element of order 4, there are six possibilities. (See Problem 10.) A little trial and error finds the right combination.

```
HomoDef(F, c, P(2,3,4,1) )
FinishHomo(F)
    'Homomorphism defined.'
```

Next we want to see that F is an isomorphism by showing that the kernel of F,

```
Kernel(F)
    {e}
```

is just the identity. Then by the pigeonhole principle, the image of F must be all of S_4, so $G \approx S_4$.

Sage can use the circle graphs on the set $\{1, 2, \ldots, n\}$ to visualize permutations. For example,

```
CircleGraph([1,2,3,4,5], P(5,4,1,2,3))
```

produces the circle graph on the left side of Figure 5.1. The solid arrows form a triangle that connects 1, 5, and 3, while the dotted "double arrow" connects 2 and 4. So this circle graph reveals some additional structure to the permutation that we will study later.

We can graph two or more permutations simultaneously. The command

```
CircleGraph([1,2,3,4,5], P(5,4,1,2,3), P(4,3,5,1,2))
```

produces the circle graph on the right of Figure 5.1. Here, the solid arrows represent the permutation $P(5, 4, 1, 2, 3)$, while the dotted arrows represent $P(4, 3, 5, 1, 2)$. If one imagines a permutation formed by traveling first through a dotted arrow, and then through a solid arrow, one obtains the permutation $P(2, 1, 3, 5, 4)$, which is $P(5, 4, 1, 2, 3) \cdot P(4, 3, 5, 1, 2)$. Note that the arrows are like functions, in that we apply the arrow of the second permutation first, and then the arrow for the first permutation.

The inverse of a permutation can be found using *Sage*.

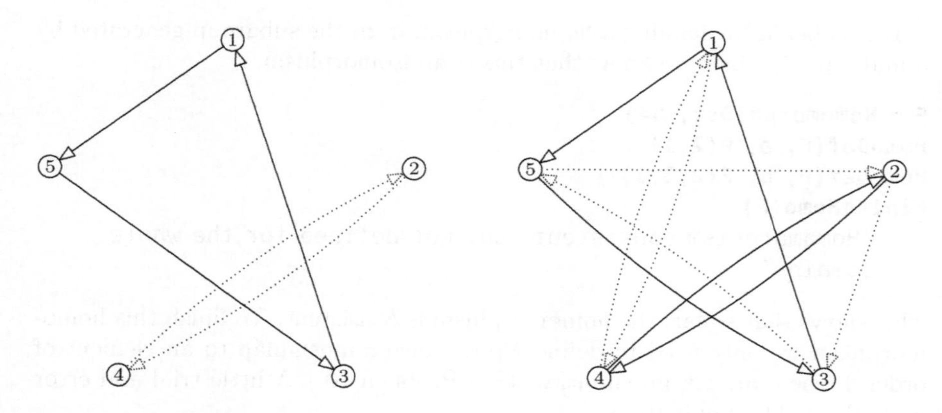

FIGURE 5.1: Circle graphs for typical permutations

```
P(5,4,1,2,3)^-1
    P(3, 4, 5, 2, 1)
```

The circle graph of the inverse permutation is similar to the circle graph of $P(5, 4, 1, 2, 3)$ except that all arrows are going in the opposite direction. The product of a permutation and its inverse of course will yield the identity element, denoted by $P(\,)$ in *Sage*.

```
P(5,4,1,2,3) * P(3,4,5,2,1)
    P()
```

Sage can also treat a permutation as a function,

```
P(5,4,1,2,3)(2)
    4
```

showing that $f(2) = 4$. In spite of the simplicity of the notations for a permutation, we will find that there is yet another notation that is even more concise. We will study this in the next section.

Problems for §5.1

For Problems **1** through **8**: Compute the following permutation products.

1 $\begin{pmatrix} 1 & 2 & 3 & 4 & 5 \\ 3 & 1 & 4 & 5 & 2 \end{pmatrix} \cdot \begin{pmatrix} 1 & 2 & 3 & 4 & 5 \\ 4 & 2 & 5 & 3 & 1 \end{pmatrix}$

2 $\begin{pmatrix} 1 & 2 & 3 & 4 & 5 \\ 4 & 2 & 5 & 3 & 1 \end{pmatrix} \cdot \begin{pmatrix} 1 & 2 & 3 & 4 & 5 \\ 3 & 1 & 4 & 5 & 2 \end{pmatrix}$

3 $\begin{pmatrix} 1 & 2 & 3 & 4 & 5 & 6 \\ 4 & 2 & 1 & 6 & 3 & 5 \end{pmatrix} \cdot \begin{pmatrix} 1 & 2 & 3 & 4 & 5 & 6 \\ 5 & 1 & 6 & 3 & 2 & 4 \end{pmatrix}$

4 $\begin{pmatrix} 1 & 2 & 3 & 4 & 5 & 6 \\ 5 & 1 & 6 & 3 & 2 & 4 \end{pmatrix} \cdot \begin{pmatrix} 1 & 2 & 3 & 4 & 5 & 6 \\ 4 & 2 & 1 & 6 & 3 & 5 \end{pmatrix}$

5 $\begin{pmatrix}1&2&3&4&5&6&7\\2&6&3&7&1&4&5\end{pmatrix}\cdot\begin{pmatrix}1&2&3&4&5&6&7\\6&1&7&2&4&3&5\end{pmatrix}$

6 $\begin{pmatrix}1&2&3&4&5&6&7\\6&1&7&2&4&3&5\end{pmatrix}\cdot\begin{pmatrix}1&2&3&4&5&6&7\\2&6&3&7&1&4&5\end{pmatrix}$

7 $\begin{pmatrix}1&2&3&4&5&6&7&8\\7&3&8&1&4&6&5&2\end{pmatrix}\cdot\begin{pmatrix}1&2&3&4&5&6&7&8\\3&7&4&2&8&1&5&6\end{pmatrix}$

8 $\begin{pmatrix}1&2&3&4&5&6&7&8\\3&7&4&2&8&1&5&6\end{pmatrix}\cdot\begin{pmatrix}1&2&3&4&5&6&7&8\\7&3&8&1&4&6&5&2\end{pmatrix}$

9 Form a multiplication table of S_3 using the permutation notation for the elements. That is, use the elements

$$S_3 = \left\{\begin{pmatrix}123\\123\end{pmatrix},\begin{pmatrix}123\\132\end{pmatrix},\begin{pmatrix}123\\213\end{pmatrix},\begin{pmatrix}123\\231\end{pmatrix},\begin{pmatrix}123\\312\end{pmatrix},\begin{pmatrix}123\\321\end{pmatrix}\right\}.$$

10 Find the six elements of S_4 that are of order 4.
Hint: All four of the numbers must move.

11 Find the eight elements of S_4 that are of order 3.
Hint: One number must map to itself.

12 Find the nine elements of S_4 that are of order 2.

13 Find a nontrivial element of S_5 that commutes with the permutation

$$x = \begin{pmatrix}1&2&3&4&5\\4&2&3&5&1\end{pmatrix}.$$

14 Find a permutation x in S_4 that solves the equation

$$x\cdot\begin{pmatrix}1&2&3&4\\1&3&4&2\end{pmatrix} = \begin{pmatrix}1&2&3&4\\4&1&3&2\end{pmatrix}\cdot x.$$

(There are in fact three different answers.)

15 Find a permutation x in S_5 that solves the equation

$$x\cdot\begin{pmatrix}1&2&3&4&5\\4&2&5&3&1\end{pmatrix} = \begin{pmatrix}1&2&3&4&5\\3&1&5&4&2\end{pmatrix}\cdot x.$$

(There are in fact four different answers.)

16 *Sage* views the permutations

$$\begin{pmatrix}1&2&3&4&5\\2&1&4&3&5\end{pmatrix}\quad\text{and}\quad\begin{pmatrix}1&2&3&4\\2&1&4&3\end{pmatrix}$$

as being the same permutation, $P(2,1,4,3)$. But are these really the same? If not, why can *Sage* use the same notation for these two elements?

Interactive Problems

For Problems **17** through **20**: Determine how the following permutations can be expressed in terms of the book rearrangements **`First`**, **`Last`**, **`Left`**, **`Right`**, and **`Rev`**.

17 $\begin{pmatrix} 1 & 2 & 3 & 4 \\ 1 & 3 & 2 & 4 \end{pmatrix}$

18 $\begin{pmatrix} 1 & 2 & 3 & 4 \\ 4 & 2 & 3 & 1 \end{pmatrix}$

19 $\begin{pmatrix} 1 & 2 & 3 & 4 \\ 3 & 1 & 4 & 2 \end{pmatrix}$

20 $\begin{pmatrix} 1 & 2 & 3 & 4 \\ 2 & 4 & 1 & 3 \end{pmatrix}$

5.2 Cycles

Although we have a notation for the elements of a permutation, it is not very convenient. We would like a way to express the permutations in a way that is easy to use, and more concise. The key to the new notation is to study the cycle structure of a permutation.

In the circle graph for the permutation $P(5,4,1,2,3)$, we saw that the arrows connecting 1, 5, and 3 were of one color, while a different colored arrow connected 2 and 4. By experimenting, we find that other permutations such as $P(4,5,2,3,1)$ have circle graphs with arrows of only one color, as in Figure 5.2.

These arrows indicate that the permutation can be expressed by a single chain

$$1 \to 4 \to 3 \to 2 \to 5 \to 1.$$

Other permutations, such as $P(2,4,1,6,5,3)$, have every *straight* arrow of the same color, even though there is one point (5) that maps to itself. We can

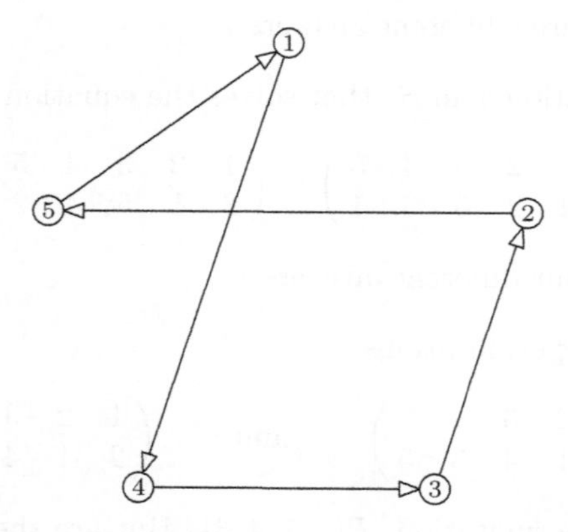

FIGURE 5.2: Circle graph of a typical cycle

still express this permutation as a single chain

$$1 \to 2 \to 4 \to 6 \to 3 \to 1,$$

if we stipulate that all numbers that are not mentioned in the chain map to themselves.

DEFINITION 5.2 Any permutation that can be expressed as a single chain is called a *cycle.* A cycle that moves exactly r of the numbers is called an *r-cycle.*

Let us introduce a concise notation for cycles. We can abbreviate a chain such as

$$1 \to 2 \to 4 \to 6 \to 3 \to 1,$$

to simply

$$(1\,2\,4\,6\,3).$$

This is called the *cycle notation* for the permutation. Each number in the cycle is mapped to the next number. The last number in the cycle is mapped to the first number. In general, the r-cycle

$$(i_1\, i_2\, i_3\, \ldots\, i_r)$$

represents the permutation that maps i_1 to i_2 , i_2 to i_3, etc., and finally i_r back to i_1. Notice that

$$(i_1\, i_2\, i_3\, \ldots\, i_r)^{-1} = (i_r\, i_{r-1}\, \ldots\, i_3\, i_2\, i_1),$$

so the inverse of an r-cycle will always be an r-cycle. The identity element can be written as the 0-cycle ( ).

A 1-cycle is actually impossible, since if one number is not fixed by a permutation, then the number that it maps to cannot be fixed. Thus, a non-identity permutation must move at least two numbers. We say that an r-cycle is a *nontrivial cycle* if $r > 1$.

Most permutations cannot be written as a single chain. This is evident from looking at the circle graph for the permutation $P(5, 4, 1, 2, 3)$. However, the two different types of arrows suggest that this permutation could be expressed as *two* cycles, one that represents the triangle from 1 to 5 to 3, and back to 1, and the other that exchanges 2 and 4. These two permutations are $P(5, 2, 1, 4, 3)$ and $P(1, 4, 3, 2, 5)$. These two cycles multiply together to give $P(5, 4, 1, 2, 3)$. In fact, this product can be done in either order. If we write these two permutations in cycle notation,

$$P(5, 2, 1, 4, 3) = (1\,5\,3), \qquad P(1, 4, 3, 2, 5) = (2\,4),$$

we notice that there are no numbers in common between these two cycles.

DEFINITION 5.3 Two cycles

$$(i_1\, i_2\, i_3\, \ldots\, i_r) \qquad \text{and} \qquad (j_1\, j_2\, j_3\, \ldots\, j_s)$$

are *disjoint* if $i_m \neq j_n$ for any m and n. That is, there are no integers in common between the two cycles.

LEMMA 5.1
Let x be an element of S_n that is not the identity. Then x can be written as a product of nontrivial disjoint cycles. This representation of x is unique up to the rearrangement of the cycles.

PROOF: Let us say that x fixes the integer i if $x(i) = i$. We will use induction on the number of integers not left fixed by x, denoted by m. Because x is not the identity, there is at least one integer not fixed by x. In fact, m must be at least 2, for the first integer must have somewhere to go.

If $m = 2$, then only two numbers i_1 and i_2 are moved. Since these are the only two integers not fixed, x must be a 2-cycle $(i_1\, i_2)$.

We now will assume by induction that the lemma is true whenever the number of integers not left fixed by x is fewer than m. Let i_1 be one integer that is not fixed, and let $i_2 = x(i_1)$. Then $x(i_2)$ cannot be i_2 for x is one-to-one, and if $x(i_2)$ is not i_1, we define $i_3 = x(i_2)$. Likewise, $x(i_3)$ cannot be either i_2 or i_3, since x is one-to-one. If $x(i_3)$ is not i_1, we define $i_4 = x(i_3)$.

Eventually this process must stop, for there are only m elements that are not fixed by x. Thus, there must be some value k such that $x(i_k) = i_1$. Define the permutation y to be the k-cycle $(i_1\, i_2\, i_3\, \ldots\, i_k)$. Then $x \cdot y^{-1}$ fixes all of the integers fixed by x, along with $i_1, i_2, i_3, \ldots, i_k$. By induction, $x \cdot y^{-1}$ can be expressed by a series of nontrivial disjoint cycles $c_1 \cdot c_2 \cdot c_3 \cdots c_t$. Moreover, the integers appearing in $c_1 \cdot c_2 \cdot c_3 \cdots c_t$ are just those that are not fixed by $x \cdot y^{-1}$. Thus, $c_1 \cdot c_2 \cdot c_3 \cdots c_t$ are disjoint from y. Finally, we have

$$x = y \cdot c_1 \cdot c_2 \cdot c_3 \cdots c_t.$$

Therefore, x can be written as a product of disjoint nontrivial cycles. By induction, every permutation besides the identity can be written as a product of nontrivial disjoint cycles.

For the uniqueness, suppose that a permutation x has two ways of being written in terms of nontrivial disjoint cycles:

$$x = c_1 \cdot c_2 \cdot c_3 \cdots c_r = d_1 \cdot d_2 \cdot d_3 \cdots d_s.$$

For any integer i_1 not fixed by x, one and only one cycle must contain i_1. Suppose that cycle is $c_j = (i_1\, i_2\, i_3\, \ldots\, i_q)$. But by the way we constructed the cycles above, this cycle must also be one of the d_k's. Thus, each cycle c_j is equal to d_k for some k. By symmetry, each d_k is equal to c_j for some

j. Thus, the two ways of writing x in terms of nontrivial disjoint cycles are merely rearrangements of the cycles. ∎

Lemma 5.1 gives us a succinct way to express permutations. *Sage* uses the notation

```
C(2,3,4,5)
```

to denote the cycle $(2\,3\,4\,5)$. We can multiply two cycles together,

```
C(2,3,4,5) * C(1,2,4)
    (1, 3, 4)(2, 5)
```

forming the answer as a product of two disjoint cycles, expressed using only parentheses. Note that when two cycles are disjoint, they are displayed without the times sign between them. We call this the *cycle decomposition* of the permutation. We can convert from the cycle notation to the permutation and vice versa in *Sage* with the commands

```
CycleToPerm( C(1,3,4) * C(2,5) )
    P(3, 5, 4, 1, 2)
PermToCycle( P(4,6,1,8,2,5,7,3) )
    (1, 4, 8, 3)(2, 6, 5)
```

We may even mix the two notations in *Sage* within an expression, such as:

```
C(1,2,3) * P(3,1,2,5,4) * C(4,5)
    ()
```

Whenever *Sage* encounters a mixture like this, it puts the answer in terms of cycles. In this case the result is the identity permutation, so *Sage* returns the 0-cycle ( ).

In *Sage*, we can create a circle graph of a cycle, or product of cycles, just as we did for permutations. We can even treat a cycle as a function, as we did for permutations. For example,

```
C(1,4,8,3)(3)
    1
```

determines where the cycle $(1\,4\,8\,3)$ sends the number 3. However, to evaluate a product of cycles at a given number, an extra pair of parentheses is needed:

```
(C(1,4,8,3)*C(2,6,5))(2)
    6
```

We mentioned that there are no permutations that move just one element, but the permutations that move exactly 2 elements will be important. We will give these 2-cycles a special name.

DEFINITION 5.4 A *transposition* is a 2-cycle $(i_1\, i_2)$, where $i_1 \neq i_2$.

Observe that i_1 can be any of the n numbers, and i_2 can be any of the remaining $n-1$ numbers, but this counts each transposition twice, since $(i_1\, i_2) = (i_2\, i_1)$. Thus, there are

$$\frac{n(n-1)}{2} = \frac{n^2 - n}{2}$$

transpositions of S_n.

LEMMA 5.2

For $n > 1$, the set of transpositions in S_n generates S_n.

PROOF: We need to show that every element of S_n can be written as a product of transpositions. The identity element can be written as $(1\,2)(1\,2)$, so we let x be a permutation that is not the identity. By Lemma 5.1, we can express x as a product of nontrivial disjoint cycles:

$$x = (i_1\, i_2\, i_3\, \ldots\, i_r) \cdot (j_1\, j_2\, \ldots\, j_s) \cdot (k_1\, k_2\, \ldots\, k_t) \cdot \cdots .$$

Now, consider the product of transpositions

$$(i_1\, i_2) \cdot (i_2\, i_3) \cdots (i_{r-1}\, i_r) \cdot (j_1\, j_2) \cdot (j_2\, j_3) \cdots (j_{s-1}\, j_s) \cdot (k_1\, k_2) \cdots (k_{t-1}\, k_t) \cdot \cdots .$$

Note that this product is equal to x. (Recall that we are working from right to left.) Therefore, we have expressed every element of S_n as a product of transpositions. □

Of course, a particular permutation can be expressed as a product of transpositions in more than one way. But an important property of the symmetric groups is that the number of transpositions used to represent a given permutation will always have the same parity; that is, even or odd. To show this, we will first prove the following lemma.

LEMMA 5.3

The product of an odd number of transpositions in S_n cannot equal the identity element.

PROOF: Since S_2 only contains one transposition, $(1\,2)$, raising this to an odd power will not be the identity element, so the lemma is true for the case $n = 2$. So by induction we can assume that the lemma is true for S_{n-1}. Suppose that there is an odd number of transpositions producing the identity in S_n. Then we can find such a product that uses the fewest number of transpositions, say k transpositions, with k odd. At least one transposition will involve moving n, since the lemma is true for S_{n-1}. Suppose that the

m^{th} transposition is the last one that moves n. If $m = 1$, then only the first transposition moves n, so the product cannot be the identity. We will now use induction on m. That is, we will assume that no product of k transpositions can be the identity for a smaller m. But then the $(m-1)^{\text{st}}$ and the mth transpositions are one of the four possibilities

$$(n\,x)(n\,x), \qquad (n\,x)(n\,y), \qquad (x\,y)(n\,x), \quad \text{or} \quad (y\,z)(n\,x)$$

for some x, y, and z. In the first case, the two transpositions cancel, so we can form a product using a fewer number of transpositions. In the other three cases, we can replace the pair with another pair,

$$(n\,x)(n\,y) = (n\,y)(x\,y); \quad (x\,y)(n\,x) = (n\,y)(x\,y); \quad (y\,z)(n\,x) = (n\,x)(y\,z);$$

for which m is smaller. Thus, there is no odd product of transpositions in S_n equaling the identity. ∎

We can use this lemma to prove the following theorem.

THEOREM 5.1: The Signature Theorem
For the symmetric group S_n, define the function

$$\sigma : S_n \to \mathbb{Z}$$

by

$$\sigma(x) = (-1)^{N(x)},$$

where $N(x)$ is the minimum number of transpositions needed to express x as a product of transpositions. Then this function, called the signature function, *is a homomorphism from S_n to the set of integers $\{-1, 1\}$.*

PROOF: By Lemma 5.2, every element of S_n can be written as a product of transpositions, so $\sigma(x)$ is well defined. Obviously this maps S_n to $\{-1, 1\}$, so we only need to establish that this is a homomorphism. Suppose that $\sigma(x \cdot y) \neq \sigma(x) \cdot \sigma(y)$. Then $N(x \cdot y) - (N(x) + N(y))$ would be an odd number. Since $N(x^{-1}) = N(x)$, we would also have $N(x \cdot y) + N(y^{-1}) + N(x^{-1})$ being an odd number. But then we would have three sets of transpositions, totaling an odd number, which when strung together produce $x \cdot y \cdot y^{-1} \cdot x^{-1} = (\,)$. This contradicts Lemma 5.3, so in fact $\sigma(x \cdot y) = \sigma(x) \cdot \sigma(y)$ for all x and y in S_n. ∎

We can compute the signature function on both permutations and products of cycles, using the **Signature** command.

```
Signature( P(4,3,5,1,2) )
    -1
Signature( C(1,4,2,7)*C(6,7,3) )
    -1
```

The signature of an r-cycle will be -1 if r is even, and 1 if r is odd.

DEFINITION 5.5 A permutation is an *alternating permutation* or an *even permutation* if the signature of the permutation is 1. A permutation is an *odd permutation* if it is not even, that is, if the signature is -1. The set of all alternating permutations of order n is written A_n.

COROLLARY 5.1

The set of all alternating permutations A_n is a normal subgroup of S_n. If $n > 1$, then S_n/A_n is isomorphic to Z_2.

PROOF: Clearly A_n is a normal subgroup of S_n, since A_n is the kernel of the signature homomorphism. Also if $n > 1$, then S_n contains at least one transposition whose signature would be -1. Thus, the image of the homomorphism is $\{-1, 1\}$. This group is isomorphic to Z_2. Then by the first isomorphism theorem (4.1), S_n/A_n is isomorphic to Z_2. □

PROPOSITION 5.1

For $n > 2$, the alternating group A_n is generated by the set of 3-cycles.

PROOF: Since every 3-cycle is a product of two transpositions, every 3-cycle is in A_n. Thus, it is sufficient to show that every element in A_n can be expressed in terms of 3-cycles. We have already seen that any element can be expressed as a product of an even number of transpositions. Suppose we group these in pairs as follows:

$$x = [(i_1\, j_1) \cdot (k_1\, l_1)] \cdot [(i_2\, j_2) \cdot (k_2\, l_2)] \cdot \cdots \cdot [(i_n\, j_n) \cdot (k_n\, l_n)].$$

If we could convert each pair of transpositions into 3-cycles, we would have the permutation x expressed as a product of 3-cycles. There are three cases to consider:

Case 1:

The integers i_m, j_m, k_m, l_m are all distinct. In this case,

$$(i_m\, j_m) \cdot (k_m\, l_m) = (i_m\, k_m\, l_m) \cdot (i_m\, j_m\, l_m).$$

Case 2:

Three of the four integers i_m, j_m, k_m, l_m are distinct. The four combinations that would produce this situation are $i_m = k_m$, $i_m = l_m$, $j_m = k_m$, or $j_m = l_m$. However, these four possibilities are essentially the same, so we only have to check one of these four combinations: $i_m = k_m$. Then we have

$$(i_m\, j_m) \cdot (i_m\, l_m) = (i_m\, l_m\, j_m).$$

Case 3:

Only two of the four integers i_m, j_m, k_m, and l_m are distinct. Then we must either have $i_m = k_m$ and $j_m = l_m$, or $i_m = l_m$ and $j_m = k_m$. In either case, we have

$$(i_m\, j_m) \cdot (k_m\, l_m) = (\,) = (1\,2\,3)(1\,3\,2).$$

In all three cases, we were able to express a pair of transpositions in terms of a product of one or two 3-cycles. Therefore, the permutation x can be written as a product of 3-cycles. ∎

Let us use this proposition to find the elements of A_4. We know that this is generated by 3-cycles, and has $4!/2 = 12$ elements. Since

```
Group( C(1,2,3), C(1,2,4) )
    {(), (1, 3, 2), (1, 2, 3), (1, 2)(3, 4), (2, 4, 3), (1, 4, 3),
    (2, 3, 4), (1, 4, 2), (1, 3)(2, 4), (1, 3, 4), (1, 2, 4),
    (1, 4)(2, 3)}
```

has 12 elements, this must be A_4. Eight of the twelve elements are 3-cycles. The other four elements form a subgroup that we have seen before.

Problems for §5.2

For Problems **1** through **4**: Find the product of the cycles without using *Sage*.

1 $(1\,5\,6) \cdot (3\,5\,2\,4) \cdot (1\,4\,3\,5)$

2 $(2\,4\,7) \cdot (1\,3\,6\,4) \cdot (1\,7\,5\,3\,6)$

3 $(1\,7\,2\,3\,8\,4) \cdot (1\,3\,5\,2\,4\,6) \cdot (2\,4\,3\,5\,8)$

4 $(1\,9\,3\,5\,2\,4\,8) \cdot (2\,7\,3\,9\,5\,4) \cdot (4\,7\,6\,8)$

5 Simplify the product of the cycles

$$(1\,2\,3)(2\,3\,4)(3\,4\,5)\cdots(n-1\ \ n\ \ n+1)(n\ \ n+1\ \ n+2)$$

for $n > 1$.

Hint: Try it with $n = 2$, $n = 3$, and $n = 4$ to see a pattern. Then prove using induction that the pattern persists.

6 Find the order of the permutations

$$(1\,2\,5)(3\,4) \qquad \text{and} \qquad (1\,2\,5)(3\,4\,6\,7).$$

7 Prove that the order of a permutation written in disjoint cycles is the least common multiple of the orders of the cycles.

8 Show that A_8 contains an element of order 15.
Hint: See Problem 7.

9 Show that if H is a subgroup of S_n, then either every member of H is an even permutation or exactly half of them are even.

10 Let S_Ω be the collection of all one-to-one and onto functions from $\mathbb{Z}^+$ to $\mathbb{Z}^+$ that only move a finite number of elements. Prove that S_Ω is a group. Show that we can write

$$S_\Omega = \bigcup_{n=1}^{\infty} S_n.$$

How should we interpret this union?

11 Let S_∞ be the collection of all one-to-one and onto functions from $\mathbb{Z}^+$ to $\mathbb{Z}^+$. Prove that S_∞ is a group. Find an element of this group that is not in S_Ω. (See Problem 10.)

12 Consider the set G of all one-to-one and onto functions $f(x)$ from $\mathbb{Z}^+$ to $\mathbb{Z}^+$ such that there is some integer M for which

$$|f(x) - x| < M \qquad \forall x \in \mathbb{Z}^+.$$

(The value of M is different for different elements of the group.) Prove that G is a group containing S_Ω. Find an element of G that is not in S_Ω. Find an element of S_∞ that is not in G. (See Problems 10 and 11.)

13 How many elements of order 5 are there in S_6?

14 A card-shuffling machine will always shuffle cards in the same way relative to the order in which they were given. All of the spades arranged in order from ace to king are put into the machine, and then the shuffled cards are re-entered into the machine again. If the cards after the second shuffle are in the order 10, 9, 4, Q, 6, J, 5, 3, K, 7, 8, 2, A, what order were the cards in after the first shuffle?

15 A subgroup H of the group S_n is called *transitive* on $B = \{1, 2, \ldots, n\}$ if for each pair i, j of elements of B, there exists an element f in H such that $f(i) = j$. Show that there exists a cyclic subgroup H of S_n that is transitive on B.

16 Let ϕ denote an r-cycle in S_n, and let x be any permutation in S_n. Show that $x \cdot \phi \cdot x^{-1}$ is an r-cycle.

17 Let ϕ and f denote two disjoint cycles in S_n, and let x be any permutation in S_n. Show that $x \cdot \phi \cdot x^{-1}$ and $x \cdot f \cdot x^{-1}$ are disjoint cycles. (See Problem 16.)

Interactive Problems

18 Use *Sage* to find a pair of 3-cycles whose product is a 3-cycle. Can there be a product of two 4-cycles that yields a 4-cycle?

19 The *cycle structure* of a permutation is the number of 2-cycles, 3-cycles, etc. it contains when written as a product of disjoint cycles. For example, $(1\,2\,3)(4\,5)$ and $(3\,4\,5)(1\,2)$ have the same cycle structure. Consider the elements

```
a = C(1, 2, 3); a
      (1, 2, 3)
b = C(1, 4, 2, 5, 6, 7); b
      (1, 4, 2, 5, 6, 7)
```

Predict the cycle structure of a^2, a^3, b^2, b^3, and b^6. Check your answers with *Sage*.

20 Calculate $a \cdot b$ from Problem 19. Predict the cycle structure of $(a \cdot b)^2$, $(a \cdot b)^3$, and $(a \cdot b)^4$, and verify your predictions with *Sage*.

21 Calculate $a \cdot b \cdot a^{-1}$ from Problem 19. Notice that it has the same cycle structure as b. Try this with other random permutations. Does $a \cdot b \cdot a^{-1}$ always have the same cycle structure as b? How do Problems 16 and 17 explain what is happening?

5.3 Cayley's Theorem

It was mentioned earlier that Galois originally defined a group as a subgroup of a permutation group. When Cayley created an abstract definition of a group, he showed that his definition was equivalent to Galois' definition. (He still only considered finite groups. See the Historical Diversion on page 169.) Today we refer to his result as Cayley's theorem, that every finite group is isomorphic to a subgroup of a permutation group.

To visualize Cayley's theorem, consider the circle graphs produced in §5.1. These demonstrate the property that every permutation is *one-to-one* and *onto*. In fact, every one-to-one and onto function on a finite set can be seen as a permutation on that set. But we also saw one-to-one and onto circle graphs in §3.1 while working with cosets. Is there a connection between these circle graphs? To demonstrate, let us work with the group Q of order 8:

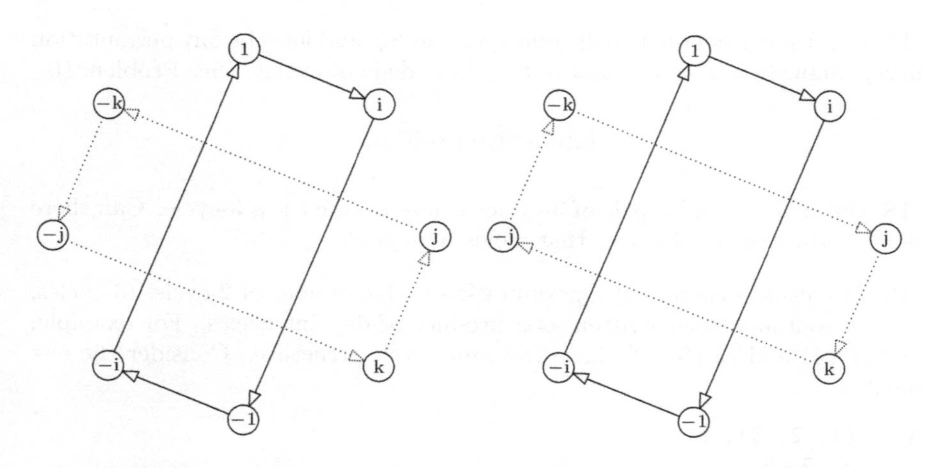

CircleGraph(Q, LeftMult(i)) **CircleGraph(Q, RightMult(i))**

FIGURE 5.3: Circle graphs for multiplying elements of Q by i

```
Q = InitQuaternions(); Q
    {1, i, j, k, -1, -i, -j, -k}
```

To find the left and right cosets of a subgroup generated by i, we use the commands

```
CircleGraph(Q, LeftMult(i))
CircleGraph(Q, RightMult(i))
```

which produce the two circle graphs in Figure 5.3.

If we number the elements of Q from 1 to 8, starting with 1 and going clockwise around the circles of Figure 5.3, we find that the left circle graph mimics the permutation $P(2, 5, 8, 3, 6, 1, 4, 7) = (1\,2\,5\,6)(3\,8\,7\,4)$, while the second graph is similar to the permutation $P(2, 5, 4, 7, 6, 1, 8, 3) = (1\,2\,5\,6)(3\,4\,7\,8)$. If we used different elements of Q in place of the i, we would have a different set of permutations. Thus, we can define two functions, $f(x)$ and $\phi(x)$, which map elements of Q to S_8. Table 5.2 shows both of these two functions.

Motivational Example 5.1

Let us use *Sage* to see if either of these is a homomorphism. Normally, in defining a homomorphism, we first determine the domain group and the target group. But in this case the target group is S_8, which has 40320 elements. Rather than having *Sage* construct all of the elements of this group, which would take an unreasonable amount of time, we can find the range of the homomorphism by determining which group is generated by $f(i)$ and $f(j)$.

TABLE 5.2: Ways to assign permutations to Q

x	$f(x)$ LeftMult(x)	$\phi(x)$ RightMult(x)
1	()	()
i	(1256)(3874)	(1256)(3478)
j	(1357)(2468)	(1357)(2864)
k	(1458)(2763)	(1458)(2367)
-1	(15)(26)(37)(48)	(15)(26)(37)(48)
$-i$	(1652)(3478)	(1652)(3874)
$-j$	(1753)(2864)	(1753)(2468)
$-k$	(1854)(2367)	(1854)(2763)

```
T = Group( C(1,2,5,6)*C(3,8,7,4), C(1,3,5,7)*C(2,4,6,8) ); T
     {(), (1, 2, 5, 6)(3, 8, 7, 4), (1, 7, 5, 3)(2, 8, 6, 4),
     (1, 8, 5, 4)(2, 3, 6, 7), (1, 5)(2, 6)(3, 7)(4, 8),
     (1, 6, 5, 2)(3, 4, 7, 8), (1, 3, 5, 7)(2, 4, 6, 8),
     (1, 4, 5, 8)(2, 7, 6, 3)}
```

We are now ready for the homomorphism.

```
F = Homomorph(Q, T)
HomoDef(F, i, C(1,2,5,6)*C(3,8,7,4) )
HomoDef(F, j, C(1,3,5,7)*C(2,4,6,8) )
HomoDef(F, k, C(1,4,5,8)*C(2,7,6,3) )
FinishHomo(F)
     (1, 2, 5, 6)(3, 8, 7, 4) * (1, 3, 5, 7)(2, 4, 6, 8) is not
     (1, 4, 5, 8)(2, 7, 6, 3)
     'Homomorphism failed'
```

So f must not be a homomorphism. Let us try seeing if ϕ is a homomorphism.

```
T = Group( C(1,2,5,6)*C(3,4,7,8), C(1,3,5,7)*C(2,8,6,4) ); T
     {(), (1, 6, 5, 2)(3, 8, 7, 4), (1, 3, 5, 7)(2, 8, 6, 4),
     (1, 8, 5, 4)(2, 7, 6, 3), (1, 5)(2, 6)(3, 7)(4, 8),
     (1, 2, 5, 6)(3, 4, 7, 8), (1, 7, 5, 3)(2, 4, 6, 8),
     (1, 4, 5, 8)(2, 3, 6, 7)}
phi = Homomorph(Q, T)
HomoDef(phi, i, C(1,2,5,6)*C(3,8,7,4) )
HomoDef(phi, j, C(1,3,5,7)*C(2,4,6,8) )
HomoDef(phi, k, C(1,4,5,8)*C(2,7,6,3) )
FinishHomo(phi)
     'Homomorphism defined'
```

This time, *Sage* found that ϕ is a homomorphism, generated by **RightMult** permutations. ∎

We can easily generalize this example to prove the following.

THEOREM 5.2: Cayley's Theorem
Every finite group of order n is isomorphic to a subgroup of S_n.

PROOF: Let G be a group of order n. For each g in G, define the mapping

$$p_g : G \to G$$

by $p_g(x) = g \cdot x$. For a given g, if $p_g(x) = p_g(y)$, then $g \cdot x = g \cdot y$, so $x = y$. Hence, p_g is a one-to-one mapping. Since G is a finite group, we can use the pigeonhole principle to show that p_g is also onto, and hence is a permutation of the elements of G.

We now can consider the mapping ϕ from G to the symmetric group $S_{|G|}$ on the elements of G, given by

$$\phi(g) = p_g$$

Now, consider two elements $\phi(g)$ and $\phi(h)$. The product of these is the mapping

$$x \to (p_g \cdot p_h)(x) = p_g(p_h(x)) = p_g(h \cdot x) = g \cdot (h \cdot x) = (g \cdot h) \cdot x.$$

Since this is the same as $\phi(g \cdot h)$, ϕ is a homomorphism.

The element g will be in the kernel of the homomorphism ϕ only if $\phi_g(x)$ is the identity permutation. This means that $g \cdot x = x$ for all elements x in G. Thus, the kernel consists just of the identity element of G, and hence ϕ is an isomorphism. Therefore, G is isomorphic to a subgroup of $S_{|G|}$. ∎

Although this theorem shows that all finite groups can be considered as a subgroup of a symmetric group, the theorem can apply to infinite groups as well. Of course we then must consider infinite symmetric groups, whose elements would be permutations on an infinite collection of objects. We might have a difficult time expressing some of the permutations! For example, if we had a library of an infinite number of books, we could not begin to express how one could rearrange the books. Some of the permutations could be expressed as one-to-one and onto functions. However, most of the permutations in an infinite symmetric group are not expressible using a finite number of words or symbols. Problems 10 through 12 of §5.2 reveal some of the unusual properties of infinite symmetric groups. Fortunately, we will mainly work with finite symmetric groups.

Although Cayley's theorem (5.2) shows that any finite group G is a subgroup of S_n, where n is the size of the group G, we often can find a smaller symmetric group that contains an isomorphic copy of G.

Historical Diversion

Arthur Cayley (1821–1895)

Author Cayley was a British mathematician, born in Richmond. He started at Trinity College at the early age of 17. By the time he was 20, he had already published 3 papers in the Cambridge Mathematical Journal. A few years later, Cayley introduced the concept of n-dimensional geometry. He graduated from Trinity winning the Senior Wrangler (top mathematician). In a competition he won a fellowship to Cambridge University for four years.

After his fellowship was over, at age 25 Cayley chose to be a lawyer. Yet he continued to work on mathematics in his spare time. Over the course of 14 years, Cayley would publish between 200 and 300 papers.

In 1863, a new position was established at Cambridge University, the Sadleirian. Cayley was elected to this position, and remained there the rest of his life. Cayley played a major role in allowing women to be admitted to Cambridge.

Although matrices have been around since antiquity, Cayley is considered to be the creator of matrix algebra, since he was the first to define the product of matrices. He showed that a square matrix satisfied its own characteristic equation, and made other huge developments in linear algebra.

One of Cayley's major contributions is the first step towards the modern definition of a group. Galois had originally defined a group as a set of permutations that is closed under multiplication. (See Historical Diversion on page 525.) In 1854 Cayley instead defined the group abstractly as a finite set, containing the identity (which he called 1), which was closed under an associative multiplication. He also insisted that the cancellation laws hold, that is, $a \cdot b = a \cdot c$ or $b \cdot a = c \cdot a$ implies that $b = c$. (From this rule, and the fact that the set is finite, one can prove that inverses exist. See Problem 19.) Cayley went on to prove that the two definitions are equivalent, the result currently called Cayley's theorem.

Cayley proceeded to make multiplication tables for the groups (now called Cayley tables) and showed how the tables revealed many important structures within the group, such as the inverse of the elements.

Unfortunately, Cayley's abstraction of the group definition went virtually unnoticed, and groups continued to mean only permutation groups for 26 more years. The idea of an infinite group did not occur until 1882.

Image source: Smithsonian Library, used by permission

Motivational Example 5.2

Consider the group D_4, whose multiplication table is given in Table 4.2.

```
InitGroup("e")
AddGroupVar("a", "b")
Define(a^4, e)
Define(b^2, e)
Define(b*a, a^3*b)
D4 = ListGroup(); D4
    {e, a, a^2, a^3, b, a*b, a^2*b, a^3*b}
```

Let us consider a *non-normal* subgroup of D_4:

```
H = Group(b)
    {e, b}
```

We saw in Cayley's theorem (5.2) that **RightMult** applied to the elements of the group derived a homomorphism. What if we applied **RightMult** to the cosets of the group? Recall that **RightMult(x)** can be thought of as a function $p_g(x) = g \cdot x$, that is, it multiplies the argument of the function to the right of g. If we apply this function to a left coset of H, we have $p_g(xH) = g \cdot xH$, which yields another left coset. (Right cosets won't work here, since $p_g(Hx) = g \cdot Hx$, which is neither a left nor right coset.) The list of left cosets is given by

```
L = LftCoset(D4, H); L
    {{e, b}, {a, a*b}, {a^2, a^2*b}, {a^3, a^3*b}}
```

If we multiply each coset to the right of a fixed element of the group, say a or $a \cdot b$, we get the circle graphs in Figure 5.4.

We see that each coset is mapped to another coset, so once again we can treat each circle graph as a permutation. By numbering the cosets in the order that they appear in **L**, we see that **RightMult(a)** acts as the permutation **P(2,3,4,1)** $= (1\,2\,3\,4)$, whereas **RightMult(b)** acts as the permutation **P(1,4,3,2)** $= (2\,4)$. *Sage* can check that this extends to a homomorphism.

```
S4 = Group( C(1,2), C(1,2,3), C(1,2,3,4) )
F = Homomorph(D4, S4)
HomoDef(F, a, C(1,2,3,4) )
HomoDef(F, b, C(2,4) )

FinishHomo(F)
    'Homomorphism defined'
Kernel(F)
    {e}
```

Since the kernel is just the identity element, we see that there is a subgroup of S_4 isomorphic to D_4. $\square$

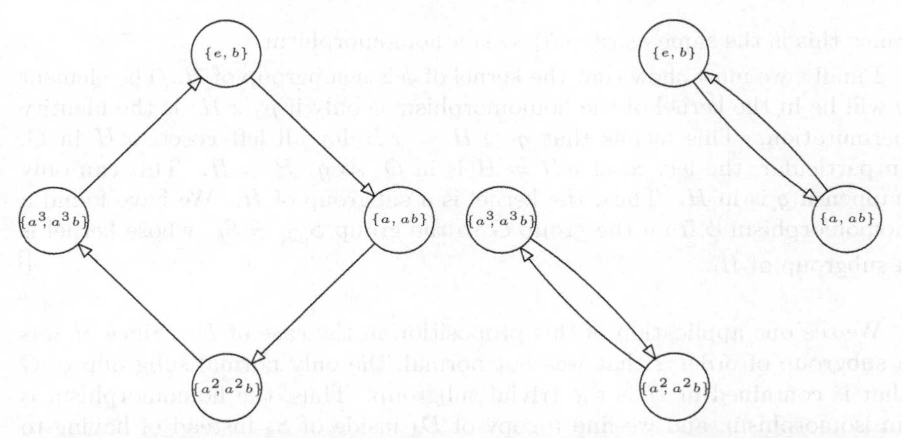

FIGURE 5.4: Circle graphs for multiplying cosets of D_4

Note that this is a much stronger result than Cayley's theorem (5.2), which only says that D_4 is isomorphic to a subgroup of the larger group S_8. We can follow this procedure to produce the following result:

THEOREM 5.3: Generalized Cayley's Theorem

Let G be a finite group of order n, and H a subgroup of order m. Then there is a homomorphism from G to S_k, with $k = n/m$, and whose kernel is a subgroup of H.

PROOF: Let Q be the set of left cosets G/H. For each g in G, define the mapping

$$p_g : Q \to Q$$

by $p_g(x\,H) = g \cdot x\,H$. Note that this is well defined, since if $x\,H = y\,H$, then $g \cdot x\,H = g \cdot y\,H$.

For a given g, if $p_g(x\,H) = p_g(y\,H)$, then $g \cdot x\,H = g \cdot y\,H$, so $x\,H = y\,H$. Hence, p_g is a one-to-one mapping. Since Q is a finite set, by the pigeonhole principle, p_g must also be onto, and hence is a permutation of the elements of Q.

We now can consider the mapping ϕ from G to the symmetric group $S_{|Q|}$ on the elements of Q, given by

$$\phi(g) = p_g.$$

Now, consider two elements $\phi(g)$ and $\phi(h)$. The product of these is the mapping

$$x\,H \to (p_g \cdot p_h)(x\,H) = p_g(p_h(x\,H)) = p_g(h \cdot x\,H) = g \cdot (h \cdot x\,H) = (g \cdot h) \cdot x\,H.$$

Since this is the same as $\phi(g \cdot h)$, ϕ is a homomorphism.

Finally, we must show that the kernel of ϕ is a subgroup of H. The element g will be in the kernel of the homomorphism ϕ only if $p_g(x\,H)$ is the identity permutation. This means that $g \cdot x\,H = x\,H$ for all left cosets $x\,H$ in Q. In particular, the left coset $e\,H = H$ is in Q, so $g \cdot H = H$. This can only happen if g is in H. Thus, the kernel is a subgroup of H. We have found a homomorphism ϕ from the group G to the group $S_{|Q|} = S_k$, whose kernel is a subgroup of H. ▯

We see one application of this proposition in the case of D_4. Since H was a subgroup of order 2 that was not normal, the only normal subgroup of G that is contained in H is the trivial subgroup. Thus, the homomorphism is an isomorphism, and we find a copy of D_4 inside of S_4 instead of having to look in the larger group S_8. This idea can be applied whenever we can find a subgroup of G that does not contain any nontrivial normal subgroups of G.

But there is another important ramification from this proposition. We can prove the existence of a normal subgroup of a group, knowing only the order of the group!

COROLLARY 5.2

Let G be a finite group, and H any subgroup of G. Then H contains a subgroup N, which is a normal subgroup of G, such that $|G|$ divides $(|G|/|H|)! \cdot |N|$.

PROOF: By the generalized Cayley's theorem (5.3), there exists a homomorphism ϕ from G to S_k, where $k = |G|/|H|$. Furthermore, the kernel is a subgroup of H. If we let N be the kernel, and let I be the image of the homomorphism, we have by the first isomorphism theorem (4.1) that

$$G/N \approx I.$$

In particular, $|G|/|N| = |I|$, and $|I|$ is a factor of $|S_k| = k!$. This means that $|G|$ is a factor of $k! \cdot |N|$. ▯

Here is an example of how we can prove the existence of a nontrivial normal subgroup, using just the order of the group. Suppose we have a group G of order 108. Suppose that G has a subgroup of order 27. (We will find in §7.4 that all groups of order 108 must have a subgroup of order 27.) Using $|G| = 108$ and $|H| = 27$, we find that G must contain a subgroup N such that 108 divides $(108/27)! \cdot |N| = 24 \cdot |N|$. But this means that $|N|$ must be a multiple of 9. Since N is a subgroup of H, which has order 27, we see that N is of order 9 or 27. Hence, we have proven that G contains a normal subgroup of either order 9 or 27. This will go a long way in finding the possible group structures of G, using only the size of the group G.

Problems for §5.3

1 Find a subgroup of S_4 that is isomorphic to Z_8^*.

Hint: Look at the proof of Cayley's theorem (5.2).

2 Find a subgroup of S_5 that is isomorphic to Z_5. (Do you really need Cayley's theorem (5.2) for this one?)

3 Follow the proof of Cayley's theorem (5.2) to find a subgroup of S_8 isomorphic to $D_4 = \{e, a, a^2, a^3, b, a \cdot b, a^2 \cdot b, a^3 \cdot b\}$, using this ordering of the elements.

4 Follow the proof of Cayley's theorem (5.2) to find a subgroup of S_8 isomorphic to $Z_{15}^* = \{1, 2, 4, 7, 8, 11, 13, 14\}$, using this ordering of the elements.

5 Follow the proof of Cayley's theorem (5.2) to find a subgroup of S_8 isomorphic to $Z_{24}^* = \{1, 5, 7, 11, 13, 17, 19, 23\}$, using this ordering of the elements.

6 According to Cayley's theorem (5.2), the quaternion group Q is isomorphic to a subgroup of S_8. Show that Q is not isomorphic to a subgroup of S_7.

Hint: Assume that a subgroup is isomorphic to Q. Is the permutation corresponding to $-1 = i^2$ odd or even? How many disjoint cycles can it contain? What possible permutations can i, j, k, $-i$, $-j$, and $-k$ be mapped to? From this, produce a contradiction.

7 In the text we found a group isomorphic to D_4 actually contained in S_4, which is a much smaller group than S_8 used by Cayley's theorem (5.2). What is the smallest symmetric group that contains a subgroup isomorphic to Z_{24}^*?

8 The function $f(x)$, which used **LeftMult** instead of **RightMult**, was seen not to be a homomorphism. Show that

$$f(x \cdot y) = f(y) \cdot f(x).$$

A function with this property is called an *anti-homomorphism*.

9 Show that if G is a group of order 35, and H is a subgroup of order 7, then H is normal.

Hint: Use Corollary 5.2.

10 Show that if G is a group of order 36, and H is a subgroup of order 9, then either H is normal, or H contains a subgroup of order 3 that is normal in G.

11 Show that if G is a group of order 200, and H is a subgroup of order 25, then either H is normal, or H contains a subgroup of order 5 that is normal in G.

12 Show that if G is a group of order 60, and H is a subgroup of order 15, then either H is normal, or H contains a subgroup of order 5 that is normal in G.

13 Show that if G is a group of order 189, and H is a subgroup of order 27, then either H is normal, or H contains a subgroup of order 3 or 9 that is normal in G.

14 Use Corollary 5.2 to show that if G is a group of order $p \cdot m$, where p is prime and $p > m$, then any subgroup of order p is normal.

15 Let G be a finite group, and H be a subgroup containing exactly 1/3 of the elements of G. Use Corollary 5.2 to show that either H is normal, or exactly half the elements of H form a normal subgroup of G.

16 Suppose G is a finite group, and let p be the smallest prime that divides $|G|$. Show that a subgroup H with order $|G|/p$ must be normal.

17 Suppose G has order p^2, where p is prime. Show that all subgroups are normal.

18 Show that in Cayley's theorem, the subgroup of S_n created is transitive in S_n. See Problem 15 from §5.2 for the definition of transitive.

19 Show that Cayley's definition of a finite group agrees with the current definition. (See Historical Diversion on page 169.) That is, show that if the cancellation laws hold for a finite set, $a \cdot b = a \cdot c$ or $b \cdot a = c \cdot a$ implies $a = c$, then inverses exist.

Interactive Problems

20 Use Cayley's theorem (5.2) to find a subgroup of S_{12} that is isomorphic to Z_{21}^*.

21 Use Cayley's theorem (5.2) to find a subgroup of S_{12} that is isomorphic to the following group:

```
InitGroup("e")
AddGroupVar("a", "b")
Define(a^3, e)
Define(b^4, e)
Define(b*a, a^2*b)
G = ListGroup(); G
    {e, a, a^2, b, a*b, a^2*b, b^2, a*b^2, a^2*b^2, b^3, a*b^3,
    a^2*b^3}
```

22 Use the generalized Cayley's theorem (5.3) to find a subgroup of S_8 that is isomorphic to the following group:

```
InitGroup("e")
AddGroupVar("a", "b")
Define(a^2, e)
Define(b^8, e)
Define(b*a, a*b^5)
G = ListGroup(); G
    {e, a, b, a*b, b^2, a*b^2, b^3, a*b^3, b^4, a*b^4, b^5,
    a*b^5, b^6, a*b^6, b^7, a*b^7}
```

Hint: Find a subgroup of order 2 that is not normal.

5.4 Numbering the Permutations

Although using cycles to denote permutations is in most cases more succinct and more readable, *Sage* works much faster using the permutation notation. Thus, for large time-consuming operations, such as checking that a function is a homomorphism, it will actually be faster using the $P(\ldots)$ notation than the $C(\ldots)$ notation. For example, we saw using Cayley's theorem that there was a copy of Q inside of S_8. It was generated by the elements

$$\phi(i) = (1\,2\,5\,6)(3\,4\,7\,8) \quad \text{and} \quad \phi(j) = (1\,3\,5\,7)(2\,8\,6\,4).$$

These two elements can be converted to the permutation notation, and we can use these to generate a subgroup of S_8. Thus, we could form a group isomorphic to Q by the command

```
Q = Group({P(2,5,4,7,6,1,8,3), P(3,8,5,2,7,4,1,6)}); Q
    {P(), P(6, 1, 8, 3, 2, 5, 4, 7), P(3, 8, 5, 2, 7, 4, 1, 6),
    P(8, 7, 2, 1, 4, 3, 6, 5), P(5, 6, 7, 8, 1, 2, 3, 4),
    P(2, 5, 4, 7, 6, 1, 8, 3), P(7, 4, 1, 6, 3, 8, 5, 2),
    P(4, 3, 6, 5, 8, 7, 2, 1)}
```

Alternatively, we could have used the cycle notation.

```
[ PermToCycle(x) for x in Q ]
    [(), (1, 6, 5, 2)(3, 8, 7, 4), (1, 3, 5, 7)(2, 8, 6, 4),
    (1, 8, 5, 4)(2, 7, 6, 3), (1, 5)(2, 6)(3, 7)(4, 8),
    (1, 2, 5, 6)(3, 4, 7, 8), (1, 7, 5, 3)(2, 4, 6, 8),
    (1, 4, 5, 8)(2, 3, 6, 7)]
```

Which method is best? For small groups, using cycles would be a good choice, because the results are easy to read. But for larger groups (say, over 100 elements, and yes, we will be working with groups that large in the next chapter), having *Sage* write out all of the elements in terms of cycles would be time-consuming and messy. It would be convenient to have a succinct way to describe each permutation using some type of abbreviation.

This section introduces a way to work with permutations that combines succinctness and speed. *Sage* has a preset order in which it lists the permutations.

$$
\begin{aligned}
1^{\text{st}} \text{ permutation} &= P(\,) \\
2^{\text{nd}} \text{ permutation} &= P(2,1) \\
3^{\text{rd}} \text{ permutation} &= P(1,3,2) \\
4^{\text{th}} \text{ permutation} &= P(3,1,2) \\
5^{\text{th}} \text{ permutation} &= P(2,3,1) \\
6^{\text{th}} \text{ permutation} &= P(3,2,1) \\
7^{\text{th}} \text{ permutation} &= P(1,2,4,3) \\
\cdots \quad & \quad \cdots \\
24^{\text{th}} \text{ permutation} &= P(4,3,2,1)
\end{aligned}
$$

Notice that the first 2 permutations give the group S_2, the first 6 give S_3, and the first 24 elements give S_4. This pattern can be extended to higher-order permutations, so that the first $n!$ permutations gives the group S_n.

The order of the permutations are designed so that *Sage* can quickly find the nth permutation on the list. For example, to find out what the 2000th permutation would be on this list, we use the **NthPerm** command.

```
NthPerm(2000)
    P(4, 1, 7, 6, 3, 2, 5)
```

We can also quickly determine the position of a given permutation on this list. The command

```
PermToInt( P(4,1,7,6,3,2,5) )
    2000
```

converts the permutation back to the number 2000.

Rather than spelling out each permutation, we can now give a single number that describes where the permutation is on the list of permutations. This will be called the *integer representation* of the permutation. Although this representation hides most of the information about the permutation, *Sage* can quickly recover the needed information to do group operations.

For example, we can multiply the 3rd permutation with the 21st on the list with the command

```
NthPerm(3) * NthPerm(21)
   P(3, 4, 2, 1)
```

If we wanted this converted back to a number, we would type

```
PermToInt(NthPerm(3) * NthPerm(21))
   23
```

Hence the 3rd permutation times the 21st permutation gives the 23rd permutation. If we had multiplied in the other order, we would get 19 instead, indicating that the group is non-abelian.

Sage provides an abbreviation to the permutations. By setting the variable **DisplayPermInt** to true, permutations will be displayed as their integer counterpart.

```
DisplayPermInt = true
```

Now, every permutation will be displayed as an integer.

```
P(4,1,7,6,3,2,5)
   2000
```

This integer representation of the permutations allows us to find other groups within the permutations easily. For example, the quaternion group was generated by the elements

$$(1\,2\,5\,6)(3\,4\,7\,8) \quad \text{and} \quad (1\,3\,5\,7)(2\,8\,6\,4).$$

Converting these to permutations will reveal their integer representation.

```
CycleToPerm( C(1,2,5,6)*C(3,4,7,8) )
   25827
CycleToPerm( C(1,3,5,7)*C(2,8,6,4) )
   14805
```

So we find that the quaternion group contains the 25827th and 14805th permutations. Now we can form the group using these two permutations as generators.

```
Q = Group(NthPerm(25827), NthPerm(14805)); Q
   {1, 7526, 14805, 16992, 23617, 25827, 32484, 39728}
```

This gives the whole group on a single line that encodes the entire structure of the group. Finally, the command **MultTable(Q)** produces Table 5.3.

This integer representation of the permutations allows us to form such a table, and has many other advantages over cyclic permutations, especially when we are working with extremely large subgroups of a symmetric group.

There are simple algorithms to convert from the permutation representation to the integer representation and back without a computer. We begin by presenting a method of converting from a permutation to an integer.

TABLE 5.3: Multiplication table for Q using integer representation

·	1	7526	14805	16992	23617	25827	32484	39728
1	1	7526	14805	16992	23617	25827	32484	39728
7526	7526	23617	16992	32484	25827	1	39728	14805
14805	14805	39728	23617	7526	32484	16992	1	25827
16992	16992	14805	25827	23617	39728	32484	7526	1
23617	23617	25827	32484	39728	1	7526	14805	16992
25827	25827	1	39728	14805	7526	23617	16992	32484
32484	32484	16992	1	25827	14805	39728	23617	7526
39728	39728	32484	7526	1	16992	14805	25827	23617

Example 5.3
Demonstrate without *Sage* that $P(4,1,7,6,3,2,5)$ is the 2000th permutation.

SOLUTION: For each number in the permutation, we count how many numbers further left are larger than that number. For example, the 4 has no numbers further left, so the count would be 0. The 3, however, has three numbers to the left of it that are larger, namely 4, 7, and 6. Here are the results of these counts.

$$\begin{array}{c} P(4,1,7,6,3,2,5) \\ 0\ 1\ 0\ 1\ 3\ 4\ 2 \end{array}$$

Next, we multiply the mth count by $(m-1)!$, and add the products together, and finally add 1. Thus,

$$0 \cdot 0! + 1 \cdot 1! + 0 \cdot 2! + 1 \cdot 3! + 3 \cdot 4! + 4 \cdot 5! + 2 \cdot 6! + 1 = 2000. \qquad \square$$

A similar algorithm reverses the procedure, and determines the nth permutation.

Example 5.4
Determine the 4000th permutation without *Sage*.

SOLUTION: We begin by subtracting 1, then using the division algorithm to successively divide by 2, 3, 4, etc., until the quotient is 0.

$$\begin{aligned} 3999 &= 2 \cdot 1999 + 1 \\ 1999 &= 3 \cdot \ 666 + 1 \\ 666 &= 4 \cdot \ 166 + 2 \\ 166 &= 5 \cdot \ \ 33 + 1 \\ 33 &= 6 \cdot \ \ \ 5 + 3 \\ 5 &= 7 \cdot \ \ \ 0 + 5 \end{aligned}$$

Since the last division was by $n = 7$, the permutation is in S_7. We will use the remainders to determine the permutation, starting from the last remainder, and working towards the first. We start with a list of numbers from 1 to n:

$$\{1, 2, 3, 4, 5, 6, 7\}$$

For each remainder m, we consider the $(m+1)^{\text{st}}$ largest number that has not been crossed out from the list. Since the last remainder is 5, we take the 6th largest number, which is 2. This eliminates 2 from the list. Here is the result after processing two more remainders.

$$\begin{array}{rcl} 3999 = 2 \cdot 1999 + 1 & & \\ 1999 = 3 \cdot 666 + 1 & & \\ 666 = 4 \cdot 166 + 2 & & \\ 166 = 5 \cdot 33 + 1 & \Rightarrow & 6 \\ 33 = 6 \cdot 5 + 3 & \Rightarrow & 4 \\ 5 = 7 \cdot 0 + 5 & \Rightarrow & 2 \end{array}$$

$$\{1, \not{2}, 3, \not{4}, 5, \not{6}, 7\}$$

The next remainder is 2, so we take the 3rd largest number that is not crossed out, which is 3. Continuing, we get the following.

$$\begin{array}{rcl} 3999 = 2 \cdot 1999 + 1 & \Rightarrow & 1 \\ 1999 = 3 \cdot 666 + 1 & \Rightarrow & 5 \\ 666 = 4 \cdot 166 + 2 & \Rightarrow & 3 \\ 166 = 5 \cdot 33 + 1 & \Rightarrow & 6 \\ 33 = 6 \cdot 5 + 3 & \Rightarrow & 4 \\ 5 = 7 \cdot 0 + 5 & \Rightarrow & 2 \end{array}$$

$$\{\not{1}, \not{2}, \not{3}, \not{4}, \not{5}, \not{6}, 7\}$$

The only number not crossed out is 7, which becomes the first number in the permutation. The rest of the permutation can be read from the new numbers from top to bottom, producing $P(7, 1, 5, 3, 6, 4, 2)$. ∎

The simple algorithms for converting permutations to integers and back make this association more natural. It also explains why *Sage* is able to convert permutations so quickly.

Problems for §5.4

For Problems **1** through **8**: Convert the following permutations to integers. Note that cycle notations must first be converted to a permutation.

1 $P(5,1,3,6,4,2)$	**4** $P(4,5,3,7,1,6,2)$	**7** $(1\,4\,3\,8)(2\,7\,6)$
2 $P(3,6,2,1,5,4)$	**5** $(1\,7\,2)(3\,6\,5)$	**8** $(1\,6\,8)(2\,5\,7\,4)$
3 $P(2,6,1,3,5,7,4)$	**6** $(1\,5\,6\,2)(4\,7)$	**9** $(1\,5)(2\,6\,7)(3\,8)$

For Problems **10** through **17**: Determine the nth permutation for the following values of n,

10 506	**12** 927	**14** 3816	**16** 6923
11 629	**13** 2593	**15** 4207	**17** 8510

18 Show that the set of elements in S_Ω is countable. See Problem 10 from §5.2, and Definition 0.12.

Interactive Problems

19 Find the elements of A_4 converted to the integer representation. Is there a pattern as to which positive integers correspond to the even permutations, and which correspond to odd? Does the pattern continue to A_5?

20 Use *Sage* to find all elements of S_7 whose square is $P(3,5,1,7,6,2,4)$. Hint: Use a "for" loop to test all of the elements of S_7:

```
for i in range(5040):
    if NthPerm(i)^2 == P(3,5,1,7,6,2,4):
        print(NthPerm(i))
```

21 Use *Sage* to find all elements of S_6 whose cube is $P(3,5,6,1,2,4)$. (See the hint for Problem 20.)

Chapter 6

Building Larger Groups from Smaller Groups

In this chapter, we will use the smaller groups that we have previously studied as building blocks to form larger groups. We will discover that *all* finite abelian groups can be constructed using just the cyclic groups Z_n. In fact, we will find that all small groups of order 15 or less can be expressed in terms of the groups we have studied.

6.1 The Direct Product

In this section we will consider the easiest way to combine two groups to form a larger group. In spite of its simplicity, we will find that all finite abelian groups can be constructed from this operation.

One way in which we can create a larger group from two smaller groups is to consider ordered pairs (g_1, g_2), in which the first component g_1 is a member of one group, and the second component g_2 is an element of a second group. We then can multiply these ordered pairs component-wise.

DEFINITION 6.1 Given two groups H and K, the *direct product* of H and K, denoted $H \times K$, is the group of ordered pairs (h, k) such that $h \in H$ and $k \in K$, with multiplication defined by

$$(h_1, k_1) \cdot (h_2, k_2) = (h_1 \cdot h_2, k_1 \cdot k_2).$$

The four group properties for the direct product are easy to verify. Certainly $H \times K$ is closed under multiplication, since the component-wise product of two ordered pairs is again an ordered pair. If e_1 is the identity element for H, and e_2 the identity element for K, then (e_1, e_2) would be the identity element of the direct product. Also, the inverse of an ordered pair (h, k) is (h^{-1}, k^{-1}). Finally, the associative law would hold for $H \times K$, since it holds for both H and K.

TABLE 6.1: Cayley table of $Z_4 \times Z_2$

	(0,0)	(0,1)	(1,0)	(1,1)	(2,0)	(2,1)	(3,0)	(3,1)
(0,0)	(0,0)	(0,1)	(1,0)	(1,1)	(2,0)	(2,1)	(3,0)	(3,1)
(0,1)	(0,1)	(0,0)	(1,1)	(1,0)	(2,1)	(2,0)	(3,1)	(3,0)
(1,0)	(1,0)	(1,1)	(2,0)	(2,1)	(3,0)	(3,1)	(0,0)	(0,1)
(1,1)	(1,1)	(1,0)	(2,1)	(2,0)	(3,1)	(3,0)	(0,1)	(0,0)
(2,0)	(2,0)	(2,1)	(3,0)	(3,1)	(0,0)	(0,1)	(1,0)	(1,1)
(2,1)	(2,1)	(2,0)	(3,1)	(3,0)	(0,1)	(0,0)	(1,1)	(1,0)
(3,0)	(3,0)	(3,1)	(0,0)	(0,1)	(1,0)	(1,1)	(2,0)	(2,1)
(3,1)	(3,1)	(3,0)	(0,1)	(0,0)	(1,1)	(1,0)	(2,1)	(2,0)

Example 6.1
Let $H = Z_4$ and $K = Z_2$. Consider the direct product $G = Z_4 \times Z_2$. Since Z_4 consists of the elements $\{0, 1, 2, 3\}$ and Z_2 consists of $\{0, 1\}$, the set of all ordered pairs (h, k) with $h \in Z_4$ and $k \in Z_2$ is

$$\{(0,0), (0,1), (1,0), (1,1), (2,0), (2,1), (3,0), (3,1)\}.$$

Thus, we will have a group of order 8. Multiplication is performed componentwise in the two groups. ∎

In order to define this group in *Sage*, we first define the groups Z_4 and Z_2.

```
Z4 = ZGroup(4)
Z2 = ZGroup(2)
G = DirectProduct(Z4, Z2); G
    {(0, 0), (0, 1), (1, 0), (1, 1), (2, 0), (2, 1), (3, 0), (3, 1)}
```

The Cayley table produced by *Sage* is given in Table 6.1.

We notice from the table that $Z_4 \times Z_2$ is commutative, has an element of order 4, yet has no element of order 8. Since we found all groups of order 8 in Chapter 4, we can use process of elimination to determine that this group must be isomorphic to Z_{15}^*.

It is not hard to show that the direct product of two abelian groups will be abelian.

PROPOSITION 6.1
Let H and K be two groups. Then $H \times K$ is commutative if, and only if, both H and K are commutative.

PROOF: First, suppose that H and K are both abelian. Then for two elements (h_1, k_1) and (h_2, k_2) in $H \times K$, we have

$$(h_1, k_1) \cdot (h_2, k_2) = (h_1 \cdot h_2, k_1 \cdot k_2) = (h_2 \cdot h_1, k_2 \cdot k_1) = (h_2, k_2) \cdot (h_1, k_1).$$

So the two elements in $H \times K$ commute. Hence, $H \times K$ is abelian.

Now suppose that $H \times K$ is commutative. We then have

$$(h_1 \cdot h_2, k_1 \cdot k_2) = (h_1, k_1) \cdot (h_2, k_2) = (h_2, k_2) \cdot (h_1, k_1) = (h_2 \cdot h_1, k_2 \cdot k_1).$$

Comparing components, we see that $h_1 \cdot h_2 = h_2 \cdot h_1$ and $k_1 \cdot k_2 = k_2 \cdot k_1$. Since this is true for all h_1 and h_2 in H, and all k_1 and k_2 in K, both H and K are abelian. □

It is easy to find the number of elements in a direct product. If H has order n, and K has order m, then the number of ordered pairs (h, k) would be $n \cdot m$.

We can generalize the direct product to a set of more than two groups. Let

$$G_1, G_2, G_3, \ldots, G_n$$

be a collection of n groups. Then we define $G_1 \times G_2 \times G_3 \times \cdots \times G_n$ to be the set of ordered n-tuples $(g_1, g_2, g_3, \ldots, g_n)$ with multiplication defined by

$$(g_1, g_2, \ldots, g_n) \cdot (h_1, h_2, \ldots, h_n) = (g_1 \cdot h_1, g_2 \cdot h_2, \ldots, g_n \cdot h_n).$$

The direct product of more than two groups can also be defined by taking the direct product of direct products. That is, given three groups G, H, and K, we could define both $(G \times H) \times K$ and $G \times (H \times K)$. But it is trivial to see that the mappings

$$f : (G \times H) \times K \to G \times H \times K$$

$$\phi : G \times (H \times K) \to G \times H \times K$$

given by

$$f(((g, h), k)) = (g, h, k) \qquad \text{and} \qquad \phi((g, (h, k))) = (g, h, k)$$

are both surjective isomorphisms. Thus,

$$(G \times H) \times K \approx G \times H \times K \approx G \times (H \times K).$$

It also should be noted that there is the natural mapping

$$\phi : H \times K \to K \times H$$

given by $\phi((h, k)) = (k, h)$. Thus, $H \times K \approx K \times H$.

This shows that the direct product between groups is a commutative operation, as well as associative. This suggests that some groups may be able to be expressed as a direct product of two or more smaller groups. If this is the case, then the order in which the smaller groups are combined would be irrelevant.

DEFINITION 6.2 Let G be a group. We say that G has a *decomposition* if $G \approx H \times K$, where neither H nor K is the trivial group.

For example, the group Z_{15}^* has a decomposition, since we saw in Example 6.1 that this group is isomorphic to $Z_4 \times Z_2$. We would like to find a way of testing whether a general group can be decomposed into smaller groups. In the case of S_3, we could use the fact that all smaller groups are abelian, along with Proposition 6.1 to show that S_3 cannot have a decomposition. But for other groups, the problem is more difficult. The following theorem gives us a way to determine whether a given group has a decomposition.

THEOREM 6.1: The Direct Product Theorem
Let G be a group with identity e, and let H and K be two subgroups of G. Suppose the following two statements are true:

1. $H \cap K = \{e\}$.

2. $h \cdot k = k \cdot h$ *for all* $h \in H$ *and* $k \in K$.

Then $H \cdot K \approx H \times K$.

PROOF: First, we show that every element in $H \cdot K$ can be *uniquely* written in the form $h \cdot k$, where $h \in H$ and $k \in K$. Suppose that

$$h_1 \cdot k_1 = h_2 \cdot k_2.$$

Then $h_2^{-1} \cdot h_1 = k_2 \cdot k_1^{-1}$. Since this element must be in both H and K, and the intersection of H and K is the identity element, we have that

$$h_2^{-1} \cdot h_1 = k_2 \cdot k_1^{-1} = e.$$

Thus, $h_1 = h_2$ and $k_1 = k_2$. Therefore, every element of $H \cdot K$ can be written uniquely as $h \cdot k$, where h is in H, and k is in K.

Next, we need to show that $H \cdot K$ is a group. Since $h \cdot k = k \cdot h$ for all $h \in H$ and $k \in K$, we have that $H \cdot K = K \cdot H$. Thus, by Lemma 4.2, $H \cdot K$ is a subgroup of G.

We can now define a mapping

$$\phi : H \cdot K \to H \times K$$

by $\phi(x) = (h, k)$, where h and k are the unique elements such that $h \in H$, $k \in K$, and $x = h \cdot k$. It is clear that ϕ is one-to-one, since the element (h, k) can only have come from $h \cdot k$. Also, ϕ is onto, for the element $h \cdot k$ maps to (h, k). All that remains to show is that $\phi(x \cdot y) = \phi(x) \cdot \phi(y)$. Let $x = h_1 \cdot k_1$, and $y = h_2 \cdot k_2$. Then

$$\begin{aligned}\phi(x \cdot y) &= \phi(h_1 \cdot k_1 \cdot h_2 \cdot k_2)\\ &= \phi(h_1 \cdot h_2 \cdot k_1 \cdot k_2)\\ &= (h_1 \cdot h_2, k_1 \cdot k_2)\\ &= (h_1, k_1) \cdot (h_2, k_2)\\ &= \phi(x) \cdot \phi(y).\end{aligned}$$

Thus, ϕ is an isomorphism, and so $H \cdot K \approx H \times K$. ∎

We can use this theorem to define the direct product of two groups in *Sage*.

Computational Example 6.2
Suppose we wish to generate the direct product $S_3 \times Z_8^*$. We first must define the two groups in *Sage* using the same identity element and different letters for the generators. The group S_3 is defined by the commands

```
InitGroup("e")
AddGroupVar("a", "b")
Define(a^3, e); Define(b^2, e); Define(b*a, a^2*b)
H = Group(a, b); H
    {e, b, a^2*b, a, a^2, a*b}
```

Now let us define Z_8^*, using c and d for the two generators.

```
AddGroupVar("c", "d")
Define(c^2, e); Define(d^2, e); Define(d*c, c*d)
K = Group(c, d); K
    {e, c, d, c*d}
```

Of course we did not use the **InitGroup** command before defining the second group, otherwise we would have cleared the first group. Notice that

```
Intersection(H, K)
    {e}
```

is just the identity element, so the first condition of the direct product theorem is satisfied.

In order for the second condition of the direct product theorem to be satisfied, every element of H must commute with every element of K. This will be true as long as all of the *generators* of H commute with all of the *generators* of K. Since there are 2 generators of H and 2 of K, we can tell *Sage* that the generators commute using $2 \cdot 2 = 4$ definitions:

```
Define(c*a, a*c); Define(c*b, b*c)
Define(d*a, a*d); Define(d*b, b*d)
```

```
H = Group(a, b); H
    {e, b, a^2*b, a, a^2, a*b}
K = Group(c, d); K
    {e, c, d, c*d}
```

Note that we were consistent in the direction of these definitions. That is, we defined an element of the form $k \cdot h$ to $h \cdot k$, where h is in H, and k is in K. Also, we recalculated the groups H and K so that the new rules will apply to the elements in this set.

According to the direct product theorem, $H \cdot K$ is now the same as $H \times K$. Here, then, is the direct product:

```
H * K
    {e, b, a^2*b, a, a^2, a*b, c, b*c, d, b*d, a^2*b*c, a*c,
    a^2*b*d, a*d, a^2*c, a*b*c, a^2*d, a*b*d, c*d, b*c*d,
    a^2*b*c*d, a*c*d, a^2*c*d, a*b*c*d}
```

We would get the same result by finding the smallest group that contains all of the generators.

```
G = Group(a, b, c, d)
len(G)
    24
```

This gives us a group of 24 elements. □

Since S_4 also has 24 elements, we could ask if the group in Example 6.2 is isomorphic to S_4. But recall that S_4 had exactly 9 elements of order 2, whereas the computation

```
RootCount(G, 2)
    16
```

reveals that G has 16 solutions to $x^2 = e$, with one being the identity. Thus, there are 15 elements of order 2, so S_4 is not isomorphic to $S_3 \times Z_8^*$.

This technique of counting elements of a certain order is an efficient way of showing that two groups are not isomorphic. Recall in §2.3 we denoted the number of solutions to $x^n = e$ by $R_n(G)$. In particular, if G is cyclic, $R_n(G) = \gcd(|G|, n)$. It is also rather easy to calculate $R_n(G)$ for direct products.

PROPOSITION 6.2

Let H and K be finite groups, and let n be a positive integer. Then

$$R_n(H \times K) = R_n(H) \cdot R_n(K).$$

PROOF: Let e_1 denote the identity element of H, and e_2 denote the identity element of K. An element $x = (h, k)$ in $H \times K$ solves the equation $x^n = (e_1, e_2)$ if and only if

$$h^n = e_1 \qquad \text{and} \qquad k^n = e_2.$$

Since there are $R_n(H)$ solutions to the former, and $R_n(K)$ solutions to the latter, there are $R_n(H) \cdot R_n(K)$ ordered pairs (h, k) that solve both of these equations. Thus, there are $R_n(H) \cdot R_n(K)$ elements of $H \times K$ for which $x^n = (e_1, e_2)$. ∎

For example, $R_2(S_3) = 4$, since there are 3 elements of order 2 in S_3, plus the identity. Also, all 4 elements of Z_8^* satisfy $x^2 = e$, so $R_2(Z_8^*) = 4$. Thus, there are 16 elements of $S_3 \times Z_8^*$ that satisfy $x^2 = e$, one of which is the identity. Thus, we quickly see that there are 15 elements of order 2.

As powerful as the direct product theorem (6.1) is, it is often difficult to check that $h \cdot k = k \cdot h$ for all $h \in H$ and $k \in K$. Here is a more convenient way of showing that a group can be expressed as a direct product of two subgroups.

COROLLARY 6.1

Let G be a group with identity e, and let H and K be two normal *subgroups of G. Then if $H \cap K = \{e\}$, $H \cdot K \approx H \times K$.*

PROOF: The first condition of the direct product theorem (6.1) is given, so we only need to show that the second condition holds. That is, we need to show that $h \cdot k = k \cdot h$ for all h in H, and k in K. Let $h \in H$ and $k \in K$.

Since K is a normal subgroup of G, $h \cdot k \cdot h^{-1}$ is in K. Thus, $h \cdot k \cdot h^{-1} \cdot k^{-1}$ is also in K.

But H is also a normal subgroup of G, so $k \cdot h^{-1} \cdot k^{-1}$ is in H. Hence, $h \cdot k \cdot h^{-1} \cdot k^{-1}$ is also in H.

We now use the fact that the only element in both H and K is e. Thus, $h \cdot k \cdot h^{-1} \cdot k^{-1} = e$, which implies $h \cdot k = k \cdot h$. Therefore, the second condition of the direct product theorem (6.1) holds, and so by this theorem, $H \cdot K \approx H \times K$. ∎

This corollary is sometimes more useful than the direct product theorem, even though for abelian groups the two are equivalent, since all subgroups of abelian groups are normal. In the next section we will continue to study the decomposition of abelian groups, and find that all finite abelian groups can be decomposed uniquely into a certain form.

Problems for §6.1

1 We have shown by process of elimination that $Z_4 \times Z_2$ is isomorphic to Z_{15}^*. Demonstrate the isomorphism by giving multiplication tables for the two groups with the same pattern.

2 Demonstrate that $Z_3 \times Z_2$ is isomorphic to Z_6.

3 Construct a multiplication table for $Z_2 \times Z_8^*$.

4 Construct a multiplication table for $Z_3 \times Z_8^*$.

5 Let $G = H \times K$, and define

$$\overline{H} = \{(h, e) \mid h \in H\}$$

and

$$\overline{K} = \{(e, k) \mid k \in K\}.$$

Prove that $G/\overline{H} \approx K$ and $G/\overline{K} \approx H$.

Hint: Use the first isomorphism theorem on an appropriate homomorphism.

For Problems **6** through **13**: Use Proposition 6.2 to find the number of elements of orders 2, 3, and 4 for the following groups.

Hint: First calculate $R_2(G)$, $R_3(G)$, and $R_4(G)$.

6 $Z_2 \times Z_6$
7 $Z_3 \times Z_4$
8 $Z_6 \times Z_8^*$
9 $S_3 \times Z_3$
10 $S_3 \times Z_4$
11 $A_4 \times Z_2$
12 $Z_2 \times Z_3 \times Z_4$
13 $Z_3 \times S_3 \times Z_4$
14 $Z_4 \times A_4 \times Z_6$

15 Show that $Z_2 \times Z_6$ is not isomorphic to Z_{12}.

16 Show that $S_3 \times Z_2$ is not isomorphic to A_4.

17 Using only the fact that $R_2(S_4) = 10$, prove that S_4 is not the decomposition of two smaller groups. You can use the result of Problem 20 in §2.3.

18 Using the fact that $R_3(A_5) = 21$ and $R_5(A_5) = 25$, prove that A_5 is not the decomposition of two smaller groups.

Interactive Problems

19 Use *Sage* to define the group $Z_2 \times Z_6$, and display the multiplication table. Then have *Sage* find the multiplication table for Z_{21}^*, and rearrange the elements to show that these groups are isomorphic.

20 Use *Sage* to define the group $Z_3 \times Z_8^*$, and display the multiplication table. Then have *Sage* find the multiplication table for Z_{36}^*, and rearrange the elements to show that these groups are isomorphic.

6.2 The Fundamental Theorem of Finite Abelian Groups

In the last section we promised that we will be able to construct any finite abelian group using as building blocks the groups that we have already learned. In this section, we will use the direct product to show that all finite abelian groups can be expressed in terms of the cyclic groups Z_n. We will even be able to find all abelian groups of a given order.

Example 6.3
Can we express the group Z_6 as the direct product of two smaller groups?
SOLUTION: By the direct product theorem, we must find two subgroups of Z_6 whose intersection is just the identity element, and whose product is the whole group. It is not hard to see that the subgroups

$$H = \{0, 3\} \qquad \text{and} \qquad K = \{0, 2, 4\}$$

satisfy these two conditions. Thus, $Z_6 \approx Z_2 \times Z_3$. This is easily verified using *Sage*.

```
Z2 = ZGroup(2); Z2
    {0, 1}
Z3 = ZGroup(2); Z3
    {0, 1, 2}
G = DirectProduct(Z2, Z3); G
    {(0, 0), (0, 1), (0, 2), (1, 0), (1, 1), (1, 2)}
RootCount(G, 2)
    2
RootCount(G, 3)
    3
```

Since we only have one element of order 2, and 2 elements of order 3, there must be an element of order 6, so the product $Z_2 \times Z_3$ must be isomorphic to Z_6. □

Observe the groups $H = \{0, 3\}$ and $K = \{0, 2, 4\}$ in this example. Notice that H consists of all of the elements such that $h^2 = 0$, and K consists of all the elements such that $k^3 = 0$. These two subgroups had only the identity element in common. We can extend this observation to general abelian groups.

LEMMA 6.1
Let G be an abelian group of order mn, where m and n are coprime. Then

$$H = \{h \in G \mid h^m = e\}$$

and

$$K = \{k \in G \mid k^n = e\}$$

are both subgroups of G*, and* $G \approx H \times K$.

PROOF: To check that H and K are indeed subgroups simply observe that since G is commutative the functions $\phi(x) = x^m$ and $f(x) = x^n$ are both homomorphisms of G. Then H and K are the kernels of the mappings ϕ and f.

To show that H and K have only the identity element in common, we consider an element x in the intersection. By the Chinese remainder theorem (0.7), there exists a non-negative number $k < m \cdot n$ such that

$$k \textbf{ mod } m = 1 \qquad \text{and} \qquad k \textbf{ mod } n = 0.$$

Then $k = 1 + mb$ for some number b. Thus,

$$x^k = x^{(1+mb)} = x \cdot (x^m)^b = x \cdot e^b = x$$

since x is in H. Yet $k = nc$ for some number c, so

$$x^k = x^{nc} = (x^n)^c = e^c = e$$

since x is in K. Thus, $x = e$, and so $H \cap K = \{e\}$. Since G is abelian, the direct product theorem (6.1) proves that

$$H \cdot K \approx H \times K.$$

All that is left to prove is that $G = H \cdot K$. Let g be an element in G. Since m and n are coprime, by the greatest common divisor theorem (0.4) there exists a and b such that

$$an + bm = \gcd(m, n) = 1.$$

Then

$$g = g^1 = g^{(an+bm)} = g^{an} \cdot g^{bm}.$$

Now, $(g^{an})^m = (g^a)^{nm} = e$, so g^{an} is in H. Likewise, g^{bm} is in K. Thus, every element of G is in $H \cdot K$, and so

$$G \approx H \times K.$$

□

This lemma tells us that if an abelian group has an order that is a product of two coprime numbers, this group can be written as a direct product of two groups. Unfortunately, the lemma does not tell us that H and K are proper subgroups. It is conceivable that either H or K from Lemma 6.1 is the whole group, and the other is just the identity element. We would still have $G = H \times K$, but this would not give a decomposition of G.

Here is an example to illustrate the possible problem that could occur. Suppose we know G is an abelian group of order 24. Since $24 = 8 \cdot 3$, we could let $m = 8$, and $n = 3$ in Lemma 6.1. Then H would consist of all elements of order 1, 2, 4, or 8, while K would consist of the elements of order 1 or 3. Then we would have $G \approx H \times K$.

But what if G had no elements of order 3? Then K would be just the identity element, and H would have to be all of G. Lemma 6.1 would hold, but since H and K are not proper subgroups, this would not give a decomposition of G. The next lemma uses induction to show that, in fact, G must have an element of order 3.

LEMMA 6.2

If G is a finite abelian group and p is a prime that divides the order of G, then G has an element of order p.

PROOF: We will proceed using induction on the order of G. If $|G|$ is a prime number, then p must be $|G|$, and G must be isomorphic to Z_p. So there would be an element of order p in G.

Suppose that the assumption is true for all groups of order less than $|G|$. If G does not have any proper subgroups, then G would be a cyclic group of prime order (which we have already covered). Thus, we may assume that G has a subgroup N that is neither G nor $\{e\}$.

Since G is abelian all subgroups are normal. Thus we could consider the quotient group G/N./ / Since $|G| = |N| \cdot |G/N|$, p must divide either $|N|$ or $|G/N|$. If p divides N, then because N is a smaller group than G, by induction N must have an element of order p, which would be in G.

If p does not divide $|N|$ it must divide $|G/N|$. Since G/N is a smaller group than G, by induction G/N must have an element of order p. This element can be written aN for some a in G.

Since aN is of order p, a cannot be in N, yet a^p must be in N. If we let $q = |N|$, we would have by Corollary 3.2 that $(a^p)^q = e$.

If $b = a^q$ is not the identity, then $b^p = e$, and so b would be the required element. But if $b = e$, then $(aN)^q = N$. But aN was of order p, and so p must divide q. But we assumed that p did not divide $q = |N|$. Hence, b is not the identity, and so G has an element of order p. ∎

This lemma is known as Cauchy's theorem for abelian groups. (See Historical Diversion on page 192.) Later on we will see that Cauchy's theorem is true for *all* groups, not just abelian groups. However, the result for abelian groups is sufficient for this chapter. This lemma guarantees that the subgroups H and K generated by Lemma 6.1 must be proper subgroups. In fact, there are times when it is possible to predict the size of the subgroups H and K.

Historical Diversion

Augustin Cauchy (1789–1857)

Augustin Cauchy was born in Paris, and by the time he was 11, both Laplace and Lagrange had recognized his potential. Lagrange told Laplace, "You see that little young man? He will supplant all of us in so far as we are mathematicians." On Lagrange's advice, Cauchy was enrolled in the best secondary school in Paris at the time, the École Centrale du Panthéon. In spite of his many awards in Latin and Humanities, Cauchy chose an engineering career.

At 21, he was given a commission as a civil engineer in Napolean's army. But during this job, Cauchy was doing mathematics on the side, submitting three manuscipts to the Première Classe. In 1812, he became ill from overwork, and returned to Paris to find a mathematical position.

By 1815 Cauchy was recognized as the leading mathematician in France, and was given an appointment at the École Polytechnique. Cauchy, along with Gauss, are considered to be the last two mathematicians to know all known mathematics as of their lifetime. Cauchy made contributions to almost every branch of mathematics. He was the first to prove Taylor's theorem using a remainder term. He was the first to define a function of a complex variable. He also worked with permutation groups, proving that if a prime p divides the order of a group, then some element is of order p. He introduced a new level of rigor in his proofs, which served as a model for future mathematicians.

During the French revolution of 1830, when Louis-Philippe succeeded Charles X, Cauchy fled to Fribourg, Switzerland, leaving his family behind. Because he refused to swear an oath of allegiance to the new regime, he lost almost all of his positions in Paris. In 1831, the King of Sardinia offered him a chair of theoretical physics in Turin. In 1833 he left Turin to go to Prague, to become a science tutor of the grandson of Charles X, the thirteen-year old Duke Henri d'Artois. Unfortunately, the Duke acquired a dislike of mathematics, and Cauchy did very little mathematics during these years. In 1834, his wife and two daughters joined Cauchy in Prague, reuniting his family.

Cauchy returned to Paris in 1838, but could not secure a position because he still refused to take an oath. In 1848,the oath of allegiance was abolished, allowing Cauchy to have an academic appointment. In 1849, he was reinstated as a professor of mathematical astronomy at the Faculté de Sciences. During these final years, until his death in 1857, Cauchy wrote over 500 research papers.

Image source: Smithsonian Library, used by permission

LEMMA 6.3

Let G be an abelian group of order $p^n \cdot k$ where p is prime, k is not divisible by p, and $n > 0$. Then there are subgroups P and K of G such that $G \approx P \times K$, where $|P| = p^n$, and $|K| = k$.

PROOF: Since p^n and k are coprime, we can use Lemma 6.1 to form the subgroups

$$P = \{x \in G \mid x^{(p^n)} = e\}$$

and

$$K = \{x \in G \mid x^k = e\}.$$

By Lemma 6.1 these two subgroups have only the identity in common, and $G \approx P \times K$. If p divided $|K|$, then by Lemma 6.2, K would contain an element of order p. But this element would then be in P as well, which contradicts the fact that only the identity element is in common between P and K. So p does not divide the order of K.

Also note that the order of every element of P is a power of p. Thus, Lemma 6.2 tells us that no other prime other than p divides $|P|$.

Finally, note that $|G| = p^n \cdot k = |P| \cdot |K|$. Since p does not divide $|K|$, we have that p^n must divide $|P|$. But no other primes can divide $|P|$, and so $|P| = p^n$. Hence, $|K| = k$. ∎

Lemma 6.3 is a tremendous help in finding the decomposition of abelian groups. To illustrate, suppose we have an abelian group G of order 24. Since $24 = 2^3 \cdot 3$, Lemma 6.3 states that G is isomorphic to a direct product of a group of order 8 and a group of order 3. Thus, G must be one of the groups

$$Z_8 \times Z_3, \qquad Z_{15}^* \times Z_3, \quad \text{or} \quad Z_{24}^* \times Z_3.$$

If we can find all abelian groups of order p^n for p a prime number, then we will in a similar manner be able to find all finite abelian groups.

Hence, our next line of attack is abelian groups of order p^n, where p is prime. If this is not a cyclic group, we can find a decomposition for this group as well.

LEMMA 6.4

Suppose P is an abelian group of order p^n, where p is a prime. Let x be an element in P that has the maximal order of all of the elements of P. Then $P \approx X \times T$, where X is the cyclic group generated by x, T is a subgroup of P, and $X \cap T = \{e\}$.

PROOF: We will use induction on n. If $n = 1$, then P is a cyclic group of order p, and hence is generated by non-identity element x in P. We then have $X = P$, so we can let $T = \{e\}$, and $P \approx X \times T$, with $X \cap T = \{e\}$.

Now suppose that the assertion is true for all powers of p less than n. Notice that the order of every element of P is a power of p. Thus, if we let x be an element with the *largest* order, say m, then the order of all elements in P must divide m. Hence, $g^m = e$ for all elements g in P.

We now let X be the subgroup generated by x. If $X = P$, then we can again let $T = \{e\}$ and we are done. If X is not P, we let y be an element of P not in X that has the *smallest* possible order. Then since the order of y^p is less than the order of y, y^p must be in X. This means that $y^p = x^q$ for some $0 \le q < m$.

Since y is in P, $y^m = e$. But

$$y^m = (y^p)^{(m/p)} = (x^q)^{(m/p)} = x^{(mq/p)}.$$

Because x is of order m, this can be the identity only if mq/p is a multiple of m. Hence, q is a multiple of p.

If we let $k = x^{-(q/p)} \cdot y$, then k is not in X because y is not, and

$$k^p = \left(x^{-(q/p)}\right)^p \cdot y^p = x^{-q} \cdot y^p = x^{-q} \cdot x^q = e.$$

Therefore, we have found an element k of order p that is not in X. If we let K be the group generated by the element k, then $X \cap K = \{e\}$.

Consider the quotient group P/K. What is the order of xK in P/K? We see that

$$(xK)^j = K \Longleftrightarrow x^j \in K \Longleftrightarrow x^j \in X \cap K \Longleftrightarrow x^j = e.$$

Therefore, the order of xK is the same as the order of x, which is m. Also note that no element of P/K can have an element of higher order since $g^m = e$ for all elements g in P.

Now we use the induction. Since the order of P/K is less than the order of P, and xK is an element of maximal order, we have by induction that

$$P/K \approx Y \times B,$$

where Y is the subgroup of P/K generated by xK, and B is a subgroup of P/K such that only the identity element K is in the intersection of Y and B.

Let ϕ be the canonical homomorphism from P to P/K given by $\phi(g) = gK$. Let $T = \phi^{-1}(B)$. Then T is a subgroup of P.

If g is in both X and T, then $\phi(g)$ is in both Y and B. Since the intersection of Y and B is the identity element, we have $\phi(g) = g \cdot K = K$. Thus, g is in the subgroup K. But $X \cap K = \{e\}$, so we have

$$X \cap T = \{e\}.$$

Thus, by the direct product theorem (6.1), we find that $X \cdot T \approx X \times T$.

We finally need to show that $P = X \cdot T$. Let u be an element in P, and since $P/K \approx Y \times B$, we can write $\phi(u)$ as $(x^c K) \cdot (kK)$ for some number c, and some kK in B. Then

$$u \in x^c \cdot k \cdot K \subseteq X \cdot T.$$

Thus, $P = X \cdot T$, and so $P \approx X \times T$. ∎

To illustrate the application of Lemma 6.4, consider the group Z_{24}^*. All non-identity elements of Z_{24}^* are of order 2, so this is the maximal order. Thus, Lemma 6.4 states that Z_{24}^* can be decomposed into Z_2 and a group of order 4. Since we have seen that $Z_4 \times Z_2 \approx Z_{15}^*$, the only other choice is $Z_2 \times Z_8^*$.

Now we apply Lemma 6.4 to Z_8^*. This is of order 4, and all elements besides the identity are of order 2, so Z_8^* can be decomposed into Z_2 and a group of order 2, which must be Z_2. Thus, $Z_8^* \approx Z_2 \times Z_2$, and so

$$Z_{24}^* \approx Z_2 \times Z_2 \times Z_2.$$

We have found a way to decompose any abelian group, to the point where each factor is a cyclic group whose order is a power of a prime. But now we want to address the issue as to whether a decomposition is *unique.* Can two different decompositions be isomorphic?

The main tool for testing whether two groups are isomorphic is to count elements of a given order. This can be accomplished by computing $R_n(G)$ for various values of n. It is natural to compute $R_n(G)$ for a decomposition of cyclic groups.

LEMMA 6.5

Let p be a prime number, and G be the direct product of cyclic groups

$$Z_{(p^{h_1})} \times Z_{(p^{h_2})} \times \cdots \times Z_{(p^{h_n})} \times Z_{k_1} \times Z_{k_2} \times \cdots \times Z_{k_m},$$

where $h_1, h_2, \ldots, h_n$ are positive integers, and $k_1, k_2, \ldots, k_m$ are coprime to p. Then if $q = p^x$,

$$R_q(G) = p^{\left(\sum_{i=1}^{n} \mathrm{Min}(h_i, x)\right)},$$

where $\mathrm{Min}(h_i, x)$ *denotes the minimum of h_i and x.*

PROOF: Since G is expressed as a direct product we can use Proposition 6.2 and find $R_q(H)$ for each factor H in the product, and multiply these numbers together. Since each factor is cyclic, we can use Corollary 2.1. For all of the factors $Z_{k_1}, Z_{k_2}, \ldots Z_{k_m}$, since $\gcd(k_i, q) = \gcd(k_i, p^x) = 1$, $R_q(H)$ would be 1. On the other hand, $R_q(Z_{(p^{h_i})})$ is

$$\gcd(p^{h_i}, q) = \gcd(p^{h_i}, p^x) = p^{\mathrm{Min}(h_i, x)}.$$

Thus, $R_q(G)$ is the product of the above for factors 1 through n of G, which gives us a grand total of

$$p^{\left(\sum_{i=1}^{n} \mathrm{Min}(h_i,x)\right)}.$$

□

We are now ready to show that *all* finite abelian groups can be represented as the direct product of cyclic groups. However, we would like to show at the same time that such a representation is unique. To this end we will use the previous lemma in conjunction with the following.

LEMMA 6.6

Let $h_1, h_2, h_3, \ldots, h_n$ be a set of positive integers, and define $f(x)$ as

$$f(x) = \sum_{i=1}^{n} \mathrm{Min}(h_i, x)$$

where $\mathrm{Min}(h_i, x)$ *denotes the minimum of h_i and x. Then the number of times that the integer x appears in the set of integers $h_1, h_2, h_3, \ldots, h_n$ is given by*

$$2f(x) - f(x-1) - f(x+1).$$

PROOF: Let us begin by observing the value of the expression

$$2\ \mathrm{Min}(h_i, x) - \mathrm{Min}(h_i, x-1) - \mathrm{Min}(h_i, x+1).$$

When $h_i < x$, then $\mathrm{Min}(h_i, x) = \mathrm{Min}(h_i, x-1) = \mathrm{Min}(h_i, x+1) = h_i$, and so the above evaluates to 0. On the other hand, if $h_i > x$, then the above expression simplifies to be

$$2x - (x-1) - (x+1) = 0.$$

However, if $h_i = x$, then $\mathrm{Min}(h_i, x) = x$, $\mathrm{Min}(h_i, x-1) = x-1$, and $\mathrm{Min}(h_i, x+1) = x$. Hence, we have

$$2\ \mathrm{Min}(h_i, x) - \mathrm{Min}(h_i, x-1) - \mathrm{Min}(h_i, x+1) = 2x - (x-1) - x = 1.$$

Thus, we see that

$$2\ \mathrm{Min}(h_i, x) - \mathrm{Min}(h_i, x-1) - \mathrm{Min}(h_i, x+1) = \begin{cases} 1 & \text{if } h_i = x \\ 0 & \text{if } h_i \neq x \end{cases}.$$

Thus, if we sum the above expression for i going from 1 to n, we will count the number of terms h_i that are equal to x. Hence this count will be

$$\sum_{i=1}^{n} 2\ \mathrm{Min}(h_i, x) - \mathrm{Min}(h_i, x-1) - \mathrm{Min}(h_i, x+1) = 2f(x) - f(x-1) - f(x+1).$$

□

We can now use Lemmas 6.3 through 6.6 to prove the following.

***THEOREM 6.2*: The Fundamental Theorem of Finite Abelian Groups**
A nontrivial finite abelian group is isomorphic to

$$Z_{(p_1^{h_1})} \times Z_{(p_2^{h_2})} \times Z_{(p_3^{h_3})} \times \cdots Z_{(p_n^{h_n})},$$

where $p_1, p_2, p_3, \ldots, p_n$ are prime numbers (not necessarily distinct). Furthermore, this decomposition is unique up to the rearrangement of the factors.

PROOF: We will proceed on induction on the order of the group. If the order of the group is 2, then the theorem is true since the group would be isomorphic to Z_2. Let G be a finite abelian group and suppose the theorem is true for all groups of order less than G. Let p be a prime that divides the order of G. By Lemma 6.3, $G \approx P \times K$, where P is the subgroup of G containing the elements of order p^m for some m.

Furthermore, if x is an element of maximal order in P, and X is the group generated by x, then by Lemma 6.4, $G \approx X \times T \times K$. Since X will be a nontrivial cyclic group, the orders of T and K will be less than $|G|$. Thus, by induction, T and K can be written as a direct product of cyclic groups whose orders are powers of primes. Since X is also a cyclic group of order p^r for some r, G can be written as a direct product of cyclic groups whose orders are powers of primes.

We next have to show that this decomposition is *unique*. We will do this by showing that the number of times $Z_{(p^x)}$ appears in the decomposition, where p is a prime, is completely determined by $R_q(G)$ for various values of q. From Lemma 6.5,

$$R_{p^x}(G) = p^{(\sum \text{ Min}(h_i, x))}$$

where the sum is taken over all i such that $p_i = p$. Thus, we see that

$$f_p(x) = \sum_{p_i = p} \text{Min}(h_i, x) = \log_p(R_{p^x}(G))$$

will be completely determined by the orders of the elements of G. But then by Lemma 6.6 the number of times that $Z_{(p^x)}$ appears in the decomposition is given by

$$2f_p(x) - f_p(x-1) - f_p(x+1).$$

Hence, the decomposition of G as a direct product of cyclic groups of the form $Z_{(p^x)}$ is unique. ∎

From this theorem, we can easily find all non-isomorphic abelian groups of a given order. For example, to find all non-isomorphic abelian groups of order 16, we note that all such groups are direct products of the cyclic groups of orders 2, 4, 8, or 16. We want to find all possible combinations of 2, 4, 8, and 16 that will multiply to give 16. With a little experimenting, we find that there are five such combinations:

$$2 \cdot 2 \cdot 2 \cdot 2, \qquad 2 \cdot 2 \cdot 4, \qquad 4 \cdot 4, \qquad 2 \cdot 8, \quad \text{and} \quad 16.$$

Thus, there are 5 possible abelian groups of order 16:

$$Z_2 \times Z_2 \times Z_2 \times Z_2, \qquad Z_2 \times Z_2 \times Z_4, \qquad Z_4 \times Z_4, \qquad Z_2 \times Z_8, \quad \text{and} \quad Z_{16}.$$

Since the fundamental theorem (6.2) also states that the representation is unique, these five groups must be non-isomorphic to each other. Notice that there is a correlation between these five groups, and the five ways we can express the number 4 as a sum of positive integers:

$$1+1+1+1=4$$
$$1+1+2=4$$
$$2+2=4$$
$$1+3=4$$
$$4=4$$

This leads us to a way of finding the number of non-isomorphic groups of order p^m for any m.

COROLLARY 6.2

Let $P(m)$ denote the number of ways in which m can be expressed as a sum of positive integers, without regard to order. Then if p is a prime number, there are exactly $P(m)$ non-isomorphic abelian groups of order p^m.

PROOF: By the fundamental theorem of abelian groups (6.2), every abelian group of order p^m must be isomorphic to

$$Z_{(p^{h_1})} \times Z_{(p^{h_2})} \times Z_{(p^{h_3})} \times \cdots \times Z_{(p^{h_n})}.$$

Also,

$$p^{h_1} \cdot p^{h_2} \cdot p^{h_3} \cdots p^{h_n} = p^m.$$

Hence $h_1 + h_2 + h_3 + \cdots + h_n = m$. Furthermore, the decomposition of the abelian group is unique up to rearrangement of the factors. Thus, there is a one-to-one correspondence between non-isomorphic abelian groups of order p^m and ways m can be written as a sum of positive integers without regard to order. ∎

We call $P(m)$ the number of *partitions* of m. We can have *Sage* count the number of partitions for us. For example, to find the number of partitions of the number 4, we can enter

```
PartitionsP(4)
    5
```

to find that there are five groups of order 2^4. The number of partitions increases exponentially with m; in fact a *Sage* plot reveals that it grows approximately like the function $e^{c\sqrt{m}}$ for some c. See Problem 21.

We can now find the number of non-isomorphic abelian groups of any order.

COROLLARY 6.3

Let $m > 1$ be an integer with prime factorization

$$p_1^{h_1} \cdot p_2^{h_2} \cdot p_3^{h_3} \cdots p_n^{h_n},$$

where $p_1, p_2, p_3, \ldots, p_n$ are distinct primes. Then the number of non-isomorphic abelian groups of order m is given by

$$P(h_1) \cdot P(h_2) \cdot P(h_3) \cdots P(h_n).$$

PROOF: We know from the fundamental theorem of abelian groups (6.2) that each such group is isomorphic to a direct product of cyclic groups whose order is a power of a prime. If we collect all factors involving the same primes together, we find that such a group is isomorphic to a direct product of a series of groups of orders $p_1^{h_1}$, $p_2^{h_2}$, $p_3^{h_3}$, $\cdots$, and $p_n^{h_n}$.

We know from Corollary 6.2 that there are exactly $P(x)$ non-isomorphic abelian groups of order p^x. Thus, there are $P(h_i)$ possible groups for the ith factor in this decomposition. Therefore, there are

$$P(h_1) \cdot P(h_2) \cdot P(h_3) \cdots P(h_n)$$

possible ways of forming a product of groups with orders

$$p_1^{h_1}, p_2^{h_2}, p_3^{h_3}, \ldots, \text{ and } p_n^{h_n}.$$

Since the fundamental theorem of abelian groups (6.2) also states that the decomposition is unique up to the rearrangement of the factors, every group thus formed is isomorphically different. So we have exactly $P(h_1) \cdot P(h_2) \cdot P(h_3) \cdots P(h_n)$ non-isomorphic abelian groups of order m. ☐

Computational Example 6.4

Suppose we wish to find the number of non-isomorphic abelian groups of order 180 billion. Since $180{,}000{,}000{,}000 = 2^{11} \cdot 3^2 \cdot 5^{10}$, we have that the number of groups is

```
PartitionsP(11) * PartitionsP(2) * PartitionsP(10)
    4704
```

giving us 4704 abelian groups of order 180 billion. ∎

From these two corollaries, we see that all finite abelian groups have been classified. One of the outstanding problems in group theory is to classify all finite groups. This is as yet an unsolved problem although much progress has been made through the use of computers. In the next two sections we will show some other ways of generating larger groups, which have become a key to some of the recent work that has been done in group theory.

Problems for §6.2

1 Let n be any integer greater than 1. Prove that $Z_n \times Z_n$ is not isomorphic to Z_{n^2}.

2 Let G be an abelian group with order mn, where m and n are coprime. Prove that $R_m(G) = m$ and $R_n(G) = n$.

Hint: Use Lemma 6.1 and the strategy of Lemma 6.3.

For Problems **3** through **11**: Find, up to isomorphism, all abelian groups of the following orders:

3 $|G| = 32$
4 $|G| = 200$
5 $|G| = 210$
6 $|G| = 300$
7 $|G| = 450$
8 $|G| = 500$
9 $|G| = 600$
10 $|G| = 675$
11 $|G| = 900$

12 What is the smallest positive integer n for which there are exactly four non-isomorphic abelian groups of order n?

13 Calculate the number of elements of order 4 in the groups

$$Z_{16}, \qquad Z_8 \times Z_2, \qquad Z_4 \times Z_4, \quad \text{and} \quad Z_4 \times Z_2 \times Z_2.$$

14 How many elements of order 25 are in $Z_5 \times Z_{25}$? (Do not do this exercise by brute force.)

15 An abelian group G of order 256 has 1 element of order 1, 7 elements of order 2, 24 elements of order 4, 96 elements of order 8, and 128 elements of order 16. Determine up to isomorphism the group G as a direct product of cyclic groups.

Hint: Use Lemma 6.5 to determine the value of the function

$$f(x) = \sum_{i=1}^{n} \text{Min}(h_i, x)$$

for $x = 1$, 2, 3, and 4. Then use Lemma 6.6 to determine how many times Z_2, Z_4, Z_8, and Z_{16} appear in the decomposition.

16 An abelian group G of order 512 has 1 element of order 1, 15 elements of order 2, 112 elements of order 4, 128 elements of order 8, and 256 elements of order 16. Determine up to isomorphism the group G as a direct product of cyclic groups. See the hint for Problem 15.

17 If an abelian group G of order 40 has exactly three elements of order 2, determine up to isomorphism the group G.

18 Classify the integers n for which the only abelian groups of order n are cyclic.

19 Recall from Problem 19 that the cycle structure of a permutation is the number of 2-cycles, 3-cycles, etc. it contains when written as a product of disjoint cycles. Show that the number of possible cycle structures in S_n is $P(n)$.

Interactive Problems

20 Use *Sage*'s **PartitionsP** command to find the number of abelian groups of order 120,000,000.

21 Notice that the logarithm of the **PartitionsP** function looks like a sideways parabola.

```
S = list_plot(ln(PartitionsP(i)) for i in range(999)]); S
```

This indicates that the **PartitionsP** function grows like $e^{c\sqrt{m}}$ for some constant c. Here is a way we can plot a sideways parabola on top of the above graph.

```
var("x")
P = plot(1.0 * sqrt(x), [x, 1, 999]); P + S
```

Try varying the constant 1.0 until the curves seem to run parallel to each other. Approximately what is this constant?

6.3 Automorphisms

There is another way to combine groups together other than the direct product. But before we can understand how this is defined, we must first consider a new group created from a single group. Many times, but not always, this new group will be larger that the original group. In fact, some very important examples stem from the groups created in this section.

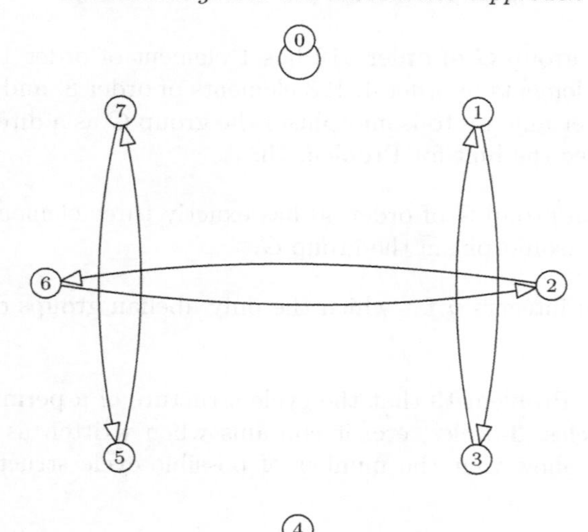

FIGURE 6.1: Circle graph for $x \to x^3$ in Z_8

We have already studied several examples of homomorphisms and isomorphisms *between* two groups, but suppose we considered a mapping from a group *to itself.*

Motivational Example 6.5

Find an isomorphism from Z_8 onto itself.

We can consider the following mapping:

```
Z8 = ZGroup(8)
CircleGraph(Z8, Pow(3))
```

which produces Figure 6.1. This mapping could be considered as the permutation $(1\,3)(2\,6)(5\,7)$ since the element 0 is left fixed. However, to make this into a homomorphism in *Sage*, we have to define a mapping that sends **Z8[1]** to **Z8[3]**.

```
F = Homomorph(Z8, Z8)
HomoDef(F, Z8[1], Z8[3])
FinishHomo(F)
    'Homomorphism defined'
```

The circle graph of **F** will be the same as Figure 6.1, which shows that in fact the homomorphism is one-to-one and onto. □

We give such a homomorphism a special name.

DEFINITION 6.3 An *automorphism* of the group G is a homomorphism from G to G that is one-to-one and onto.

If we study the above automorphism f on Z_8, we discover why this works. Recall that the operation of this group is addition modulo 8. Hence the mapping $x \to x^3$ in Z_8 will send each number x to $(3x)$ **mod** 8. Therefore,

$$f(x \cdot y) = f((x+y) \textbf{ mod } 8) = (3(x+y)) \textbf{ mod } 8 = (3x+3y) \textbf{ mod } 8 = f(x) \cdot f(y).$$

By observing this pattern, we can find another automorphism of Z_8 by sending x to x^5 instead of x^3. In fact, it is possible to define the product of two automorphisms as follows: If f and ϕ are both automorphisms of G, then $f \cdot \phi$ is the mapping $x \to f(\phi(x))$. This leads us to the proof of the following.

PROPOSITION 6.3

Given a group G, the set of all automorphisms on G forms a group, denoted $\text{Aut}(G)$. *In fact,* $\text{Aut}(G)$ *is a subgroup of the group of permutations on the elements of G.*

PROOF: The mapping $i(x) = x$ for all x in G is obviously an automorphism on G, so the set of all automorphisms on G is non-empty. Also, each automorphism is a permutation on the elements of G. Suppose ϕ and f are two automorphisms on G. Then $\phi(f(x))$ is a one-to-one and onto mapping from G to G.

Furthermore,

$$\phi(f(x \cdot y)) = \phi(f(x) \cdot f(y)) = \phi(f(x)) \cdot \phi(f(y)).$$

So $\phi(f(x))$ is a homomorphism on G, so $\phi \cdot f$ is an automorphism of G.

Also, since f is one-to-one and onto, f^{-1} exists on G, and

$$f\left(f^{-1}(x) \cdot f^{-1}(y)\right) = f\left(f^{-1}(x)\right) \cdot f\left(f^{-1}(y)\right) = x \cdot y.$$

Taking f^{-1} of both sides of the equation gives us

$$f^{-1}(x) \cdot f^{-1}(y) = f^{-1}(x \cdot y).$$

So f^{-1} is a homomorphism. Hence both f^{-1} and $\phi \cdot f^{-1}$ are automorphisms of G. Therefore by Proposition 2.2, $\text{Aut}(G)$ is a subgroup of the group of permutations on the elements of G. ∎

Example 6.6

Find the automorphism group for Z_8.

SOLUTION: The element 1 must be mapped by an automorphism to an element of order 8. Thus, 1 is mapped to either 1, 3, 5, or 7. But since 1 is a generator of Z_8, this would completely define the automorphism. Thus, there are at most four elements of $\text{Aut}(Z_8)$. But besides the identity mapping, we can easily find three other automorphisms:

$$x \to x^3, \qquad x \to x^5, \quad \text{and} \quad x \to x^7.$$

So we have exactly four automorphisms of Z_8. By converting these mappings to permutations on the non-zero elements of Z_8, we can express the automorphism group as

$$\{P(), P(3,6,1,4,7,2,5), P(5,2,7,4,1,6,3), P(7,6,5,4,3,2,1)\}.$$

This automorphism group can quickly be seen to be isomorphic to Z_8^*. ☐

It is not hard to generalize this result.

PROPOSITION 6.4

$$\text{Aut}(Z_n) \approx Z_n^*.$$

PROOF: Consider the mapping

$$\psi : Z_n^* \to \text{Aut}(Z_n)$$

given by $\psi(j) = f_j$, where $f_j(x) = (jx)$ **mod** n. Then given two elements j and k in Z_n^*, we have that

$$f_j(f_k(x)) = (j \cdot (k \cdot x)) \textbf{ mod } n = ((j \cdot k) \cdot x) \textbf{ mod } n = f_{j \cdot k}(x).$$

So

$$\psi(j) \cdot \psi(k) = f_j(f_k) = f_{j \cdot k} = \psi(j \cdot k).$$

Hence, ψ is a homomorphism from Z_n^* to $\text{Aut}(Z_n)$. To see that ψ is one-to-one, we note that $f_j(1) = j$, and so $f_j = f_k$ only if $j = k$.

To see that ψ is onto, we can use the pigeon-hole principle. If we consider a general automorphism f of Z_n, then $f(1)$ must be a generator of Z_n, since 1 is a generator. But f will be completely determined by knowing $f(1)$. Thus, the number of automorphisms is at most the number of generators of Z_n, which is $\phi(n)$. Since $|Z_n^*| = \phi(n)$, we know the function is one-to-one, so it must also be onto. ☐

So far, the automorphism group is smaller than the original group, but the goal of this chapter is to form larger groups. Let us consider a non-cyclic group.

Example 6.7

Find the automorphism group of the group Z_8^*, which has the following Cayley table.

·	1	3	5	7
1	1	3	5	7
3	3	1	7	5
5	5	7	1	3
7	7	5	3	1

SOLUTION: A good strategy for finding all of the automorphisms is to first determine an upper bound for the number of automorphisms. Suppose f is an automorphism. Then $f(1) = 1$, but all other elements are of order 2. Hence, any of the other elements might map to each other in any way. For example, $f(3)$ might be 3, 5, or 7. Once we know where 3 is mapped, $f(5)$ might be either of the other two elements. However, once we know $f(3)$ and $f(5)$, then $f(7)$ must be $f(3) \cdot f(5)$. Thus, there are at most $3 \cdot 2 = 6$ elements of $\mathrm{Aut}(Z_8^*)$. If we find that there are indeed this many automorphisms, then $\mathrm{Aut}(Z_8^*)$ would be larger than Z_8^*.

Here is one possible automorphism.

$$f(1) = 1$$
$$f(3) = 5$$
$$f(5) = 3$$
$$f(7) = 7$$

This can be represented as a transposition $(3\,5)$. Note that here, we are using the cycle notation with *elements* in place of numbers. We can test to see if this is an automorphism by constructing the Cayley table with the new ordering, and see if it has the same "color pattern." The new table is on the left side.

·	1	5	3	7
1	1	5	3	7
5	5	1	7	3
3	3	7	1	5
7	7	3	5	1

·	1	3	7	5
1	1	3	7	5
3	3	1	5	7
7	7	5	1	3
5	5	7	3	1

We can also ask whether there is an automorphism that sends 3 to 3, but exchanges 5 to 7, giving us the transposition $(5\,7)$. The new Cayley table is shown above on the right. Both of these tables preserve the color pattern of the original Cayley table, so both are automorphisms. These two automorphism will generate a copy of S_3, which gives 6 automorphisms. Since we established that this is the maximum number of automorphisms for Z_8^*, we have found the entire automorphism group. Hence $\mathrm{Aut}(Z_8^*) \approx S_3$. □

For non-commutative groups, there is a quick way to find many of the automorphisms. Let G be a non-commutative group, and let x be any element in G. The mapping $f_x : G \to G$ defined by

$$f_x(y) = x \cdot y \cdot x^{-1}$$

will always be an automorphism, for

$$f_x(y \cdot z) = x \cdot y \cdot z \cdot x^{-1} = (x \cdot y \cdot x^{-1}) \cdot (x \cdot z \cdot x^{-1}) = f_x(y) \cdot f_x(z).$$

So $f_x(y)$ is a homomorphism. Since the inverse homomorphism can easily be found,

$$y \in f_x^{-1}(v) \Longleftrightarrow x \cdot y \cdot x^{-1} = v \Longleftrightarrow y = x^{-1} \cdot v \cdot x \Longleftrightarrow y = f_{x^{-1}}(v),$$

we have that $f_x(y)$ is one-to-one and onto, therefore $f_x(y)$ is an automorphism.

DEFINITION 6.4 An automorphism $\phi(y)$ of a group G is called an *inner automorphism* if there is an element x in G such that

$$\phi(y) = x \cdot y \cdot x^{-1} \qquad \text{for all } y \in G.$$

The set of inner automorphisms of G is denoted $\text{Inn}(G)$.

Example 6.8
Find the inner automorphisms of the quaternion group

$$Q = \{1, i, j, k, -1, -i, -j, -k\}.$$

SOLUTION: Let us begin by determining an upper bound for the number of automorphisms. If f is an automorphism of Q, then $f(1) = 1$, but also $f(-1)$ must be -1, since this is the only element of order 2. All of the other elements are of order 4, so $f(i)$ could be any one of the remaining six elements. Once $f(i)$ is determined, we have that $f(-i) = f(i)^3$. Then $f(j)$ could be one of the remaining four elements. Since i and j generate Q, f will be determined by knowing $f(i)$ and $f(j)$. Thus, there is a maximum of $6 \cdot 4 = 24$ automorphisms.

It is fairly easy to find the inner automorphisms on Q. If we choose $x = i$, we have the mapping

$$\begin{aligned} f(1) &= i \cdot 1 \cdot (-i) = 1 & f(-1) &= i \cdot (-1) \cdot (-i) = -1 \\ f(i) &= i \cdot i \cdot (-i) = i & f(-i) &= i \cdot (-i) \cdot (-i) = -i \\ f(j) &= i \cdot j \cdot (-i) = -j & f(-j) &= i \cdot (-j) \cdot (-i) = j \\ f(k) &= i \cdot k \cdot (-i) = -k & f(-k) &= i \cdot (-k) \cdot (-i) = k \end{aligned}$$

We can express this automorphism in terms of cycles: $(j, -j)(k, -k)$. If we use $x = j$ or $x = k$ instead of $x = i$, we get the automorphisms $(i, -i)(k, -k)$ and $(i, -i)(j, -j)$. These three automorphisms, along with the identity automorphism, form a group. These are the only 4 inner automorphisms. ∎

Although we were able to find the inner automorphisms by hand, we will need *Sage*'s help to find the rest of the automorphisms.

Computational Example 6.9

Determine the automorphism group of Q.

With a bit of trial and error, we can come up with a new automorphism.

```
Q = InitQuaternions(); Q
    {1, i, j, k, -1, -i, -j, -k}
X = Homomorph(Q, Q)
HomoDef(X, i, i)
HomoDef(X, j, k)
FinishHomo(X)
    'Homomorphism defined'
```

This homomorphism from Q to itself can be shown to be one-to-one and onto. In fact, it can be represented by the cycle $(j, k, -j, -k)$. Also, the commands

```
Y = Homomorph(Q, Q)
HomoDef(Y, i, k)
HomoDef(Y, j, j)
FinishHomo(Y)
    'Homomorphism defined'
```

show that there is yet another automorphism on Q, which can be represented by $(i, k, -i, -k)$. These two automorphisms, along with the group of 4 inner automorphisms, generate a total of 24 automorphisms.

```
A = Group( C(j, -j)*C(k, -k), C(i, -i)*C(k, -k), C(j, k, -j, -k),
C(i, k, -i, -k) ); A
    {(), (-i, -j)(-k, k)(j, i), (-i, -j, -k)(k, i, j),
    (-i, -j, k)(-k, i, j), (-i, -j, i, j), (-i, -k)(-j, j)(k, i),
    (-i, -k, -j)(k, j, i), (-i, -k, j)(-j, i, k), (-i, -k, i, k),
    (-i, k)(-j, j)(-k, i), (-i, k, -j)(-k, j, i),
    (-i, k, j)(-j, i, -k), (-i, k, i, -k), (-i, j)(-j, i)(-k, k),
    (-i, j, -k)(-j, k, i), (-i, j, k)(-j, -k, i), (-i, j, i, -j),
    (-i, i)(-j, -k)(k, j), (-i, i)(-j, k)(-k, j), (-i, i)(-j, j),
    (-i, i)(-k, k), (-j, -k, j, k), (-j, k, j, -k),
    (-j, j)(-k, k)}
```

Notice that *Sage* allows group elements inside of cycles. We can see that the inner automorphisms are embedded in this list. What is this group isomorphic to?

In fact, $\text{Aut}(Q) \approx S_4$, as can be seen by Figure 6.2. Each rotation of the octahedron represents an automorphism of Q. For example, rotating the front face 120° clockwise corresponds to the automorphism

$$(i, j, k)(-i, -j, -k).$$

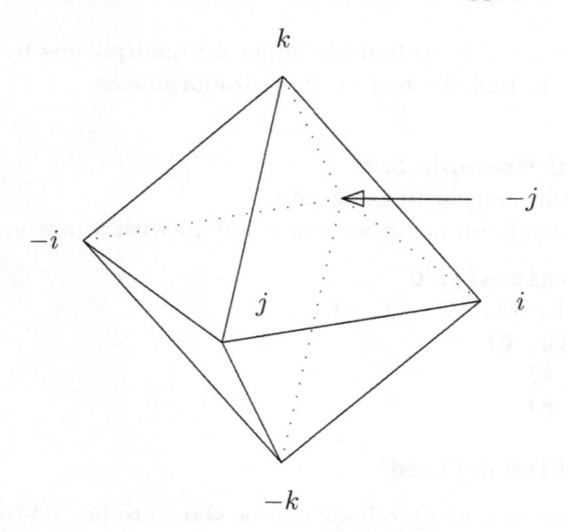

FIGURE 6.2: Labeling the octahedron to show $\text{Aut}(Q) \approx S_4$

So the automorphism group is isomorphic to the octahedral group, which we saw was isomorphic to S_4. ∎

Although the inner automorphisms did not produce the full automorphism group, this set of inner automorphisms turns out to be a very important subgroup of the automorphism group. Let us discover the first main property of this subgroup.

PROPOSITION 6.5

Let G be a group. Then $\text{Inn}(G)$ *is a normal subgroup of* $\text{Aut}(G)$.

PROOF: First we need to show that $\text{Inn}(G)$ is a subgroup. Let

$$f_x(y) = x \cdot y \cdot x^{-1}$$

be an inner automorphism. The inverse can be easily found by observing

$$y \in f_x^{-1}(v) \Longleftrightarrow x \cdot y \cdot x^{-1} = v \Longleftrightarrow y = x^{-1} \cdot v \cdot x \Longleftrightarrow y = f_{(x^{-1})}(v),$$

so the inverse of f_x is also an inner automorphism.

If we consider two inner automorphisms f_x and f_y, then

$$(f_x \cdot f_y)(v) = f_x(f_y(v)) = x \cdot (y \cdot v \cdot x^{-1}) \cdot y^{-1} = (x \cdot y) \cdot v \cdot (x \cdot y)^{-1} = f_{(x \cdot y)}(v).$$

Thus the product of two inner automorphisms is also an inner automorphism. So by Proposition 2.2, $\text{Inn}(G)$ is a subgroup of $\text{Aut}(G)$.

Finally, we need to show that Inn(G) is normal in Aut(G). Let ϕ be any automorphism and let $f_x = x \cdot y \cdot x^{-1}$ be an inner automorphism. Then

$$(\phi \cdot f_x \cdot \phi^{-1})(v) = \phi(f_x(\phi^{-1}(v))) = \phi(x \cdot (\phi^{-1}(v)) \cdot x^{-1}).$$

Since ϕ is a homomorphism, this will simplify.

$$\begin{aligned}\phi(x \cdot (\phi^{-1}(v)) \cdot x^{-1}) &= \phi(x) \cdot \phi(\phi^{-1}(v)) \cdot \phi(x^{-1}) \\ &= \phi(x) \cdot v \cdot [\phi(x)]^{-1} = f_{\phi(x)}(v).\end{aligned}$$

So $\phi \cdot f_x \cdot \phi^{-1}$ is an inner automorphism of G. Therefore, by Proposition 3.4, Inn(G) is a normal subgroup of Aut(G). □

For example, we found four inner automorphisms of Q. All of them but the identity were of order 2. Thus, we see that Inn(Q) $\approx Z_8^*$.

Because the inner automorphism group is always a normal subgroup, we could consider the quotient group.

DEFINITION 6.5 We define the *outer automorphism group* to be the quotient group

$$\mathrm{Out}(G) = \mathrm{Aut}(G)/\mathrm{Inn}(G).$$

The outer automorphism group of Q must contain six elements, and with some experimenting in *Sage*, one finds that Out(Q) is non-abelian. Therefore, Out(Q) $\approx S_3$.

If G is an abelian group, then the only inner automorphism is the identity automorphism. Thus, for abelian groups,

$$\mathrm{Inn}(G) \approx \{e\} \qquad \text{and} \qquad \mathrm{Out}(G) \approx \mathrm{Aut}(G).$$

Let us look at one last example, which will create a huge group.

Computational Example 6.10

Find the automorphism group of Z_{24}^*.

SOLUTION: Rather than using **ZStar(24)**, we will consider this group as $Z_2 \times Z_2 \times Z_2$ so we can see the relationship with the generators. We can load this group into *Sage* with the following commands:

```
InitGroup("e")
AddGroupVar("a", "b", "c")
Define(a^2, e); Define(b^2, e); Define(c^2, e)
Define(b*a, a*b); Define(c*a, a*c); Define(c*b, b*c)
Y = ListGroup(); Y
    {e, a, b, a*b, c, a*c, b*c, a*b*c}
```

Once again, we will begin by determining an upper bound for the number of automorphisms. Suppose $\phi(x)$ is an automorphism of Z_{24}^*. Naturally $\phi(e) = e$, but $\phi(a)$ could be any of the seven remaining elements of order 2. Also, $\phi(b)$ could be any one of the remaining six elements. Then we would have $\phi(a \cdot b) = \phi(a) \cdot \phi(b)$, so four elements will be accounted for. But $\phi(c)$ could be any of the four elements left over. Since the group is generated by $\{a, b, c\}$, there are at most $7 \cdot 6 \cdot 4 = 168$ possible automorphisms.

One possible automorphism would be to send a to b, b to c, and c back to a.

```
F = Homomorph(Y, Y)
HomoDef(F, a, b)
HomoDef(F, b, c)
HomoDef(F, c, a)
FinishHomo(F)
    'Homomorphism defined'
```

which *Sage* verifies is an automorphism. Another automorphism, given by

```
G = Homomorph(Y, Y)
HomoDef(G, a, a)
HomoDef(G, b, a*b)
HomoDef(G, c, c)
FinishHomo(G)
    'Homomorphism defined'
```

indicates that there may indeed be many automorphisms.

It would be more concise if we could use permutations for a group this large. If we number the non-identity elements in the order they appear in **ListGroup**, we have $a = 1$, $b = 2$, $a \cdot b = 3$, $c = 4$, $a \cdot c = 5$, $b \cdot c = 6$, and $a \cdot b \cdot c = 7$. With this ordering we can convert F and G to standard permutations $(1\,2\,4)(3\,5\,6)$ and $(2\,3)(6\,7)$. That is, F maps element 1 (a) to element 2 (b), which is mapped to element 4 (c), etc. Likewise, G exchanges the 2nd and 3rd elements, and exchanges the 6th and 7th elements of Z_{24}^*. Once we have all of the elements as permutations, we can use the integer notation feature to list them.

```
f = CycleToPerm( C(1,2,4)*C(3,6,5) ); f
    P(2, 4, 6, 1, 3, 5)
g = CycleToPerm( C(2,3)*C(6,7) ); g
    P(1, 3, 2, 4, 5, 7, 6)
DisplayPermInt = true
A = Group(f, g); A
    {1, 27, 61, 87, 122, 149, 187, 231, 244, 270, 331, 357, 374,
    404, 437, 467, 496, 548, 558, 593, 640, 670, 684, 714, 723,
    745, 783, 805, 844, 870, 931, 957, 962, 989, 1027, 1071,
    1096, 1148, 1158, 1193, 1214, 1244, 1277, 1307, 1366, 1384,
```

```
    1410, 1428, 1445, 1466, 1509, 1549, 1566, 1588, 1653, 1675,
    1681, 1707, 1741, 1767, 1822, 1862, 1889, 1902, 1966, 1984,
    2010, 2028, 2054, 2084, 2117, 2147, 2166, 2188, 2253, 2275,
    2285, 2306, 2349, 2389, 2403, 2425, 2463, 2485, 2566, 2584,
    2610, 2628, 2662, 2702, 2729, 2742, 2780, 2798, 2843, 2861,
    2897, 2927, 2954, 2984, 3018, 3071, 3076, 3110, 3144, 3185,
    3206, 3220, 3288, 3306, 3328, 3346, 3361, 3387, 3421, 3447,
    3487, 3517, 3531, 3561, 3618, 3671, 3676, 3710, 3737, 3767,
    3794, 3824, 3888, 3906, 3928, 3946, 3984, 4025, 4046, 4060,
    4083, 4105, 4143, 4165, 4213, 4231, 4257, 4275, 4362, 4392,
    4402, 4432, 4488, 4506, 4528, 4546, 4577, 4607, 4634, 4664,
    4703, 4721, 4760, 4778, 4809, 4839, 4849, 4879, 4935, 4953,
    4975, 4993}
len(A)
    168
```

Since this gives us 168 elements, we know we have all of the automorphisms. Notice that *Sage* orders the numbers, making it easier to find a particular element. In particular, the elements f and g are found to be

```
f
    187
g
    723
```

So the group $\text{Aut}(Z_{24}^*)$ is generated by the 187th and 723rd permutations. This group has special properties we will explore in the next chapter. □

We have now seen several examples where the group of automorphisms is larger than the original group. But this group of automorphisms can also be used as a tool for connecting two groups to form an even larger group, in much the same way that two groups formed the direct product. The next section will explore this methodology.

Problems for §6.3

For Problems **1** through **6**: Determine an upper bound for the size of the automorphism group for the following groups. It helps to first determine how many elements there are of each order.

1 S_3
2 D_4
3 Z_{15}^*
4 $Z_6 \times Z_2$
5 $Z_3 \times Z_3$
6 $Z_2 \times Z_2 \times Z_2 \times Z_2$

7 Prove that if G is a finite group of order n, then $\text{Aut}(G)$ is isomorphic to a subgroup of S_{n-1}.

8 Prove that if G is non-abelian, then there is an inner automorphism that is not trivial.

9 Prove that if G is abelian, and there is an element of G with an order greater than 2, then $\phi(x) = x^{-1}$ is a non-trivial automorphism.

10 Prove that any finite group of order greater than 2 has at least two automorphisms.

Hint: The only groups not covered by Problems 8 and 9 are isomorphic to $Z_2 \times Z_2 \times \cdots \times Z_2$.

11 Prove that if G is not abelian, then $\text{Aut}(G)$ is not cyclic.

12 Find all of the inner automorphisms of S_3. Use cycle notation for the automorphisms, as we did for Example 6.8. The multiplication table for S_3 is on page 71.

13 Find all of the inner automorphisms of D_4. Use cycle notation for the automorphisms, as we did for Example 6.8. The multiplication table for D_4 is on page 124.

14 Show that for the group D_4, there is an automorphism with $\phi(a) = a$ and $\phi(b) = a \cdot b$. Show that the multiplication table with the new ordering of elements created by the automorphism has the same "color pattern" as the table on page 124.

15 Find the automorphism group of S_3. See Problem 12.

16 Find the automorphism group of D_4. See Problems 13 and 14.

17 Find $\text{Aut}(\mathbb{Z})$.

18 Find two non-isomorphic groups G and M for which $\text{Aut}(G) \approx \text{Aut}(M)$.

Interactive Problems

For Problems **19** through **21**: Find all of the automorphisms of the following groups.

19 Z_{15}^* **20** Z_{21}^* **21** D_5

22 Find all of the automorphisms of the group $Z_3 \times Z_3$. Because of the large number of automorphisms, it is useful to number the non-identity elements of the group as we did for $\text{Aut}(Z_{24}^*)$ in Example 6.10.

6.4 Semi-Direct Products

We have already seen one way to combine two groups H and K to form the direct product $H \times K$. In this section we will see another way to combine to groups H and K, very similar to the direct product, but with a twist. Once again the larger group will have isomorphic copies of H and K as subgroups, but only *one* of the two subgroups will be a normal subgroup.

Note that this section is more advanced than previous sections. Although some examples will later refer to this section, particularly in §7.4, many readers may want to skip this section and go on to the next chapter.

Suppose that H and K are any two groups, and suppose that we have a homomorphism $\phi : H \to \text{Aut}(K)$. Because the function ϕ returns another function, we will write ϕ_h instead of $\phi(h)$. The expression $\phi_h(k)$ represents the automorphism ϕ_h evaluated at the element k. That is, if h_1 and h_2 are two elements of H, then $\phi_{h_1}(k)$ and $\phi_{h_2}(k)$ will be two automorphisms of K, and also $\phi_{h_1 \cdot h_2}(k) = (\phi_{h_1} \cdot \phi_{h_2})(k) = \phi_{h_1}(\phi_{h_2}(k))$. (Recall that $\phi_{h_1} \cdot \phi_{h_2}$ means we do ϕ_{h_2} first, then do ϕ_{h_1}.)

There will always be at least one homomorphism from H to $\text{Aut}(K)$, the trivial homomorphism. However, there will often be several nontrivial homomorphisms from H to $\text{Aut}(K)$. For each such homomorphism, we can define a product of H and K.

DEFINITION 6.6 Let K and H be two groups, and let G be the set of all ordered pairs (k, h), where k is in K and h is in H. Let ϕ be a nontrivial homomorphism from H to $\text{Aut}(K)$. Then the *semi-direct product of K with H through ϕ*, denoted $K \rtimes_\phi H$, is the set G with multiplication defined by

$$(k_1, h_1) \cdot (k_2, h_2) = (k_1 \cdot \phi_{h_1}(k_2),\ h_1 \cdot h_2).$$

PROPOSITION 6.6
The semi-direct product of K with H through ϕ is a group.

PROOF: It is clear that the product of two ordered pairs in G is an ordered pair in G. If we let e_1 denote the identity element of K, and e_2 denote the identity element of H, then

$$\phi_{e_2}(k) = k,$$

since ϕ must map e_2 to the identity automorphism of K. Thus

$$(k_1, h_1) \cdot (e_1, e_2) = (k_1 \cdot \phi_{h_1}(e_1), h_1 \cdot e_2) = (k_1, h_1),$$

and

$$(e_1, e_2) \cdot (k_2, h_2) = (e_1 \cdot \phi_{e_2}(k_2), e_2 \cdot h_2) = (k_2, h_2).$$

So (e_1, e_2) acts as the identity element of G.

Next we note that the element (k, h) has an inverse $(\phi_{h^{-1}}(k^{-1}), h^{-1})$, since

$$\begin{aligned}(\phi_{h^{-1}}(k^{-1}), h^{-1}) \cdot (k, h) &= (\phi_{h^{-1}}(k^{-1}) \cdot \phi_{h^{-1}}(k), h^{-1} \cdot h) \\ &= (\phi_{h^{-1}}(k^{-1} \cdot k), e_2) = (\phi_{h^{-1}}(e_1), e_2) = (e_1, e_2),\end{aligned}$$

and

$$\begin{aligned}(k, h) \cdot (\phi_{h^{-1}}(k^{-1}), h^{-1}) &= (k \cdot \phi_h(\phi_{h^{-1}}(k^{-1})), h \cdot h^{-1}) \\ &= (k \cdot \phi_{e_2}(k^{-1}), e_2) = (k \cdot k^{-1}, e_2) = (e_1, e_2).\end{aligned}$$

The final thing we need to check is that the multiplication on G is associative. Note that

$$\begin{aligned}[(k_1, h_1) \cdot (k_2, h_2)] \cdot (h_3, k_3) &= (k_1 \cdot \phi_{h_1}(k_2), h_1 \cdot h_2) \cdot (k_3, h_3) \\ &= (k_1 \cdot \phi_{h_1}(k_2) \cdot \phi_{h_1 \cdot h_2}(k_3), (h_1 \cdot h_2) \cdot h_3),\end{aligned}$$

while

$$\begin{aligned}(k_1, h_1) \cdot [(k_2, h_2) \cdot (k_3, h_3)] &= (k_1, h_1) \cdot (k_2 \cdot \phi_{h_2}(k_3), h_2 \cdot h_3) \\ &= (k_1 \cdot \phi_{h_1}(k_2 \cdot \phi_{h_2}(k_3)), h_1 \cdot (h_2 \cdot h_3)) \\ &= (k_1 \cdot \phi_{h_1}(k_2) \cdot \phi_{h_1}(\phi_{h_2}(k_3)), (h_1 \cdot h_2) \cdot h_3) \\ &= (k_1 \cdot \phi_{h_1}(k_2) \cdot \phi_{h_1 \cdot h_2}(k_3), (h_1 \cdot h_2) \cdot h_3).\end{aligned}$$

Hence the multiplication on G is associative and so G forms a group. □

Example 6.11

Find a semi-direct product $Z_3 \rtimes_\phi Z_2$.

SOLUTION: First we find $\text{Aut}(Z_3) \approx Z_3^* \approx Z_2$. Hence, there is only one non-trivial automorphism on Z_3, which is $x \to x^{-1}$. To get a non-trivial automorphism from Z_2 to $\text{Aut}(Z_3)$, we must have 0 map to the identity automorphism, and 1 map to the other automorphism. That is, $\phi_0(x) = x$ and $\phi_1(x) = x^{-1}$. Thus,

$$(2, 1) \cdot (1, 0) = (2 \cdot \phi_1(1), 1 \cdot 0) = (2 \cdot 2, 1 \cdot 0) = (1, 1).$$

The multiplication table is given in Table 6.2. This is a non-abelian group of order 6, so this is isomorphic to S_3. □

A semi-direct product of two groups acts in many ways like the direct product. One property that is in common is that there are copies of the two original groups within the product. In fact, we have the following:

TABLE 6.2: Cayley table of $Z_3 \rtimes_\phi Z_2$

	(0,0)	(0,1)	(1,0)	(1,1)	(2,0)	(2,1)
(0,0)	(0,0)	(0,1)	(1,0)	(1,1)	(2,0)	(2,1)
(0,1)	(0,1)	(0,0)	(2,1)	(2,0)	(1,1)	(1,0)
(1,0)	(1,0)	(1,1)	(2,0)	(2,1)	(0,0)	(0,1)
(1,1)	(1,1)	(1,0)	(0,1)	(0,0)	(2,1)	(2,0)
(2,0)	(2,0)	(2,1)	(0,0)	(0,1)	(1,0)	(1,1)
(2,1)	(2,1)	(2,0)	(1,1)	(1,0)	(0,1)	(0,0)

LEMMA 6.7
Let $G = K \rtimes_\phi H$ be the semi-direct product of K with H through the homomorphism ϕ. Suppose that e_1 is the identity element of K, and e_2 is the identity element of H. Then

$$\overline{H} = \{(e_1, h) \mid h \in H\}$$

is a subgroup of G, and

$$\overline{K} = \{(k, e_2) \mid k \in K\}$$

is a normal subgroup of G. Furthermore, $\overline{H} \approx H$, $\overline{K} \approx K$, and $\overline{H} \cap \overline{K}$ is the identity element of G.

PROOF: We will use Proposition 2.2 and observe that

$$(e_1, h)^{-1} = (\phi_{h^{-1}}(e_1^{-1}), h^{-1}) = (e_1, h^{-1}),$$

so

$$(e_1, h_1)\cdot(e_1, h_2)^{-1} = (e_1, h_1)\cdot(e_1, h_2^{-1}) = (e_1 \cdot \phi_{h_1}(e_1), h_1 \cdot h_2^{-1}) = (e_1, h_1 \cdot h_2^{-1}).$$

Thus, whenever a and b are in $\overline{H}$, $a \cdot b^{-1}$ is in $\overline{H}$. So $\overline{H}$ is a subgroup.

The mapping $f : G \to H$ given by

$$f((k, h)) = h$$

is a homomorphism, since

$$f((k_1, h_1)\cdot(k_2, h_2)) = f((k_1 \cdot \phi_{h_1}(k_2), h_1 \cdot h_2)) = h_1 \cdot h_2 = f((k_1, h_1))\cdot f((k_2, h_2)).$$

The kernel of this homomorphism is $\overline{K}$, so $\overline{K}$ is a normal subgroup of G. By restricting the function f to the set $\overline{H}$, we find that it is one-to-one and onto. Thus, $\overline{H} \approx H$. A similar function $g : K \to \overline{K}$, given by

$$g(k) = (k, e_2)$$

can show that $\overline{K} \approx K$. This function is clearly one-to-one and onto, and

$$g(k_1) \cdot g(k_2) = (k_1, e_2) \cdot (k_2, e_2) = (k_1 \cdot \phi_{e_2}(k_2), e_2) = (k_1 \cdot k_2, e_2) = g(k_1 \cdot k_2).$$

Finally, it is clear that the intersections of the two groups give $\{(e_1, e_2)\}$. ◻

Since the semi-direct product contains copies of the two smaller groups within itself, the natural question is whether an arbitrary group G can be expressed as a semi-direct product of two of its subgroups. The conditions for when this happens is set forth in the following theorem.

THEOREM 6.3: The Semi-Direct Product Theorem
Suppose that a group G has two subgroups N and H whose intersection is the identity element. Then if N is a normal subgroup of G and H is not a normal subgroup of $N \cdot H$, then there exists a nontrivial homomorphism ϕ from H to Aut(N) *such that*

$$N \cdot H \approx N \rtimes_{\phi} H.$$

PROOF: Note that since H is a subgroup of G, and N is a normal subgroup, we have by Lemma 4.3 that $N \cdot H$ is a subgroup of G. We next want to define the homomorphism ϕ. For each h in H, we define

$$\phi_h(k) = h \cdot k \cdot h^{-1}$$

for all $k \in N$. We first need to show that ϕ_h is an automorphism on N for each h in H, and then we need to show that ϕ itself is a nontrivial homomorphism. Note that

$$\phi_h(k_1 \cdot k_2) = h \cdot k_1 \cdot k_2 \cdot h^{-1} = (h \cdot k_1 \cdot h^{-1}) \cdot (h \cdot k_2 \cdot h^{-1}) = \phi_h(k_1) \cdot \phi_h(k_2).$$

So ϕ_h is a homomorphism from N to N. Since

$$y \in \phi_h^{-1}(k) \iff h \cdot y \cdot h^{-1} = k \iff y = h^{-1} \cdot k \cdot h$$

we see that ϕ_h is a one-to-one and onto function. Thus, ϕ_h is an automorphism of N.

Next, we need to see that ϕ itself is a homomorphism from H to Aut(N). Note that

$$\begin{aligned}(\phi_{h_1} \cdot \phi_{h_2})(k) &= \phi_{h_1}(\phi_{h_2}(k)) \\ &= \phi_{h_1}(h_2 \cdot k \cdot h_2^{-1}) \\ &= h_1 \cdot h_2 \cdot k \cdot h_2^{-1} \cdot h_1^{-1} \\ &= (h_1 \cdot h_2) \cdot k \cdot (h_1 \cdot h_2)^{-1} = \phi_{h_1 \cdot h_2}(k).\end{aligned}$$

So $\phi_{h_1} \cdot \phi_{h_2} = \phi_{(h_1 \cdot h_2)}$ and we see that ϕ is a homomorphism. In fact, the homomorphism must be nontrivial, because if $\phi_h(k) = k$ for all h and k, then since $\phi_h(k) = h \cdot k \cdot h^{-1} = k$ we have that $k \cdot h = h \cdot k$ for all h in H, and k in N. This would indicate that H is a *normal* subgroup of $N \cdot H$, which contradicts our original assumption. Thus, ϕ is a nontrivial homomorphism.

We can now proceed in a way similar to how we proved the direct product theorem (6.1). As before, we will begin by showing that every element in $N \cdot H$ can be *uniquely* written in the form $k \cdot h$, where $k \in N$ and $h \in H$.

Suppose that we have

$$k_1 \cdot h_1 = k_2 \cdot h_2.$$

Then $k_2^{-1} \cdot k_1 = h_2 \cdot h_1^{-1}$. Since this element is in both N and H, which has just the identity element in the intersection, we must have

$$k_2^{-1} \cdot k_1 = h_2 \cdot h_1^{-1} = e.$$

Therefore, $k_1 = k_2$ and $h_1 = h_2$. Thus, we have shown that every element of $N \cdot H$ is written uniquely as $k \cdot h$, where k is in N, and h is in H.

We now want to create a mapping

$$f : N \cdot H \to N \rtimes_{\phi} H$$

defined by

$$f(x) = (k, h),$$

where k and h are the unique elements such that $k \in N$, $h \in H$, and $x = k \cdot h$. The function f is one-to-one since the element (k, h) can only come from $k \cdot h$. Also, the element $k \cdot h$ maps to (k, h) so f is onto.

The final step is to show that f is a homomorphism. Let $x = k_1 \cdot h_1$, and $y = k_2 \cdot h_2$. Then

$$x \cdot y = k_1 \cdot h_1 \cdot k_2 \cdot h_2 = (k_1 \cdot h_1 \cdot k_2 \cdot h_1^{-1}) \cdot (h_1 \cdot h_2).$$

Since N is a normal subgroup, $h_1 \cdot k_2 \cdot h_1^{-1}$ is in N, and so $k_1 \cdot h_1 \cdot k_2 \cdot h_1^{-1}$ is in N while $h_1 \cdot h_2$ is in H. Thus,

$$\begin{aligned} f(x \cdot y) &= f((k_1 \cdot h_1 \cdot k_2 \cdot h_1^{-1}) \cdot (h_1 \cdot h_2)) \\ &= (k_1 \cdot h_1 \cdot k_2 \cdot h_1^{-1}, h_1 \cdot h_2) \\ &= (k_1 \cdot \phi_{h_1}(k_2), h_1 \cdot h_2) \\ &= (k_1, h_1) \cdot (k_2, h_2) = f(x) \cdot f(y). \end{aligned}$$

So f is an isomorphism, and we have $N \cdot H \approx N \rtimes_{\phi} H$. ∎

Note that if both H and N are normal subgroups of $H \cdot N$, we have by Corollary 6.1 that $H \cdot N \approx H \times N$.

We will use the semi-direct product theorem to define this product in *Sage*. After defining the two groups H and N using the same identity element, we must find the homomorphism ϕ from H to $\mathrm{Aut}(N)$. As in the case of the direct product, we will want to express every element of the form $k \cdot h$, where k is in N, and h is in H. From the definition, we see that

$$(k, e_2) \cdot (e_1, h) = (k \cdot \phi_{e_2}(e_1), e_2 \cdot h) = (k, h),$$

Thus, we see that $k \cdot h$ can represent the ordered pair (k, h). We need to tell *Sage* how to handle expressions of the form $h \cdot k$.

For each generator a of N, and each generator b of H, we can calculate how **b * a** should be defined by evaluating $(e_1, b) \cdot (a, e_2) = (\phi_b(a), b)$. Thus we make a definition in *Sage* of the form

Define(b*a, $\phi_b(a)$ *** b)**

where we replace the expression $\phi_b(a)$ with its element of N.

Computational Example 6.12

Use *Sage* to find a semi-direct product of Z_5 with Z_2.

SOLUTION: We first must define Z_5 and Z_2 into *Sage* using the same identity but different generators.

```
InitGroup("e")
AddGroupVar("a", "b")
Define(a^5, e)
Define(b^2, e)
Z5 = Group(a); Z5
    {e, a^4, a, a^3, a^2}
Z2 = Group(b); Z2
    {e, b}
```

After loading the groups Z_5 and Z_2, we want to find a nontrivial homomorphism ϕ from Z_2 to $\mathrm{Aut}(Z_5)$. But $\mathrm{Aut}(Z_5) \approx Z_5^* \approx Z_4$. Since the element b is of order 2, ϕ_b must be of order 2 to keep the homomorphism from being trivial. But it is easy to find the one element of $\mathrm{Aut}(Z_5)$ of order 2:

$$\phi(k) = k^{-1}.$$

In fact, this will always be an automorphism whenever N is an abelian group. As long as N has an element that is not its own inverse, this automorphism will be of order 2. If we let $\phi_b(k) = k^{-1}$, then $\phi_b(a) = a^{-1} = a^4$. Thus, the definition

```
Define(b*a, a^4*b)
```

completes the definition of the semi-direct product.

```
G = ListGroup(); G
    {e, a, a^2, a^3, a^4, b, a*b, a^2*b, a^3*b, a^4*b}
```

TABLE 6.3: Multiplication table for D_5

$\cdot$	e	a	a^2	a^3	a^4	b	$a\cdot b$	$a^2\cdot b$	$a^3\cdot b$	$a^4\cdot b$
e	e	a	a^2	a^3	a^4	b	$a\cdot b$	$a^2\cdot b$	$a^3\cdot b$	$a^4\cdot b$
a	a	a^2	a^3	a^4	e	$a\cdot b$	$a^2\cdot b$	$a^3\cdot b$	$a^4\cdot b$	b
a^2	a^2	a^3	a^4	e	a	$a^2\cdot b$	$a^3\cdot b$	$a^4\cdot b$	b	$a\cdot b$
a^3	a^3	a^4	e	a	a^2	$a^3\cdot b$	$a^4\cdot b$	b	$a\cdot b$	$a^2\cdot b$
a^4	a^4	e	a	a^2	a^3	$a^4\cdot b$	b	$a\cdot b$	$a^2\cdot b$	$a^3\cdot b$
b	b	$a^4\cdot b$	$a^3\cdot b$	$a^2\cdot b$	$a\cdot b$	e	a^4	a^3	a^2	a
$a\cdot b$	$a\cdot b$	b	$a^4\cdot b$	$a^3\cdot b$	$a^2\cdot b$	a	e	a^4	a^3	a^2
$a^2\cdot b$	$a^2\cdot b$	$a\cdot b$	b	$a^4\cdot b$	$a^3\cdot b$	a^2	a	e	a^4	a^3
$a^3\cdot b$	$a^3\cdot b$	$a^2\cdot b$	$a\cdot b$	b	$a^4\cdot b$	a^3	a^2	a	e	a^4
$a^4\cdot b$	$a^4\cdot b$	$a^3\cdot b$	$a^2\cdot b$	$a\cdot b$	b	a^4	a^3	a^2	a	e

The multiplication table is given in Table 6.3, which shows that this is a non-abelian group of order 10. ∎

We can ask *Sage* what this group is, using the **`StructureDescription()`** command. This command analyzes the last group which was defined using the **`InitGroup`** and **`Define`** commands.

```
StructureDescription()
   D5
```

This shows that the group is the dihedral group D_5. We can generalize this example as follows:

DEFINITION 6.7 Let $n > 2$, and let ϕ be the homomorphism from $Z_2 = \{e, b\}$ to $\mathrm{Aut}(Z_n)$ given by

$$\phi_e(k) = k, \qquad \phi_b(k) = k^{-1}.$$

Then the semi-direct product $Z_n \rtimes_\phi Z_2$ is called the *dihedral group of order* $2n$. It is denoted D_n, and is a non-abelian group of order $2n$.

The commands

```
InitGroup("e")
AddGroupVar("a", "b")
Define(a^n, e)
Define(b^2, e)
Define(b*a, a^-1*b)
Dn = Group(a, b)
```

define the group D_n. The symbol n must be replaced with an integer before executing these commands. When $n = 3$, we get a non-abelian group of order 6, so $D_3 \approx S_3$. We also have already seen D_4, since this was one of the 5 groups of order 8 that we found in Chapter 4.

Note that the semi-direct product may greatly depend on the choice of the homomorphism ϕ.

Computational Example 6.13

Consider finding the semi-direct products of Z_8 with Z_2. Since $\text{Aut}(Z_8) \approx Z_8^*$ has three elements of order 2, there are three nontrivial homomorphisms from Z_2 to $\text{Aut}(Z_8)$. One of these produces the dihedral group D_8 above, but the other two homomorphisms produce different groups. If we let $\phi_b(a) = a^3$, we get the following.

```
InitGroup("e")
AddGroupVar("a", "b")
Define(a^8, e)
Define(b^2, e)
Define(b*a, a^3*b)
G = ListGroup(); G
    {e, a, a^2, a^3, a^4, a^5, a^6, a^7, b, a*b, a^2*b, a^3*b,
    a^4*b, a^5*b, a^6*b, a^7*b}
StructureDescription()
    QD16
```

Sage calls this group "QD16", since it is the *quasidihedral group* of order 16, written QD_{16}. If we let $\phi_b(a) = a^5$ instead, we get

```
InitGroup("e")
AddGroupVar("a", "b")
Define(a^8, e)
Define(b^2, e)
Define(b*a, a^5*b)
M = ListGroup(); M
    {e, a, a^2, a^3, a^4, a^5, a^6, a^7, b, a*b, a^2*b, a^3*b,
    a^4*b, a^5*b, a^6*b, a^7*b}
StructureDescription()
    Z8 : Z2
```

Even though the list of elements look the same for the two groups, the structure description is different. *Sage* uses a colon for the semi-direct product symbol $\rtimes$, so *Sage* recognized that the last group was of the form $Z_8 \rtimes Z_2$, but otherwise there is no special name for this group. $\square$

Another way of showing that the three groups are different is by having *Sage* display the multiplication tables, and counting the number of times the identity element appears along the diagonals. We find that $R_2(D_8) = 10$, $R_2(QD_{16}) = 6$, and $R_2(M) = 4$, where M is the last group of the form $Z_8 \rtimes Z_2$.

We see that the semi-direct product $Z_8 \rtimes_\phi Z_2$ depends on the choice of the homomorphism ϕ. In fact, even though the three elements of $\mathrm{Aut}(Z_8)$ of order 2 are essentially equivalent (since the automorphisms of Z_8^* included all permutations of these three elements), we see that the three elements produced three different semi-direct products.

This example is really more of an exception rather than a rule. Part of what makes this example unusual is that the automorphism group Z_8^* is abelian, and hence does not have any nontrivial *inner* automorphisms. If two homomorphisms ϕ and f from H to $\mathrm{Aut}(N)$ are related through an inner automorphism of $\mathrm{Aut}(N)$, then the corresponding semi-direct products will if fact be isomorphic.

PROPOSITION 6.7

Let ϕ be a homomorphism from a group H to the group $\mathrm{Aut}(N)$. *Suppose that f is another homomorphism such that*

$$f_h(k) = w(\phi_h(w^{-1}(k))),$$

where $w(k)$ is an automorphism of N. Then $N \rtimes_f H \approx N \rtimes_\phi H$.

PROOF: Let us write $G = N \rtimes_\phi H$, and $M = N \rtimes_f H$. These are two different groups, even though they are both written using ordered pairs. Let us define a mapping

$$v : G \to M$$

defined by

$$v((k, h)) = (w(k), h).$$

Because $w(k)$ is one-to-one and onto, certainly v is one-to-one and onto. All we would have to check is that

$$v((k_1, h_1)) \cdot v((k_2, h_2)) = v((k_1, h_1) \cdot (k_2, h_2)).$$

We have that

$$\begin{aligned} v((k_1, h_1)) \cdot v((k_2, h_2)) &= (w(k_1), h_1) \cdot (w(k_2), h_2) \\ &= (w(k_1) \cdot f_{h_1}(w(k_2)), h_1 \cdot h_2) \\ &= (w(k_1) \cdot w(\phi_{h_1}(w^{-1}(w(k_2)))), h_1 \cdot h_2) \\ &= (w(k_1) \cdot w(\phi_{h_1}(k_2)), h_1 \cdot h_2). \end{aligned}$$

On the other hand,

$$\begin{aligned} v((k_1, h_1) \cdot (k_2, h_2)) &= v((k_1 \cdot \phi_{h_1}(k_2), h_1 \cdot h_2)) \\ &= (w(k_1 \cdot \phi_{h_1}(k_2)), h_1 \cdot h_2) \\ &= (w(k_1) \cdot w(\phi_{h_1}(k_2)), h_1 \cdot h_2). \end{aligned}$$

Since these are equal, we have an isomorphism. □

It is also clear that whenever two homomorphisms ϕ and f are related through an automorphism of H, the semi-direct products must be isomorphic since we are merely relabeling the elements of H. As a result there will be many instances in which there will be only one non-isomorphic semi-direct product of N by H. In this case, we can denote *the* semi-direct product as $N \rtimes H$, without having to specify the homomorphism ϕ.

We will find that we can describe many groups in terms of semi-direct products that would be hard to describe in any other way. With *Sage*, the structure of these semi-direct products can easily be studied.

Problems for §6.4

For Problems **1** through **6**: Let $\phi : Z_8^* \to \mathrm{Aut}(Z_8^*)$ be defined as follows: $\phi_1 = \phi_3 = ()$, $\phi_5 = \phi_7 = (3\,5)$, where we used the cycle notation for the automorphisms. Compute the following in $Z_8^* \rtimes_\phi Z_8^*$:

1 $(5,3)\cdot(3,5)$ **3** $(7,5)^{-1}$ **5** $(1,5)\cdot(3,7)\cdot(5,3)$

2 $(3,5)\cdot(5,3)$ **4** $(5,7)^{-1}$ **6** $(5,3)\cdot(3,7)\cdot(1,5)$

7 Show that there is only one semi-direct product $Z_8^* \rtimes Z_2$, and form a Cayley table. Which of the five groups of order 8 is this isomorphic to?

Hint: Use Proposition 6.7.

8 Show that there is only one semi-direct product of the form $Z_8^* \rtimes Z_3$. Form a Cayley table of this group. You have seen this group before. Do you recognize it?

9 Form a Cayley table of the only semi-direct product of the form $Z_3 \rtimes Z_4$.

10 Show that there is only one semi-direct product of the form $\mathbb{Z} \rtimes Z_2$. Describe this group.

11 Show that there is only one semi-direct product of the form $\mathbb{Z} \rtimes \mathbb{Z}$. Describe this group.

12 Let G be any group, and let i be the identity mapping from $\mathrm{Aut}(G)$ to itself. We can define the semi-direct product $H = G \rtimes_i \mathrm{Aut}(G)$. The group H is called the *holomorph* of G. Show that every automorphism of G is the restriction of some inner automorphism of the holomorph H.

13 Let G be a group, and n a positive integer. We will let G^n denote the direct product of G with itself n times, or the set of n-tuples in G. If $\sigma \in S_n$, we can define

$$\psi_\sigma : G^n \to G^n \qquad \text{by} \qquad \psi_\sigma(g_1, g_2, \ldots g_n) = (g_{\sigma^{-1}(1)}, g_{\sigma^{-1}(2)}, \ldots g_{\sigma^{-1}(n)}).$$

Show that ψ_σ is an automorphism of G^n.

14 Let G^n and ψ be defined as in Problem 13. Show that if σ and τ are two elements of S_n, then $\psi_\tau(\psi_\sigma(x)) = \psi_{\tau\cdot\sigma}(x)$.

Hint: Think of an n-tuple as a function f from the set $1 \le i \le n$ to G, with $f(i)$ being the ith component of the n-tuple. Then $\phi_\sigma(f)$ sends $f(i)$ to $f(\sigma^{-1}(i))$.

15 Let G be a group, and H a subgroup of S_n. We define the *wreath product*

$$G \text{ Wr } H$$

as the semi-direct product $G^n \rtimes_\psi H$, where G^n and ψ are defined as in Problem 13. Show that if G is a finite group, the wreath product is a finite group of size $|G|^n \cdot |H|$.

16 Form the multiplication table of Z_2 Wr S_2. See Problem 15.

Interactive Problems

17 Use *Sage* to find the only semi-direct product $Z_8^* \rtimes Z_8^*$. Is this group isomorphic to any of the three groups of order 16 found by considering $Z_8 \rtimes_\phi Z_2$?

18 From Problems 16, and 19 from §6.1, Problem 9, and Definition 6.7, we have found six groups of order 12: Z_{12}, $Z_2 \times Z_6$, A_4, D_6, $S_3 \times Z_2$, and $Z_3 \rtimes Z_4$. Yet Table 4.4 indicates that there are only five non-isomorphic groups of order 12. Which two of these groups are isomorphic? Use *Sage* to show the isomorphism.

19 Use *Sage* to define the wreath product Z_3 Wr S_2. Then use **`StructureDescription()`** to determine what group this is. See Problem 15.

20 Use *Sage* to define the wreath product Z_2 Wr A_3. Then use **`StructureDescription()`** to determine what group this is. See Problem 15.

21 Use *Sage* to define the wreath product Z_2 Wr S_3. Then use **`StructureDescription()`** to determine what group this is. See Problem 15.

Chapter 7

The Search for Normal Subgroups

7.1 The Center of a Group

We saw several instances in the last chapter in which the structure of a group hinges on its normal subgroups. Thus, we will want to develop techniques for finding *all* of the normal subgroups of a given group G. We will discover in the process that some of the normal groups have additional properties. We will naturally concentrate our attention on non-abelian groups, since every subgroup of an abelian group is normal.

In this section we will consider a simple way of constructing a normal subgroup from a given group. In fact, the definition was suggested in Problem 8 of § 3.3. However, we will find that this particular normal subgroup, called the *center* of the group, has some very important properties.

Motivational Example 7.1

Let us begin by considering the dihedral group D_4. Table 7.1 gives us the Cayley table of this group.

There are five elements of order 2 in this group, but one of these, a^2, has another important property. Notice that the locations of the a^2 in Table 7.1 form a symmetrical pattern reflected along the main diagonal, even though the entire table is not symmetric. This indicates that whenever $x \cdot y = a^2$,

TABLE 7.1: *Sage*'s multiplication table for D_4

$\cdot$	e	a	a^2	a^3	b	$a \cdot b$	$a^2 \cdot b$	$a^3 \cdot b$
e	e	a	a^2	a^3	b	$a \cdot b$	$a^2 \cdot b$	$a^3 \cdot b$
a	a	a^2	a^3	e	$a \cdot b$	$a^2 \cdot b$	$a^3 \cdot b$	b
a^2	a^2	a^3	e	a	$a^2 \cdot b$	$a^3 \cdot b$	b	$a \cdot b$
a^3	a^3	e	a	a^2	$a^3 \cdot b$	b	$a \cdot b$	$a^2 \cdot b$
b	b	$a^3 \cdot b$	$a^2 \cdot b$	$a \cdot b$	e	a^3	a^2	a
$a \cdot b$	$a \cdot b$	b	$a^3 \cdot b$	$a^2 \cdot b$	a	e	a^3	a^2
$a^2 \cdot b$	$a^2 \cdot b$	$a \cdot b$	b	$a^3 \cdot b$	a^2	a	e	a^3
$a^3 \cdot b$	$a^3 \cdot b$	$a^2 \cdot b$	$a \cdot b$	b	a^3	a^2	a	e

then $y \cdot x = a^2$ in D_4. Hence $y = x^{-1} \cdot a^2 = a^2 \cdot x^{-1}$ for all elements x. In order for this to happen, a^2 must commute with *all* of the elements of D_4. □

DEFINITION 7.1 Given a group G, the *center* of G is defined to be the set of elements x for which $x \cdot y = y \cdot x$ for all elements $y \in G$. The center of a group G is customarily denoted $Z(G)$ because of the German word for center, *zentrum.* [1, p. 150]

From this definition, we see that $a^2 \in Z(D_4)$. It is also clear that $e \in Z(G)$ for all groups, since $e \cdot y = y \cdot e$. By examining Table 7.1 we find that there are no other elements of D_4 in $Z(D_4)$, so $Z(D_4) = \{e, a^2\}$. This is obviously a subgroup, but it turns out to be a normal subgroup because of the following proposition.

PROPOSITION 7.1
Given a group G, then $Z(G)$ is a normal subgroup of G.

PROOF: First, we need to show that $Z(G)$ is a subgroup of G. If x and y are in $Z(G)$, and a is any element in G, then

$$x \cdot y \cdot a = x \cdot a \cdot y = a \cdot x \cdot y.$$

So $x \cdot y$ commutes with all of the elements of G. Thus, $x \cdot y$ is in $Z(G)$.
Also, we have

$$x^{-1} \cdot a = (a^{-1} \cdot x)^{-1} = (x \cdot a^{-1})^{-1} = a \cdot x^{-1}.$$

So x^{-1} must also be in $Z(G)$. Thus, by Proposition 2.2, $Z(G)$ is a subgroup of G.
Next, we can see that

$$a \cdot x \cdot a^{-1} = x \cdot a \cdot a^{-1} = x.$$

So $a \cdot x \cdot a^{-1}$ is in $Z(G)$ whenever x is in $Z(G)$ and a is in G. Thus, by Proposition 3.4, $Z(G)$ is a normal subgroup of G. □

We use the command **GroupCenter** to find the center of a group in *Sage*. For example, the command

```
Z = GroupCenter(D4); Z
    {e, a^2}
```

verifies our earlier observation that $Z(D_4) = \{e, a^2\}$.
Although the center always produces a normal subgroup, this subgroup is not always interesting.

Example 7.2
Show that the center of the group $S_3 = \{(), (1\,2), (1\,3), (2\,3), (1\,2\,3), (1\,3\,2)\}$ is just the identity element.
SOLUTION: Since $(1\,2) \cdot (2\,3) = (1\,2\,3) \neq (2\,3) \cdot (1\,2) = (1\,3\,2)$, neither $(1\,2)$ nor $(2\,3)$ are in the center. Also, $(1\,3) \cdot (1\,2\,3) = (1\,2) \neq (1\,2\,3) \cdot (1\,3) = (2\,3)$, so neither $(1\,3)$ nor $(1\,2\,3)$ are in the center. Finally, $(1\,3\,2)$ cannot be in the center, since we have established that $(1\,3\,2)^2 = (1\,2\,3)$ is not in the center. Thus, only $()$ is in the center. □

Whenever the center is just the identity element, we say the group is *centerless.* In fact, all of the permutation groups S_n bigger than S_3 are centerless. Since the proof involves an even permutation, we will find the center of A_n at the same time.

PROPOSITION 7.2
If $n > 3$, then the groups S_n and A_n are centerless.

PROOF: Suppose that ϕ is an element of S_n or A_n which is not the identity. We need to show that ϕ cannot be in the center of either S_n or A_n, which amounts to finding an element of A_n that does not commute with ϕ.

Since ϕ is not the identity, there is some number x that is not fixed by ϕ, say x is mapped to y. Since $n > 3$, there is at least one number not in the list $\{x, y, \phi(y)\}$. Let z be one of these remaining numbers. Finally, we let f be the 3-cycle $(x\ y\ z)$.

Since f is an even permutation, f is in A_n. Then $f \cdot \phi$ sends x to z, but $\phi \cdot f$ sends x to $\phi(y) \neq z$. Thus, $f \cdot \phi \neq \phi \cdot f$, and ϕ is not in the center of either A_n or S_n. □

The other extreme is if $Z(G)$ is the entire group G. This happens if, and only if, the group G is abelian.

Since $Z(G)$ is a normal subgroup of G, what is the quotient group? The answer is rather interesting.

PROPOSITION 7.3
If G is a group, then $G/Z(G) \approx \text{Inn}(G)$.

PROOF: We begin by observing that the mapping

$$\phi : G \to \text{Inn}(G)$$

given by

$$\phi_x(y) = x \cdot y \cdot x^{-1}$$

is a homomorphism, as we saw in the proof of the semi-direct product theorem (6.3). By the definition of the inner automorphisms, this mapping is surjective.

However, this mapping is not necessarily injective. Let us determine the kernel of ϕ.

Suppose that ϕ_x is the identity homomorphism. Then $\phi_x(y) = y$ for all y in G. This means that $x \cdot y \cdot x^{-1} = y$, or $x \cdot y = y \cdot x$, for all y in G. Thus, x is in the center of G.

Now, suppose x is in $Z(G)$. Then $\phi_x(y) = x \cdot y \cdot x^{-1} = y \cdot x \cdot x^{-1} = y$, so ϕ_x is the identity homomorphism. Thus the kernel of ϕ is precisely the center of G. Therefore, by the first isomorphism theorem (4.1), we have

$$G/Z(G) \approx \text{Inn}(G).$$

□

The center of a group possesses a characteristic that is even stronger than that of a normal subgroup. To illustrate this characteristic, consider the next proposition.

PROPOSITION 7.4

Let N be a normal subgroup of a group G. Then $Z(N)$ is a normal subgroup not only of N, but also of G.

PROOF: Let g be an element of G, and z an element of $Z(N)$. We need to show that $g \cdot z \cdot g^{-1}$ is in $Z(N)$. Since N is a normal subgroup of G, we certainly know that $g \cdot z \cdot g^{-1}$ is in N, so the way to test that it is in $Z(N)$ is to show that it commutes with every element of N.

Let n be an element of N. We want to show that $g \cdot z \cdot g^{-1} \cdot n = n \cdot g \cdot z \cdot g^{-1}$. Let $h = g^{-1} \cdot n \cdot g$. Then h is in N, since N is normal in G. Also, $n = g \cdot h \cdot g^{-1}$, so

$$\begin{aligned} g \cdot z \cdot g^{-1} \cdot n &= (g \cdot z \cdot g^{-1}) \cdot (g \cdot h \cdot g^{-1}) = g \cdot z \cdot h \cdot g^{-1} = g \cdot h \cdot z \cdot g^{-1} \\ &= (g \cdot h \cdot g^{-1}) \cdot (g \cdot z \cdot g^{-1}) = n \cdot g \cdot z \cdot g^{-1}. \end{aligned}$$

Hence, $g \cdot z \cdot g^{-1}$ commutes with every element n in N, so $g \cdot z \cdot g^{-1}$ is in $Z(N)$. By Proposition 3.4, we have that $Z(N)$ is a normal subgroup of G. □

This proposition demonstrates a rather unusual property of a center of a group. In general, the normal subgroup of a normal subgroup is not necessarily a normal subgroup. Consider $M = \{(\,), (12)(34), (13)(24), (14)(23)\}$, which is a normal subgroup of S_4, and $H = \{(\,), (12)(34)\}$, which is a normal subgroup of M.

```
S4 = Group( C(1,2), C(1,2,3), C(1,2,3,4) )
M = Group( C(1,2)*C(3,4), C(1,3)*C(2,4) ); M
    {(), (1, 2)(3, 4), (1, 3)(2, 4), (1, 4)(2, 3)}
H = Group( C(1,2)*C(3,4) ); H
    {(), (1, 2)(3, 4)}
```

We find that H is *not* a normal subgroup of S_4.

```
LftCoset(S4, H)
    {{(), (1, 2)(3, 4)}, {(1, 2), (3, 4)}, {(2, 3), (1, 3, 4, 2)},
    {(1, 3, 2), (2, 3, 4)}, {(1, 2, 3), (1, 3, 4)},
    {(1, 3), (1, 2, 3, 4)}, {(2, 4, 3), (1, 4, 2)},
    {(1, 4, 3, 2), (2, 4)}, {(1, 2, 4, 3), (1, 4)},
    {(1, 4, 3), (1, 2, 4)}, {(1, 3)(2, 4), (1, 4)(2, 3)},
    {(1, 4, 2, 3), (1, 3, 2, 4)}}
RtCoset(S4, H)
    {{(), (1, 2)(3, 4)}, {(1, 2), (3, 4)}, {(2, 3), (1, 2, 4, 3)},
    {(1, 3, 2), (1, 4, 3)}, {(1, 2, 3), (2, 4, 3)}
    {(1, 3), (1, 4, 3, 2)}, {(2, 3, 4), (1, 2, 4)},
    {(1, 3, 4, 2), (1, 4)}, {(2, 4), (1, 2, 3, 4)},
    {(1, 4, 2), (1, 3, 4)}, {(1, 3)(2, 4), (1, 4)(2, 3)},
    {(1, 4, 2, 3), (1, 3, 2, 4)}}
```

Contrast this situation to the center of a group. We found that the center of a group $Z(N)$ is a normal subgroup of G, even though $Z(N)$ contains no information about the larger group G. Any group that contains N as a normal subgroup, such as a semi-direct product of N by another group, will have $Z(N)$ as a normal subgroup.

Problems for §7.1

1 Find the center of the group Q.

2 Find the center of the group D_5.

3 Find the center of the group $Z_3 \rtimes Z_4$. See Problem 9 from §6.4.

4 Find the center of the group $Z_8^* \rtimes Z_8^*$ from Problems 1 through 6 of §6.4.

5 Must the center of a group be abelian?

6 Let G be a group and $Z(G)$ the center of G. Prove that G is abelian if, and only if, $G/Z(G)$ is cyclic.
Hint: Use Proposition 7.3.

7 Show that if A and B are two groups, then $Z(A \times B) \approx Z(A) \times Z(B)$.

8 Prove that if a group only has one element of order 2, then that element must be in the center.

9 Prove that if H is a normal subgroup of G, and $|H| = 2$, then $H \in Z(G)$.

10 Let G be a group, and H be a transitive subgroup of S_n. (See Problem 15 of §5.2.) Show that $Z(G \text{ Wr } H) \approx Z(G)$. See Problem 15 of §6.4 for the definition of G Wr H.

11 Let ϕ be an automorphism on the group G, and let $z \in Z(G)$. Prove that $\phi(z) \in Z(G)$.

12 A *characteristic* subgroup of G is a subgroup H such that $\phi(h) \in H$ for all $h \in H$ and all automorphisms ϕ of G. Problem 11 shows that $Z(G)$ is a characteristic subgroup of G. Prove that all characteristic subgroups are also normal subgroups.

13 Let H be the only subgroup of G of size $|H|$. Prove that H is a characteristic subgroup of G. See Problem 12.

14 Let G be an abelian group, and let H be the subgroup of size $R_k(G)$ given by

$$\{x \in G \mid x^k = e\}.$$

Prove that H is a characteristic subgroup of G. See Problem 12.

15 Prove that all subgroups of a cyclic group are characteristic.
Hint: See Problems 12 and 13.

16 Prove that if N is a characteristic subgroup of G, and H is a characteristic subgroup of N, then H is a characteristic subgroup of G. Note this statement is not true if "characteristic" is replaced with "normal." See Problem 12.

17 Prove that if N is a normal subgroup of G, and H is a characteristic subgroup of N, then H is a normal subgroup of G. This generalizes Proposition 7.4, since the center is a characteristic subgroup. See Problem 12.

Interactive Problems

18 Use *Sage* to find the center of the group D_6. This can be loaded by the commands:

```
InitGroup("e")
AddGroupVar("a", "b")
Define(a^6, e); Define(b^2, e); Define(b*a, a^5*b)
D6 = ListGroup(); D6
    {e, a, a^2, a^3, a^4, a^5, b, a*b, a^2*b, a^3*b, a^4*b, a^5*b}
```

What familiar group is the quotient group $D_6/Z(D_6)$ isomorphic to?

19 In Problem 22 of §6.3, we computed the group $G = \text{Aut}(Z_3 \times Z_3)$. Find the center of this group. What familiar group is $G/Z(G)$ isomorphic to?

20 Find the centers of the groups D_3, D_4, D_5, D_6, D_7, and D_8. Do you see any patterns?

7.2 The Normalizer and Normal Closure Subgroups

In the last section, we found a subgroup of N that was not only normal, but also was normal in any group G for which N was a normal subgroup. In this section, we will essentially turn the question around: Given a subgroup H of G, can we find a subgroup N of G for which H lies inside of N as a normal subgroup? In the process of answering this question, we will produce a powerful tool that can be used to identify the possible normal subgroups. In addition, we will consider a related question, finding the smallest normal subgroup of G that contains the subgroup H.

DEFINITION 7.2 Let S be a *subset* of a group G. We define the *normalizer of S by G,* denoted $N_G(S)$, to be the set

$$N_G(S) = \{g \in G \mid g \cdot S \cdot g^{-1} = S\}.$$

Notice that this definition allows for S to be merely a *subset* of G, not necessarily a subgroup. We will later find uses for having a more generalized definition. For now, let us show that the normalizer has some of the properties that we are looking for.

PROPOSITION 7.5
Let S be a subset of the group G. Then $N_G(S)$ is a subgroup of G.

PROOF: Suppose x and y are in $N_G(S)$. Then both $x \cdot S \cdot x^{-1} = S$, and $y \cdot S \cdot y^{-1} = S$. Thus, $S = y^{-1} \cdot S \cdot y$, and so

$$(x \cdot y^{-1}) \cdot S \cdot (x \cdot y^{-1})^{-1} = x \cdot (y^{-1} \cdot S \cdot y) \cdot x^{-1} = x \cdot S \cdot x^{-1} = S.$$

Thus, $x \cdot y^{-1}$ is in $N_G(S)$, and so by Proposition 2.2, $N_G(S)$ is a subgroup of G. ∎

Example 7.3
Consider the group $Q = \{1, i, j, k, -1, -i, -j, -k\}$. Find the normalizer of the single element $\{i\}$.
SOLUTION: We want to find the elements such that $g \cdot i \cdot g^{-1} = i$, which clearly contains i. Since we know from Proposition 7.5 that the normalizer is a subgroup, $\{1, i, -1, -i\}$ is in the normalizer. But j is not in the normalizer, so $N_G(\{i\}) = \{1, i, -1, -i\}$. ∎

If, in addition, S is a subgroup of G, then the normalizer lives up to its name.

PROPOSITION 7.6
Let H be a subgroup of the group G. Then $N_G(H)$ is the largest subgroup of G that contains H as a normal subgroup.

PROOF: First, we must check that H is a normal subgroup of $N_G(H)$. But this is obvious, since $g \cdot H \cdot g^{-1} = H$ for all g in $N_G(H)$.

Next, we must see that $N_G(H)$ is the largest such group. Suppose that Y is another subgroup of G that contained H as a normal subgroup. Then $y \cdot H \cdot y^{-1} = H$ for all $y \in Y$. Thus, $Y \subseteq N_G(H)$.

Since any subgroup of G that contains H as a normal subgroup is itself contained in $N_G(H)$, we have that $N_G(H)$ is the largest such group. ▯

Example 7.4
Find the normalizer of the subgroup $[i] = \{1, i, -1, -i\}$ of Q.
SOLUTION: Since this is a normal subgroup of Q, the normalizer is all of Q, since it is the largest group for which $[i]$ is normal. In general, the normalizer of a *normal* subgroup by G will produce the whole group G. ▯

The *Sage* command **Normalizer(G, H)** finds the normalizer $N_G(H)$ of the set H in G. We can verify the last two examples.

```
Q = InitQuaternions(); Q
    {1, i, j, k, -1, -i, -j, -k}
H = Normalizer(Q, i)
    {1, i, -1, -i}
Normalizer(Q, H)
    {1, i, j, k, -1, -i, -j, -k}
```

Note that if the set is a single element, we do not have to enclose the element in brackets. We can find the normalizer of any subset, even one that is not a subgroup. For example, the normalizer of the subset $\{i, j\}$ is

```
Normalizer(Q, [i, j])
    {1, -1}
```

which contains neither i nor j. In general though, all we can say is that the normalizer will be a subgroup, which this example illustrates.

There is one other case in which we can say that the normalizer will contain H. Notice that in the example we did where H was a single element, the normalizer contained that element. In fact, $N_G(\{g\})$ will consist of all elements of G that commute with g. It should be noted that $N_G(\{g\})$ is not the same thing as $N_G([g])$, the normalizer of the group generated by g. The former is the set of elements that commute with g, and the latter is the largest subgroup that contains $[g]$ as a normal subgroup.

We have seen that the normalizer of a subgroup H by G finds the largest subgroup of G that contains H as a normal subgroup. What if we asked for the *smallest* subgroup containing H that is a normal subgroup of G? Whether H is a subgroup or a subset, we can use the following proposition.

PROPOSITION 7.7

Let S be a subset of a group G. Then the smallest group containing S that is a normal subgroup of G is given by

$$N^* = \bigcap_{N \in L} N,$$

where L denotes the collection of normal subgroups of G that contain S.

PROOF: The group G itself is in the collection L, so this collection is not empty. Thus, by Proposition 2.3, N^* is a subgroup of G.

Also, since each N in the collection contained the set S, the intersection will also contain S. All that needs to be shown is that N^* is normal.

If n is an element of N^*, and g is an element of G, then since each N is a normal subgroup of G, and n would be in all of the groups N,

$$g \cdot n \cdot g^{-1} \in N \quad \text{for all} \quad N \in L.$$

Thus, $g \cdot n \cdot g^{-1}$ is in the intersection of all of the N's, which is N^*. Hence, by Proposition 3.4, N^* is a normal subgroup of G. ∎

We will call this subgroup the *normal closure* of S. The *Sage* command **NormalClosure(G, S)** computes this subgroup. With this command, we can systematically find *all* of the normal subgroups of a given group. Note that if S contains a single element, we can use the element instead of a set.

Computational Example 7.5

Find all of the normal subgroups of S_3, using the generators a and b.

SOLUTION: We would like to see if there are any other normal subgroups besides the two trivial groups. Since a proper subgroup must contain one of the elements $\{a, b, a \cdot b, b^2, a \cdot b^2\}$, we have five groups to try.

```
InitGroup("e")
AddGroupVar("a", "b")
Define(a^2, e); Define(b^b, e); Define(b*a, a*b^2)
S3 = ListGroup(); S3
    {e, a, b, a*b, b^2, a*b^2}
NormalClosure(S3, a)
    {e, a, b, a*b, b^2, a*b^2}
NormalClosure(S3, b)
    {e, b, b^2}
NormalClosure(S3, a*b)
```

```
    {e, a, b, a*b, b^2, a*b^2}
NormalClosure(S3, b^2)
    {e, b, b^2}
NormalClosure(S3, a*b^2)
    {e, a, b, a*b, b^2, a*b^2}
```

We see that using b and b^2 produces the normal subgroup of order 3, A_3. The other elements produced the whole group. In fact, if we considered a normal subgroup generated by *two* elements, it is obvious that this would have to contain a normal subgroup already found. But the smallest found was A_3, and no larger subgroup could still be proper. Thus, we have used *Sage* to *prove* that the only proper normal subgroup of S_3 is A_3. ☐

This method of exhaustion works well for small groups, but one can imagine that this method would be time consuming for larger groups. In the next section, we will find a short-cut so that we will not have to try every element of the group, but rather just a handful of elements.

Problems for §7.2

1 For each element g in D_4, find the normalizer $N_{D_4}(\{g\})$.

2 For each element g in D_5, find the normalizer $N_{D_5}(\{g\})$.

3 There are five subgroups of D_4 of order 2: $\{e, a^2\}$, $\{e, b\}$, $\{e, a\cdot b\}$, $\{e, a^2\cdot b\}$, and $\{e, a^3 \cdot b\}$. For each subgroup, find $N_{D_4}(H)$.

4 There are five subgroups of D_5 of order 2: $\{e, b\}$, $\{e, a \cdot b\}$, $\{e, a^2 \cdot b\}$, $\{e, a^3 \cdot b\}$, and $\{e, a^4 \cdot b\}$. For each subgroup, find $N_{D_5}(H)$.

5 Must the normalizer of an element $N_G(\{g\})$ be abelian?

6 Let G be any group. Prove that

$$Z(G) = \bigcap_{g\in G} N_G(\{g\}).$$

7 Let G be a group, and let g be an element of G. Prove that

$$N_G(\{g\}) = N_G(\{g^{-1}\}).$$

8 Let G be a group, and let g be an element of G, and k be any integer. Prove that

$$N_G(\{g\}) \subseteq N_G(\{g^k\}).$$

9 Let G be a group. Prove that for any subset S,

$$Z(G) \subseteq N_G(S).$$

10 Let G be a group. Prove that $N_G(\{g\}) = G$ if, and only if, $g \in Z(G)$.

For Problems **11** through **16**: Find the normal closure of the following sets in D_4.

11 $\{a\}$
12 $\{a^2\}$
13 $\{b\}$
14 $\{a \cdot b\}$
15 $\{a^2, b\}$
16 $\{b, a \cdot b\}$

For Problems **17** through **20**: Find the normal closure of the following sets in D_5.

17 $\{a\}$
18 $\{a^2\}$
19 $\{b\}$
20 $\{a \cdot b\}$

Interactive Problems

21 Use *Sage* to find the normalizer $N_{D_6}(\{x\})$ for each of the 12 elements of the group D_6 listed in Problem 18 of §7.1. For which elements is the normalizer the same subgroup?

22 Use *Sage*'s `NormalClosure` command to find all of the normal subgroups of the group D_6 given in Problem 18 of §7.1.

7.3 Conjugacy Classes and Simple Groups

We have already seen how the cosets of a subgroup can be used to partition the group into disjoint sets. As a result we proved Lagrange's theorem, which had far-reaching consequences. In this section, we will find another way to partition the group into disjoint sets, and as a result we will find a much faster way of determining *all* of the normal subgroups. In fact, we will find some groups that do not have any non-trivial normal subgroups at all.

In the last section, we used the *Sage* command `NormalClosure(G, S)` to find the smallest group containing the subset S that was a normal group of G. Let us look closely at how this command works. We know that if the element a is in this normal subgroup, then $g \cdot a \cdot g^{-1}$ must also be in the group for all g in G. Many of the elements that must be in the normal subgroup can be found in this way.

DEFINITION 7.3 Let G be a group. We say that the element u is *conjugate* to the element v if there exists an element g in G such that $u = g \cdot v \cdot g^{-1}$.

Note that every element is conjugate to itself, for we can let g be the identity element. Also note that if u is conjugate to v, then v is also conjugate to u, since

$$v = (g^{-1}) \cdot u \cdot (g^{-1})^{-1}.$$

Finally, if u is conjugate to v, and v in turn is conjugate to w, we can see that u is conjugate to w. This is easy to see, since there is a g and h such that $u = g \cdot v \cdot g^{-1}$ and $v = h \cdot w \cdot h^{-1}$. Then

$$u = g \cdot v \cdot g^{-1} = g \cdot (h \cdot w \cdot h^{-1}) \cdot g^{-1} = (g \cdot h) \cdot w \cdot (g \cdot h)^{-1}.$$

Recall that in Definition 1.3, we defined an equivalence relationship as any relationship having three properties:

1. Every element u is equivalent to itself.

2. If u is equivalent to v, then v is equivalent to u.

3. If u is equivalent to v, and v in turn is equivalent to w, then u is equivalent to w.

These were called the reflexive, symmetric, and transitive properties. We used the equivalence relationships of cosets in §3.4 to form a partition of the group, which gave us the quotient groups. In the same way, we can use the equivalence relationship of conjugates to form a different partition of the group, called *conjugacy classes.* Unlike cosets, though, the conjugacy classes will not be all the same size. The conjugacy class containing the element u is given by

$$\{g \cdot u \cdot g^{-1} \mid g \in G\}$$

Computational Example 7.6

Find all of the conjugacy classes of S_4.

SOLUTION: The *Sage* command for finding all of the conjugacy classes of a group G is **ConjugacyClasses(G)**. Let us find the conjugacy classes of S_4, which are generated by the cycles (1 2) and (2 3 4).

```
S4 = Group( C(1,2), C(2,3,4) ); S4
    {(), (1, 2), (2, 3), (1, 3, 2), (1, 2, 3), (1, 3), (3, 4),
    (1, 2)(3, 4), (2, 4, 3), (1, 4, 3, 2), (1, 2, 4, 3), (1, 4, 3),
    (2, 3, 4), (1, 3, 4, 2), (2, 4), (1, 4, 2), (1, 3)(2, 4),
    (1, 4, 2, 3), (1, 2, 3, 4), (1, 3, 4), (1, 2, 4), (1, 4),
    (1, 3, 2, 4), (1, 4)(2, 3)}
ConjugacyClasses(S4)
    {{()}, {(1, 2), (2, 3), (1, 3), (3, 4), (2, 4), (1, 4)},
    {(1, 3, 2), (1, 2, 3), (2, 4, 3), (1, 4, 3), (2, 3, 4),
    (1, 4, 2), (1, 3, 4), (1, 2, 4)}, {(1, 2)(3, 4), (1, 3)(2, 4),
    (1, 4)(2, 3)}, {(1, 4, 3, 2), (1, 2, 4, 3), (1, 3, 4, 2),
    (1, 4, 2, 3), (1, 2, 3, 4), (1, 3, 2, 4)}}
```

The identity element is in a class by itself since $g \cdot e \cdot g^{-1}$ will always produce e. But the cycle notation reveals an interesting fact about the other four classes: one contains all of the transpositions, one contains all of the 3-cycles, one contains all of the 4-cycles, and one conjugacy class contains the products of two disjoint transpositions. Problems 16 and 17 of §5.2 may help shed some light on why this happens. ∎

The conjugacy classes are very useful for finding normal subgroups, since whenever one element of a conjugacy class is in a normal subgroup of G, the entire conjugacy class must be in the normal subgroup. Thus, in order to find *all* normal subgroups of S_4 we only have to try the unions of different combinations of the conjugacy classes. Furthermore, the identity element is guaranteed to be in every subgroup.

Example 7.7

Use Example 7.6 to find all of the normal subgroups of S_4.

SOLUTION: It would be helpful if we label the conjugacy classes.

$$\begin{aligned} A &= \{(12),(13),(14),(23),(24),(34)\} \\ B &= \{(12)(34),(13)(24),(14)(23)\} \\ C &= \{(123),(124),(132),(134),(142),(143),(234),(243)\} \\ D &= \{(1234),(1243),(1324),(1342),(1423),(1432)\} \\ E &= \{()\} \end{aligned}$$

Then a non-trivial normal subgroup would have to be one of the following unions of conjugacy classes.

$E \cup A$	7 elements
$E \cup B$	4 elements
$E \cup C$	9 elements
$E \cup D$	7 elements
$E \cup A \cup B$	10 elements
$E \cup A \cup C$	15 elements
$E \cup A \cup D$	13 elements
$E \cup B \cup C$	12 elements
$E \cup B \cup D$	10 elements
$E \cup C \cup D$	15 elements
$E \cup A \cup B \cup C$	18 elements
$E \cup A \cup B \cup D$	16 elements
$E \cup A \cup C \cup D$	21 elements
$E \cup B \cup C \cup D$	18 elements

Of course, the last combination $E \cup A \cup B \cup C \cup D$ would give us the whole group. We actually can test all of these combinations without the help of

Sage. This table also includes the number of elements in the subsets, and we can eliminate almost all of these combinations with Lagrange's theorem (3.1). Only the second and eighth combinations have the number of elements divide 24. The combination

$$E \cup B = \{(), (12)(34), (13)(24), (14)(23)\}$$

we have seen before, so we recognize this is the normal subgroup which is isomorphic to Z_8^*. The other combination, $E \cup B \cup C$, contains the even permutations of S_4, so this is the normal subgroup A_4. Hence, we can use conjugacy classes to prove that there are precisely two non-trivial normal subgroups of S_4. ∎

Computational Example 7.8

Use *Sage* to find all of the normal subgroups of A_5.

SOLUTION: This group is generated by the cycles $(1\,2\,3)$ and $(3\,4\,5)$, so the conjugacy classes are as follows:

```
A5 = Group( C(1,2,3), C(3,4,5) )
ConjugacyClasses(A5)
    {{()}, {(1, 3, 2), (1, 2, 3), (2, 4, 3), (1, 4, 3), (2, 3, 4),
    (1, 4, 2), (1, 3, 4), (1, 2, 4), (3, 5, 4), (2, 5, 4),
    (1, 5, 4), (3, 4, 5), (2, 5, 3), (1, 5, 3), (2, 4, 5),
    (2, 3, 5), (1, 5, 2), (1, 4, 5), (1, 3, 5), (1, 2, 5)},
    {(1, 2)(3, 4), (1, 3)(2, 4), (1, 4)(2, 3), (1, 2)(4, 5),
    (2, 3)(4, 5), (1, 3)(4, 5), (1, 2)(3, 5), (2, 4)(3, 5),
    (1, 4)(3, 5), (1, 3)(2, 5), (2, 5)(3, 4), (1, 4)(2, 5),
    (1, 5)(2, 3), (1, 5)(3, 4), (1, 5)(2, 4)}, {(1, 5, 4, 3, 2),
    (1, 3, 5, 4, 2), (1, 3, 2, 5, 4), (1, 2, 4, 5, 3),
    (1, 2, 5, 3, 4), (1, 5, 3, 2, 4), (1, 4, 5, 2, 3),
    (1, 4, 3, 5, 2), (1, 5, 2, 4, 3), (1, 2, 3, 4, 5),
    (1, 4, 2, 3, 5), (1, 3, 4, 2, 5)}, {(1, 2, 5, 4, 3),
    (1, 5, 4, 2, 3), (1, 2, 3, 5, 4), (1, 4, 5, 3, 2),
    (1, 5, 3, 4, 2), (1, 4, 2, 5, 3), (1, 3, 4, 5, 2),
    (1, 3, 5, 2, 4), (1, 5, 2, 3, 4), (1, 3, 2, 4, 5),
    (1, 2, 4, 3, 5), (1, 4, 3, 2, 5)}}
```

This group also has only five conjugacy classes, so it should be no more difficult to find the normal subgroups than S_4. We can pick a representative element from each of the non-trivial conjugacy classes: $(1\,2\,3)$, $(1\,2)(3\,4)$, $(1\,2\,3\,4\,5)$, and $(1\,2\,3\,5\,4)$. From this point we can proceed as in the S_4 example to show that there are *no* normal subgroups of A_5. (See Problem 9.) However, we can use *Sage* to speed up the process.

```
len(NormalClosure(A5, C(1,2,3) ))
    60
```

```
len(NormalClosure(A5, C(1,2)*C(3,4) ))
    60
len(NormalClosure(A5, C(1,2,3,4,5) ))
    60
len(NormalClosure(A5, C(1,2,3,5,4) ))
    60
```

This shows that if any of the 4 representative elements are in a non-trivial normal subgroup of A_5, the subgroup would have to be all 60 elements of A_5. Hence, there can be no nontrivial normal subgroups of A_5. ∎

We will see that this is a rather unusual property for a group to have, so we will give this a special name.

DEFINITION 7.4 A group is said to be *simple* if it contains no normal subgroups besides itself and the identity subgroup.

The groups Z_p, for p a prime number, are the first examples we have seen of simple groups. We now have seen an example of a non-cyclic simple group, A_5. In fact this is the *smallest* non-cyclic simple group! (See Problem 19 of §7.4.)

Let us find other simple groups. The natural place to look is higher order alternating groups. Let us use *Sage*'s help to find the sizes of the conjugacy classes of A_6. This group is generated by the cycles $(1\,2\,3)$ and $(2\,3\,4\,5\,6)$.

```
A6 = Group(C(1, 2, 3), C(2, 3, 4, 5, 6))
len(A6)
    360
S = ConjugacyClasses(A6)
[ len(x) for x in S ]
    [1, 40, 45, 72, 72, 90, 40]
```

Thus, we see that there are 7 conjugacy classes of A_6, one of size 1 (the identity), two of size 40, two of size 72, one of size 45, and one of size 90.

Example 7.9

Use the above result to show that A_6 is simple.

SOLUTION: If there were a non-trivial subgroup N, its size would be a factor of 360, hence $|N| = 180, 120, 90, 72, 60$, or 45. Note it cannot be 40 or smaller, since it must contain the identity and at least one other conjugacy class. Clearly $|N| \neq 45$, since there is no conjugacy class of size 44. Thus, $|N|$ is even, so we must include both odd conjugacy classes, 1 and 45, plus at least one other. Hence, $|N| \geq 86$. At this point we see that $|N|$ is a multiple of 5, so both conjugacy classes of size 72 must be included to get the sum to be a multiple of 5. At this point $|N| \geq 190$, which is impossible. So A_6 is a simple group. ∎

Our goal is to show that A_n is simple for all $n > 4$. We begin by showing that all 3-cycles are in one conjugacy class.

LEMMA 7.1

If $n > 4$, any two 3-cycles are conjugate in A_n. Furthermore, the conjugate of a 3-cycle is again a 3-cycle.

PROOF: We begin by showing that the conjugate of a 3-cycle is again a 3-cycle. Let $(a\,b\,c)$ be a 3-cycle, and let ϕ be any permutation in A_n. Suppose that $x = \phi(a)$, $y = \phi(b)$, and $z = \phi(c)$. Then we can compute

$$\phi \cdot (a\,b\,c) \cdot \phi^{-1} = (x\,y\,z).$$

Thus the conjugate of a 3-cycle is another 3-cycle.

Next we will show that any 3-cycle is conjugate to the element $(1\,2\,3)$ in A_n. Let $(u\,v\,w)$ be a 3-cycle. Since $n > 4$ there must be at least two numbers not mentioned in this 3-cycle, so we will call two of them x and y. Consider the permutation

$$\phi = \begin{pmatrix} 1 & 2 & 3 & 4 & 5 & \cdots \\ u & v & w & x & y & \cdots \end{pmatrix}.$$

Here, the dots indicate that when $n > 5$, we can complete the permutation in any way so that the numbers on the bottom row will be a permutation of the numbers 1 through n.

Now ϕ will either be an even permutation or an odd permutation. If ϕ is an odd permutation, we can consider instead the permutation

$$\phi = \begin{pmatrix} 1 & 2 & 3 & 4 & 5 & \cdots \\ u & v & w & y & x & \cdots \end{pmatrix}.$$

So we may assume that ϕ is an even permutation. Thus ϕ is in A_n, and we can compute

$$\phi \cdot (1\,2\,3) \cdot \phi^{-1} = (u\,v\,w).$$

Therefore, any 3-cycle is conjugate to $(1\,2\,3)$, and so any two 3-cycles are conjugate to each other in A_n whenever $n > 4$. ∎

With this lemma, we can show that A_n will be a simple group whenever $n > 4$. This was originally proved by Abel using a long case-by-case argument. Since *Sage* helped us show that A_5 and A_6 are simple, the argument is greatly simplified.

THEOREM 7.1: Abel's Theorem

The alternating group A_n is simple for all $n > 4$.

Historical Diversion
Niels Abel (1802–1829)

Niels Abel was born in Norway at a time when the country was experiencing extreme poverty and hunger due to the Napoleonic wars. His father, Sören Georg Abel, had degrees in philosophy and theology, and served as a Protestant minister at Gjerstad. Abel was the second of seven children, and was taught by his father until he was 13 years old. The family's poverty was intensified since Abel's father was often drunk, and his mother was accused of lacking morals.

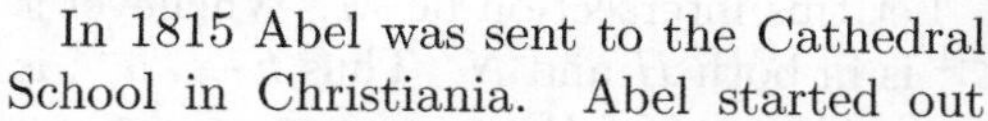

In 1815 Abel was sent to the Cathedral School in Christiania. Abel started out uninspired, but in 1817, a new mathematics teacher, Bernt Holmboë, joined the school, and took an interest in Abel. Within a year, Abel was reading the works of Euler, Newton, d'Alembert, Lagrange, and Laplace.

But in 1820, Abel's father died, and it was up to Abel to support his mother and family. Holmboë worked to raise money from his colleagues to allow Abel to enter the University of Christiania in 1821. During Abel's final year in school, he worked on the quintic equation, unsolved for 250 years.

$$ax^5 + bx^4 + cx^3 + dx^2 + ex + f = 0.$$

Abel believed he had solved the equation by radicals, and submitted a paper to the Danish mathematician Ferdinand Degen. Degen asked Abel to provide an example, and as Abel worked out the example, he found an error in his paper. Degen advised Abel to work instead on elliptic integrals, a new field that had promising consequences for both analysis and mechanics.

Abel took Degen's advise, and wrote several papers on functional equations and integrals. On a visit to see Degen, Abel met Christine Kemp, who later became his fiancée. Returning to Christiania, he again worked on the quintic equation, and in 1824 he proved the impossibility of solving the general equation in radicals. He send his proof to Gauss, who dismissed him as a crank, and never read the proof. Abel continued to work on elliptic functions in competition with Carl Jacobi. By this time Abel had become famous among the mathematical centers, and efforts were made to secure him a suitable position.

In 1828, Abel became seriously ill from tuberculosis, and his condition intensified due to Abel's sled trip to visit his fiancée. In spite of a reprieve long enough for the couple to spend Christmas together, he died soon after on April 6, 1829, just 2 days before word arrived that he was appointed as a professor at the University of Berlin.

Image source: Wikimedia Commons

PROOF: Suppose that N is a proper normal subgroup of A_n, and let ϕ be an element of N besides the identity. By Proposition 7.2, A_n is centerless. Since Proposition 5.1 tells us that A_n is generated by 3-cycles, there must be at least one 3-cycle that does not commute with ϕ, say $(a\,b\,c)$. Thus, $\phi \cdot (a\,b\,c)$ is not equal to $(a\,b\,c) \cdot \phi$, or equivalently, $(a\,b\,c) \cdot \phi \cdot (a\,c\,b) \cdot \phi^{-1}$ is not the identity element.

Since N is a normal subgroup, $(a\,b\,c) \cdot \phi \cdot (a\,c\,b)$ must be in N. Therefore, $(a\,b\,c) \cdot \phi \cdot (a\,c\,b) \cdot \phi^{-1}$ must also be in N. But $\phi \cdot (a\,c\,b) \cdot \phi^{-1}$ is the conjugate of a 3-cycle, so by Lemma 7.1 this is also a 3-cycle, say $(x\,y\,z)$. Thus, N contains a product of two 3-cycles, $(a\,b\,c) \cdot (x\,y\,z)$, which is not the identity. In essence we can say that there is a non-identity element of N that moves at most six numbers, labeled a, b, c, x, y, and z. If there are duplicates in this list, we can add arbitrary numbers so that we have six different numbers.

Here's where we can take advantage of the fact that A_6 is known to be simple. Consider the subgroup H of A_n consisting of all even permutations of the six numbers a, b, c, x, y, and z. We have just showed that there is a nontrivial intersection of N and H. Let this intersection be M. Whenever x is in M and h is in H, then $h \cdot x \cdot h^{-1}$ is in both H and N. Thus $h \cdot x \cdot h^{-1}$ is in M. Hence M is a nontrivial normal subgroup of H.

But H is isomorphic to A_6, which we have proven to be a simple group. Thus M must be all of H. In particular M contains a 3-cycle, and so N contains a 3-cycle. By Lemma 7.1 all 3-cycles of A_n are conjugate, so N contains all 3-cycles of A_n. Finally, by Proposition 5.1 the 3-cycles generate A_n, so N must be all of A_n. Therefore, A_n is simple whenever $n > 4$. □

This theorem has an immediate application to the permutation groups S_n.

COROLLARY 7.1

If $n > 4$ then the only proper normal subgroup of S_n is A_n.

PROOF: Suppose that there were another normal subgroup, N. Then the intersection of N with A_n would be another normal subgroup of S_n, and so would be a normal subgroup of A_n. Since A_n is simple for $n > 4$, this intersection must either be the identity or all of A_n.

Suppose that the intersection is all of A_n. Then N contains A_n, and if N is not equal to A_n, N would contain more than half of the elements of S_n. But this would contradict Lagrange's theorem (3.1) unless $N = S_n$.

Suppose that the intersection of N and A_n is just the identity element. Then since both N and A_n are normal subgroups, we have by Corollary 6.1,

$$N \cdot A_n \approx N \times A_n.$$

If N is not just the identity element, this quickly leads to a contradiction, for N could have order of at most 2, telling us that S_n was isomorphic to $Z_2 \times A_n$. But this is ridiculous, for we saw in Proposition 7.2 that S_n was centerless, whereas $Z_2 \times A_n$ clearly has both $(0, (\,))$ and $(1, (\,))$ in its center.

Therefore, the only normal subgroups of S_n for $n > 4$ are S_n itself, A_n, and the identity element. □

We now have found two sequences of simple groups, namely Z_p for p being a prime number, and A_n for all $n > 4$. Are any of the other groups that we have looked at simple groups?

Computational Example 7.10

Find the normal subgroups of the group $\mathrm{Aut}(Z_{24}^*)$, the group of order 168 generated by the 187th and 723rd permutation elements.

```
DisplayPermInt = true
A = Group( NthPerm(187), NthPerm(723) ); A
    {1, 27, 61, 87, 122, 149, 187, 231, 244, 270, 331, 357, 374,
    404, 437, 467, 496, 548, 558, 593, 640, 670, 684, 714, 723,
    745, 783, 805, 844, 870, 931, 957, 962, 989, 1027, 1071,
    1096, 1148, 1158, 1193, 1214, 1244, 1277, 1307, 1366, 1384,
    1410, 1428, 1445, 1466, 1509, 1549, 1566, 1588, 1653, 1675,
    1681, 1707, 1741, 1767, 1822, 1862, 1889, 1902, 1966, 1984,
    2010, 2028, 2054, 2084, 2117, 2147, 2166, 2188, 2253, 2275,
    2285, 2306, 2349, 2389, 2403, 2425, 2463, 2485, 2566, 2584,
    2610, 2628, 2662, 2702, 2729, 2742, 2780, 2798, 2843, 2861,
    2897, 2927, 2954, 2984, 3018, 3071, 3076, 3110, 3144, 3185,
    3206, 3220, 3288, 3306, 3328, 3346, 3361, 3387, 3421, 3447,
    3487, 3517, 3531, 3561, 3618, 3671, 3676, 3710, 3737, 3767,
    3794, 3824, 3888, 3906, 3928, 3946, 3984, 4025, 4046, 4060,
    4083, 4105, 4143, 4165, 4213, 4231, 4257, 4275, 4362, 4392,
    4402, 4432, 4488, 4506, 4528, 4546, 4577, 4607, 4634, 4664,
    4703, 4721, 4760, 4778, 4809, 4839, 4849, 4879, 4935, 4953,
    4975, 4993}
```

SOLUTION: As large as this group is, *Sage* can still quickly find the conjugacy classes.

```
ConjugacyClasses(A)
    {{1}, {27, 61, 87, 122, 270, 404, 593, 640, 714, 723, 745,
    1566, 1681, 2306, 2425, 3110, 3421, 3767, 4143, 4488, 4528},
    {149, 187, 244, 357, 374, 467, 548, 558, 844, 989, 1148,
    1307, 1366, 1384, 1410, 1428, 1445, 1588, 1653, 1741, 1767,
    1889, 2028, 2147, 2188, 2285, 2389, 2463, 2485, 2566, 2702,
    2798, 2984, 3071, 3076, 3220, 3306, 3361, 3387, 3531, 3671,
    3737, 3824, 3928, 3984, 4083, 4105, 4213, 4362, 4392, 4402,
    4432, 4634, 4703, 4839, 4975}, {231, 331, 437, 496, 670, 684,
    783, 805, 870, 962, 1193, 1244, 1466, 1675, 1707, 1822, 2010,
```

```
    2054, 2166, 2349, 2403, 2584, 2742, 2861, 2927, 3018, 3206,
    3346, 3447, 3517, 3710, 3794, 3888, 4025, 4165, 4257, 4506,
    4546, 4607, 4760, 4849, 4935}, {931, 1071, 1096, 1277, 1509,
    1902, 1984, 2084, 2275, 2610, 2662, 2843, 2954, 3185, 3288,
    3487, 3618, 3946, 4046, 4275, 4577, 4778, 4879, 4953},
    {957, 1027, 1158, 1214, 1549, 1862, 1966, 2117, 2253, 2628,
    2729, 2780, 2897, 3144, 3328, 3561, 3676, 3906, 4060, 4231,
    4664, 4721, 4809, 4993}}
```

So we have six conjugacy classes of this group, one of which is just the identity. The other five classes can be represented by the first element in each list, which are the 27th, 149th, 231st, 931st, and 957th permutations. This list alone can be used to show that A is simple (see Problem 10), but we can also verify that the normal closure of each of these five elements yields the whole group.

```
len(NormalClosure(A, NthPerm(27) ))
    168
len(NormalClosure(A, NthPerm(149) ))
    168
len(NormalClosure(A, NthPerm(231) ))
    168
len(NormalClosure(A, NthPerm(931) ))
    168
len(NormalClosure(A, NthPerm(957) ))
    168
```

Thus, any proper normal subgroup cannot contain any of these five elements; we have shown that there are no proper normal subgroups, so $\mathrm{Aut}(Z_{24}^*)$ is a simple group. □

This is the second largest non-cyclic simple group. (A_5 is the smallest and A_6 is the third smallest.) See Problems 11 through 17 for more examples of simple groups.

We can have *Sage* give us a structure description of a permutation group by including the integer representation of a set of generators for the arguments.

```
StructureDescription(187, 723)
    PSL(3,2)
```

So one of the official names for the group $\mathrm{Aut}(Z_{24}^*)$ is $L_3(2)$. This group is the beginning of yet another infinite family of simple groups, called the Chevalley groups. We will not go into all of the ways this group can be generalized to produce these other groups, but we will mention an important result that has taken place during the 20th century. It was once thought that *all* finite simple groups were either the cyclic groups of prime order, the alternating groups, or one of the Chevalley or twisted Chevalley groups. (One of these groups turns out to be not quite simple. Yet taking half of the elements forms a new simple group, just as we took half of the elements of S_n to form the simple groups A_n.) But there were several other simple groups that were discovered, called *sporadic* groups. In the 1960s and 1970s it was proved that there are exactly 26 sporadic groups, ranging in size from a mere 7,920 elements to the monstrous

808,017,424,794,512,875,886,459,904,961,710,757,005,754,368,000,000,000

elements! These 26 sporadic groups are listed in [13]. Because these have been proven to be the only sporadic groups, all finite simple groups are now known.

Problems for §7.3

1 Find all of the conjugacy classes of the group D_4.

2 Find all of the conjugacy classes of the quaternion group Q. (See Table 4.3 in Chapter 4 for the multiplication table of Q.)

3 Find all of the conjugacy classes of the group D_5.

4 Find the conjugacy classes of the group $Z_3 \rtimes Z_4$. See Problem 9 from §6.4.

5 Find the conjugacy classes of the group $Z_8^* \rtimes Z_8^*$ from Problems 1 through 6 of §6.4.

6 Show that the conjugacy class for an element x has only one element if, and only if, x is in the center of the group.

7 Show that if G is a finite group of odd order, and $x \in G$ is not the identity, then x^{-1} is not in the conjugacy class of x.

8 Show that if G is a finite group, and x and y are in the same conjugacy class, then $|N_G(\{x\})| = |N_G(\{y\})|$.

9 *Sage* showed that the group A_5 had conjugacy classes of orders 1, 12, 12, 15, and 20. Using just this information, without using Abel's theorem (7.1), prove that A_5 is simple. Use Example 7.9 as a guide.

10 *Sage* showed that the group $\text{Aut}(Z_{24}^*)$ had conjugacy classes of orders 1, 21, 24, 24, 42, and 56. Using just this information, prove that $\text{Aut}(Z_{24}^*)$ is simple.

11 The group $L_2(8)$ has 504 elements, and has nine conjugacy classes of orders 1, 56, 56, 56, 56, 63, 72, 72, and 72. Prove that $L_2(8)$ is simple. This is another example of a Chevalley group.

12 The group $L_2(11)$ has 660 elements, and has eight conjugacy classes of orders 1, 55, 60, 60, 110, 110, 132, and 132. Prove this group is simple. This group is related to the group $\mathrm{Aut}(Z_{11} \times Z_{11})$.

13 The group $L_2(13)$ has 1092 elements, and has nine conjugacy classes of orders 1, 84, 84, 91, 156, 156, 156, 182, and 182. Prove this group is simple. This group is related to the group $\mathrm{Aut}(Z_{13} \times Z_{13})$.

14 The group $L_2(17)$ has 2448 elements, and has eleven conjugacy classes of orders 1, 144, 144, 153, 272, 272, 272, 272, 306, 306, and 306. Prove this group is simple. This group, the seventh smallest non-cyclic simple group, is related to the group $\mathrm{Aut}(Z_{17} \times Z_{17})$.

15 Looking at the pattern of the last 3 problems, determine the eighth smallest non-cyclic simple group.

16 The group M_{11} has order 7920, and has 10 conjugacy classes of orders 1, 165, 440, 720, 720, 990, 990, 990, 1320, and 1584. Prove that M_{11} is simple. This is the smallest of the 26 sporadic simple groups.

17 The group $L_3(4)$ has 20160 elements, and has 10 conjugacy classes of orders 1, 315, 1260, 1260, 1260, 2240, 2880, 2880, 4032, and 4032. Prove that this group is simple. Show that even though A_8 is a simple group with the same order, these two groups are not isomorphic.

Hint: How many 3-cycles are in A_8? What does Lemma 7.1 say about the 3-cycles?

18 Find a representative element for each of the seven conjugacy classes of the group A_6. The number of elements in each conjugacy class is given in Example 7.9.

Hint: Are $(1\,2\,3\,4\,5)$ and $(1\,2\,3\,5\,4)$ in the same conjugacy class? Why are $(1\,2)(3\,4\,5\,6)$ and $(1\,2)(3\,4\,6\,5)$ in the same conjugacy class?

19 Using the counting methods that were used to estimate the 168 elements of $\mathrm{Aut}(Z_{24}^*)$, find the maximum number of elements of $\mathrm{Aut}(Z_2 \times Z_2 \times Z_2 \times Z_2)$. This group is in fact simple, and contains the number of elements predicted by this estimate. Are there any other simple groups that we have seen of this order?

Interactive Problems

20 The following commands load a group of order 20 into *Sage*.

```
InitGroup("e")
AddGroupVar("a", "b")
Define(a^5, e); Define(b^4, e); Define(b*a, a^2*b)
M = ListGroup()
```

Find the conjugacy classes of this group, and use this to find all of the normal subgroups of M.

21 The following commands load a group of order 24 into *Sage*.

```
DisplayPermInt = true
G = Group(NthPerm(2374), NthPerm(6212)); G
    {1, 2374, 4517, 6212, 6841, 9929, 11637, 13016, 13698, 15367,
    18454, 19853, 21239, 21896, 24132, 25315, 28226, 28986,
    30928, 31590, 33108, 37381, 38807, 39487}
StructureDescription(2374, 6212)
    SL(2,3)
```

Find the conjugacy classes of this group, and use this to find all of the normal subgroups of G.

7.4 The Class Equation and Sylow's Theorems

Just as Lagrange's theorem had far-reaching implications, the partition of a group into conjugacy classes also has important consequences if we consider the size of these classes. As a result, we will be able to prove three important theorems first discovered by Peter Sylow. (See the Historical Diversion on page 252.) Using these three theorems, we will be able to classify all groups up to order 15. This shows just how powerful these theorems can be.

In working with the conjugacy classes from the last section, we may have noticed a pattern in the *size* of each of the conjugacy classes. For example, the conjugacy classes of S_4 are given by

```
S4 = Group( C(1,2), C(2,3,4) )
ConjugacyClasses(S4)
    {{()}, {(1, 2), (2, 3), (1, 3), (3, 4), (2, 4), (1, 4)},
    {(1, 3, 2), (1, 2, 3), (2, 4, 3), (1, 4, 3), (2, 3, 4),
    (1, 4, 2), (1, 3, 4), (1, 2, 4)}, {(1, 2)(3, 4), (1, 3)(2, 4),
    (1, 4)(2, 3)}, {(1, 4, 3, 2), (1, 2, 4, 3), (1, 3, 4, 2),
    (1, 4, 2, 3), (1, 2, 3, 4), (1, 3, 2, 4)}}
```

The first class has only the identity element, the class with the transpositions has exactly 6 elements, while the other classes are of orders 8, 3, and 6. Immediately we see that the number of elements in the classes may be different. We have the obvious relationship

$$1 + 6 + 8 + 3 + 6 = 24,$$

the order of the group, since every element in the group belongs to one and only one conjugacy class. Is there another pattern? Let us compare this with the conjugacy classes of $\text{Aut}(Z_{24}^*)$. There were six conjugacy classes of size 1, 21, 42, 56, 24, and 24. We can check that

$$1 + 21 + 56 + 42 + 24 + 24 = 168.$$

But another pattern is becoming clear that is akin to Lagrange's theorem (3.1). Notice that the number of elements in each class is always a *divisor of the order of the group*. Let's see if we can discover why this pattern exists.

LEMMA 7.2
Let G be a finite group, and let g be an element of G. Then the number of elements of G that are conjugate to g is given by

$$\frac{|G|}{|N_G(\{g\})|},$$

where $N_G(\{g\})$ denotes the normalizer of the single element $\{g\}$.

PROOF: We saw in Proposition 7.5 that $N_G(\{g\})$ is a subgroup of G. We want to determine all possible conjugates of the element g. Note that if u and v are two elements of G, then $u \cdot g \cdot u^{-1}$ and $v \cdot g \cdot v^{-1}$ will represent the same element if, and only if,

$$\begin{aligned} u \cdot g \cdot u^{-1} = v \cdot g \cdot v^{-1} &\Longleftrightarrow v^{-1} \cdot u \cdot g \cdot u^{-1} \cdot v = g \\ &\Longleftrightarrow (v^{-1} \cdot u) \cdot g \cdot (v^{-1} \cdot u)^{-1} = g \\ &\Longleftrightarrow v^{-1} \cdot u \in N_G(\{g\}) \\ &\Longleftrightarrow u \in v \cdot N_G(\{g\}) \\ &\Longleftrightarrow u \cdot N_G(\{g\}) = v \cdot N_G(\{g\}). \end{aligned}$$

Thus $u \cdot g \cdot u^{-1}$ and $v \cdot g \cdot v^{-1}$ represent the same element if, and only if, u and v belong to the same left coset of $N_G(\{g\})$. Therefore, to count all of the possible conjugates of g, we merely count the number of left cosets of $N_G(\{g\})$, which is

$$\frac{|G|}{|N_G(\{g\})|}.$$

∎

We have already observed that the sum of the number of elements in each of the conjugacy classes must give the number of elements in the group. Since we now know how many elements are in each conjugacy class, we can derive what is called the *class equation.*

THEOREM 7.2: The Class Equation Theorem
Let G be a finite group. Then

$$|G| = \sum_g \frac{|G|}{|N_G(\{g\})|},$$

where the sum runs over one g from each conjugacy class.

PROOF: We simply observe that every element of G appears in exactly one of the conjugacy classes. Thus, $|G|$ is the sum of the sizes of all of the conjugacy classes. We have by Lemma 7.2 that the size of each conjugacy class is

$$\frac{|G|}{|N_G(\{g\})|}$$

where g is a representative element of the conjugacy class. Thus we get the class equation. ∎

It is helpful to give an example of the class equation, to understand the notation. The group $\mathrm{Aut}(Z_{24}^*)$ had 6 conjugacy classes, represented by the 1st, 27th, 149th, 231st, 931st, and 957th permutations. We can find the size of the normalizers for each of these elements.

```
A = Group( NthPerm(187), NthPerm(723) )
len(Normalizer(A, NthPerm(1) ))
    168
len(Normalizer(A, NthPerm(27) ))
    8
len(Normalizer(A, NthPerm(149) ))
    3
len(Normalizer(A, NthPerm(231) ))
    4
len(Normalizer(A, NthPerm(931) ))
    7
len(Normalizer(A, NthPerm(957) ))
    7
```

Finally, we form the sum

$$\sum_g \frac{|G|}{|N_G(\{g\})|} = \frac{168}{168} + \frac{168}{8} + \frac{168}{3} + \frac{168}{4} + \frac{168}{7} + \frac{168}{7} = 168.$$

We will see many very important applications of this equation, but let us begin by learning what this has to say about groups whose order is a power of a prime.

COROLLARY 7.2

If G is a group of order p^n where p is a prime and n is a positive integer, then G is not centerless.

PROOF: First we observe that an element g is in the center of G if, and only if, $y \cdot g \cdot y^{-1} = g$ for all y in G, which would happen if, and only if, the conjugacy class of g consists of just g by itself.

Now suppose G is centerless. Then the only conjugacy class that contains just one element would be the class $\{e\}$. All other conjugacy classes would have a size that is a divisor of p^n, so the number of elements in the other conjugacy classes would be a power of p. But this is impossible since the sum on the right-hand side of the class equation (7.2) would be congruent to 1 (mod p), while the left-hand side of the class equation would be p^n, which is congruent to 0 (mod p). Therefore, G is not centerless. ∎

This corollary is useful in finding all non-isomorphic groups of order p^n, where p is a prime. For example, we can easily find all non-isomorphic groups of order p^2.

COROLLARY 7.3

If p is a prime then there are exactly two non-isomorphic groups of order p^2, namely Z_{p^2} and $Z_p \times Z_p$.

PROOF: If G is a group of order p^2, then by Corollary 7.2, G has a nontrivial center. Since the number of elements of $Z(G)$ must divide p^2, so $|Z(G)|$ is either equal to p or p^2.

Suppose that $|Z(G)| = p$. Then there exists an element g not in $Z(G)$. Then $N_G(\{g\})$ denotes the set of elements that commute with g. Certainly

$$Z(G) \subseteq N_G(\{g\}),$$

and also

$$g \in N_G(\{g\}),$$

so $N_G(\{g\})$ contains at least $p + 1$ elements. But this is a subgroup of G, so the number of elements must divide p^2. Hence, $N_G(\{g\})$ contains all of G, but this would say that g is in the center $Z(G)$, which contradicts our assumption. Thus, there are p^2 elements in $Z(G)$ and hence G is an abelian group.

Finally, we can use the fundamental theorem of finite abelian groups (6.2) to say that G must be isomorphic to the direct product of cyclic groups. It is easy to see that there are exactly two possibilities for such a product to have p^2 elements, namely Z_{p^2} and $Z_p \times Z_p$. ∎

In particular we can use Corollary 7.3 to see that there are only two non-isomorphic groups of order 9, Z_9 and $Z_3 \times Z_3$.

One of the keys for finding all groups of a certain order is knowing whether there is a normal subgroup of a certain order. The next proposition will allow us to know that there will be a normal subgroup *without knowing the structure of the group.*

PROPOSITION 7.8

Let G be a group of order p^n. Then G contains a normal subgroup of order p^{n-1}.

PROOF: We will proceed by using induction on n. Note that if $n = 1$, then there is obviously a normal subgroup of order $p^{1-1} = p^0 = 1$, namely the trivial subgroup $\{e\}$.

Suppose that we know that every group of order p^{n-1} has a normal subgroup of order p^{n-2}. Let G be a group of order p^n. Then by Corollary 7.2, the center of G is not just the identity element. Since p would then divide the order of $Z(G)$, by Lemma 6.2 there is an element of $Z(G)$ of order p, say x. Then the group generated by x would be of order p, and since x is in the center, all elements of G would commute with x. Thus, $X = [x]$ would be a normal subgroup of G.

We then can consider the quotient group G/X. This would have order p^{n-1}, and we would have the canonical homomorphism

$$\phi : G \to G/X$$

whose kernel is the subgroup X. By the induction hypothesis, G/X is a group of order p^{n-1}, and so has a normal subgroup of order p^{n-2}, say Y.

We will now "lift" the subgroup Y back to the original group. Since $\phi^{-1}(Y)$ is the inverse image of a normal subgroup, by Corollary 4.2, this is a normal subgroup of G. Note Y is a set of cosets, and that $g \in \phi^{-1}(Y)$ if, and only if, g is contained in one of the cosets of Y. Since each of the cosets of Y contains p elements, it is clear that the size of $\phi^{-1}(Y)$ is $p \cdot p^{n-2} = p^{n-1}$. Therefore, we have proved by induction that there is a normal subgroup of G of order p^{n-1}. □

We now are ready to start finding normal subgroups of a more general group, knowing only the group's order. The most important set of theorems that tackle this problem are by a Norwegian high school teacher named Ludwig Sylow (1832–1918) [1, p. 324]. Before we work on finding normal subgroups let us see if we can find a subgroup of a given order within a group.

THEOREM 7.3: The First Sylow Theorem

Suppose that G is a group of order $p^n \cdot m$, where p is a prime, and m is coprime to p. Then G has a subgroup of order p^n.

PROOF: We will proceed by using induction on the size of G. That is, we will assume that the theorem is true for all groups smaller than G.

If p^n divided $|H|$ for some proper subgroup H of G, then by our induction hypothesis, H would have a subgroup of order p^n, which would be a subgroup of G for which we are searching. So we may assume that p^n does not divide the order of any proper subgroup of G.

Historical Diversion
Peter Ludwig Sylow (1832–1918)

Peter Sylow (SEE-loh) was born in Christiania, Norway, which is now Oslo. He was raised to be modest and be hard-working, but this later would result in him settling for a lowly position rather that aspiring to the position he deserved. Sylow attended Christiania Cathedral School, graduating in 1850. He then attended Christiania University, and won a gold medal in a mathamatics contest in 1853. In 1855 he became a high school teacher, because there was no university post available for him at the time. This was unfortunate for both Sylow and the high school, because Sylow would have made an outstanding university professor, but he was not a particularly good high school teacher.

In spite of the long hours required by his teaching duties, Sylow found time to study the papers of Niels Abel. (See Historical Diversion on page 241.) He soon found that Abel discovered many more results in the theory of equations than his published papers indicated. He tried to publish some of Abel's unpublished results, but the journal's editor, Leopold Kronecker, had derived these results himself, and had no wish to admit that his results were in fact proven by Abel long before him. Since Abel's unpublished works were eventually published, these results are indeed credited to Abel rather than Kronecker.

In 1862 Sylow received a temporary appointment at Chistiania University, and taught Galois theory and permutation groups. It was during this year that Sylow proposed questions about p-groups. After demonstrating Cauchy's theorem to the class (a group of order divisible by a prime p has a subgroup of order p), Sylow conjectured whether this can be generalized to a power of p. One of his students was Sophus Lie, who would later assist Sylow in publishing Abel's unpublished papers.

Ten years after his conjecture, Sylow proved his famous theorems in a 10-page paper, *Théorèmes sur les groupes de substitutions*. Almost all work done on finite groups uses these three Sylow's theorems.

In 1898, Lie had a special position created for Sylow at Christiania University. Even though he was 65 years old when he finally became a university professor, he held on to the position for 20 years until his death in 1918.

Image source: Wikimedia Commons

In particular, if g is not in the center of G, then $N_G(\{g\})$ will not be all of G. Hence, p^n does not divide $|N_G(\{g\})|$. But since p^n does divide $|G|$, we have from Lemma 7.2 that the number of conjugates of g is $|G|/|N_G(\{g\})|$, which must be a multiple of p.

Now we can use the argument that we used in Corollary 7.2. The class equation theorem (7.2) states that

$$|G| = \sum_g \frac{|G|}{|N_G(\{g\})|},$$

where the sum runs over one g from each conjugacy class. For those g in the center of G, $|G|/|N_G(\{g\})|$ will be 1, while for all other terms, $|G|/|N_G(\{g\})|$ will be a multiple of p. Since the sum is $p^n \cdot m$, which is a multiple of p, the number of elements in $Z(G)$ must be a multiple of p.

Since $Z(G)$ is an abelian group and p divides $Z(G)$, we have by Lemma 6.2 that there is an element of $Z(G)$ of order p, say x. We now can proceed in the same way as we did in Proposition 7.8. Since x is in the center, all elements of G would commute with x, and so $X = [x]$ would be a normal subgroup of order p.

The quotient group G/X would then have order $p^{n-1} \cdot m$, and we would have the canonical homomorphism

$$\phi : G \to G/X$$

whose kernel is the subgroup X. By the induction hypothesis, G/X is smaller than G, and so has a subgroup of order p^{n-1}, say Y. We can then lift Y back to the original group. Since $\phi^{-1}(Y)$ is the inverse image of a subgroup, by Corollary 4.2, this is a subgroup of G. But the kernel of the homomorphism is of order p, so the size of $\phi^{-1}(Y)$ is $p \cdot p^{n-1} = p^n$. Therefore, we have proved by induction that there is a subgroup of G of order p^n. □

The first Sylow theorem generalizes Cauchy's theorem, which states that if p divides $|G|$, then G has a subgroup of order p. See Problem 8.

Since the first Sylow theorem guarantees the existance of at least one subgroup of order p^n for a group of size $p^n \cdot m$, we will give a name to these subgroups.

DEFINITION 7.5 If G is a group of order $p^n \cdot m$, where m is coprime to the prime p, then a subgroup of order p^n is called a *p-Sylow subgroup.*

Let us give a quick application of the first Sylow theorem (7.3). Suppose we have a group G of order 10. There is guaranteed to be a 5-Sylow subgroup, say K, and a 2-Sylow subgroup, say H. Obviously,

$$K \approx Z_5 \quad \text{and} \quad H \approx Z_2.$$

Furthermore, the intersection of K and H must just be the identity element, since Z_5 does not have any elements of order 2. Also, K is a subgroup of G with index 2, so by Proposition 3.5, K is a normal subgroup of G. If H is also normal, we have by the direct product theorem (6.1) that

$$K \cdot H \approx K \times H \approx Z_5 \times Z_2 \approx Z_{10}.$$

On the other hand, if H is not a normal subgroup, then by the semi-direct product theorem (6.3)

$$K \cdot H \approx K \rtimes_{\phi} H$$

for some nontrivial homomorphism ϕ from H to $\text{Aut}(K)$. But in Chapter 6, we found that there was only one nontrivial homomorphism, yielding the dihedral group D_5. In either case, $K \cdot H$ is of order 10, so G is either isomorphic to Z_{10} or D_5.

Even though Sylow's first theorem (7.3) guarantees that there will be at least one p-Sylow subgroup, there may be more than one. The next of Sylow's theorems shows that any two p-Sylow subgroups are related.

THEOREM 7.4: The Second Sylow Theorem

If H and K are two p-Sylow subgroups of G, then there exists an element u in G such that $H = u \cdot K \cdot u^{-1}$.

PROOF: Let G be a group of order $p^n \cdot m$, where m is coprime to the prime p. We begin by showing that whenever K is a p-Sylow subgroup of G then $u \cdot K \cdot u^{-1}$ will also be a p-Sylow subgroup for all u in G. Note that the number of elements in $u \cdot K \cdot u^{-1}$ is also p^n, and if $u \cdot k_1 \cdot u^{-1}$ and $u \cdot k_2 \cdot u^{-1}$ are two elements of $u \cdot K \cdot u^{-1}$, then

$$(u \cdot k_1 \cdot u^{-1}) \cdot (u \cdot k_2 \cdot u^{-1})^{-1} = u \cdot k_1 \cdot u^{-1} \cdot (u \cdot k_2^{-1} \cdot u^{-1}) = u \cdot (k_1 \cdot k_2^{-1}) \cdot u^{-1},$$

which is in $u \cdot K \cdot u^{-1}$. So by Proposition 2.2, $u \cdot K \cdot u^{-1}$ is a p-Sylow subgroup of G.

If there is only one p-Sylow subgroup of G there is nothing to prove. Suppose H and K are two subgroups of order p^n. Let us call two elements u and v of G to be "related" if $u = h \cdot v \cdot k$ for some h in H and k in K. Note that every element is related to itself, for $u = e \cdot u \cdot e$, and e is in both H and K. Also, if u is related to v, then v is related to u, for

$$u = h \cdot v \cdot k \Longleftrightarrow v = h^{-1} \cdot u \cdot k^{-1}.$$

Finally, if u is related to v, and v is related to w, then $u = h_1 \cdot v \cdot k_1$ and $v = h_2 \cdot w \cdot k_2$, and so

$$u = h_1 \cdot (h_2 \cdot w \cdot k_2) \cdot k_1 = (h_1 \cdot h_2) \cdot w \cdot (k_2 \cdot k_1),$$

so u and w are related. Therefore, we can partition the group G into "families," where each family consists of all elements related to one element.

Now suppose that there are j families, and we select one element u_i from each family. Each of the families can be described as $H \cdot u_i \cdot K$. Hence, we can write

$$G = (H \cdot u_1 \cdot K) \cup (H \cdot u_2 \cdot K) \cup \cdots \cup (H \cdot u_j \cdot K).$$

Since each of the families have no elements in common, we have

$$|G| = |H \cdot u_1 \cdot K| + |H \cdot u_2 \cdot K| + \cdots + |H \cdot u_j \cdot K|.$$

How many elements are in each family? We note that $H \cdot u_i \cdot K$ has the same number of elements as $H \cdot u_i \cdot K \cdot u_i^{-1}$. We saw that $u_1 \cdot K \cdot u_i^{-1}$ is a group, and so even though the product of two groups was not always a group, Proposition 4.9 gave us the number of elements in the set to be

$$|H \cdot u_i \cdot K| = |H \cdot u_i \cdot K \cdot u_i^{-1}| = \frac{|H| \cdot |u_i \cdot K \cdot u_i^{-1}|}{|H \cap (u_i \cdot K \cdot u_i^{-1})|} = \frac{p^n \cdot p^n}{|H \cap (u_i \cdot K \cdot u_i^{-1})|}.$$

If we plug this formula into the equation above it, we have that $|G| =$

$$p^n \cdot m = \frac{p^n \cdot p^n}{|H \cap (u_1 \cdot K \cdot u_1^{-1})|} + \frac{p^n \cdot p^n}{|H \cap (u_2 \cdot K \cdot u_2^{-1})|} + \cdots + \frac{p^n \cdot p^n}{|H \cap (u_j \cdot K \cdot u_j^{-1})|}.$$

Note that the intersection of two groups is a subgroup of both the groups, and so the denominators will all be powers of p. Dividing both sides of the equation by p^n, we have

$$m = \frac{p^n}{|H \cap (u_1 \cdot K \cdot u_1^{-1})|} + \frac{p^n}{|H \cap (u_2 \cdot K \cdot u_2^{-1})|} + \cdots + \frac{p^n}{|H \cap (u_j \cdot K \cdot u_j^{-1})|}.$$

Since m is not a multiple of p, there must be some term on the right-hand side of this equation that is not a multiple of p. But this can happen only if one of the denominators is p^n, that is,

$$|H \cap (u_i \cdot K \cdot u_i^{-1})| = |H|$$

for some i. Since H and $u_i \cdot K \cdot u_i^{-1}$ both have p^n elements, we must have $H = u_i \cdot K \cdot u_i^{-1}$. Therefore, for any two p-Sylow subgroups of G, there is a u such that $H = u \cdot K \cdot u^{-1}$. ∎

The second Sylow theorem (7.4) allows us to know exactly when a p-Sylow subgroup is normal.

COROLLARY 7.4

The group G has only one p-Sylow subgroup for a given prime p if, and only if, G has a p-Sylow subgroup that is normal.

PROOF: Suppose that H is the only p-Sylow subgroup of G. Then for any element u in G, $u \cdot H \cdot u^{-1}$ will be a p-Sylow subgroup of G. But since there is only one p-Sylow subgroup, we have $u \cdot H \cdot u^{-1} = H$ for all u in G. Hence, H is a normal subgroup.

Now suppose that H is a normal p-Sylow subgroup of G. By the second Sylow theorem (7.4) every other p-Sylow subgroup is of the form $u \cdot H \cdot u^{-1}$. But since H is normal, $u \cdot H \cdot u^{-1} = H$. Therefore, H is the only p-Sylow subgroup. ☐

The natural question that Corollary 7.4 raises is, "How do we know if there is only one p-Sylow subgroup?" The next lemma allows us to find the number of p-Sylow subgroups in terms of the size of the normalizer. In fact it allows us to find the number of p-Sylow subgroups of a certain type.

LEMMA 7.3

Let G be a group of order $p^n \cdot m$, and let P be a p-Sylow subgroup of G. Let H be any other subgroup of G. Then the number of p-Sylow subgroups that can be written as $u \cdot P \cdot u^{-1}$ with u an element of H is given by

$$\frac{|H|}{|N_G(P) \cap H|}.$$

PROOF: Since P is a subgroup of G, $N_G(P)$ is a subgroup of G, so the intersection of $N_G(P)$ and H will be a subgroup of H. We can use the same argument as Lemma 7.2, and note that if u and v are two elements of H, then $u \cdot P \cdot u^{-1}$ and $v \cdot P \cdot v^{-1}$ will represent the same p-Sylow subgroup if, and only if,

$$\begin{aligned} u \cdot P \cdot u^{-1} = v \cdot P \cdot v^{-1} &\Longleftrightarrow v^{-1} \cdot u \cdot P \cdot u^{-1} \cdot v = P \\ &\Longleftrightarrow (v^{-1} \cdot u) \cdot P \cdot (v^{-1} \cdot u)^{-1} = P \\ &\Longleftrightarrow v^{-1} \cdot u \in N_G(P) \cap H \\ &\Longleftrightarrow u \in v \cdot (N_G(P) \cap H) \\ &\Longleftrightarrow u \cdot (N_G(P) \cap H) = v \cdot (N_G(P) \cap H). \end{aligned}$$

Thus, $u \cdot P \cdot u^{-1}$ and $v \cdot P \cdot v^{-1}$ represent the same p-Sylow subgroup if, and only if, $u \cdot (N_G(P) \cap H)$ and $v \cdot (N_G(P) \cap H)$ are the same left cosets of $N_G(P) \cap H$ over H. Therefore, the number of p-Sylow subgroups that can be expressed as $u \cdot P \cdot u^{-1}$, with u an element of H, is

$$\frac{|H|}{|N_G(P) \cap H|}.$$

☐

We now are ready to prove the last of Sylow's theorems, which in many cases will tell us the number of p-Sylow subgroups of a group.

THEOREM 7.5: The Third Sylow Theorem
Suppose that the number of p-Sylow subgroups of G is k. Then k divides $|G|$, and $k \equiv 1 \pmod{p}$.

PROOF: Suppose that we label the p-Sylow subgroups of G as $P_0, P_1, P_2, \ldots, P_{k-1}$. Let us partition all of the p-Sylow subgroups of G into different categories where two p-Sylow subgroups P_i and P_j are in the same category if there is an element u in P_0 such that

$$P_j = u \cdot P_i \cdot u^{-1}.$$

Note that P_0 would be in its own category while the number of p-Sylow subgroups in the other categories would be, according to Lemma 7.3,

$$\frac{|P_0|}{|N_G(P_i) \cap P_0|}$$

where P_i is one p-Sylow subgroup in the category.

Recall that the normalizer of each P_i contains P_i as a normal subgroup, so $N_G(P_i)$ is divisible by p^n, and hence by Corollary 7.4 the only p-Sylow subgroup of $N_G(P_i)$ is P_i. Thus, $N_G(P_i)$ cannot contain P_0, lest there would be 2 p-Sylow subgroups of $N_G(P_i)$. This makes $|N_G(P_i) \cap P_0| < p^n$, so we have that the number of p-Sylow subgroups in each category, besides the category containing just P_0, is a power of p, and hence is a multiple of p.

Therefore, the total number of p-Sylow subgroups is one more than a multiple of p, so $k \equiv 1 \pmod{p}$.

Finally, if we let $H = G$ in Lemma 7.3, we find that the number of conjugates of P_0 is

$$\frac{|G|}{|N_G(P_0)|}.$$

By the second Sylow theorem (7.4), this would give us all of the p-Sylow subgroups. Therefore, k is also a divisor of the order of the group G. ∎

These three theorems of Sylow provide a means of finding normal subgroups of a group G just from knowing the order of G. For example, suppose that a group is of order 45. Since 3^2 divides 45, there is a 3-Sylow subgroup of order 9. We also know that the number of 3-Sylow subgroups divides 45, so this number must be 1, 3, 5, 9, 15, or 45. However, the number must be congruent to 1 (mod 3). Thus, the only possibility is that there is only one subgroup of order 9, say H. But then this subgroup is normal.

We can use the same argument to find a normal subgroup of order 5. Again, the number of 5-Sylow subgroups must be 1, 3, 5, 9, 15, or 45. But this number must also be congruent to 1 (mod 5), so there is only one subgroup of order 5, and this group must also be normal.

Although the Sylow theorems are powerful tools, when combined with the tools of semi-direct products and the computational power of *Sage*, we can determine most of the groups of a given order.

Example 7.11
Find all of the groups of order 12.
SOLUTION: The divisors of 12 are 1, 2, 3, 4, 6, and 12, so by the third Sylow theorem there are either one or four 3-Sylow subgroups and there are either one or three 2-Sylow subgroups. Let H be a 3-Sylow subgroup, and let K be a 2-Sylow subgroup (which will be of order 4). Certainly the intersection of H and K is just the identity element since K cannot contain an element of order 3.

Let us show that either H or K is normal. Suppose that H is not normal. Then there must be four 3-Sylow subgroups of G. Each of these 3-Sylow groups contains two different elements of order 3, so G would have eight elements of order 3. But that would leave only four elements left over, and so K must be composed of all of those four elements. Then there would be only one 2-Sylow subgroup, which would be normal.

By the direct product theorem (6.1) and the semi-direct product theorem (6.3), $H \cdot K$ would have to be of one of the following forms:

1. $H \cdot K \approx Z_3 \times Z_4 \approx Z_{12}$,

2. $H \cdot K \approx Z_3 \times Z_8^* \approx Z_3 \times Z_2 \times Z_2$,

3. $H \cdot K \approx Z_4 \rtimes_\phi Z_3$,

4. $H \cdot K \approx Z_8^* \rtimes_\phi Z_3$,

5. $H \cdot K \approx Z_3 \rtimes_\phi Z_4$,

6. $H \cdot K \approx Z_3 \rtimes_\phi Z_8^*$.

In all six cases $H \cdot K$ contains 12 elements, and so $G = H \cdot K$. Let us work these six cases separately. The first two give the two possible abelian groups of order 12. Case 3 is actually impossible, since $\text{Aut}(Z_4) \approx Z_4^*$ has only two elements, and therefore has no elements of order 3. Therefore, there is no nontrivial homomorphism from Z_3 to $\text{Aut}(Z_4)$. The other three cases are as follows:

Case 4

An element of order 3 in Z_3 must map to an element of order 3 in $\text{Aut}(Z_8^*)$, which is isomorphic to S_3. There are two elements of order 3 in S_3, and these two elements are conjugates. By Proposition 6.7, it does not matter which element of Z_3 maps to which elements in $\text{Aut}(Z_8^*)$, so the semi-direct product $Z_3 \rtimes_\phi Z_8^*$ is unique up to isomorphisms. But A_4 is a group of order 12, has a normal subgroup isomorphic to Z_8^*, and does not have a normal subgroup of order 3. Thus, A_4 must be this unique semi-direct product $Z_3 \rtimes Z_8^*$.

Case 5

The homomorphism ϕ must map a generator of Z_4 to a nontrivial element of Aut(Z_3). But Aut(Z_3) has only two elements, so this homomorphism is uniquely determined. The group is generated by following the *Sage* commands.

```
InitGroup("e")
AddGroupVar("a", "b")
Define(a^3, e); Define(b^4, e); Define(b*a, a^2*b)
M = ListGroup(); M
    {e, a, a^2, b, a*b, a^2*b, b^2, a*b^2, a^2*b^2, b^3, a*b^3,
    a^2*b^3}
RootCount(M, 2)
    2
```

This non-abelian group has only one element of order 2. Thus, it is not isomorphic to any group we have seen before. If we ask *Sage* for the description of the structure,

```
StructureDescription()
    Z3 : Z4
```

we find that this is considered to be the group $Z_3 \rtimes Z_4$. This is how we will identify this group.

Case 6

The homomorphism ϕ maps Z_8^* to Aut(Z_3). Since Aut(Z_3) contains only two elements, the homomorphism ϕ is completely determined by its kernel. The kernel of ϕ cannot be just the identity, since there is not an isomorphic copy of Z_8^* in Aut(Z_3). On the other hand, the kernel of a nontrivial homomorphism cannot be all of Z_8^*. Thus, the kernel contains exactly two elements, and because there are automorphisms of Z_8^* mapping one subgroup of order 2 to any other, it will not matter which subgroup of order 2 we pick. Thus, there is a unique semi-direct product $z_3 \rtimes Z_8^*$.

The obvious group of order 12 that we have yet to consider is $Z_2 \times S_3$. This has a normal subgroup of order 3, so by process of elimination must be $Z_3 \rtimes Z_8^*$. In summary, we have found five possible groups of order 12:

$$Z_{12}, \qquad A_4 \qquad Z_2 \times Z_2 \times Z_3 \qquad Z_2 \times S_3 \quad \text{and} \quad Z_4 \rtimes Z_3. \qquad \square$$

Let us summarize our findings formally with a proposition.

PROPOSITION 7.9

There are exactly 28 non-isomorphic groups of order less than 16.

PROOF: The trivial group is the only group of order 1, and since 2, 3, 5,

7, 11, and 13 are prime, we have only one non-isomorphic group of each of these orders.

In Chapter 4 we found that the only non-isomorphic groups of order 4 were

$$Z_4 \quad \text{and} \quad Z_8^*,$$

the only non-isomorphic groups of order 6 were

$$Z_6 \quad \text{and} \quad S_3,$$

and the only non-isomorphic groups of order 8 were

$$Z_8, \quad Z_{15}^*, \quad Z_{24}^*, \quad Q, \quad \text{and} \quad D_4.$$

By Corollary 7.3 the only two non-isomorphic groups of order 9 are

$$Z_9 \quad \text{and} \quad Z_3 \times Z_3.$$

We have already used the first Sylow theorem (7.3) to find all of the non-isomorphic groups of order 10:

$$Z_{10} \quad \text{and} \quad D_5.$$

We just found all of the groups of order 12:

$$Z_{12}, \quad A_4, \quad Z_2 \times Z_2 \times Z_3, \quad Z_2 \times S_3, \quad \text{and} \quad Z_4 \rtimes Z_3.$$

We can use the same argument to find all of the non-isomorphic groups of order 14. If $|G| = 14$, there must be a 7-Sylow subgroup of G, say K. Since K contains half the elements, by Proposition 3.5, K is normal. We also must have a 2-Sylow subgroup, H. Since K cannot have an element of order 2, H and K have only the identity element in common. If H is normal, then $K \cdot H \approx K \times H \approx Z_7 \times Z_2 \approx Z_{14}$. If H is not normal, by the semi-direct product theorem (6.3),

$$K \cdot H \approx K \rtimes_{\phi} H$$

for some homomorphism ϕ from H to $\text{Aut}(K)$. In either case $K \cdot H$ has 14 elements, and so $G = K \cdot H$. Also, ϕ is determined by where the non-identity element of H is mapped. Since this must be an element of $\text{Aut}(K)$ of order 2, and since

$$\text{Aut}(K) \approx \text{Aut}(Z_7) \approx Z_7^* \approx Z_6$$

has only one element of order 2, there can only be one such homomorphism. Since D_7 is a non-abelian group of order 14, this must be the one semi-direct product that we found. Thus, the only two groups of order 14 are

$$Z_{14} \quad \text{and} \quad D_7.$$

Let us move on to find all groups of order 15. Suppose $|G| = 15$. Then the number of 3-Sylow subgroups and the number of 5-Sylow subgroups must both divide 15, so both of these numbers must be one of $1, 3, 5$, or 15. But 1 is the only number in this set that is congruent to 1 (mod 5). So there is only one 5-Sylow subgroup, K. Likewise, 1 is the only number in the set that is congruent to 1 (mod 3). So there is only one 3-Sylow subgroup, H. By Corollary 7.4, both K and H are normal subgroups of G, and the intersection must be just the identity element. Thus, by Corollary 6.1,

$$K \cdot H \approx K \times H \approx Z_5 \times Z_3 \approx Z_{15}.$$

Since this has all 15 elements, this must be all of G, and so there is only one non-isomorphic group of order 15, namely Z_{15}.

Therefore, counting all of the groups of order less than 16, we find that there are exactly 28 of them. ∎

Unfortunately, finding all the groups of order 16 is a difficult problem. Even though Proposition 7.8 tells us that there must be a normal subgroup K of order 8, there is no guarantee that there would be a subgroup H of order 2 such that $H \cdot K$ gives the whole group. Thus, we would not be able to use the semi-direct product theorem (6.3) to find *all* of the groups of order 16 (although we can find many of them, as we did in the last chapter).

Problems for §7.4

For Problems **1** through **6**: Determine whether or not the following lists could represent the sizes of the conjugacy classes for some finite group.

Hint: Add up the numbers to determine the size of the group.

1 $\{2, 2, 4, 4, 4\}$

2 $\{1, 3, 6, 8\}$

3 $\{1, 2, 2, 5\}$

4 $\{1, 1, 3, 3, 3, 11, 11\}$

5 $\{1, 1, 1, 1, 1, 5, 5, 5, 5\}$

6 $\{1, 3, 6, 6, 8\}$

7 If G has order p^n for some prime p, show that every subgroup of order p^{n-1} is a normal subgroup of G.

8 Use the class equation to prove Cauchy's theorem: if p is a prime that divides $|G|$, then G has a subgroup of order p.

Hint: Use induction on $|G|$. You can use the fact that we proved it for the abelian groups in Lemma 6.2.

9 If H is a subgroup of G, and H has order p^i for some prime p, show that H is contained in a p-Sylow subgroup of G.

Hint: Mimic the proof of the second Sylow theorem (7.4).

10 Use Sylow's theorem to show that all groups of order 33 are cyclic.

11 Prove that no group of order 56 is simple.

12 Show that if p is an odd prime, then any group with $2p$ elements is isomorphic to either Z_{2p} or D_p.

13 Determine all non-isomorphic groups of order 99.

14 Show that there are exactly four non-isomorphic groups of order 66:

$$Z_{66}, \qquad D_{33}, \qquad D_{11} \times Z_3, \quad \text{and} \quad D_3 \times Z_{11}.$$

Hint: Use Sylow's theorems along with Problem 10.

15 Show that all groups of order 255 are cyclic.
Hint: Use Lemma 4.3.

16 Let $|G| = p \cdot q$, where $p > q$ are both primes. Show that G has a normal subgroup of order p.

17 If $|G| = p^2 \cdot q$, where p and q are different primes, show that G must contain a normal subgroup of either size p^2 or q.
Hint: Generalize the case $|G| = 12$ done in the text.

18 Show that a group of order $p^3 \cdot q$, where p and q are different primes, cannot be simple.
Hint: Use Corollary 5.2 for the case $|G| = 24$. Then do the case $q < p$. With these out of the way, you can assume that $q > p + 1$.

19 Use the results of Problems 16 through 18 to show that no non-cyclic group of order less than 60 is simple.

Interactive Problems

20 Use *Sage* to find all of the 2-Sylow and 5-Sylow subgroups of the group M defined in Problem 20 of §7.3. How many of the subgroups are there? Does this agree with the prediction given by the third Sylow theorem?

21 Using *Sage*, find all non-isomorphic groups of order 21.
Hint: What can you determine from Sylow's theorems? Which semi-direct products are possible?

Chapter 8

Solvable and Insoluble Groups

In this chapter we will study the concept of *solvable* groups. Every group will be classified either as solvable or insoluble, and in fact most of the groups we have looked at so far turn out to be solvable.

Solvable groups play a key role in studying polynomial equations. Whether or not a given polynomial can be solved in terms of radicals (square roots, cube roots, etc.) depends on whether a certain group corresponding to that polynomial is a solvable group. In fact this application is the origin of the term "solvable group."

8.1 Subnormal Series and the Jordan-Hölder Theorem

We will eventually define a solvable group as one that can be broken down into smaller, cyclic pieces. But before we can formally define a solvable group, we must first show that if we break the group down in two different ways, the pieces will essentially be the same. This is not at all obvious, and involves a result known as the Jordan-Hölder theorem.

We must first make some preliminary definitions. We have already encountered situations in which we had a normal subgroup of a normal subgroup, such as in the third isomorphism theorem. But suppose we have a whole series of subgroups of a group G, each one fitting inside of the previous one like Russian dolls.

DEFINITION 8.1 A *subnormal series* for a group G is a sequence $G_0, G_1, G_2, \ldots G_n$ of subgroups of G such that

$$G = G_0 \supseteq G_1 \supseteq G_2 \supseteq \cdots \supseteq G_n = \{e\},$$

where each G_i is a normal subgroup of G_{i-1} for $i = 1, 2, \cdots n$.

A subnormal series is called a *normal series* if it satisfies the stronger condition that all of the groups G_i are normal subgroups of the original group G. We will be mainly interested in subnormal series, but there are a few of the exercises regarding normal series.

Motivational Example 8.1
The group S_4, has a normal subgroup of order 4, namely

$$K = \{(\,), (1\,2)(3\,4), (1\,3)(2\,4), (1\,4)(2\,3)\}.$$

The identity element is of course a normal subgroup of K, so we can write

$$S_4 \supseteq K \supseteq \{(\,)\}$$

which would be a subnormal series of length $n = 2$. Is there a way that we can make a longer series out of this one? Because A_4 is also a normal subgroup of S_4, and K is a normal subgroup of A_4, we can slip this group into our series. Also, the group K contains the subgroup

$$H = \{(\,), (1\,2)(3\,4)\},$$

which is a normal subgroup of K since K is abelian. Therefore, we have a longer subnormal series of length 4:

$$S_4 \supseteq A_4 \supseteq K \supseteq H \supseteq \{(\,)\}.$$

We say that this new subnormal series is a *refinement* of the first subnormal series. □

DEFINITION 8.2 We say that a subnormal (or normal) series

$$G = H_0 \supseteq H_1 \supseteq H_2 \supseteq \cdots \supseteq H_k = \{e\}$$

is a *refinement* of the subnormal (or normal) series

$$G = G_0 \supseteq G_1 \supseteq G_2 \supseteq \cdots \supseteq G_n = \{e\}$$

if each subgroup G_i appears as H_j for some j.

Is there a way that we can refine our subnormal series to produce an even longer chain? Our definition did not exclude the possibility of two groups in the series being the same, so we could consider

$$S_4 \supseteq A_4 \supseteq A_4 \supseteq K \supseteq H \supseteq H \supseteq H \supseteq \{(\,)\}.$$

Although this is a longer subnormal series, it is usually pointless to repeat the same subgroup in the series.

DEFINITION 8.3 A *composition series* of a group G is a subnormal series

$$G = G_0 \supset G_1 \supset G_2 \supset \cdots \supset G_n = \{e\}$$

for which each subgroup is smaller than the proceeding subgroup, and for which there is no refinement that includes additional subgroups.

Using this definition, we see that

$$S_4 \supseteq A_4 \supseteq K \supseteq H \supseteq \{()\}$$

is a composition series. Clearly no subgroups are repeated, and there simply is not enough room between two of these subgroups to slip in another subgroup. For example, A_4 is half of S_4, so any subgroup containing more than A_4 must be all of S_4. In fact, we can easily test to see whether a subnormal series is a composition series.

PROPOSITION 8.1

The subnormal series

$$G = G_0 \supseteq G_1 \supseteq G_2 \supseteq \cdots \supseteq G_n = \{e\}$$

is a composition series if, and only if, all of the quotient groups G_{k-1}/G_k are nontrivial simple groups.

PROOF: Note that if there are no repeated subgroups in the subnormal series, then G_{i-1}/G_i must contain at least two elements. Likewise, if G_{i-1}/G_i is nontrivial, then G_{i-1} is not equal to G_i. So the quotient groups are nontrivial if, and only if, there are no repeated subgroups in the subnormal series.

Suppose that the subnormal series is not a composition series yet does not repeat any subgroups. Then there must be an additional group H that we can add between G_{k-1} and G_k, so that

$$G_{k-1} \supseteq H \supseteq G_k,$$

where H is a normal subgroup of G_{k-1} and G_k is a normal subgroup of H. Then by Lemma 4.6, H/G_k will be a normal subgroup of G_{k-1}/G_k, and since H is neither G_{k-1} nor G_k, we have a proper normal subgroup of G_{k-1}/G_k.

Now suppose that there is a proper normal subgroup N of G_{k-1}/G_k. Can we then lift N to find a suitable subgroup H to fit between G_{k-1} and G_k? If we consider the canonical homomorphism ϕ from G_{k-1} to the quotient group G_{k-1}/G_k we can take $H = \phi^{-1}(N)$. Then since N is a normal subgroup of G_{k-1}/G_k, by Corollary 4.2, H will be a normal subgroup of G_{k-1}. Also, G_k will be a normal subgroup of H, for H is in G_{k-1}. Because N has at least two elements, H will be strictly larger than the kernel of ϕ, yet since N is not the entire image of ϕ, H will be strictly smaller than G_k. Therefore, the subnormal series is not a composition series.

Thus, a subnormal series is a composition series if, and only if, the quotient groups G_{k-1}/G_k are nontrivial simple groups. □

The quotient groups G_{k-1}/G_k in a composition series for G are called the *composition factors* of the composition series.

For example, the composition factors for the composition series

$$S_4 \supseteq A_4 \supseteq K \supseteq H \supseteq \{()\}$$

are

$$S_4/A_4 \approx Z_2, \qquad A_4/K \approx Z_3, \qquad K/H \approx Z_2, \quad \text{and} \quad H/\{()\} \approx Z_2.$$

It is certainly possible for a group to have more than one composition series. For example, we could have picked the subgroup $B = \{(), (1,4)(2,3)\}$ instead of H, producing the composition series

$$S_4 \supseteq A_4 \supseteq K \supseteq B \supseteq \{()\}.$$

Even though this is a different composition series, the composition factors are isomorphically the same. Our goal for this section is to prove that this happens all of the time. However, we have yet to see why two composition series must have the same length. Even if we can prove that the composition series are the same length, the composition factors may not appear in the same order. For example, the group Z_{12} has the following two subnormal series:

$$Z_{12} \supseteq \{0, 3, 6, 9\} \supseteq \{0\}.$$

$$Z_{12} \supseteq \{0, 2, 4, 6, 8, 10\} \supseteq \{0, 4, 8\} \supseteq \{0\}.$$

No matter how we refine these series, the quotient group isomorphic to Z_3 in the first series will come before any other non-trivial quotient groups, yet any refinement of the second series will have the last non-trivial quotient group isomorphic to Z_3.

It helps if we use a diagram to demonstrate the strategy that we will be using. Suppose that we have a group G with two subnormal series, one of length 2, and one of length 3, as pictured in Figure 8.1.

$$G = A_0 \supseteq A_1 \supseteq A_2 = \{e\}, \qquad G = B_0 \supseteq B_1 \supseteq B_2 \supseteq B_3 = \{e\}.$$

It is immediately clear that $A_0 = B_0$ and $A_2 = B_3$, but A_1 does not have to be either B_1 or B_2.

The goal is to refine both of the subnormal series by adding two subgroups within each gap of the A series, and one subgroup within each gap in the B series. Here, we will allow the possibility of duplicate subgroups in the refinements. Nonetheless, both series will have length 6, which we can express as follows:

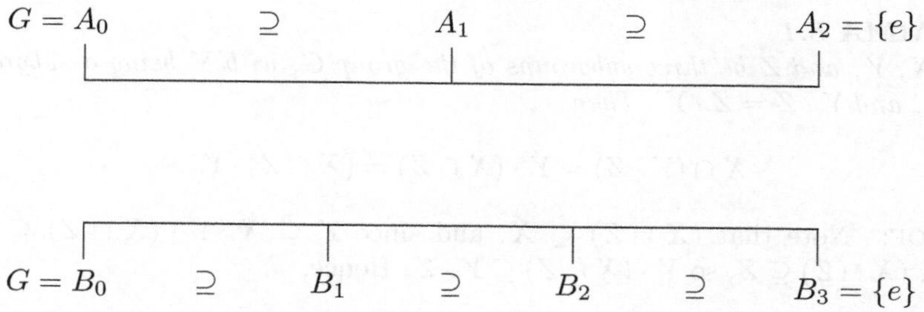

$G = A_0 \quad \supseteq \quad A_1 \quad \supseteq \quad A_2 = \{e\}$

$G = B_0 \quad \supseteq \quad B_1 \quad \supseteq \quad B_2 \quad \supseteq \quad B_3 = \{e\}$

FIGURE 8.1: Example of two subnormal series of different lengths

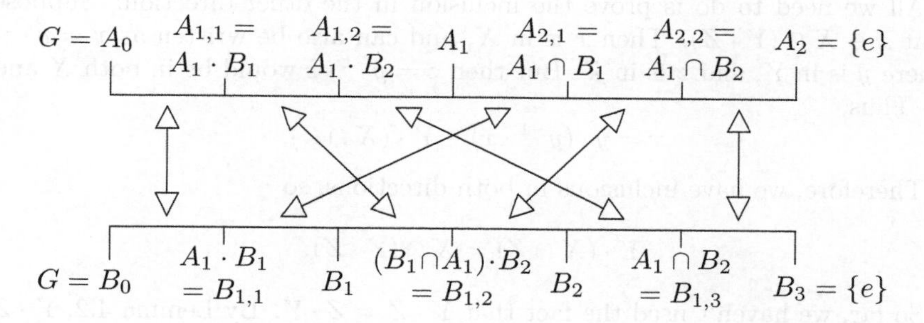

FIGURE 8.2: Diagram showing the strategy of the refinement theorem

$$G = A_0 \supseteq A_{1,1} \supseteq A_{1,2} \supseteq A_1 \supseteq A_{2,1} \supseteq A_{2,2} \supseteq A_0 = \{e\},$$
$$G = B_0 \supseteq B_{1,1} \supseteq B_1 \supseteq B_{1,2} \supseteq B_2 \supseteq B_{1,3} \supseteq B_0 = \{e\}.$$

Figure 8.2 shows these set inclusions, and also gives a hint on how we are to define these intermediate subgroups.

The next step will be to show that the quotient groups for each interval of the A series is isomorphic to a quotient group for an interval of the B series, as shown by the arrows in Figure 8.2. Note that this scrambles the order of the quotient groups, so that the ith subinterval of the jth interval in the A series corresponds to the jth subinterval of the ith interval of the B series.

Although it is clear that

$$G \supseteq A_1 \cdot B_1 \supseteq A_1 \cdot B_2 \supseteq A_1 \supseteq A_1 \cap B_1 \supseteq A_1 \cap B_2 \supseteq \{e\}, \qquad \text{and}$$
$$G \supseteq A_1 \cdot B_1 \supseteq B_1 \supseteq (B_1 \cap A_1) \cdot B_2 \supseteq B_2 \supseteq A_1 \cap B_2 \supseteq \{e\},$$

it is not at all clear that each is a normal subgroup of the previous group, or even that all of these sets are subgroups of G. Before we show this, we will need the following lemma.

LEMMA 8.1
Let X, Y, and Z be three subgroups of the group G, with Y being a subgroup of X, and $Y \cdot Z = Z \cdot Y$. Then

$$X \cap (Y \cdot Z) = Y \cdot (X \cap Z) = (X \cap Z) \cdot Y.$$

PROOF: Note that $(X \cap Z) \subseteq X$, and since $Y \subseteq X$, $Y \cdot (X \cap Z) \subseteq X$. Also, $(X \cap Z) \subseteq Z$, so $Y \cdot (X \cap Z) \subseteq Y \cdot Z$. Hence,

$$Y \cdot (X \cap Z) \subseteq X \cap (Y \cdot Z).$$

All we need to do is prove the inclusion in the other direction. Suppose that $x \in X \cap (Y \cdot Z)$. Then x is in X, and can also be written as $x = y \cdot z$, where y is in Y, and z is in Z. But then $z = y^{-1} \cdot x$ would be in both X and Z. Thus,

$$x = y \cdot (y^{-1} \cdot x) \in Y \cdot (X \cap Z).$$

Therefore, we have inclusions in both directions, so

$$Y \cdot (X \cap Z) = X \cap (Y \cdot Z).$$

So far, we haven't used the fact that $Y \cdot Z = Z \cdot Y$. By Lemma 4.2, $Y \cdot Z$ is a subgroup of G, and so the intersection of X with $Y \cdot Z$ is a subgroup of G. So by Lemma 4.2 again, we have

$$Y \cdot (X \cap Z) = (X \cap Z) \cdot Y.$$

□

We will need one more lemma that will help us to show the isomorphisms indicated by the arrows in Figure 8.2.

LEMMA 8.2
Let X, Y, and Z be three subgroups of the group G, with Y being a normal subgroup of X, and Z a normal subgroup of G. Then $Y \cdot Z$ is a normal subgroup of $X \cdot Z$, and

$$(X \cdot Z)/(Y \cdot Z) \approx X/(X \cap (Y \cdot Z)).$$

PROOF: Since Z is a normal subgroup of G, both $Y \cdot Z$ and $X \cdot Z$ are subgroups of G by Lemma 4.3. If we let $y \cdot z$ be in $Y \cdot Z$, and $x \cdot w$ be in $X \cdot Z$, then

$$\begin{aligned}(x \cdot w) \cdot (y \cdot z) \cdot (x \cdot w)^{-1} &= x \cdot (y \cdot x^{-1} \cdot x \cdot y^{-1}) \cdot w \cdot y \cdot z \cdot w^{-1} \cdot x^{-1} \\ &= (x \cdot y \cdot x^{-1}) \cdot (x \cdot (y^{-1} \cdot w \cdot y) \cdot z \cdot w^{-1} \cdot x^{-1}).\end{aligned}$$

Now, $x \cdot y \cdot x^{-1}$ is in Y, since Y is a normal subgroup of X. Likewise, $y^{-1} \cdot w \cdot y$ is in Z, since y is in G. Then $(y^{-1} \cdot w \cdot y) \cdot z \cdot w^{-1}$ is in Z, and so $x \cdot (y^{-1} \cdot w \cdot$

$y) \cdot z \cdot w^{-1} \cdot x^{-1}$ is in Z, since x is in G. Therefore, $(x \cdot w) \cdot (y \cdot z) \cdot (x \cdot w)^{-1}$ is in $Y \cdot Z$, and so $Y \cdot Z$ is a normal subgroup of $X \cdot Z$.

We now can use the second isomorphism theorem (4.2), using $K = Y \cdot Z$. We have that $X \cdot K = X \cdot Y \cdot Z = X \cdot Z$ since Y is a subgroup of X. So

$$(X \cdot Z)/(Y \cdot Z) = (X \cdot K)/K \approx X/(X \cap K) = X/(X \cap (Y \cdot Z)). \qquad \square$$

We are now ready to put the pieces together, and show any two subnormal series can be refined in such a way that the quotient groups are isomorphic.

THEOREM 8.1: The Refinement Theorem

Suppose that there are two subnormal series for a group G. That is, there are subgroups A_i and B_j such that

$$G = A_0 \supseteq A_1 \supseteq A_2 \supseteq \cdots \supseteq A_n = \{e\},$$

and

$$G = B_0 \supseteq B_1 \supseteq B_2 \supseteq \cdots \supseteq B_m = \{e\},$$

where each A_i is a normal subgroup of A_{i-1}, and each B_j is a normal subgroup of B_{j-1}. Then it is possible to refine both series by inserting the subgroups

$$A_{i-1} = A_{i,0} \supseteq A_{i,1} \supseteq A_{i,2} \supseteq \cdots \supseteq A_{i,m} = A_i, \qquad i = 1, 2, \ldots n,$$

$$B_{j-1} = B_{j,0} \supseteq B_{j,1} \supseteq B_{j,2} \supseteq \cdots \supseteq B_{j,n} = B_j, \qquad j = 1, 2, \ldots m$$

in such a way that

$$A_{i,j-1}/A_{i,j} \approx B_{j,i-1}/B_{j,i}.$$

PROOF: We let

$$A_{i,j} = (A_{i-1} \cap B_j) \cdot A_i \quad \text{and} \quad B_{j,i} = (B_{j-1} \cap A_i) \cdot B_j.$$

To see that these fit the conditions we need, we first want to show that these are groups. Note that both

$$X = (A_{i-1} \cap B_{j-1}) \quad \text{and} \quad Y = (A_{i-1} \cap B_j)$$

are subgroups of A_{i-1}, Y is a subgroup of X, and $Z = A_i$ is a normal subgroup of A_{i-1}.

So by Lemma 4.3, both $A_{i,j-1} = X \cdot Z$ and $A_{i,j} = Y \cdot Z$ are subgroups of A_{i-1}. We can now use Lemma 8.2, using $G = A_{i-1}$. Since B_j is a normal subgroup of B_{j-1}, Y is a normal subgroup of X, so by Lemma 8.2, $Y \cdot Z$ is a normal subgroup of $X \cdot Z$, and

$$A_{i,j-1}/A_{i,j} = (X \cdot Z)/(Y \cdot Z) \approx X/(X \cap (Y \cdot Z)).$$

Now Lemma 8.1 comes into use. Since Y is a subgroup of X,

$$\begin{aligned} X \cap (Y \cdot Z) = Y \cdot (X \cap Z) &= (A_{i-1} \cap B_j) \cdot (A_{i-1} \cap B_{j-1} \cap A_i) \\ &= (A_{i-1} \cap B_j) \cdot (A_i \cap B_{j-1}) \\ &= (A_i \cap B_{j-1}) \cdot (A_{i-1} \cap B_j). \end{aligned}$$

Thus,

$$A_{i,j-1}/A_{i,j} \approx (A_{i-1} \cap B_{j-1})/[(A_{i-1} \cap B_j) \cdot (A_i \cap B_{j-1})].$$

By switching the roles of the two series we find by the exact same argument that

$$B_{j,i-1}/B_{j,i} \approx (B_{j-1} \cap A_{i-1})/[B_{j-1} \cap A_i) \cdot (B_j \cap A_{i-1})].$$

Notice that these are exactly the same thing, so

$$A_{i,j-1}/A_{i,j} \approx B_{j,i-1}/B_{j,i}.$$

∎

If we now apply the refinement theorem to two composition series we find that the composition factors will be the same.

THEOREM 8.2: The Jordan-Hölder Theorem
Let G be a finite group, and let

$$G = A_0 \supset A_1 \supset A_2 \supset \cdots \supset A_n = \{e\}$$

and

$$G = B_0 \supset B_1 \supset B_2 \supset \cdots \supset B_m = \{e\}$$

be two composition series for G. Then $n = m$, and the composition factors A_{i-1}/A_i are isomorphic to the composition factors B_{j-1}/B_j in some order.

PROOF: By the refinement theorem (8.1), there is a refinement of both composition series such that the quotient groups of the two subnormal series are isomorphic to each other in some order. In particular, the nontrivial quotient groups of one subnormal series are isomorphic to the nontrivial quotient groups of the other. But these are composition series, so any refinements merely repeat a subgroup a number of times. Thus, by eliminating these repetitions, we eliminate the trivial quotient groups and produce the original two composition series. Thus, the quotient groups A_{i-1}/A_i are isomorphic to the quotient groups B_{j-1}/B_j in some order. The fact that $n = m$ merely comes from the one-to-one correspondence of the nontrivial quotient groups. ∎

The Jordan-Hölder theorem (8.2) shows that the composition factors do not depend on the composition series, but rather the finite group G. This is reminiscent of the unique factorization of integers, where every integer greater than one can be written as a unique product of prime numbers. Since the composition factors are always nontrivial simple groups, in a sense the simple groups play the same role in group theory that prime numbers play in number theory. The correspondence is heightened by the fact that Z_p is a nontrivial simple group if, and only if, p is a prime number. However, we have seen that there are other simple groups, such as $\mathrm{Aut}(Z_{24}^*)$ and A_n for $n > 4$. Since these groups are rather large (at least 60 elements), they will only show up as composition factors for very large groups.

For example, a composition series for S_5 is given by

$$S_5 \supset A_5 \supset \{()\}, \qquad S_5/A_5 \approx Z_2, \quad \text{and} \quad A_5/\{()\} \approx A_5.$$

Since Z_2 and A_5 are both simple groups, this is a composition series, and so the composition factors of S_5 are Z_2 and A_5.

The composition series will play a vital role in determining whether groups are solvable or not. However, we will hold off on the definition of a solvable group until we have defined another tool in group theory, the derived group.

Problems for §8.1

1 Let

$$G = Z_{12} \supseteq A_1 = \{0, 3, 6, 9\} \supseteq \{0\}$$

and

$$G = Z_{12} \supseteq B_1 = \{0, 2, 4, 6, 8, 10\} \supseteq B_2 = \{0, 4, 8\} \supseteq \{0\}$$

be two subnormal series for Z_{12}. Find all of the subgroups shown in Figure 8.2, and show that the quotient groups indicated by the arrows are indeed isomorphic.

For Problems **2** through **10**: Write out a composition series for the group.

2 Z_{15}^*

3 Z_{24}^*

4 Z_{21}^*

5 $Z_{12} \times Z_{18}$

6 The quaternion group Q

7 D_4

8 D_5

9 D_6

10 S_6

11 Show that there are exactly three possible composition series for A_4.

12 Find an example of two non-isomorphic groups for which the composition factors are isomorphic.

13 Find two groups of the same order with composition series of different lengths.

14 Find a non-simple group for which all of the composition factors are non-cyclic.

15 Find a simple group for which all of the composition factors are cyclic.

16 Find a non-abelian group with cyclic composition factors for which there is only one composition series.

17 Prove that if the refinement theorem (8.1) is applied to two *normal* series, the resulting series will be normal. That is, if A_u and B_v are such that

$$G = A_0 \supseteq A_1 \supseteq A_2 \supseteq \cdots \supseteq A_n = \{e\},$$

and

$$G = B_0 \supseteq B_1 \supseteq B_2 \supseteq \cdots \supseteq B_m = \{e\},$$

where each A_i and B_j is a normal subgroup of G (not just the previous group), then the $A_{i,j}$ and $B_{j,i}$ given by the refinement theorem will all be normal subgroups of G.

Hint: Use the result of Problem 17 from §4.3.

18 A *chief* series is a normal series for which no refinements produce normal series. Show that the Jordan-Hölder theorem (8.2) applies to chief series as well as to composition series. That is, show that if

$$G = A_0 \supseteq A_1 \supseteq A_2 \supseteq \cdots \supseteq A_n = \{e\}$$

and

$$G = B_0 \supseteq B_1 \supseteq B_2 \supseteq \cdots \supseteq B_m = \{e\}$$

are two chief series, then $n = m$, and the quotient groups of the first series are isomorphic to the quotient groups of the second in some order. (Use the result from Problem 17.)

Interactive Problems

19 Use *Sage* to find a composition series for the following group of order 20:

```
InitGroup("e")
AddGroupVar("a", "b")
Define(a^5, e); Define(b^4, e); Define(b*a, a^2*b)
M = ListGroup()
```

20 Use *Sage* to find a composition series for the following group:

```
DisplayPermInt = true
G = Group(NthPerm(2374), NthPerm(6212)); G
    {1, 2374, 4517, 6212, 6841, 9929, 11637, 13016, 13698, 15367,
    18454, 19853, 21239, 21896, 24132, 25315, 28226, 28986,
    30928, 31590, 33108, 37381, 38807, 39487}
```

8.2 Derived Group Series

Although we could define a finite solvable group in terms of the composition factors, we will instead use a definition that will apply for both finite and infinite groups. This definition of solubility hinges on the concept of the derived group. Hence, we must first define the derived group, and then we can formulate a general definition of a solvable group.

In the process, we will find a method for producing a composition series that is easily implemented using *Sage*. This method hinges on the concept of a "commutator."

DEFINITION 8.4 Given two elements x and y of a group G, the *commutator* of x and y is the element $x^{-1} \cdot y^{-1} \cdot x \cdot y$, and is written $[x, y]$.

Notice that if G is an abelian group the commutator will always give the identity element. We can also consider the commutator of two subgroups of G. If H and K are two subgroups, then consider the set

$$\{x^{-1} \cdot y^{-1} \cdot x \cdot y \quad | \quad x \in H \quad \text{and} \quad y \in K\}.$$

Unfortunately, this set will not always form a group.

Example 8.2
Consider the two subgroups of S_4:

$$H = \{(\,), (1\,2)\}, \qquad K = \{(\,), (2\,3\,4), (2\,4\,3)\}.$$

Then the set

$$\{x^{-1} \cdot y^{-1} \cdot x \cdot y \quad | \quad x \in H \quad \text{and} \quad y \in K\}$$

can be found by making a table for possible values of x and y.

$x^{-1} \cdot y^{-1} \cdot x \cdot y$	$(\,)$	$(2\,3\,4)$	$(2\,4\,3)$
$(\,)$	$(\,)$	$(\,)$	$(\,)$
$(1\,2)$	$(\,)$	$(1\,4\,2)$	$(1\,3\,2)$

So we get $\{(\,), (1\,3\,2), (1\,4\,2)\}$, which is not a subgroup. However, we can consider the subgroup *generated* by all of the commutators, which of course will make a subgroup. ∎

DEFINITION 8.5 Given two subgroups H and K of a group G, we define the *mutual commutator subgroup of H and K*, denoted $[H, K]$, to be the subgroup generated by the elements

$$\{x^{-1} \cdot y^{-1} \cdot x \cdot y \quad | \quad x \in H \quad \text{and} \quad y \in K\}.$$

We can find the mutual commutator with the *Sage* command **MutualCommutator**.

```
H = Group( C(1,2) ); H
    {(), (1, 2)}
K = Group( C(2,3,4) ); K
    {(), (2, 4, 3), (2, 3, 4)}
MutualCommutator(H, K)
    {(), (1, 3, 2), (1, 2, 3), (1, 2)(3, 4), (2, 4, 3), (1, 4, 3),
    (2, 3, 4), (1, 4, 2), (1, 3)(2, 4), (1, 3, 4), (1, 2, 4),
    (1, 4)(2, 3)}
```

So the commutator $[H, K]$ in this case is A_4. Note that whenever an element u is in $[H, K]$, we cannot say that $u = x^{-1} \cdot y^{-1} \cdot x \cdot y$ for some $x \in H$ and $y \in K$. Rather, we must write

$$u = u_1 \cdot u_2 \cdot \cdots \cdot u_n,$$

where either u_i or u_i^{-1} is $x_i^{-1} \cdot y_1^{-1} \cdot x_i \cdot y_i$. In spite of this difficulty, we will be able to discover some important properties with the mutual commutator groups.

PROPOSITION 8.2

If H and K are normal subgroups of G, then $[H, K]$ is a normal subgroup of G.

PROOF: Let u be an element of $[H, K]$, and g an element of G. Then $u = u_1 \cdot u_2 \cdots u_n$, where either u_i or u_i^{-1} is $x_i^{-1} \cdot y_i^{-1} \cdot x_i \cdot y_i$. Then

$$g \cdot u \cdot g^{-1} = (g \cdot u_1 \cdot g^{-1}) \cdot (g \cdot u_2 \cdot g^{-1}) \cdots (g \cdot u_n \cdot g^{-1}),$$

and

$$\begin{aligned} g \cdot x_i^{-1} \cdot y_i^{-1} \cdot x_i \cdot y_i \cdot g^{-1} &= \\ (g \cdot x_i^{-1} \cdot g^{-1}) \cdot (g \cdot y_i^{-1} \cdot g^{-1}) \cdot (g \cdot x_i \cdot g^{-1}) \cdot (g \cdot y_i \cdot g^{-1}) & \\ = \left[g \cdot x_i \cdot g^{-1},\ g \cdot y_i \cdot g^{-1}\right]. & \end{aligned}$$

If H and K are both normal subgroups of G, then $g \cdot x_i \cdot g^{-1}$ is in H, and $g \cdot y_i \cdot g^{-1}$ is in K. Thus, $[g \cdot x_i \cdot g^{-1},\ g \cdot y_i \cdot g^{-1}]$ is in $[H, K]$. Since $(g \cdot u_i \cdot g^{-1})^{-1} = (g \cdot u_i^{-1} \cdot g^{-1})$, if one of these is in $[H, K]$, they both are. Hence $g \cdot u_i \cdot g^{-1}$ is in $[H, K]$ for every u_i, and $g \cdot u \cdot g^{-1} \in [H, K]$. By proposition 3.4, $[H, K]$ is a normal subgroup of G. ∎

Many times one of the two groups H or K will be the whole group G. We call the subgroup $[G, H]$ the *commutator subgroup of H in G*. In this case *Sage* can find the commutator subgroup faster with the simplified command **Commutator(G, H)**, which takes advantage of the fact that H is a subgroup of G. In fact, *Sage* will correctly find the commutator subgroup if only the *generators* of H are specified. For example, suppose we wish to find the commutator $[S_4, A_4]$. By using only the generators of A_4, we get

```
S4 = Group( C(1,2), C(1,2,3,4) )
A4 = Group( C(1,2,3), C(2,3,4) )
Commutator(S4, [C(1,2,3), C(2,3,4)])
    {(), (1, 3, 2), (1, 2, 3), (1, 2)(3, 4), (2, 4, 3), (1, 4, 3),
    (2, 3, 4), (1, 4, 2), (1, 3)(2, 4), (1, 3, 4), (1, 2, 4),
    (1, 4)(2, 3)}
```

which gives us A_4 again. The commutator $[S_4, S_4]$ is given by

```
Commutator(S4, [C(1,2), C(1,2,3,4)])
    {(), (1, 3, 2), (1, 2, 3), (1, 2)(3, 4), (2, 4, 3), (1, 4, 3),
    (2, 3, 4), (1, 4, 2), (1, 3)(2, 4), (1, 3, 4), (1, 2, 4),
    (1, 4)(2, 3)}
```

which is also A_4. However, the commutator $[A_4, A_4]$ is

```
K = Commutator(A4, [C(1,2,3), C(2,3,4)]); K
    {(), (1, 2)(3, 4), (1, 3)(2, 4), (1, 4)(2, 3)}
```

which gives a subgroup with only four elements. This is exactly the subgroup K from the last section.

DEFINITION 8.6 We define the commutator subgroup of G with itself, $[G, G]$, to be the *derived group* of G, denoted G'.

Since G is a normal subgroup of itself, Proposition 8.2 states that the derived group will be a normal subgroup of G. Since the commutator of any two elements in an abelian group is e, $[G, G]$ will be the trivial group whenever G is abelian.

We can denote the derived group of the derived group G' as G''. Likewise, the derived group of G'' will be denoted G''', and so on. Because each of these groups is a normal subgroup of the previous one, we have the series

$$G \supseteq G' \supseteq G'' \supseteq G''' \supseteq \cdots.$$

This is called the *derived series* for the group G. The derived series is in fact a subnormal series as long as the groups keep getting smaller and smaller until they finally get to the trivial subgroup.

From our experiments in *Sage*, we can find the derived series of the group S_4. Since $[S_4, S_4] = A_4$ and $[A_4, A_4] = K$, we have that $G' = A_4$, $G'' = K$, and $G''' = \{()\}$, since K is abelian. So we produce the series

$$S_4 \supseteq A_4 \supseteq K \supseteq \{()\}.$$

However, if we start with the group A_5, then $[A_5, A_5]$ must be a normal subgroup of the simple group A_5. Since the derived group is not the identity element, we see that the derived group must be all of A_5.

Thus, the derived series for A_5 is

$$A_5 \supseteq A_5 \supseteq A_5 \supseteq A_5 \supseteq \cdots$$

which never gets to the trivial subgroup.

DEFINITION 8.7 A group G is called *solvable* if the derived series

$$G \supseteq G' \supseteq G'' \supseteq G''' \supseteq \cdots$$

includes the trivial group in a finite number of steps. If the derived series never reaches the trivial group, G is said to be *insoluble*.

By our experiments, we see that S_4 is a solvable group, whereas A_5 is not. Whenever we have a solvable group G, the derived series is in fact a subnormal series for G. So it is natural that the derived series would shed some light as to what the composition factors of G are. First we will need the following lemma, which characterizes the derived group.

LEMMA 8.3

Let G be a group. Then the derived group G' is the smallest normal subgroup for which the quotient group is abelian.

PROOF: First we need to show that G/G' is abelian. Consider the canonical homomorphism ϕ from G onto G/G'. Then for x and y in G, $x^{-1} \cdot y^{-1} \cdot x \cdot y$ is in G', and so $\phi(x^{-1} \cdot y^{-1} \cdot x \cdot y)$ is the identity element in G/G'. But then

$$\phi(x^{-1} \cdot y^{-1} \cdot x \cdot y) = \phi(x)^{-1} \cdot \phi(y)^{-1} \cdot \phi(x) \cdot \phi(y) = e,$$

so $\phi(x) \cdot \phi(y) = \phi(y) \cdot \phi(x)$. Since ϕ is surjective, we see that G/G' is abelian.

Now suppose that N is another normal subgroup of G for which G/N is abelian. To show that G' is a smaller group, we will show that N contains G'.

For any x and y in G, note that $x^{-1} \cdot y^{-1} \cdot x \cdot y$ is certainly contained in $x^{-1} \cdot N \cdot y^{-1} \cdot N \cdot x \cdot N \cdot y \cdot N$. But since the quotient group G/N is abelian, we have

$$x^{-1} \cdot N \cdot y^{-1} \cdot N \cdot x \cdot N \cdot y \cdot N = x^{-1} \cdot N \cdot x \cdot N \cdot y^{-1} \cdot N \cdot y \cdot N = N \cdot N = N.$$

Thus, $x^{-1} \cdot y^{-1} \cdot x \cdot y$ is in N for all x and y in G. Since G' is generated by all such elements, G' is contained in N. ∎

We now can express a relationship between the composition factors of a group and the derived series of a group.

THEOREM 8.3: The Solvability Theorem
Let G be a finite group. Then G is solvable if, and only if, the composition factors of G are cyclic groups of prime order.

PROOF: Suppose that the composition factors of G are all cyclic groups of prime order. Then there exists a composition series for G:

$$G = G_0 \supset G_1 \supset G_2 \supset \cdots \supset G_n = \{e\}.$$

Since G_0/G_1 is an abelian group, we have from Lemma 8.3 that G' is contained in G_1. But since G_1/G_2 is also abelian, by Lemma 8.2 we have G_1' is in G_2, and so

$$G'' \subseteq G_1' \subseteq G_2.$$

Proceeding in this way we find that the nth derived group, $G^{(n)}$, must be contained in $G_n = \{e\}$. Thus, the derived series produced the trivial group in at most n steps, so G is solvable.

Now suppose that G is solvable and finite, and so the derived series can be written

$$G \supseteq G' \supseteq G'' \supseteq G''' \supseteq \cdots \supseteq G^{(n)} = \{e\}.$$

If $G^{(n)}$ is the first term in the derived series equal to $\{e\}$, then this subnormal series can never repeat any two subgroups. Because this is a finite group, there are only a finite number of ways this series could be refined without repeating subgroups. Thus, by the refinement theorem, we can refine this to produce a composition series. Because each of the quotient groups of the derived series is abelian, the quotient groups of the refinement must also be abelian. But by Proposition 8.1, the quotient groups of the composition series must be nontrivial simple groups. The only nontrivial simple groups that are abelian are the cyclic groups of prime order. Thus, the quotient groups for this composition series are cyclic groups of prime order. By the Jordan-Hölder theorem (8.2), all composition series have the same composition factors, so all composition series have quotient groups that are cyclic of prime order. ∎

Historically, solvability was defined in terms of the composition factors. But this definition could only be used on finite groups, since infinite groups do not have a composition series. By defining solvability in terms of the derived series, we allow for infinite groups to be solvable.

For example, the group of integers $\mathbb{Z}$ is abelian, so the derived group is the trivial group. Hence, $\mathbb{Z}$ is solvable, yet there are no composition series for $\mathbb{Z}$. This is because every proper subgroup of $\mathbb{Z}$ is isomorphic to $\mathbb{Z}$.

From the solvability theorem we see that for finite groups, solvability can be defined in terms of the composition factors. Does this hold true for infinite groups as well? That is, is an infinite group solvable as long as there is no non-abelian simple group (finite or infinite) lurking somewhere within its structure, either as a subgroup or as a quotient group? To shed some light on this problem, we will first need the following lemma.

LEMMA 8.4

If N is a normal subgroup of G, and H is a subgroup of G, then

$$(H \cdot N/N)' = (H' \cdot N)/N.$$

PROOF: We first note that since N is a normal subgroup of G, $H \cdot N$ is a subgroup of G, and so N is a normal subgroup of $H \cdot N$. Two typical elements of $H \cdot N/N$ are $h \cdot n \cdot N$ and $k \cdot m \cdot N$, where h and k are in H, and n and m are in N. Then $(H \cdot N/N)'$ is generated from the elements of the form

$$(h \cdot n \cdot N)^{-1} \cdot (k \cdot m \cdot N)^{-1} \cdot (h \cdot n \cdot N) \cdot (k \cdot m \cdot N) = h^{-1} \cdot k^{-1} \cdot h \cdot k \cdot N.$$

But these elements are also in $(H' \cdot N)/N$. In fact, $(H' \cdot N)/N$ is generated by the elements of the form $h^{-1} \cdot k^{-1} \cdot h \cdot k \cdot N$. Therefore, the groups $(H \cdot N/N)'$ and $(H' \cdot N)/N$ are equal. □

With this lemma we will be able to show the relationship with a solvable group to its subgroups and quotient groups.

PROPOSITION 8.3

Suppose that G is a group and H is a normal subgroup of G. Then G is solvable if, and only if, both H and G/H are solvable.

PROOF: We begin by showing that if G is solvable, and H is a subgroup of G, normal or not, then H is solvable. Since H is contained in G, we have

$$H' \subseteq G' \Longrightarrow H'' \subseteq G'' \Longrightarrow H''' \subseteq G''' \cdots.$$

Thus, since $G^{(n)} = \{e\}$ for some n, $H^{(n)} = \{e\}$, and H is solvable.

Next we want to show that if H is normal, then G/H is solvable. Since $G = G \cdot H$ we can use Lemma 8.4 to find $(G/H)' = (G' \cdot H)/H$. But since G' is a subgroup, we can continue to use Lemma 8.4 to find

$$(G/H)'' = (G' \cdot H/H)' = (G'' \cdot H)/H,$$
$$(G/H)''' = (G'' \cdot H/H)' = (G''' \cdot H)/H, \cdots.$$

Since G is a solvable group, $G^{(n)} = \{e\}$ for some n. Thus

$$(G/H)^{(n)} = (G^{(n)} \cdot H)/H$$

would be the identity group H/H. Therefore, G/H is a solvable group.

Now suppose that both H and G/H are solvable. Then $(G/H)^n$ is the identity for some n, so $(G^{(n)} \cdot H)/H$ is the identity. Thus, $G^{(n)}$ is a subgroup of H, and since H is solvable, $G^{(n)}$ must be solvable. Therefore, $G^{(n+m)}$ is the identity for some m, and so G is a solvable group. ☐

From this proposition, we see that for an infinite solvable group there cannot be any non-abelian simple groups within its structure whether as a subgroup, a quotient group, a subgroup of a quotient group, etc. Thus the current definition of solvability for infinite groups agrees with the historical notion of a group that does not contain non-abelian simple groups in the composition factors.

Why do we want to know whether a group is solvable or not? Notice that most of the *solvable* groups could be entered into *Sage* using the **InitGroup** and **Define** commands, whereas the *insoluble* groups, such as $\mathrm{Aut}(Z_{24}^*)$, had to be considered as a subgroup of a symmetric group. In the next section, we will show why the solvable groups were the only groups that could be entered into *Sage* using the **Define** commands.

Problems for §8.2

1 Show that any group of order p^n, where p is prime, is solvable.
Hint: See Corollary 7.2.

2 Show that S_n is solvable for $n < 5$, but is insoluble for $n > 4$.

3 Show that $[z \cdot x \cdot z^{-1}, z \cdot y \cdot z^{-1}] = z \cdot [x, y] \cdot z^{-1}$.

4 Let G be the group from Example 1.8 in §1.3, the group of *linear functions* of the form $f(x) = mx + b$, with $m, b \in \mathbb{R}$, $m \neq 0$. By finding the derived group G', show that this group is solvable.

5 Show that if G is a non-cyclic simple group, then $G' = G$. Is it true that if $G' = G$, then G must be simple?

6 Let H and K be two subgroups of G. Prove that the mutual commutator $[H, K]$ is a normal subgroup of the group generated by the elements of H and K.

For Problems **7** through **10**, find the derived series of the group.

7 S_3 **8** D_4 **9** D_5 **10** Q

11 Find the derived series for the group $G = Z_3 \rtimes Z_4$. See Problem 9 from §6.4.

12 Show that the group G' is a characteristic subgroup of G. See Problem 12 of §7.1 for the definition of a characteristic subgroup.

13 Show that if a group G is solvable, then the derived series is in fact a *normal* series for G.

Hint: Use Problem 12 and the key property of characteristic subgroups found in Problem 16 of §7.1.

14 If G is a group, define the sequence $G_1 = [G, G]$, $G_2 = [G, G_1]$, $G_3 = [G, G_2], \ldots$. G is said to be *nilpotent* if $|G_n| = 1$ for some n. Prove that if G is nilpotent, then G is solvable.

Hint: Prove that G_n contains the nth derived group of G.

15 Find a solvable group that is not nilpotent. (See Problem 14.)

16 Show that a group of order p^n, where p is prime, is nilpotent. (See Problem 14 and Corollary 7.2.)

17 A group is called *supersolvable* if there is a chief series with cyclic factors. Show that if G is supersolvable, then G' is nilpotent. (See Problem 18 from §8.1 and Problem 14.)

Interactive Problems

18 Use *Sage* to find the derived series of the group Q:

```
Q = InitQuaternions(); Q
    {1, i, j, k, -1, -i, -j, -k}
```

Add any subgroups necessary to make this series a composition series.

19 Use *Sage*'s **Commutator** command as an alternative way to show that $\text{Aut}(Z_{24}^*)$ is insoluble. Load this group with the commands

```
DisplayPermInt = true
A = Group( NthPerm(187), NthPerm(723) ); A
    {1, 27, 61, 87, 122, 149, 187, 231, 244, 270, 331, 357, 374,
    404, 437, 467, 496, 548, 558, 593, 640, 670, 684, 714, 723,
    745, 783, 805, 844, 870, 931, 957, 962, 989, 1027, 1071,
    1096, 1148, 1158, 1193, 1214, 1244, 1277, 1307, 1366, 1384,
    1410, 1428, 1445, 1466, 1509, 1549, 1566, 1588, 1653, 1675,
    1681, 1707, 1741, 1767, 1822, 1862, 1889, 1902, 1966, 1984,
    2010, 2028, 2054, 2084, 2117, 2147, 2166, 2188, 2253, 2275,
    2285, 2306, 2349, 2389, 2403, 2425, 2463, 2485, 2566, 2584,
    2610, 2628, 2662, 2702, 2729, 2742, 2780, 2798, 2843, 2861,
    2897, 2927, 2954, 2984, 3018, 3071, 3076, 3110, 3144, 3185,
    3206, 3220, 3288, 3306, 3328, 3346, 3361, 3387, 3421, 3447,
    3487, 3517, 3531, 3561, 3618, 3671, 3676, 3710, 3737, 3767,
    3794, 3824, 3888, 3906, 3928, 3946, 3984, 4025, 4046, 4060,
    4083, 4105, 4143, 4165, 4213, 4231, 4257, 4275, 4362, 4392,
```

```
    4402, 4432, 4488, 4506, 4528, 4546, 4577, 4607, 4634, 4664,
    4703, 4721, 4760, 4778, 4809, 4839, 4849, 4879, 4935, 4953,
    4975, 4993}
```

and find A'. Note that *Sage* can find the derived group fairly quickly.

20 Find the derived group series of the following group:

```
DisplayPermInt = true
G = Group(NthPerm(2374), NthPerm(6212)); G
    {1, 2374, 4517, 6212, 6841, 9929, 11637, 13016, 13698, 15367,
    18454, 19853, 21239, 21896, 24132, 25315, 28226, 28986,
    30928, 31590, 33108, 37381, 38807, 39487}
```

What group is G' isomorphic to? Is G a semi-direct product of two familiar groups?

8.3 Polycyclic Groups

In this section we will find an efficient method for constructing any finite solvable group. We will find that the composition series is usually overkill, since there is often a shorter subnormal series that will do the job. In the end, we will have a method of entering even large finite groups into *Sage* using a *polycyclic format* that will take advantage of *Sage*'s ability.

Throughout this book, we used *Sage*'s **InitGroup** and **Define** commands to produce many of the groups we have been studying. Only occasionally did we have to use permutations to represent groups, such as the groups A_5 and $\text{Aut}(Z_{24}^*)$. However, the method for converting a finite group into a set of *Sage* commands has never been fully explained. We know that the groups can be represented by a small number of generators. Why was S_4 defined in *Sage* with three generators when only two generators would generate the group?

The method for defining a group G in *Sage* using a set of generators stems from the composition series for a solvable group G. However, a composition series is actually more than we need. We will still insist that the factors of a series be cyclic, but not necessarily of prime order.

DEFINITION 8.8 A subnormal series

$$G = G_0 \supseteq G_1 \supseteq G_2 \supseteq \cdots \supseteq G_n = \{e\}$$

is a *polycyclic series* if the quotient groups G_{i-1}/G_i are all cyclic groups. The number n is called the *length* of the polycyclic series.

It is obvious that a group with a polycyclic series must be solvable, since the cyclic quotient groups would be solvable. Although any finite solvable group has a polycyclic series, it should be noted that an *infinite* solvable group may not always have a polycyclic series. The groups that have a polycyclic series are called *polycyclic groups.*

The first step in expressing a finite group in *Sage* is to find a polycyclic series for the group, preferably with the smallest possible length. For example, the quaternion group Q has a normal subgroup of order 4: $\{1, i, -1, -i\}$. This is of course cyclic, and the quotient group is isomorphic to Z_2, which is also cyclic. Thus, there is a polycyclic series of Q of length 2. The length of the polycyclic series will be the number of generators required for expressing the group in *Sage*, so naturally the shorter the polycyclic series, the less work the definition entails.

Our strategy will be to work inductively on the length of the series. That is, given a polycyclic series for a polycyclic group,

$$G = G_0 \supseteq G_1 \supseteq G_2 \supseteq \cdots \supseteq G_n = \{e\},$$

we will begin by defining G_n, the trivial group, and then define G_{i-1} in terms of G_i. Thus, after n steps, we will have defined the group G into *Sage*.

Defining G_n is easy, since this is the trivial group. This group is defined by the single command

```
InitGroup("e")
```

where e is the name of the identity element. Next we add the variables for the group, in the order of the polycyclic series.

AddGroupVar("g_1**", "**g_2**", ... "**g_n**")**

We will suppose that the group G_i is defined inductively by *Sage* in terms of the elements $g_{i+1}, g_{i+2}, \ldots, g_n$. Since G_{i-1}/G_i is cyclic, there is a generator of this quotient group, which we will call $g_i \cdot G_i$, where g_i is in the subgroup G_{i-1}. If this quotient group is of order m_i, then $(g_i \cdot G_i)^{m_i} = G_i$, so $g_i^{m_i} \in G_i$. So *Sage* can represent $g_i^{m_i}$ in terms of the elements $g_{i+1}, g_{i+2}, \ldots, g_n$. Thus, we can make the definition

Define(g_i**^**m_i**,** b**)**

where b is the *Sage* representation of $g_i^{m_i}$ in terms of the previously defined elements $g_{i+1}, g_{i+2}, \ldots, g_n$.

Our goal will be to have *Sage* express every element in terms of generators that are multiplied in increasing order. We consider the generators $g_1, g_2, g_3, \ldots$ as "letters," going in alphabetical order, then we need definitions that would find a way of expressing any element of the group as a product of generators such that the generators are in alphabetical order. That is, the expression $g_2 \cdot g_3 \cdot g_1$ will need to be reduced, whereas $g_1 \cdot g_2 \cdot g_3 \cdot g_3$ does not. As such, we have to program *Sage* to unravel expressions that are "in the wrong order." Any expression in the wrong order will contain a sequence $g_k \cdot g_i$, where $k > i$. Since G_i is a normal subgroup of G_{i-1}, $g_i^{-1} \cdot g_k \cdot g_i = y_k$ is in G_i, and hence can be expressed in *Sage* in terms of the elements $g_{i+1}, g_{i+2}, \ldots, g_n$. We can perform the following sequence of commands to force *Sage* to always put the generators in the proper order.

```
Define(gi^-1*gi+1*gi, yi+1)
Define(gi^-1*gi+2*gi, yi+2)
   .........
Define(gi^-1*gn*gi, yn)
```

With these definitions, *Sage* will continue to process a given element until the generators are arranged in order. Although it is not clear right now that a given element can be expressed as a product of generators arranged in order, we will see that the group generated by the elements $\{g_i, g_{i+1}, \cdots g_n\}$ will be isomorphic to G_{i-1}. Thus, we will be able to construct the group G inductively. Note that *Mathematica* uses a different format for entering polycyclic groups. See the *Mathematica* notebook for details.

Computational Example 8.3

Use a polycyclic series to define the group Q in *Sage*.

We could use the series

$$G_0 = Q \supseteq G_1 = \{1, i, -1, -i\} \supseteq G_2 = \{1\}.$$

Since this is a series of length 2, we will need 2 generators. To find g_2, we need to find a generator of the group G_1/G_2. Certainly $\{i\}$ or $\{-i\}$ would work, so we let g_2 be one of these elements. We might as well pick $g_2 = i$.

We then observe that G_0/G_1 is of order 2, and a generator is $\{j, k, -j, -k\}$. We pick g_1 to be any one of these elements, say j. To make this a true polycyclic group, it is important that we add the generator names in the left-to-right order of the polycyclic series.

```
InitGroup("e")
AddGroupVar("j", "i")
```

We are now ready to define G_1. Since G_1/G_2 is cyclic of order 4, we have that i^4 is in G_2, which is the identity. So we can define

```
Define(i^4, e)
```

This defines G_1. We now observe that G_0/G_1 is of order 2, so j^2 must be in G_1, and in fact $j^2 = -1 = i^2$. So we can define

```
Define(j^2, i^2)
```

Finally, we note that $j^{-1} \cdot i \cdot j$ must be in G_1. In fact, $j^{-1} \cdot i \cdot j = -i = i^3$. We will use a rather unusual **Define** command this time to enter this information into Sage. This will allow us to display the group.

```
Define(j^-1*i*j, i^3)
Group(j, i)
    {e, j, j^3, i, j^2*i, j^2, j*i, j^3*i}
```

This puts the elements in "alphabetical" order because $g_1 = j$ is considered to be before $g_2 = i$. To create a more conventional ordering, we could have planned ahead and chosen $G_1 = \{1, j, -1, -j\}$. See Problem 16. □

The purpose of defining the group the way we did in *Sage* is that we can convert the group to a polycyclic format. This is done with the command **ToPolycyclic()**.

```
Q = ToPolycyclic()
    Group converted to the Polycyclic format.
Q
{e, i, i^2, i^3, j, j*i, j*i^2, j*i^3}
```

Now all products are simplified to a standardized form.

```
j^4 * i^2 * j
    j*i^2
```

This is like having the best parts of **ReducedMultiplication** and **ListGroup** at the same time. This makes working with polycyclic groups much easier and faster than groups defined the standard way.

Computational Example 8.4

Use the polycyclic series for S_4,

$$G_0 = S_4 \supseteq G_1 = A_4 \supseteq G_2 = K \supseteq G_3 = H \supseteq G_4 = \{(\,)\}$$

to enter this group into *Sage* as a polycyclic group.

SOLUTION: Since there are four cyclic quotient groups, we will need four generators g_1, g_2, g_3, g_4 such that $g_i G_i$ is a generator of G_{i-1}/G_i. Some natural choices are $g_1 = (1\,2)$, $g_2 = (1\,2\,3)$, $g_3 = (1\,3)(2\,4)$, and $g_4 = (1\,2)(3\,4)$.

Next, $g_i^{m_i} \in G_i$, where m_i is the order of G_{i-1}/G_i. Looking at the polycyclic series for S_4, we find that $m_1 = 2$, $m_2 = 3$, $m_3 = 2$, and $m_4 = 2$. Hence we calculate $g_1^2 = (\,)$, $g_2^3 = (\,)$, $g_3^2 = (\,)$, and $g_4^2 = (\,)$. In this case, all of these turned out to be the identity element, but we are only promised that $g_i^{m_i}$ will be in G_i, and hence expressible in terms of $g_{i+1}, \ldots g_n$.

Finally, we calculate $g_i^{-1} \cdot g_j \cdot g_1 \in G_i$ for each combination $j > i$, and express each of these in terms of $g_{i+1}, \ldots g_n$. We find that $g_1^{-1} \cdot g_2 \cdot g_1 = (1\,3\,2) = g_2^2$,

$g_1^{-1} \cdot g_3 \cdot g_1 = (1\,4)(2\,3) = g_3 \cdot g_4$, $g_1^{-1} \cdot g_4 \cdot g_1 = (1\,2)(3\,4) = g_4$, $g_2^{-1} \cdot g_3 \cdot g_2 = (1\,4)(2\,3) = g_3 \cdot g_4$, $g_2^{-1} \cdot g_4 \cdot g_2 = (1\,3)(2\,4) = g_3$, and $g_3^{-1} \cdot g_4 \cdot g_3 = (1\,2)(3\,4) = g_4$.

We are now ready to enter this into *Sage* as a polycyclic group. We can use a, b, c, and d as the four generators.

```
InitGroup("e")
AddGroupVar("a", "b", "c", "d")
Define(a^2, e); Define(b^3, e); Define(c^2, e); Define(d^2, e)
Define(a^-1*b*a, b^2); Define(a^-1*c*a, c*d)
Define(a^-1*d*a, d); Define(b^-1*c*b, c*d)
Define(b^-1*d*b, c); Define(c^-1*d*c, d)
Group(a,b,c,d)
    {e, b*a, a*b, a*b*a, b, a, b*a*c, c, b*c, a*c, a*b*c,
    a*b*a*c, b*a*c*a*b, c*a*b, b*c*a*b, a*c*a*b, a*b*c*a*b,
    a*b*a*c*a*b, c*a, b*a*c*a, a*b*c*a, a*b*a*c*a, b*c*a, a*c*a}
```

Sage is expressing each element as a product of generators, but not always in alphabetical order. But since we used a polycyclic series to define this group, we can convert it to a polycyclic form.

```
S4 = ToPolycyclic()
    Group converted to the Polycyclic format.
S4
    {e, d, c, c*d, b, b*d, b*c, b*c*d, b^2, b^2*d, b^2*c,
    b^2*c*d, a, a*d, a*c, a*c*d, a*b, a*b*d, a*b*c, a*b*c*d,
    a*b^2, a*b^2*d, a*b^2*c, a*b^2*c*d}
```

We now see every element in a form where the generators are in alphabetical order. We can verify that we have indeed defined S_4.

```
StructureDescription()
    S4
```

□

Computational Example 8.5

Table 8.1 shows a multiplication table for a non-abelian group that we will simply call A. Enter this group into *Sage* as a polycyclic group.

Because there are no elements of order 8, this cannot be one of the groups of the form $Z_8 \rtimes_\phi Z_2$ studied in §6.4.

Finding a polycyclic series is not hard, but finding a short series of length 2 is a little trickier. We find that $\{1, Z, Y, X\}$ is a normal subgroup isomorphic to Z_4, and the quotient group is also cyclic. Thus, the series

$$G_0 = A \supset G_1 = \{1, Z, Y, X\} \supset G_2 = \{1\}$$

is a polycyclic series of length 2. By using this series, we need only two generators, a and b. Since G_1/G_2 has two generators, $\{Z\}$ and $\{X\}$, we can let b represent either element, say $b = Z$. Then $b^4 = Z^4$ must be in $G_2 = \{1\}$, so

TABLE 8.1: Multiplication table for the mystery group A

·	1	Z	Y	X	W	V	U	T	S	R	Q	P	O	N	M	L
1	1	Z	Y	X	W	V	U	T	S	R	Q	P	O	N	M	L
Z	Z	Y	X	1	T	W	V	U	R	Q	P	S	L	O	N	M
Y	Y	X	1	Z	U	T	W	V	Q	P	S	R	M	L	O	N
X	X	1	Z	Y	V	U	T	W	P	S	R	Q	N	M	L	O
W	W	V	U	T	S	R	Q	P	O	N	M	L	1	Z	Y	X
V	V	U	T	W	P	S	R	Q	N	M	L	O	X	1	Z	Y
U	U	T	W	V	Q	P	S	R	M	L	O	N	Y	X	1	Z
T	T	W	V	U	R	Q	P	S	L	O	N	M	Z	Y	X	1
S	S	R	Q	P	O	N	M	L	1	Z	Y	X	W	V	U	T
R	R	Q	P	S	L	O	N	M	Z	Y	X	1	T	W	V	U
Q	Q	P	S	R	M	L	O	N	Y	X	1	Z	U	T	W	V
P	P	S	R	Q	N	M	L	O	X	1	Z	Y	V	U	T	W
O	O	N	M	L	1	Z	Y	X	W	V	U	T	S	R	Q	P
N	N	M	L	O	X	1	Z	Y	V	U	T	W	P	S	R	Q
M	M	L	O	N	Y	X	1	Z	U	T	W	V	Q	P	S	R
L	L	O	N	M	Z	Y	X	1	T	W	V	U	R	Q	P	S

```
InitGroup("e")
AddGroupVar("a", "b")
Define(b^4, e)
```

defines $b = Z$ in *Sage*. Next, we note that both $\{W, V, U, T\}$ and $\{O, N, M, L\}$ are generators of G_0/G_1. Thus, we can let a be any of these eight elements, say $a = W$. Then $a^4 = W^4$ must be in G_1, and in fact the table shows that $a^4 = e$.

```
Define(a^4, e)
```

Finally, we need to let *Sage* know how to handle the combination $b \cdot a$. Using the multiplication table, we have that $a^{-1} \cdot b \cdot a = W^{-1} \cdot Z \cdot W = X = b^3$. So $a^{-1} \cdot b \cdot a = b^3$. Let us add this fact into *Sage*.

```
Define(a^-1*b*a, b^3)
```

We now have the group entered into *Sage*.

```
A = ListGroup(); A
    {e, a, a^2, a^3, b, a*b, a^2*b, a^3*b, b^2, a*b^2, a^2*b^2,
    a^3*b^2, b^3, a*b^3, a^2*b^3, a^3*b^3}
```

Because we used a polycyclic series to define the group, we can convert it to a polycyclic group.

```
A = ToPolycyclic()
    Group converted to the Polycyclic format.
A
    {e, b, b^2, b^3, a, a*b, a*b^2, a*b^3, a^2, a^2*b, a^2*b^2,
    a^2*b^3, a^3, a^3*b, a^3*b^2, a^3*b^3}
```

We see that each element is denoted in the same way as the command **ListGroup()**, although they are listed in a different order. □

We still have not identified this group in terms of the groups that we are familiar with. We can have Sage determine what the group is:

```
StructureDescription()
    Z4 : Z4
```

We see that this group is a semi-direct product of Z_4 with itself. In fact, it is the only such semi-direct product, so we can refer to this group as $Z_4 \rtimes Z_4$.

Problems for §8.3

For Problems **1** through **9**: Find the shortest possible polycyclic series for the following solvable groups. (There may be more than one solution.)

1 D_4 **2** D_5 **3** Z_{15}^* **4** Z_{24}^* **5** Z_{26}^* **6** $Z_3 \rtimes Z_4$ **7** $Z_2 \times Z_3 \times Z_4$ **8** $Z_2 \times Z_4 \times Z_8$ **9** $Z_2 \times Z_2 \times Z_2 \times Z_3 \times Z_3$

10 We saw that S_4's shortest polycyclic series was of length 4. What other group has its shortest polycyclic series being of length 4?

11 Suppose a group has two polycyclic series of minimum length. Must the quotient groups of the series be isomorphic?

12 Suppose a group has two polycyclic series of maximum length. Must the quotient groups of the series be isomorphic?

13 Throughout this course, we have encountered a number of groups of order 16. Here is a list of some of these groups:

$$Z_{16}, \quad Z_8 \times Z_2, \quad Z_4 \times Z_4, \quad Z_4 \times Z_2 \times Z_2, \quad Z_2 \times Z_2 \times Z_2 \times Z_2,$$

three groups of the form $Z_2 \rtimes_\phi Z_8$ in §6.4 (one is D_8),

$$Z_2 \times Q, \quad Z_2 \times D_4, \quad Z_4 \rtimes Z_4 \text{ studied in this section,}$$

and three mystery groups B, C, and D found in Problems 18, 19, and 20. Show that these 14 groups are all non-isomorphic. (In fact, these are all of the non-isomorphic groups of order 16.)

Hint: Find $R_2(G)$ by counting the number of identity elements along the diagonals. Note that group B has only 1's and L's along its diagonal, whereas group C has three different elements along its diagonal.

TABLE 8.2: Mystery group B used in Problem 18

·	1	I	J	K	L	M	N	O	P	Q	R	S	T	U	V	W
1	1	I	J	K	L	M	N	O	P	Q	R	S	T	U	V	W
I	I	L	K	N	M	1	O	J	Q	T	S	V	U	P	W	R
J	J	O	L	I	N	K	1	M	R	W	T	Q	V	S	P	U
K	K	J	M	L	O	N	I	1	S	R	U	T	W	V	Q	P
L	L	M	N	O	1	I	J	K	T	U	V	W	P	Q	R	S
M	M	1	O	J	I	L	K	N	U	P	W	R	Q	T	S	V
N	N	K	1	M	J	O	L	I	V	S	P	U	R	W	T	Q
O	O	N	I	1	K	J	M	L	W	V	Q	P	S	R	U	T
P	P	Q	R	S	T	U	V	W	L	M	N	O	1	I	J	K
Q	Q	T	S	V	U	P	W	R	M	1	O	J	I	L	K	N
R	R	W	T	Q	V	S	P	U	N	K	1	M	J	O	L	I
S	S	R	U	T	W	V	Q	P	O	N	I	1	K	J	M	L
T	T	U	V	W	P	Q	R	S	1	I	J	K	L	M	N	O
U	U	P	W	R	Q	T	S	V	I	L	K	N	M	1	O	J
V	V	S	P	U	R	W	T	Q	J	O	L	I	N	K	1	M
W	W	V	Q	P	S	R	U	T	K	J	M	L	O	N	I	1

14 Show that there is a group of order 24 for which there are two elements x and y that generate the group such that $x^3 = y^6 = e$, and $y \cdot x = x^2 \cdot y^2$. This problem is referred to in the *Mathematica* notebook.

Hint: What are the orders of the elements $x \cdot y$ and $y \cdot x$? Determine the subgroup generated by these two elements.

15 Let G be an infinite group such that every element besides the identity has order 2. Show that G is solvable, yet G does not have a polycyclic series.

Interactive Problems

16 Redo Example 8.3 with the subgroup $\{1, j, -1, -j\}$. How does *Sage* rename the elements of the group?

17 Use a polycyclic series of A_4 to enter this group into *Sage*. Then use **`ToPolycyclic`** to convert the group to a polycyclic form.

18 Find a polycyclic series of group B of order 16 given in Table 8.2, and use this to enter the group into *Sage*. Then use **`ToPolycyclic`** to convert the group to a polycyclic form.

19 Find a polycyclic series of group C of order 16 given in Table 8.3, and use this to enter the group into *Sage*. Then use **`ToPolycyclic`** to convert the group to a polycyclic form.

TABLE 8.3: Mystery group C used in Problem 19

·	1	F	G	H	I	J	K	L	M	N	O	P	Q	R	S	T
1	1	F	G	H	I	J	K	L	M	N	O	P	Q	R	S	T
F	F	1	H	G	J	I	L	K	N	M	P	O	R	Q	T	S
G	G	H	1	F	K	L	I	J	O	P	M	N	S	T	Q	R
H	H	G	F	1	L	K	J	I	P	O	N	M	T	S	R	Q
I	I	K	J	L	M	O	N	P	Q	S	R	T	1	G	F	H
J	J	L	I	K	N	P	M	O	R	T	Q	S	F	H	1	G
K	K	I	L	J	O	M	P	N	S	Q	T	R	G	1	H	F
L	L	J	K	I	P	N	O	M	T	R	S	Q	H	F	G	1
M	M	N	O	P	Q	R	S	T	1	F	G	H	I	J	K	L
N	N	M	P	O	R	Q	T	S	F	1	H	G	J	I	L	K
O	O	P	M	N	S	T	Q	R	G	H	1	F	K	L	I	J
P	P	O	N	M	T	S	R	Q	H	G	F	1	L	K	J	I
Q	Q	S	R	T	1	G	F	H	I	K	J	L	M	O	N	P
R	R	T	Q	S	F	H	1	G	J	L	I	K	N	P	M	O
S	S	Q	T	R	G	1	H	F	K	I	L	J	O	M	P	N
T	T	R	S	Q	H	F	G	1	L	J	K	I	P	N	O	M

20 Find a polycyclic series of group D of order 16 given in Table 8.4, and use this to enter the group into *Sage*. Then use `ToPolycyclic` to convert the group to a polycyclic form.

8.4 Solving the Pyraminx™

We will close this chapter by returning to a problem introduced in §2.3—the Rubik's Pyraminx™. This example is included because it is a perfect illustration of how several of the many techniques that we have learned apply to an actual problem. Although the Rubik's Pyraminx is just a toy, there are important applications to the complex groups produced, such as cryptography. Thus, this example acts as a springboard into applying the principles of group theory to real-world applications.

The Pyraminx™ group was described by four generators, r, l, b, and f, which rotated the right, left, back, or front corners 120° clockwise. The size of the group (933120 elements) makes it infeasible to list the elements in *Sage*, but we still can use the tools we have learned to analyze this group.

Does the group have a nontrivial center? Notice that the four corner pieces

TABLE 8.4: Mystery group D used in Problem 20

·	1	L	M	N	O	P	Q	R	S	T	U	V	W	X	Y	Z
1	1	L	M	N	O	P	Q	R	S	T	U	V	W	X	Y	Z
L	L	M	N	O	P	Q	R	1	T	U	V	W	X	Y	Z	S
M	M	N	O	P	Q	R	1	L	U	V	W	X	Y	Z	S	T
N	N	O	P	Q	R	1	L	M	V	W	X	Y	Z	S	T	U
O	O	P	Q	R	1	L	M	N	W	X	Y	Z	S	T	U	V
P	P	Q	R	1	L	M	N	O	X	Y	Z	S	T	U	V	W
Q	Q	R	1	L	M	N	O	P	Y	Z	S	T	U	V	W	X
R	R	1	L	M	N	O	P	Q	Z	S	T	U	V	W	X	Y
S	S	Z	Y	X	W	V	U	T	O	N	M	L	1	R	Q	P
T	T	S	Z	Y	X	W	V	U	P	O	N	M	L	1	R	Q
U	U	T	S	Z	Y	X	W	V	Q	P	O	N	M	L	1	R
V	V	U	T	S	Z	Y	X	W	R	Q	P	O	N	M	L	1
W	W	V	U	T	S	Z	Y	X	1	R	Q	P	O	N	M	L
X	X	W	V	U	T	S	Z	Y	L	1	R	Q	P	O	N	M
Y	Y	X	W	V	U	T	S	Z	M	L	1	R	Q	P	O	N
Z	Z	Y	X	W	V	U	T	S	N	M	L	1	R	Q	P	O

will never change location in the puzzle. The sequence of moves

```
InitPuzzle()
RotatePuzzle(f,r,f,r,r,f,r,f,r,r)
```

rotates one of these corner pieces, returning all other pieces to their original positions. It is clear that this sequence would commute with all other sequences performed on the puzzle. Since the four corners act independently, we would find at least $3^4 = 81$ elements in the center of the group. Let us call this subgroup K.

Are there elements in the center besides those in K? The sequence

```
InitPuzzle()
RotatePuzzle(l,l,b,f,l,l,b,f,l,l,b,f)
```

returns the four corner pieces to their place, while putting all the edge pieces in the right position, but reversed. If a further sequence of moves was performed from this position rather than the original position, the difference in the end positions would be that all six edges would be reversed. Thus, the above sequence of order 2 will commute with all other elements of the group. It is clear that there can be no more elements in the center, for such an element would have to keep the edge pieces in place. Hence, the center is a normal subgroup isomorphic to the group $Z_2 \times Z_3 \times Z_3 \times Z_3 \times Z_3$.

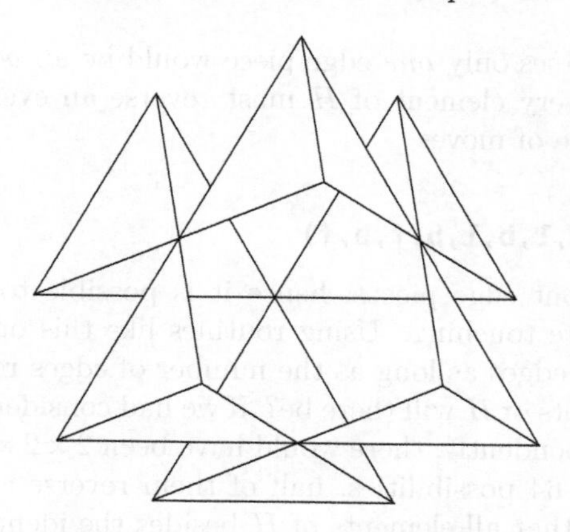

FIGURE 8.3: The Pyraminx^{TM} without the corners

Suppose we consider the subgroup E of actions that return all of the *corners* to their original place. If x is an element of E, and y is a general element, say y rotates the front corner n degrees. Then $y \cdot x \cdot y^{-1}$ rotates the front corner $n + 0 + (-n) = 0$ degrees, so the front corner would return to its original position. Since the same is true for the other three corners, we see that E is a normal subgroup.

The intersection of E and K would be the only element that leaves both the edges and the corners fixed, the identity element. Since both E and K are normal (since K is in the center), by the direct product theorem, $E \cdot K$ is isomorphic to $E \times K$. Yet any action on the Pyraminx^{TM} can be performed by first moving all of the edge pieces, and then moving all of the corners. Thus, the entire group is in $E \cdot K$, and so the Pyraminx^{TM} group is isomorphic to

$$E \times K \approx E \times Z_3 \times Z_3 \times Z_3 \times Z_3.$$

To find the structure of the subgroup E, we analyze the puzzle without the corners, as in Figure 8.3 created by *Sage*'s **`HideCorners`** command.

Since there are only 12 triangles remaining, it is clear that each action could be described as a permutation of the 12 triangles. In fact, notice that turning one corner 120° moves 6 triangles—two sets of 3 triangles rotate places. Thus, each turn produces an *even* permutation of the 12 triangles, so E is a subgroup of A_{12}.

Let us now try to find a normal subgroup of E. What if we considered the subgroup of actions that returns the edge pieces to their place, but may reverse some of them? Let us call this subgroup H./ / Let x be an element of H, and y an element of E. The action $y^{-1} \cdot x \cdot y$ may temporarily move an edge piece out of position, but will return it to its proper place after possibly flipping it. Therefore, H will be a normal subgroup of E.

Let us determine the structure of H. At first one might think that each edge piece can be reversed independently of all of the others, but this is not true.

An action that reverses only *one* edge piece would be an *odd* permutation of the triangles. So every element of H must reverse an even number of edge pieces. The sequence of moves

```
InitPuzzle()
RotatePuzzle(l,f,l,b,l,b,f,b,f)
```

reverses the two front edge pieces, hence it is possible to reverse two edge pieces when they are touching. Using routines like this one, we can reverse any combination of edges as long as the number of edges reversed is even.

How many elements of H will there be? If we had considered the edge pieces to be reversed independently, there would have been $2 \times 2 \times 2 \times 2 \times 2 \times 2 = 64$ elements. Of these 64 possibilities, half of them reverse an even number of edges. By noticing that all elements of H besides the identity are of order 2, we find that the 32 elements of H are isomorphic to $Z_2 \times Z_2 \times Z_2 \times Z_2 \times Z_2$. The quotient group E/H can now be visualized by ignoring whether the six edge pieces are reversed. Certainly this would be a subgroup of the permutations of the six edges. But again we can only consider even permutations, for the edges are moved three at a time. Thus E/H must be isomorphic to a subgroup of A_6. It is fairly clear that we can position four of the six edges in any position, so $E/H \approx A_6$.

Is E isomorphic to a semi-direct product of H with A_6? To see that it is, we need to find a copy of A_6 inside of E that contains no elements of H besides the identity. Such a subgroup is generated by the three actions

```
RotatePuzzle(f)
RotatePuzzle(b)
RotatePuzzle(r,f,f,r,r,f)
```

so the group M generated by these three sequences is isomorphic to A_6. Since it is impossible to reverse any edges with the elements of M, the intersection of M and H is the identity. Every arrangement of the edges can be obtained by first putting all of the edges into position, and then reversing several edges. Thus, $E = M \cdot H$. Therefore by the semi-direct product theorem (6.3), E is isomorphic to a semi-direct product of H with M. If we let ϕ represent the homomorphism from M to $\mathrm{Aut}(H)$, we have that

$$E \approx (Z_2 \times Z_2 \times Z_2 \times Z_2 \times Z_2) \rtimes_{\phi} A_6.$$

Unfortunately, this representation of E depends on the homomorphism ϕ from A_6 to $\mathrm{Aut}(Z_2 \times Z_2 \times Z_2 \times Z_2 \times Z_2)$. Let us try to find a representation of E that does not require additional knowledge. So instead, we will express E in terms of the *wreath product.*

DEFINITION 8.9 *Let G be a group, and H a subgroup of S_n. We define the* wreath product

$$G \text{ Wr } H$$

as the semi-direct product $G^n \rtimes_\psi H$, where G^n denotes the direct product of G with itself n times, and for $\sigma \in H$, we define $\psi_\sigma : G^n \to G^n$ by

$$\psi_\sigma(g_1, g_2, \ldots g_n) = (g_{\sigma^{-1}(1)}, g_{\sigma^{-1}(2)}, \ldots g_{\sigma^{-1}(n)}).$$

We explored the wreath product in Problems 13 through 16 of §6.4. In these problems, we demonstrated that ψ is a homomorphism from H to $\mathrm{Aut}(G^n)$, so the wreath product is a group of size $|G|^n \cdot |H|$.

Consider the wreath product Z_2 Wr A_6. This would be a semi-direct product

$$(Z_2 \times Z_2 \times Z_2 \times Z_2 \times Z_2 \times Z_2) \rtimes_\phi A_6,$$

which is similar to, but twice as large, as E. In fact, it is not too hard to see the correlation. If we considered the 6 edges of the puzzle being able to be flipped independently, then the group of all edge flips would be $(Z_2)^6$. But then we can permute the edges with any even permutation. The wreath product combines the actions of the edge permutations with the edge flips. Thus, E is isomorphic to a subgroup of Z_2 Wr A_6. In fact, it is a normal subgroup, since it contains half of the elements.

How can we specify this subgroup? Consider the derived subgroup of Z_2 Wr A_6. This subgroup is generated by elements of the form $x^{-1} \cdot y^{-1} \cdot x \cdot y$, which clearly would flip an even number of edges. There is a natural subgroup of Z_2 Wr A_6 that is isomorphic to A_6, and since $(A_6)' = A_6$, this subgroup would be in the derived subgroup. Also, if x flips two of the six edges, and y permutes three edges without flipping any, moving only one of the two edges flipped by x, then $x^{-1} \cdot y^{-1} \cdot x \cdot y$ will flip two edges, returning them to their original position. Thus, we see that $E \approx (Z_2 \text{ Wr } A_6)'$, and hence the Pyraminx group is isomorphic to

$$(Z_2 \text{ Wr } A_6)' \times Z_3 \times Z_3 \times Z_3 \times Z_3.$$

We can use *Sage* to analyze this group, by analyzing $(Z_2 \text{ Wr } A_6)'$. First we consider the subgroup H, which is the subgroup of flipping an even number of edges. We can represent the edges by disjoint transpositions.

```
H = Group( C(1,2)*C(3,4), C(3,4)*C(5,6), C(5,6)*C(7,8),
   C(7,8)*C(9,10), C(9,10)*C(11,12) )
len(H)
   32
```

Next, we consider the subgroup generated from even permutations of the cycles, without flipping them. This subgroup is generated by a 3-cycle and 5-cycle of edges.

```
M = Group( C(1,3,5)*C(2,4,6), C(3,5,7,9,11)*C(4,6,8,10,12) )
len(M)
   360
```

TABLE 8.5: Orders of the elements for $(Z_2 \text{ Wr } A_6)'$

1	element of order 1,
391	elements of order 2,
800	elements of order 3,
2520	elements of order 4,
2304	elements of order 5,
1760	elements of order 6,
1440	elements of order 8,
2304	elements of order 10,
11520	elements total.

We now can combine these subgroups, to form the whole group.

```
G = H * M
len(G)
    11520
```

This group is far too large to display, even with integer representation. However, we can determine how many elements there are of a given order, by computing $R_k(G)$ for various k.

```
RootCount(G, 2)
    392
```

This shows the group has 391 elements of order 2. By changing the value of k, we can find the number of elements of any given order, summarized in Table 8.5. This table, along with the fact that the Pyraminx group is

$$(Z_2 \text{ Wr } A_6)' \times Z_3 \times Z_3 \times Z_3 \times Z_3,$$

allows us to analyze the Pyraminx group.

Knowing the structure of the group allows us to solve the puzzle! Here is the strategy based on this decomposition of the group.

1. First put all of the edge pieces in place. We can begin with the bottom, then rotate the front and back corners until the back two edges are in the right place (they may be reversed). Finally, rotate the front corner until all six edges are in place.

2. At this point, an even number of edges will be reversed. We can find routines that will flip two, four, or six of the edges. These may rotate corners in the process.

3. Now only the four corner pieces are out of position. We can find routines to rotate these into position.

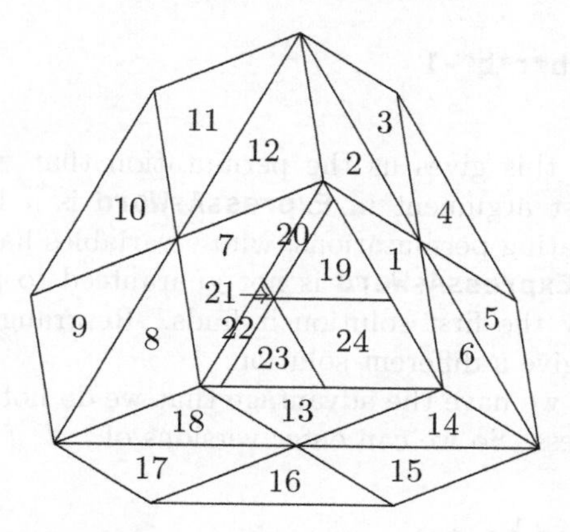

FIGURE 8.4: The Pyraminx™ with numbered faces

To find a combination of the four moves f, b, r, and l that will accomplish these goals, we can have *Sage* help us. First we can number the 24 triangles, as in Figure 8.4. Since we consider the product of several rotations to be done from left to right, we need to convert the rotations to permutations the way that we converted book rearrangements. That is, for each number, we consider what new number will be in that position after the rotation. Thus the permutation (4 14 23)(5 15 24)(6 16 19) can represent r, l = (8 21 16)(9 22 17)(10 23 18), f = (1 7 13)(2 8 14)(6 12 18), and finally b = (2 19 10)(3 20 11)(4 21 12). We can then enter the Pyraminx™ group as a subgroup of S_{24}.

```
r = C(4,14,23)*C(5,15,24)*C(6,16,19)
l = C(8,21,16)*C(9,22,17)*C(10,23,18)
f = C(1,7,13)*C(2,8,14)*C(6,12,18)
b = C(2,19,10)*C(3,20,11)*C(4,21,12)
```

Now that these rotations are entered into Sage as permutations, the natural question is how to express any given permutation in the group generated by these elements in terms of f, b, r, and l in the most efficient way. For example, suppose we want to find an efficient way to rotate just the right corner piece clockwise, which is the permutation (5 15 24). Sage can do this with the **ExpressAsWord** command.

```
ExpressAsWord(["r", "l", "f", "b"], C(5,15,24) )
    'r*b*r^-2*b^-1*r*b*r*b^-1'
```

This returns a string that describes one of the fastest ways to reach the target permutation from the permutations given. If we evaluate the contents of the string,

```
r*b*r^-2*b^-1*r*b*r*b^-1
    (5, 15, 24)
```

we see that indeed this gives us the permutation that we are looking for. Notice that the first argument in **ExpressAsWord** is a list of strings that represent the generating permutations, whose variables have been previously set up. Note that **ExpressAsWord** is not guaranteed to produce the *shortest* solution, merely the first solution it finds. Rearranging the generating permutations may give a different solution.

In flipping edges, we have the advantage that we do not care if corners are rotated in the process. So we can enter versions of r, l, f, and b that ignore the corner pieces.

```
r = C(4,14,23)*C(6,16,19)
l = C(8,21,16)*C(10,23,18)
f = C(2,8,14)*C(6,12,18)
b = C(2,19,10)*C(4,21,12)
```

By ignoring corners, we reduce the number of puzzle positions down to 11520, so it should be easy to find combinations that produce the right flips. For example, to flip the top and front left edges, we need the permutation $(2\,12)(8\,18)$.

```
ExpressAsWord(["r", "l", "f", "b"], C(2,12)*C(8,18) )
    'r*l^-1*b^-1*l*r^-1*f^-1'
r*l^-1*b^-1*l*r^-1*f^-1
    (2, 12)(8, 18)
```

We summarize the necessary moves in Tables 8.6 and 8.7.

By applying these four routines once or twice, we can get all four corners into position, and solve the puzzle!

Notice that our three steps can be expressed in terms of a subnormal series for the Pyraminx™ group:

$$\begin{aligned}((Z_2 \text{ Wr } A_6)' \times Z_3 \times Z_3 \times Z_3 \times Z_3) &\supset \\ (Z_2 \times Z_2 \times Z_2 \times Z_2 \times Z_2 \times Z_3 \times Z_3 \times Z_3 \times Z_3) &\supset \\ (Z_3 \times Z_3 \times Z_3 \times Z_3) &\supset \{e\}.\end{aligned}$$

This same type of analysis can be used to solve other puzzles, such as the Rubik's Cube®. Several problems in the homework relate to this puzzle. Thus, we can see a practical application of the properties of groups that we have studied throughout the course.

But not all applications of groups are fun and games. Group theory has also become the backbone of modern mathematics, and many important proofs, such as the impossibility of finding solutions to fifth-degree polynomials, hinge

TABLE 8.6: Flipping the edges of the Pyraminx™

$l^{-1} \cdot b \cdot f \cdot l^{-1} \cdot b \cdot f \cdot l^{-1} \cdot b \cdot f$	flip all six edges
$f \cdot b \cdot r^{-1} \cdot l \cdot r \cdot b^{-1}$	flip two front edges
$b \cdot l \cdot b \cdot r \cdot l \cdot r^{-1} \cdot l^{-1} \cdot b$	flip top & bottom edges
$f \cdot r \cdot l^{-1} \cdot b \cdot l \cdot r^{-1}$	flip top & front left edges
$r \cdot l^{-1} \cdot b \cdot l \cdot r^{-1} \cdot f$	flip top & front right edges
$r \cdot b \cdot r \cdot l \cdot b \cdot l^{-1} \cdot b^{-1} \cdot r$	flip left rear & front right edges
$l \cdot r \cdot l \cdot b \cdot r \cdot b^{-1} \cdot r^{-1} \cdot l$	flip right rear & front left edges
$r \cdot b \cdot l^{-1} \cdot f \cdot l \cdot b^{-1}$	flip bottom & front right edges
$l \cdot b \cdot f^{-1} \cdot r \cdot f \cdot b^{-1}$	flip bottom & front left edges
$b \cdot r \cdot f^{-1} \cdot l \cdot f \cdot r^{-1}$	flip top & left rear edges
$b \cdot l \cdot r^{-1} \cdot f \cdot r \cdot l^{-1}$	flip top & right rear edges
$b \cdot f \cdot l^{-1} \cdot r \cdot l \cdot f^{-1}$	flip rear two edges
$l \cdot f \cdot r^{-1} \cdot b \cdot r \cdot f^{-1}$	flip bottom & left rear edges
$r \cdot f \cdot b^{-1} \cdot l \cdot b \cdot f^{-1}$	flip bottom & right rear edges
$l \cdot r \cdot b^{-1} \cdot f \cdot b \cdot r^{-1}$	flip two left hand edges
$r \cdot l \cdot f^{-1} \cdot b \cdot f \cdot l^{-1}$	flip two right hand edges

TABLE 8.7: Rotating the corners of the Pyraminx™

$f \cdot r \cdot f \cdot r^{-1} \cdot f \cdot r \cdot f \cdot r^{-1}$	rotate front corner 120° clockwise
$l \cdot r \cdot l \cdot r^{-1} \cdot l \cdot r \cdot l \cdot r^{-1}$	rotate left corner 120° clockwise
$r \cdot b \cdot r \cdot b^{-1} \cdot r \cdot b \cdot r \cdot b^{-1}$	rotate right corner 120° clockwise
$b \cdot r \cdot b \cdot r^{-1} \cdot b \cdot r \cdot b \cdot r^{-1}$	rotate back corner 120° clockwise

entirely on finite groups. The theory of finite groups also has applications in quantum physics and inorganic chemistry and crystallography. Therefore, the material presented in this course has many applications beyond mathematics.

Problems for §8.4

For Problems **1** through **8**: Compute the following products in Z_2 Wr A_6.

1 $(0,0,0,0,0,0,P(2,4,3,1,5,6)) \cdot (1,0,1,0,0,1,P(1,2,3,4,5,6))$
2 $(1,1,1,1,1,1,P(2,3,6,4,1,5)) \cdot (0,1,1,0,0,1,P(5,1,3,2,6,4))$
3 $(0,1,1,0,0,1,P(1,6,3,5,4,2)) \cdot (1,0,0,1,1,0,P(6,4,3,2,5,1))$
4 $(1,1,0,1,1,0,P(5,2,1,3,6,4)) \cdot (1,1,0,0,0,1,P(6,2,4,1,3,5))$
5 $(0,0,1,0,1,1,(1\,2\,4)) \cdot (1,1,1,0,0,0,(1\,3)(5\,6))$
6 $(0,1,0,1,0,1,(1\,6\,3\,2\,5)) \cdot (0,1,1,0,0,1,(2\,5\,3))$
7 $(1,0,1,0,1,1,(2\,4\,5)) \cdot (0,1,0,0,1,0,(1\,5\,3\,4)(2\,6))$
8 $(0,0,0,1,0,1,(1\,4\,3)(2\,5\,6)) \cdot (1,1,1,0,0,1,(1\,4)(2\,5))$

9 Using the orders of the subgroup E of the Pyraminx™ group given in Table 8.5, determine the number of elements of the Pyraminx™ group that are of order 1, 2, 3, 4, 5, 6, 8, 10, 12, 15, 24, and 30. Verify that the sum of these numbers totals 933,120.

10 Using Problem 10 in §7.1, we see that the center Z of Z_2 Wr A_6 is isomorphic to Z_2. Is $(Z_2 \text{ Wr } A_6)/Z$ isomorphic to $(Z_2 \text{ Wr } A_6)'$?

11 Using Table 8.5, determine how many 5-Sylow subgroups there are in $(Z_2 \text{ Wr } A_6)'$.

12 Using Table 8.5, determine how many 2-Sylow subgroups there are in $(Z_2 \text{ Wr } A_6)'$. Note that from Problem 9 in §7.4, every element of order 2^k is contained in a 2-Sylow subgroup.

13 Using Table 8.5, determine how many 3-Sylow subgroups there are in $(Z_2 \text{ Wr } A_6)'$.

Hint: The normalizer of the group generated by $(0,0,0,0,0,0,(1\,2\,3))$ and $(0,0,0,0,0,0,(4\,5\,6))$ has a subgroup of order 4, generated by $(1,1,1,0,0,0,(\,))$ and $(0,0,0,1,1,1,(\,))$.

14 Consider a $2\times2\times2$ Rubik's Cube®, consisting of just eight corner pieces. Determine the size of the group of actions on this cube. Express the group of actions in terms of a wreath product.

Hint: It is impossible to rotate just one corner, and leave the others in place. Is it possible to move just two of the corners?

15 Consider a standard Rubik's Cube®. What is the size of the group of actions? What is the center of this group?

16 Let $a = (1\,2\,3\,4\,5)$ and $b = (1\,2\,4)$ be two elements of A_5. Find a way to express the element $(1\,2)(4\,5)$ in terms of a and b. There is more than one correct answer.

Hint: Try different combinations of a and b to find another 3-cycle.

17 Let $a = (1\,2\,3\,4\,5)$ and $b = (1\,2\,4)$ be two elements of A_5. Find a way to express the element $(1\,4)(2\,5)$ in terms of a and b. There is more than one correct answer.

Interactive Problems

18 First show that S_7 is generated by the elements $a = (2\,6\,3\,7\,4)$ and $b = (1\,5\,4\,2)$. Then use **ExpressAsWord** to find a way to express $(1\,2)$ in terms of a and b. This problem is not available in *Mathematica*.

19 First show that A_7 is generated by the elements $a = (1\,6\,7)(2\,5\,4)$ and $b = (1\,3\,7\,2)(4\,6)$. Then use **ExpressAsWord** to find a way to express $(1\,2\,3)$ in terms of a and b. This problem is not available in *Mathematica*.

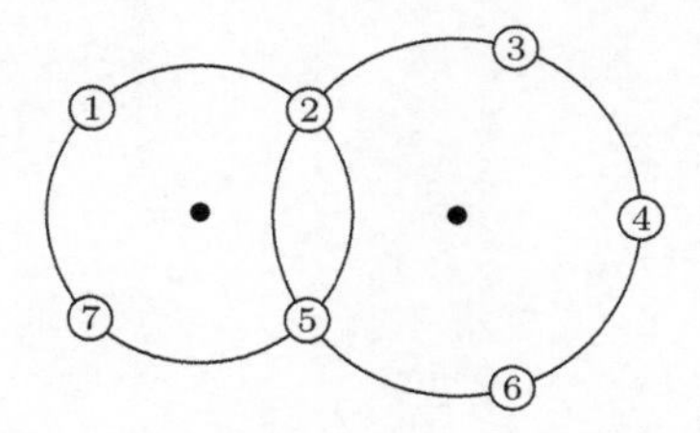

FIGURE 8.5: Simple puzzle with two wheels, used for Problem 20

20 Consider the puzzle in Figure 8.5, with 7 disks on 2 wheels. The action L turns the left wheel 90° clockwise, taking the disks with it. The action R turns the right wheel 72° clockwise, again taking the disks with it. The goal is to swap disks 5 and 6, so the disks are in consecutive order. Use *Sage*'s `ExpressAsWord` to solve this puzzle. A few brave souls might try to solve this puzzle without *Sage*'s help. This problem is not available in *Mathematica*.

Chapter 9

Introduction to Rings

This section presents the concept of a *ring*, which is a generalization of the addition and multiplication operations of standard numbers. The term *ring* was first coined by David Hilbert in 1892, although he only referred to a particular type of ring. It wasn't until 1920 that Emmy Noether gave an abstract definition of a ring, which would apply to the "hyper-complex" number systems developed earlier by William Hamilton and Hermann Grassmann. (See the Historical Diversion on page 306.) This abstraction can apply to polynomials, infinite series, matrices, and even functions. Hence, ring theory has become a valuable tool for almost every other branch of mathematics.

9.1 The Definition of a Ring

While studying the previous chapters on groups, we discovered different patterns in the group's structure by which we could project and prove many useful properties. However, many of the examples of groups we studied possess some additional structure that we have yet to take advantage of. Some of the groups had not just one, but two operations that we could define on the elements. Our goal for this section is to study these examples, and like Noether did, determine the simplest definition that would apply to all of the examples.

The simplest example to consider is the group of integers, $\mathbb{Z}$. This is a group under the operation of addition, in fact an abelian group with the identity element being 0. However, we can also multiply two integers together, always forming another integer. Is $\mathbb{Z}$ a group using multiplication instead of addition? No, because most elements do not have an inverse. However, this extra operation gives $\mathbb{Z}$ a much richer structure than standard groups.

Subgroups of $\mathbb{Z}$ can also be considered. A typical example would be the set of even integers. Once again, we have both addition and multiplication defined on this set, since both the sum and the product of two even integers yield even integers.

Likewise, the group of rationals $\mathbb{Q}$ and real numbers $\mathbb{R}$ have two operations. Although these are both abelian groups under addition, they are *al-*

TABLE 9.1: $(\cdot)$ **mod** 6

$\cdot$	0	1	2	3	4	5
0	0	0	0	0	0	0
1	0	1	2	3	4	5
2	0	2	4	0	2	4
3	0	3	0	3	0	3
4	0	4	2	0	4	2
5	0	5	4	3	2	1

most groups under multiplication as well. The multiplicative inverse exists for all elements except 0. If we considered the remaining elements $\mathbb{Q} - \{0\}$ or $\mathbb{R} - \{0\}$, we have the multiplicative groups denoted $\mathbb{Q}^*$ and $\mathbb{R}^*$.

Not only do $\mathbb{Z}$, $\mathbb{Q}$, and $\mathbb{R}$ allow for an additional operation to be defined on them, but also some groups from Chapter 1. Take for example the groups formed by modular arithmetic, such as $Z_6 = \{0, 1, 2, 3, 4, 5\}$. The group operation on Z_6 is addition modulo 6. A natural second operation would be multiplication modulo 6, shown in Table 9.1. Note that this table does not possess the "Latin square" property we have seen in the group tables. However, there is no reason for the second operation to have this familiar property.

Motivational Example 9.1

The following command produces the quaternion group Q of order 8, which we studied in Chapter 4:

```
Q = InitQuaternions(); Q
    {1, i, j, k, -1, -i, -j, -k}
```

We have seen the multiplication table before, in Table 4.3. The quaternion elements are reminiscent of the cross product between two three-dimensional vectors. That is,

$$i \cdot j = k \qquad j \cdot k = i, \quad \text{and} \quad k \cdot i = j.$$

This suggests that we can also *add* multiples of these elements together like vectors, forming such elements as

```
i - 2*j - k
    i - 2*j - k
```

which would represent the vector $\langle 1, 2, -1 \rangle$. Two vectors can be added together in the standard way.

```
(i - 2*j - k) + (3*i + j - 2*k)
    4*i - j - 3*k
```

producing the vector $\langle 4, -1, -3\rangle$. Unfortunately, as we multiply these "vectors" together using the distributive laws, we find elements of the form

```
(i - 2*j - k) * (3*i + j - 2*k)
    -3 + 5*i - j + 7*k
```

which would represent the *four*-dimensional vector $\langle -3, 5, -1, 7\rangle$. (This extra dimension could represent *time.*) However, we find that the product of any two four-dimensional vectors would give us a product in the form $a+bi+cj+dk = \langle a, b, c, d\rangle$. In fact, we are able to find the inverse of a four-dimensional vector.

```
(i - 2*j - k)^-1
     (-1/6)*i + 1/3*j + 1/6*k
```

This suggests we should explore the special properties of these vectors. ∎

PROPOSITION 9.1

The set of nonzero four-dimensional vectors forms a non-abelian group using the multiplication table for the quaternion group Q.

PROOF: If

$$x = a + bi + cj + dk$$

is nonzero, then

$$x^{-1} = \frac{a}{a^2+b^2+c^2+d^2} + \frac{-b}{a^2+b^2+c^2+d^2}\,i + \frac{-c}{a^2+b^2+c^2+d^2}\,j + \frac{-d}{a^2+b^2+c^2+d^2}\,k$$

forms a multiplicative inverse, since it is a simple exercise to show that $x \cdot x^{-1} = x^{-1} \cdot x = 1$, the multiplicative identity. (See Problem 10.) Note that since $x \neq 0$, the common denominator $a^2 + b^2 + c^2 + d^2 > 0$. It is easy to see that multiplication is closed. The only hard part is to show that the associative law holds, which is best done in *Sage*. See Problem 21. Given that the associative law holds, it is easy to see that the product of two nonzero vectors must be nonzero. If $x \cdot y = 0$, and $x \neq 0$, then

$$y = (x^{-1} \cdot x) \cdot y = x^{-1} \cdot (x \cdot y) = x^{-1} \cdot 0 = 0.$$

Thus, if both $x \neq 0$ and $y \neq 0$, then $x \cdot y \neq 0$. ∎

We call the group of four-dimensional vectors of the form $a + bi + cj + dk$ the *quaternions*, denoted by $\mathbb{H}$ after their discoverer, William Hamilton.

We have seen many examples of groups that exhibit not one but two operations defined on them. One of these operations is represented with the plus sign, and the other is usually denoted with a dot. Our goal will be to come up with a definition that unites these examples. Let us consider which properties these examples have in common. Table 9.2 organizes our findings, indicating which of the 6 groups that we looked at satisfy various properties.

We want to pay special attention to the properties that hold for *all* of the above examples. In fact, let us define a *ring* as a group possessing all of these properties. In this way, we allow all six of the above examples to be rings.

DEFINITION 9.1 A *ring* is an abelian group with the operation $(+)$ on which a second associative operation $(\cdot)$ is defined such that the following two distributive laws hold for all a, b, and c in the ring:

$$(a + b) \cdot c = (a \cdot c) + (b \cdot c) \qquad \text{and}$$
$$a \cdot (b + c) = (a \cdot b) + (a \cdot c).$$

For any ring we will use the symbol 0 to denote the additive identity of a ring, and the notation $-x$ for the additive inverse of x.

Even though we defined a ring such that all six of the groups in Table 9.2 are rings, many of these rings possessed additional properties. We will give names to rings with some of these extra properties.

DEFINITION 9.2 A ring for which $x \cdot y = y \cdot x$ for all elements x and y is called a *commutative ring.*

DEFINITION 9.3 A ring for which there is an element e such that

$$x \cdot e = e \cdot x = x$$

for all elements x in the ring is called a *unity ring* or *ring with identity.* The element e is called the *unity* or *multiplicative identity* of the ring, to distinguish it from the additive identity 0.

The next definition will deal with rings for which $x \cdot y = 0$ implies that either x or y must be 0. However, it is reasonable to first prove the following lemma:

LEMMA 9.1
If x is any element in a ring, then $0 \cdot x = x \cdot 0 = 0$, where 0 is the additive identity.

PROOF: This proof is just a little tricky because there are no other propositions to rely on. Thus, every step must directly use one of the nine properties of rings. (The temptation is to rely on some property we suspect is true, but haven't yet proven.)

TABLE 9.2: Property checklist for several groups

Property	$\mathbb{Z}$	Even Integers	$\mathbb{Q}$	Reals	Z_6	Quaternions
Closed under Addition	√	√	√	√	√	√
Closed under Multiplication	√	√	√	√	√	√
$(a+b)+c = a+(b+c)$	√	√	√	√	√	√
$(a \cdot b) \cdot c = a \cdot (b \cdot c)$	√	√	√	√	√	√
Additive Identity (0)	√	√	√	√	√	√
Multiplicative Identity (1)	√	×	√	√	√	√
Additive Inverses Exist	√	√	√	√	√	√
Multiplicative Inverses Exist Except for 0	×	×	√	√	×	√
$a+b=b+a$	√	√	√	√	√	√
$a \cdot b = b \cdot a$	√	√	√	√	√	×
$a \cdot b = 0$ only if a or $b = 0$	√	√	√	√	×	√
$(a+b) \cdot c = a \cdot c + b \cdot c$	√	√	√	√	√	√
$a \cdot (b+c) = a \cdot b + a \cdot c$	√	√	√	√	√	√

Historical Diversion

Emmy Noether (1882–1935)

Emmy Noether was a Jewish woman from a mathematically talented family. Her father, Max Noether, was a prominent mathematics professor at the University of Erlangen, and played a large part in founding the field of algebraic geometry. Her brother Fritz would also become a mathematics professor at Breslau. However, nothing in her early years would indicate her true mathematical genius.

From 1900 to 1902, she attended the University of Erlangen, studying mathematics and languages. But because she was a woman, she could not formally enroll in the courses, but only audit the lectures with the permission of the instructor, which was often denied. She was, however, allowed to take and pass the final university exams that led to a degree.

Noether moved to Göttingen to audit classes from the mathematical giants of her day, Felix Klein and David Hilbert. Hilbert specialized in an axiomatic approach to number theory. But in 1904 she returned to Erlangen, since they relaxed the rules and allowed women to register for classes. She completed her dissertation in 1907, and continued to teach, without pay, at Erlangen. In 1915, Klein and Hilbert tried to get Noether a faculty position at Göttingen, but their efforts were blocked since she was a woman. Finally, in 1919, she obtained formal admission as an academic lecturer.

Noether revealed her true genius in 1920, when she published a paper on the theory of ideals, in which she defined the left and right ideals of a ring. Noether incorporated Hilbert's axiomatic approach to abstract algebra to be the first person to give a modern definition of the ring, although her work focused on commutative rings. The following year she published *Idealtheorie in Ringbereichen,* which analyzed the ascending chain conditions among ideals. Today, we refer to a ring as a *Noetherian ring* if every ascending chain of ideals

$$I_1 \subseteq I_2 \subseteq I_3 \subseteq \cdots \subseteq I_n \subseteq \cdots$$

must eventually stop increasing in size, that is, there is some I_k such that $I_m = I_k$ for all $m > k$.

When Hilter rose to power in 1933, Noether and other Jewish professors at Göttingen were dismissed. Noether fled to the United States, to Bryn Mawr College, to be a visiting professor of mathematics. In 1935 she died at age 53 from an infection resulting from an operation to remove a uterine tumor.

Image source: Wikimedia Commons

Note that

$$(0 \cdot x + 0 \cdot x) = (0 + 0) \cdot x = 0 \cdot x,$$

so

$$(0 \cdot x + 0 \cdot x) + (-(0 \cdot x)) = 0 \cdot x + (-(0 \cdot x)) = 0.$$

Hence

$$0 \cdot x + (0 \cdot x + (-(0 \cdot x))) = 0,$$

so

$$0 \cdot x + 0 = 0 \cdot x = 0.$$

Similarly,

$$(x \cdot 0 + x \cdot 0) = x \cdot (0 + 0) = x \cdot 0,$$

so

$$(x \cdot 0 + x \cdot 0) + (-(x \cdot 0)) = x \cdot 0 + (-(x \cdot 0)) = 0.$$

Hence

$$x \cdot 0 + (x \cdot 0 + (-(x \cdot 0))) = 0,$$

so

$$x \cdot 0 + 0 = x \cdot 0 = 0. \qquad \square$$

This proof shows that we can get the equivalent of subtraction by adding the additive inverse. But although we can add, subtract, and multiply elements in a ring, we cannot, in general, divide elements. In fact, we can find some rings for which the product of two nonzero elements produces 0, such as $3 \cdot 2 = 0$ in the ring Z_6.

DEFINITION 9.4 If x is a nonzero element of a ring such that either $x \cdot y = 0$ or $y \cdot x = 0$ for a nonzero element y, then x is called a *zero divisor* of the ring. If a ring has no zero divisors, it is called a *ring without zero divisors.*

We see from this definition that 2 and 3 are zero divisors of the ring Z_6, since $3 \cdot 2 = 0$ in this ring. A related definition stems from the product of two elements equaling the multiplicative identity.

DEFINITION 9.5 If, for the element x in a unity ring, there is an element y such that

$$x \cdot y = y \cdot x = e,$$

we say that x has a multiplicative inverse, or is *invertible.*

Just because an element is not a zero divisor does not mean that it is invertible. For example, 2 is not a zero divisor of the ring $\mathbb{Z}$, yet 2 is not invertible in this ring.

The smallest possible ring is the *trivial ring*, which is defined by the *Sage* commands

```
G = ZRing(1); G
    {0}
AddTable(G)
MultTable(G)
```

+	0
0	0

·	0
0	0

This ring is rather unusual because the multiplicative identity is 0. Also, 0 is actually invertible in this ring, because $0^{-1} = 0$. These two facts are true for no other ring.

DEFINITION 9.6 A ring for which every nonzero element has a multiplicative inverse is called a *division ring.*

PROPOSITION 9.2
A division ring always has a unity and has no zero divisors.

PROOF: We just saw that the trivial ring has a unity and has no zero divisors, so we may assume that the ring has a nonzero element y. Then y has a multiplicative inverse z, so we have $y \cdot z = e$, the unity. Thus, every division ring must have a unity.

Now suppose that $x \cdot y = 0$ in a division ring, with both x and y nonzero. Then y has a multiplicative inverse z, so that $y \cdot z = e$. But then

$$x = x \cdot e = x \cdot (y \cdot z) = (x \cdot y) \cdot z = 0 \cdot z = 0,$$

which contradicts the fact that x is nonzero. Thus, a division ring has no zero divisors. ∎

DEFINITION 9.7 A non-trivial division ring for which $x \cdot y = y \cdot x$ for all x and y is called a *field.* A division ring for which multiplication is not commutative is called a *skew field.*

We can now classify each possible type of ring. For example, the ring $\mathbb{Z}$ is a commutative unity ring without zero divisors. The ring of even integers, however, has no unity element, so we would call this a commutative ring without zero divisors. Both $\mathbb{Q}$ and $\mathbb{R}$ satisfied all 13 properties, so these two rings are fields. The ring Z_6 has zero divisors, so we would call this a commutative unity ring. The quaternions $\mathbb{H}$ have all the properties of a field except that multiplication is not commutative, so this is an example of a skew field.

Problems for §9.1

For Problems **1** through **6**: Prove the following statements for arbitrary x, y, and z in a ring R, using the properties of rings, and Lemma 9.1. You can use the result of a previous problem. Note that $x - y$ is defined to be $x + (-y)$, and $x^2 = x \cdot x$.

1 $(-x) \cdot y = -(x \cdot y)$
2 $x \cdot (-y) = -(x \cdot y)$
3 $(-x) \cdot (-y) = x \cdot y$
4 $x \cdot (y - z) = x \cdot y - x \cdot z$
5 $(x - y) \cdot z = x \cdot z - y \cdot z$
6 $(x + y) \cdot (x - y) = (x^2 - y^2) + (y \cdot x - x \cdot y)$

7 If a and b are elements of a ring R, and $a \cdot b$ is a zero divisor, prove that either a or b is a zero divisor.

8 For the quaternions, $\mathbb{H}$, we define the *conjugate* of an element $x = a + bi + cj + dk$ to be $\overline{x} = a - bi - cj - dk$. Prove that $\overline{x_1} + \overline{x_2} = \overline{x_1 + x_2}$ for all x_1 and x_2 in $\mathbb{H}$.

9 Prove or disprove: $\overline{x_1} \cdot \overline{x_2} = \overline{x_1 \cdot x_2}$ for all x_1 and x_2 in $\mathbb{H}$. (See Problem 8.)

10 Prove that for x in $\mathbb{H}$, $x \cdot \overline{x} = \overline{x} \cdot x = a^2 + b^2 + c^2 + d^2$. (See Problem 8.)

11 For all x in $\mathbb{H}$, we define the *absolute value* of x to be $|x| = \sqrt{x \cdot \overline{x}}$. Prove that $|x_1 \cdot x_2| = |x_1|\,|x_2|$. (See Problem 8.)

12 Prove or disprove: For all x in the quaternions $\mathbb{H}$, $(x+1) \cdot (x-1) = x^2 - 1$.

13 Prove or disprove: For all x in the quaternions $\mathbb{H}$, $(x+i) \cdot (x-i) = x^2 + 1$.

14 Let

$$\mathbb{Z}[\sqrt{2}] = \{x + y\sqrt{2} \mid x, y \in \mathbb{Z}\}.$$

Prove that $\mathbb{Z}[\sqrt{2}]$ is a ring under the ordinary addition and multiplication of real numbers.

15 Consider the set

$$\{x + y\sqrt[3]{2} \mid x, y \in \mathbb{Z}\}.$$

Is this set a ring under the ordinary addition and multiplication of real numbers?

16 Prove that a ring can have at most one multiplicative identity.

17 Suppose that G is an abelian group with respect to addition. Define a multiplication on G by $x \cdot y = 0$ for all x and y in G. Show that G forms a ring.

18 Define new operations of addition and multiplication in $\mathbb{Z}$ by $x \oplus y = x + y - 1$ and $x \otimes y = x + y - xy$. Verify that $\mathbb{Z}$ forms a ring with respect to these new operations.

19 Fill in the remaining spaces in these addition and multiplication tables so that the resulting set forms a ring.

Hint: Use the Latin square property to fill in the addition table. Then use the distributive laws to determine the multiplication table.

$+$	0	a	b	c
0				
a				
b				c
c				

$\cdot$	0	a	b	c
0				
a				
b				c
c				

Interactive Problems

20 We saw that the the ring Z_6 had zero divisors. We can enter this ring in *Sage* with the command

```
R = ZRing(6); R
    {0, 1, 2, 3, 4, 5}
```

Try this with Z_5, Z_7, Z_8, Z_9, Z_{10}, Z_{11}, and Z_{12}, and form the multiplication tables of these rings. Which ones have zero divisors? Which ones are fields?

21 Use *Sage* to show that quaternion multiplication is associative. That is, if we define

```
Q = InitQuaternions()
var("a1 a2 a3 b1 b2 b3 c1 c2 c3 d1 d2 d3")
x = a1 + b1*i + c1*j + d1*k
y = a2 + b2*i + c2*j + d2*k
z = a3 + b3*i + c3*j + d3*k
```

then show that $(x \cdot y) \cdot z = x \cdot (y \cdot z)$.

9.2 Entering Finite Rings into *Sage*

Although we have seen a few examples of rings, we would like to expand our repertoire so that we can study properties which would be true for all rings. In particular, we want to find more examples of *finite* rings. It turns out that there are many more finite rings than groups of the same size. (There are 52 rings of order 8, as opposed to only 5 groups of this order.) Verifying the associate and distributive laws for even a small ring can be a cumbersome task, so we will let a computer do the hard part for us.

In the first eight chapters, we entered finite groups into *Sage* by using the generators of the group. If we consider a finite ring simply as an abelian group under addition, we can find a set of generators B for this group (ignoring the multiplicative structure). For each element in B we determine the additive order of the element. That is, for each generator x we want to find the smallest number n such that

$$\underbrace{x + x + \cdots + x + x}_{n \text{ times}} = 0.$$

DEFINITION 9.8 If n is a positive integer, and x is any element in a ring, we define nx inductively by letting $1x = x$, and

$$nx = (n-1)x + x.$$

We also define $(-n)x$ to be $-(nx)$ for n a positive integer. Finally, we define $0x = 0$.

Because "multiplication by an integer" is merely a shorthand for repeated addition, we immediately see that

$$(m+n)x = mx + nx \qquad \text{and} \qquad (mn)x = m(nx)$$

for any element x and any integers n and m. See Problems 13, 14, and 15.

LEMMA 9.2
Let x and y be any two elements in a ring, and let n be an integer. Then

$$(nx) \cdot y = n(x \cdot y) = x \cdot (ny).$$

PROOF: We will proceed by induction. The statement is certainly true for $n = 0$ or $n = 1$. Suppose that the statement is true for the previous case $n - 1$. But then

$$((n-1)x) \cdot y + x \cdot y = (n-1)(x \cdot y) + x \cdot y = x \cdot ((n-1)y) + x \cdot y.$$

Hence, by the distributive law,

$$((n-1)x + x) \cdot y = ((n-1) + 1)(x \cdot y) = x \cdot ((n-1)y + y),$$

and so

$$(nx) \cdot y = n(x \cdot y) = x \cdot (ny).$$

Hence, the statement is true for all positive integers.

For negative integers, we can merely show that

$$(nx) \cdot y + ((-n)x) \cdot y = (nx + (-n)x) \cdot y = ((n-n)x) \cdot y = 0 \cdot y = 0.$$

$$n(x \cdot y) + (-n)(x \cdot y) = (n - n)(x \cdot y) = 0(x \cdot y) = 0.$$
$$x \cdot (ny) + x \cdot ((-n)y) = x \cdot (ny + (-n)y) = x \cdot ((n - n)y) = x \cdot 0 = 0.$$

Thus,$((-n)x) \cdot y$, $(-n)(x \cdot y)$, and $x \cdot ((-n)y)$ are the additive inverses of $(nx) \cdot y$, $n(x \cdot y)$, and $x \cdot (ny)$, respectively. But since these latter three are equal for positive n, we have

$$((-n)x) \cdot y = (-n)(x \cdot y) = x \cdot ((-n)y).$$

Hence the lemma is proven for all integers n. ∎

We can now use this notation within *Sage* to generate a finite ring. To define a ring whose additive group is isomorphic to

$$Z_{15}^* = \{1, 2, 4, 7, 8, 11, 13, 14\},$$

we find two elements that generate this group, such as $a = 2$ and $b = 14$. Since

$$2^4 \equiv 1 \pmod{15} \quad \text{and} \quad 14^2 \equiv 1 \pmod{15},$$

we see that $a^4 = 1$ and $b^2 = 1$ in the group Z_{15}^*. But using ring notation, we write $4a = 0$ and $2b = 0$, since 0 is the additive identity of the ring.

To define this group in *Sage*, we begin by declaring that a and b will be the variables. We do this as we initialize the ring, putting the variables in quotations, just as we did with **AddGroupVar** for groups.

```
InitRing("a", "b")
```

Next, we need to tell *Sage* what the additive order of these elements would be, expressed as a list. Since a is of order 4, and b is of order 2, this could be writen as the list **[4, 2]**. The following command gives us the additive group that we want.

```
DefineRing([4, 2], [[0,0],[0,0]])
```

We will explain the meaning of the second parameter **[[0,0],[0,0]]** later. For now, this is sufficient to define the group structure of the ring. The eight elements of the group are denoted as follows:

```
R = ListRing(); R
    {0*a, a, 2*a, -a, b, a+b, 2*a+b, -a+b}
```

We notice several things from this list. First of all, the zero element is listed as **0*a**, not just 0. *Sage* interprets 0 to mean only the integer 0, so the zero element of a ring needs a different notation. Of course $0 \cdot a$ would give us the zero element for any generator a, so *Sage* picks the first generator mentioned. Also, we see that **3*a** is simplified to **-a**. *Sage* tries to find the simplest way to express the elements of the ring. We combine two elements of this group with a plus sign rather than the dot that we used for groups. For example, here is the sum of two elements:

TABLE 9.3: Addition table for a particular ring **R**

+	$0a$	a	$2a$	$-a$	b	$a+b$	$2a+b$	$-a+b$
$0a$	$0a$	a	$2a$	$-a$	b	$a+b$	$2a+b$	$-a+b$
a	a	$2a$	$-a$	$0a$	$a+b$	$2a+b$	$-a+b$	b
$2a$	$2a$	$-a$	$0a$	a	$2a+b$	$-a+b$	b	$a+b$
$-a$	$-a$	$0a$	a	$2a$	$-a+b$	b	$a+b$	$2a+b$
b	b	$a+b$	$2a+b$	$-a+b$	$0a$	a	$2a$	$-a$
$a+b$	$a+b$	$2a+b$	$-a+b$	b	a	$2a$	$-a$	$0a$
$2a+b$	$2a+b$	$-a+b$	b	$a+b$	$2a$	$-a$	$0a$	a
$-a+b$	$-a+b$	b	$a+b$	$2a+b$	$-a$	$0a$	a	$2a$

```
(-a+b) + (2*a)
    a+b
```

The addition table can be displayed using **AddTable(R)**, producing Table 9.3.

Notice that there are several differences between defining a group and defining the group structure of a ring. The obvious difference is that we use the plus sign instead of the dot for our operation. Also, when we defined a group, we began by telling *Sage* the identity element. But for a ring, the additive identity is always denoted **0*a**, and the multiplicative identity may not exist. So the first statement tells *Sage* the generators for the ring. Finally, all of the **Define** commands are combined into a single **DefineRing** command, which gives all of the necessary information about the ring.

Note that in *Mathematica*, rings are defined differently. See the *Mathematica* notebook for details.

Although this defines the additive group very quickly, we must be selective in choosing the generators. Suppose we had instead chosen the generators $a = 2$ and $b = 7$. These two elements generate the group Z_{15}^*, but both are of order 4. So the *Sage* commands for entering these two generators would be

```
InitRing("a", "b")
DefineRing([4, 4], [[0,0],[0,0]]
R = ListRing(); R
    {0*a, a, 2*a, -a, b, a+b, 2*a+b, -a+b, 2*b, a+2*b, 2*a+2*b,
    -a+2*b, -b, a-b, 2*a-b, -a-b}
```

This gives 16 elements instead of 8. The problem is that *Sage* is not using the identity $2a = 2b$, which is true since $2^2 \equiv 7^2 \pmod{15}$. Trying to add an additional *Sage* command defining $2a = 2b$ woul produce some potential problems later on. A better solution is simply to make the following restriction on the set of generators.

DEFINITION 9.9 Let G be an abelian group. A *basis* is a set $B = \{x_1, x_2, x_3, \ldots x_k\}$ that generates the group such that the only way in which

$$n_1x_1 + n_2x_2 + n_3x_3 + \cdots + n_kx_k = 0$$

for integers $n_1, n_2, n_3, \ldots n_k$ is if

$$n_1x_1 = n_2x_2 = n_3x_3 = \cdots = n_kx_k = 0.$$

For a finite group, it is clear that if we have a basis, then every combination of the form

$$n_1x_1 + n_2x_2 + n_3x_3 + \cdots + n_kx_k,$$

where each n_i is non-negative and less than the order of x_i, forms a distinct element. Also, every element of G could be put in that form. Thus, the product of the orders of all the elements of B equals the order of the group.

It should be noted that *any* finite abelian group has a basis, using the fundemental theorem of finite abelian groups (6.2). See Problem 17.

Once we have found a basis for the additive group, and have defined the additive structure into *Sage*, we are ready to consider the multiplicative definitions. If we have two generators $\{a, b\}$, we will need to define $2^2 = 4$ multiplications: $a \cdot a$, $a \cdot b$, $b \cdot a$, and $b \cdot b$. These four products could be defined to be any of the elements of the ring. Thus, for ring with the additive structure of Z_{15}^*, there are up to $8^4 = 4096$ ways to finish defining the ring! However, very few of these ways of defining the products will satisfy both the distributive laws and the associative law. For example, $b \cdot b$ cannot be defined to be a, otherwise we have the contradiction

$$2a = a + a = b \cdot b + b \cdot b = (b + b) \cdot b = (2b) \cdot b = 0 \cdot b = 0.$$

An example of a ring definition that does not produce such a contradiction comes from defining $a^2 = a$, $b^2 = b$, and $a \cdot b = b \cdot a = 0$. All other products in the ring can be determined from these using the distributive law. For example,

$$(2a + b) \cdot (a + b) = 2a^2 + b \cdot a + 2a \cdot b + b^2 = 2a + 0 + 0 + b = 2a + b.$$

The second argument of the **DefineRing** command allows us to tell *Sage* all of the possible products of two generators. These are entered as an array, using the same ordering as the original basis elements were ordered. For example, if a and b are the two generators, then the array would consist of

$$[[a \cdot a, a \cdot b], [b \cdot a, b \cdot b]].$$

To define the ring described in the above paragraph, we can use

```
InitRing("a", "b")
DefineRing([4, 2], [[a, 0],[0, b]])
R = ListRing(); R
    {0*a, a, 2*a, -a, b, a+b, 2*a+b, -a+b}
```

TABLE 9.4: Multiplication table for a particular ring **R**

$\cdot$	$0a$	a	$2a$	$-a$	b	$a+b$	$2a+b$	$-a+b$
$0a$	$0a$	$0a$	$0a$	$0a$	$0a$	$0a$	$0a$	$0a$
a	$0a$	a	$2a$	$-a$	$0a$	a	$2a$	$-a$
$2a$	$0a$	$2a$	$0a$	$2a$	$0a$	$2a$	$0a$	$2a$
$-a$	$0a$	$-a$	$2a$	a	$0a$	$-a$	$2a$	a
b	$0a$	$0a$	$0a$	$0a$	b	b	b	b
$a+b$	$0a$	a	$2a$	$-a$	b	$a+b$	$2a+b$	$-a+b$
$2a+b$	$0a$	$2a$	$0a$	$2a$	b	$2a+b$	b	$2a+b$
$-a+b$	$0a$	$-a$	$2a$	a	b	$-a+b$	$2a+b$	$a+b$

The addition table was given above in Table 9.3, while the multiplication table is given by

```
MultTable(R)
```

producing Table 9.4.

We still have not *proven* that this is a ring, since we have not verified the distributive laws and the associativity law for multiplication. The tedious task of verifying these laws can be handled by the *Sage* command

```
CheckRing()
    This is a ring.
```

Sage checks the ring most recently defined, and finds that both the distributive and associative laws hold, so this is a ring. Since R is obviously commutative from the multiplication table, the next question is whether R has a unity. *Sage* can search the ring for a unity element with the command

```
FindUnity(R)
    a+b
```

Even though we did not use the unity to construct the ring, *Sage* found one.

The multiplication table shows that many elements of R do not have inverses. Hence, this is not a division ring. Nonetheless, *Sage* can try to take inverses of some of the elements.

```
(-a+b)^-1
    -a+b
(2*a+b)^-1
    fail
```

Example 9.2

Try to define a non-commutative ring using Z_{15}^* as the additive group.

SOLUTION: If $a \cdot b = b$, yet $b \cdot a = 2a$, then the ring will not be commutative. Here is one attempt to define such a ring.

```
InitRing("a", "b")
DefineRing([4, 2],[[0, b],[2*a, 0]])
CheckRing()
    Associative law does not hold.
```

This attempt failed, so we must replace the two **0**'s with other elements of the ring.

It would seem as though there would be 64 possibilities to check, but we can narrow the search by using the associative property. For example, $(a \cdot b) \cdot a$ must be $a \cdot (b \cdot a)$, so $2a = 2a^2$. This forces a^2 to be either a or $-a$. Also, $(b \cdot a) \cdot b$ must be $b \cdot (a \cdot b)$, so $0 = b^2$.

We now have enough information to try the ring again.

```
InitRing("a", "b")
DefineRing([4, 2],[[a, b],[2*a, 0]])
CheckRing()
    This is a ring.
```

In this case, there is no unity element.

```
R = ListRing(); R
    {0*a, a, 2*a, -a, b, a+b, 2*a+b, -a+b}
FindUnity(R)
    No identity element found.
```

In fact, every nonzero element turns out to be a zero divisor. □

Since we have seen an example of a non-communitive ring without unity, can we find a non-communitive unity ring? The following proposition shows that we will not be able to use Z_{15}^* for the additive group.

PROPOSITION 9.3

If a ring with unity has an additive structure that can be generated with less than three elements, then the ring is commutative.

PROOF: Suppose that x and y are two elements of the ring that generate the group under addition. That is, every element can be expressed as $mx + ny$ for integers m and n. In particular, the unity

$$e = mx + ny$$

for some integers m and n. Since e commutes with both x and y, we have

$$mx \cdot x + ny \cdot x = (mx + ny) \cdot x = e \cdot x = x \cdot e = mx \cdot x + nx \cdot y,$$

so $ny \cdot x = nx \cdot y$.

Likewise,

$$mx \cdot y + ny \cdot y = (mx + ny) \cdot y = e \cdot y = y \cdot e = my \cdot x + ny \cdot y,$$

so $mx \cdot y = my \cdot x$.

By the greatest common divisor theorem (0.4), there are integers u and v such that

$$um + vn = \gcd(m, n).$$

If we let c denote the greatest common divisor of m and n, then

$$c(x \cdot y - y \cdot x) = (um + vn)(x \cdot y - y \cdot x) = u(mx \cdot y - my \cdot x) + v(nx \cdot y - ny \cdot x) = 0.$$

What we need to show is that $(x \cdot y - y \cdot x) = 0$. The tempting thing to do is divide by c, but this operation is not allowed in rings. Instead, we will again utilize the unity. Since $c = \gcd(m, n)$ there are integers a and b such that $m = ac$ and $n = bc$. Then

$$\begin{aligned} x \cdot y - y \cdot x &= e \cdot (x \cdot y - y \cdot x) = (acx + bcy) \cdot (x \cdot y - y \cdot x) \\ &= (ax + by) \cdot (c(x \cdot y - y \cdot x)) = (ax + by) \cdot 0 = 0. \end{aligned}$$

So $x \cdot y = y \cdot x$, and the ring is commutative. ∎

If we were to find a non-commutative unity ring, we need an additive group that requires more than two generators to define. The smallest such group is Z_{24}^*. We may suppose that the additive group is generated by the unity e, along with two other elements a and b. Suppose that $a \cdot b = a$, while $b \cdot a = b$. This would make the ring non-commutative. We still need to discern what a^2 and b^2 should be. But $a^2 = (a \cdot b) \cdot a = a \cdot (b \cdot a) = a \cdot b = a$, and $b^2 = (b \cdot a) \cdot b = b \cdot (a \cdot b) = b \cdot a = b$. So the *Sage* command for defining this ring would be

```
InitRing("e", "a", "b")
DefineRing([2, 2, 2], [[e, a, b], [a, a, a], [b, b, b]])
CheckRing()
    This is a ring.
T8 = ListRing(); T8
    {0*e, e, a, e+a, b, e+b, a+b, e+a+b}
FindUnity(T8)
    e
```

The multiplication table is given in Table 9.5. Because we will refer back to this ring often we will call this ring T_8.

It is easy to see that any finite ring can be quickly entered into *Sage*. In fact many infinite rings, such as the quaternions, can also be explored with *Sage*. This will allow us to experiment with many different rings, and find properties that are common to all rings. In the next section we will look at some of the basic relationships between rings.

TABLE 9.5: Multiplication for a non-commutative unity ring

$\cdot$	$0*e$	e	a	$e{+}a$	b	$e{+}b$	$a{+}b$	$e{+}a{+}b$
$0*e$	$0*e$	$0*e$	$0*e$	$0*e$	$0*e$	$0*e$	$0*e$	$0*e$
e	$0*e$	e	a	$e{+}a$	b	$e{+}b$	$a{+}b$	$e{+}a{+}b$
a	$0*e$	a	a	$0*e$	a	$0*e$	$0*e$	a
$e{+}a$	$0*e$	$e{+}a$	$0*e$	$e{+}a$	$a{+}b$	$e{+}b$	$a{+}b$	$e{+}b$
b	$0*e$	b	b	$0*e$	b	$0*e$	$0*e$	b
$e{+}b$	$0*e$	$e{+}b$	$a{+}b$	$e{+}a$	$0*e$	$e{+}b$	$a{+}b$	$e{+}a$
$a{+}b$	$0*e$	$a{+}b$	$a{+}b$	$0*a$	$a{+}b$	$0*e$	$0*e$	$a{+}b$
$e{+}a{+}b$	$0*e$	$e{+}a{+}b$	b	$e{+}a$	a	$e{+}b$	$a{+}b$	e

Problems for §9.2

For Problems **1** through **10**: Given the few properties of the generators of a ring, determine the array of products $[[a^2, a \cdot b], [b \cdot a, b^2]]$ that would be used to define the ring in *Sage*.

Hint: Use the associate law to fill in the missing information.

1 $a \cdot b = b, \quad b \cdot a = a$
2 $a \cdot b = a, \quad b \cdot a = 0, \quad b^2 = b$
3 $a^2 = b, \quad a \cdot b = a$
4 $a^2 = a + b, \quad b \cdot a = 0$
5 $a \cdot b = a, \quad b^2 = a + b$
6 $a \cdot b = a + b, \quad b^2 = a + b$
7 $a \cdot b = a, \quad b^2 = a$
8 $a \cdot b = 2b, \quad b \cdot a = a, \quad 3b = 0$
9 $a \cdot b = b, \quad b \cdot a = 3a, \quad 4a = 0$
10 $a^2 = b, \quad b^2 = a, \quad 2a = 2b = 0$

11 If $a^2 = a + b + c$, $a \cdot b = c$, $b \cdot c = a$, and $c \cdot a = a \cdot c = b$, determine b^2, c^2, $b \cdot a$, and $c \cdot b$.

12 Prove that a ring with a cyclic additive group must be commutative.

13 Prove that for m a positive integer, and x and y elements of a ring, then $m(x + y) = mx + my$.

14 Prove that for m and n positive integers, and x an element of a ring, then $(m + n)x = mx + nx$.

15 Prove that for m and n positive integers, and x an element of a ring, then $(mn)x = m(nx)$.

16 Prove that if n is an integer, and x is an element of a ring, then $n(-x) = -(nx)$.

17 Use the fundamental theorem of abelian groups (6.2) to show that every finite abelian group has a basis.

Interactive Problems

18 Use *Sage* to define a ring of order 2 that has no identity element. Show both the addition table and the multiplication table.

19 Use *Sage* to find a non-commutative ring of order 8, for which the additive group is isomorphic to Z_{24}^*, formed from the basis $\{a, b, c\}$, and for which $a \cdot b = a$, $b \cdot a = b$, $a \cdot c = c$, and $c \cdot a = a$.

Hint: Using the associative law, determine what a^2, b^2, and c^2 must be. Then show that $c \cdot b$ must commute with a. Use trial and error to determine $b \cdot c$.

20 Use *Sage* to find a non-commutative ring of order 8, for which the additive group is isomorphic to Z_{24}^*, formed from the basis $\{a, b, c\}$, and for which $a^2 = a + c$, $a \cdot b = b + c$, $b \cdot a = b$, and $c \cdot b = c$.

21 Define in *Sage* a non-commutative ring of order 4.

Hint: By Problem 12, the additive group must by isomorphic to Z_8^*.

9.3 Some Properties of Rings

In this section, we will explore some basic properties that are true for all rings. In particular, we want to study in what circumstances a multiplicative inverse will exist. We can use the finite rings created in *Sage* to help us determine a pattern between zero divisors and invertible elements.

One of the simplest rings to study are the rings Z_n for $n > 1$. We have already learned how to define the additive structure in *Sage* with a **ZGroup** command, and the multiplication can be defined using a **ZStar** command. We actually can define both of these at once with the command

```
Z15 = ZRing(15)
```

This defines both the addition and multiplication operations at the same time. The elements of Z_{15} are

```
Z15
    {0, 1, 2, 3, 4, 5, 6, 7, 8, 9, 10, 11, 12, 13, 14}
```

We can perform simple operations in Z_{15} such as

```
Z15[9] + Z15[7]
    1
Z15[9] * Z15[7]
    3
Z15[9] / Z15[7]
    12
```

This last operation shows that we can take multiplicative inverses of some of the elements. Even though multiplicative inverses are not guaranteed to exist for rings, some elements may be invertible.

LEMMA 9.3
Let x be an element in a ring with identity. Then if x has a multiplicative inverse, the inverse is unique. We denote the multiplicative inverse of x by x^{-1}.

PROOF: Suppose that y and z are two inverses of x. Then

$$y = y \cdot e = y \cdot (x \cdot z) = (y \cdot x) \cdot z = e \cdot z = z,$$

which is a contradiction. □

PROPOSITION 9.4
If R is a unity ring, then the invertible elements of R form a group under multiplication. This group is denoted R^*.

PROOF: Since the unity element is invertible, R^* is non-empty. Also, if x is invertible, then $(x^{-1})^{-1} = x$, so x^{-1} is also in R^*. Finally, if x and y are both invertible, then since

$$(x \cdot y) \cdot (y^{-1} \cdot x^{-1}) = x \cdot x^{-1} = e,$$

we see that $x \cdot y$ is invertible. The associative law comes from the associative multiplication of the ring. So the set of invertible elements forms a group. □

From this, we can find out when Z_n is in fact a field. The first step is to determine when Z_n will have zero divisors.

PROPOSITION 9.5
For $n > 1$, the ring Z_n has no zero divisors if, and only if, n is prime.

PROOF: First suppose that n is not prime. Then we can express $n = ab$, where a and b are less then n. If e represents the identity element of Z_n, we would then have

$$(ae) \cdot (be) = (ab)(e \cdot e) = (ab)e = ne = 0.$$

But since a and b are both less than n, ae and be are both nonzero. Hence, these would both be zero divisors in Z_n.

Now suppose that n is prime, and that there are two nonzero elements ae and be such that $(ae) \cdot (be) = 0$. Then

$$(ae) \cdot (be) = (ab)(e \cdot e) = (ab)e = 0.$$

This would imply that ab is a multiple of n. But since n is prime, we would have to conclude that either a or b is a multiple of n. But this contradicts the fact that both ae and be are nonzero. Thus, if n is prime, there are no zero divisors in Z_n. ∎

Even if n is not prime, one of the observations that can be made while studying Z_n is that the zero divisors were precisely the nonzero elements that did not have an inverse. This is true for many of the rings we have studied.

LEMMA 9.4

Let a, b, and c be elements of a ring. If a is nonzero, and is not a zero divisor, and

$$a \cdot b = a \cdot c,$$

then $b = c$. Likewise, if

$$b \cdot a = c \cdot a$$

for a nonzero and not a zero divisor, then $b = c$. This is called the cancellation law for multiplication.

PROOF: The tempting thing to do is to multiply both sides of the equation by a^{-1}. But the inverse of a may not exist, so we have to use the properties of rings instead.

If $a \cdot b = a \cdot c$ then we have

$$0 = a \cdot b - a \cdot c = a \cdot (b - c).$$

But since a is not a zero-divisor and is nonzero, we must have that $b - c = 0$. Hence $b = c$.

Likewise, if $b \cdot a = c \cdot a$, then

$$0 = b \cdot a - c \cdot a = (b - c) \cdot a$$

and since a is nonzero and not a zero divisor, $b - c = 0$, and so $b = c$. ∎

We are now ready to show a relationship between zero divisors and invertable elements. Notice that in the ring $\mathbb{Z}$, the element 2 is not invertible, but neither is it a zero divisor. This example seems to break the pattern that we have been observing, but also notice that $\mathbb{Z}$ is an *infinite* ring. Perhaps if we consider only *finite* rings we will be able to prove a relationship between zero divisors and invertible elements.

PROPOSITION 9.6

Let R be a finite ring. If b is a nonzero element of R which is not a zero divisor, then R has a unity element and b has a multiplicative inverse in R. Hence, every nonzero element in R is either a zero divisor or is invertible.

PROOF: To utilize the fact that R is finite, let us construct a sequence of powers of b:

$$\{b^1, b^2, b^3, b^4, \ldots\}.$$

Since R is finite, two elements of this sequence must be equal, say $b^m = b^n$ for $m < n$. Using the law of cancellation, we have $b^{m-1} = b^{n-1}$. Continuing this way, we eventually get $b = b^{n-m+1}$. (It is tempting to use Lemma 9.4 one more time to get $e = b^{n-m}$, but unfortunately we have yet to prove that R has a unity.)

If we now let $a = n - m + 1$, we have that $a > 1$ and $b^a = b$.

Next, let us show that b^{a-1} is a unity element in R. For any element x in R, we have

$$x \cdot b^a = x \cdot b,$$

and since b is nonzero and not a zero divisor, we can use the law of cancellation to get

$$x \cdot b^{a-1} = x.$$

Likewise, since $b^a \cdot x = b \cdot x$, we have that $b^{a-1} \cdot x = x$. Hence, there is a unity element in R, namely b^{a-1}.

Finally, we need to construct an inverse for the element b. If $a = 2$, then we have just shown that $b = e$, and hence b is its own inverse. If $a > 2$, consider the element b^{a-2}. We have that

$$b^{a-2} \cdot b = b^{a-1} = e \qquad \text{and} \qquad b \cdot b^{a-2} = b^{a-1} = e.$$

So b^{a-2} is the multiplicative inverse of b. □

COROLLARY 9.1
Every finite ring without zero divisors is a division ring.

PROOF: The trivial ring is already considered to be a division ring, so we may assume that the ring is nontrivial. Then there exists a nonzero element that is not a zero divisor, so by Proposition 9.6, the ring has a unity. Also by Proposition 9.6, every nonzero element will have a multiplicative inverse, so the ring is a division ring. □

We finally can determine which Z_n are fields.

COROLLARY 9.2
The ring Z_n is a field if, and only if, n is prime.

PROOF: If $n = 1$, then the ring $Z_n = Z_1$ is the trivial ring, which we did not consider to be a field. We may suppose that $n > 1$. If n is prime, then by Proposition 9.5 Z_n has no zero divisors, and so by Corollary 9.1 Z_n is a division ring. Since Z_n is obviously commutative, this tells us that Z_n is a field.

TABLE 9.6: The non-commutative ring T_4

$+$	0	a	b	c
0	0	a	b	c
a	a	0	c	b
b	b	c	0	a
c	c	b	a	0

$\cdot$	0	a	b	c
0	0	0	0	0
a	0	a	a	0
b	0	b	b	0
c	0	c	c	0

TABLE 9.7: The smallest non-commutative unity ring T_8

$+$	0	e	a	b	c	d	f	g
0	0	e	a	b	c	d	f	g
e	e	0	d	f	g	a	b	c
a	a	d	0	c	b	e	g	f
b	b	f	c	0	a	g	e	d
c	c	g	b	a	0	f	d	e
d	d	a	e	g	f	0	c	b
f	f	b	g	e	d	c	0	a
g	g	c	f	d	e	b	a	0

$\cdot$	0	e	a	b	c	d	f	g
0	0	0	0	0	0	0	0	0
e	0	e	a	b	c	d	f	g
a	0	a	a	a	0	0	0	a
b	0	b	b	b	0	0	0	b
c	0	c	c	c	0	0	0	c
d	0	d	0	c	c	d	f	f
f	0	f	c	0	c	d	f	d
g	0	g	b	a	c	d	f	e

Now suppose that $n > 1$ and n is not prime. By Proposition 9.5, Z_n has zero divisors, which cannot exist in a field according to Proposition 9.2. Therefore Z_n is a field if, and only if, n is prime. □

To conclude this chapter, let us find an example of each of the 11 different types of rings that could exist. First we define the rings T_4 in Table 9.6, and we will rewrite the elements of T_8 into a more compact form in Table 9.7. Then every ring will fall into one of the categories given in Table 9.8.

Problems for §9.3

1 Show that the non-commutative ring T_4 given by Table 9.6 has two elements r such that $x \cdot r = x$ for all x in the ring, yet has no element for which $r \cdot x = x$ for all x in the ring.

2 Let x be an element of a commutative ring R that has an inverse x^{-1}. Let y be another element of R such that $y^2 = 0$. Prove that $x + y$ has an inverse in R.

3 Let x be an element of a commutative ring R that has an inverse x^{-1}. Let y be another element of R such that $y^3 = 0$. Prove that $x + y$ has an inverse in R.

TABLE 9.8: Examples for the 11 possible types of rings

Type	Name	Example(s)
I	The trivial ring	Only one such ring, $\{0\}$.
II	Fields	$\mathbb{R}$, $\mathbb{Q}$, Z_p with p prime.
III	Skew fields	$\mathbb{H} =$ the quaternions.
IV	Commutative unity rings w/o zero divisors, but that are not fields	$\mathbb{Z}$, polynomials. These rings are called *integral domains*.
V	Non-commutative unity rings w/o zero divisors, but that are not skew fields	Integer quaternions: $a + bi + cj + dk$, with $a, b, c, d \in \mathbb{Z}$.
VI	Commutative rings w/o unity and w/o zero divisors	Even integers, multiples of n, $n > 1$.
VII	Non-commutative rings w/o unity and w/o zero divisors	Even quaternions.
VIII	Commutative unity rings w/ zero divisors	Z_n whenever $n > 1$ and n is not prime.
IX	Non-commutative unity rings w/ zero divisors	T_8 in table 9.7.
X	Commutative rings w/o unity and w/ zero divisors	The subset $\{0, 2, 4, 6\}$ of Z_8.
XI	Non-commutative rings w/o unity and w/ zero divisors	T_4 in table 9.6.

4 Find a specific example of two elements x and y in a ring R such that $x \cdot y = 0$, but $y \cdot x$ is nonzero.

Hint: Which of the 11 types of rings would R have to be?

5 Consider the subset $\{0, 2, 4, 6, 8\}$ of Z_{10}. Form addition and multiplication tables of this set. Is this a ring? Which of the 11 types of ring is this?

6 Let R be a ring for which $x^2 = x$ for all x in the ring. Prove that $-x = x$ for all elements x. Such rings are called *Boolean* rings.

7 Let R be a ring for which $x^2 = x$ for all x in the ring. Prove that the ring R is commutative. (See Problem 6.)

8 Let R be a ring for which $x^3 = x$ for all x in the ring. Prove that $6x = 0$ for all x in the ring.

9 An element a in a ring R is *idempotent* if $a^2 = a$. Prove that a nontrivial division ring must contain exactly two idempotent elements.

10 Let a be an idempotent element in a ring with unity. Show that $e - a$ is also an idempotent element. See Problem 9.

11 Show that if R is a commutative ring, and x and y are elements of R, then

$$(x + y)^2 = x^2 + 2xy + y^2$$

and

$$(x + y)^3 = x^3 + 3x^2y + 3xy^2 + y^3.$$

12 Let R be a commutative ring. Define the *binomial coefficient*

$$\binom{n}{k} = \frac{n \cdot (n-1) \cdot (n-2) \cdots (n-k+1)}{1 \cdot 2 \cdot 3 \cdots k}, (0 \le k \le n).$$

Using induction, prove the *binomial theorem* in R:

$$(x + y)^n = x^n + \binom{n}{1} x^{n-1} y + \binom{n}{2} x^{n-2} y^2 + \cdots + \binom{n}{n} y^n.$$

13 Determine all elements of T_8 in Table 9.7 that have a multiplicative inverse.

14 Determine all elements of the ring defined by Tables 9.3 and 9.4 in Chapter 9 that have a multiplicative inverse.

15 An *irreducible* element p of a ring R is a non-invertible element for which the only way for $p = a \cdot b$ is for either a or b to have a multiplicative inverse. Determine the irreducible elements of the ring defined by Tables 9.3 and 9.4.

Hint: Cross out the rows and columns corresponding to the invertible elements. Which elements are no longer in the interior of the table?

16 Does T_4 or T_8 in Tables 9.6 and 9.7 have any irreducible elements? (See Problem 15.)

17 A *prime* element $p \neq 0$ of a ring R is a non-invertible element such that, whenever $a \cdot b$ is a multiple of p, either a or b is a multiple of p. (A multiple of p would be any element that can be expressed as either $x \cdot p$ or $p \cdot x$.) Find a prime element of the ring T_8 in Table 9.7.

Hint: To determine if p is prime, first find all the multiples of p. Then cross out the rows and columns of the multiplication table corresponding to those elements. If there are no more multiples of p remaining, then p is prime.

18 Find a prime element of the ring defined by Tables 9.3 and 9.4 that is not irreducible. (See Problems 15 and 17.)

Interactive Problems

19 Define in *Sage* the smallest non-commutative ring, T_4 defined by Table 9.6. Use a and c and the generators.

20 Define in *Sage* the smallest non-commutative unity ring T_8 defined by Table 9.7.

Hint: The basis can be chosen to be e, a, and b.

Chapter 10

The Structure within Rings

Just as we can have subgroups, normal subgroup, quotient groups, and homomorphisms between groups, we can have similar structures within rings. In fact, ring theory runs almost parallel with the study of groups. In this chapter we will demonstrate the similarities between the two theories. These similarities are startling, since a "normal subring" is defined totally differently than a normal subgroup.

10.1 Subrings

It is natural to ask whether we can have smaller rings within a larger ring, just as we saw smaller groups inside of a larger group. This suggests the following definition.

DEFINITION 10.1 Let R be a ring. A non-empty subset S is a *subring* if S is a ring with respect to the addition $(+)$ and multiplication $(\cdot)$ of R.

We have already seen some examples of subrings. For example, the set of even integers is a ring contained in the ring of integers, which is contained in the ring of rational numbers, which in turn is contained in the ring of real numbers. The next proposition gives us a quick way to determine if a subset is indeed a subring.

PROPOSITION 10.1

A non-empty subset S is a subring of a ring R if, and only if, whenever x and y are in S, $x - y$ and $x \cdot y$ are in S.

PROOF: Certainly if S is a subring, then $x - y$ and $x \cdot y$ would be in S whenever x and y are in S. So let us suppose that S is non-empty, and is closed with respect to subtraction and multiplication. If x is any element in S, then $x - x = 0$ is in S, so S contains an additive identity. Also, $0 - x = -x$ would also be in S, so S contains additive inverses of all of its elements. Then whenever x and y are in S, $x - (-y) = x + y$ is in S, so S is closed with respect to addition. The commutative and associative properties of addition, as well as the associative and two distributive laws for multiplication, come from the original ring R. Finally, S is closed with respect to multiplication, so S is a subring. ∎

Notice that from the definition every nontrivial ring R will contain at least two subrings: the trivial ring $\{0\}$ will be a subring, as well as the entire ring R. These two subrings are called the *trivial subrings.*

Example 10.1
Consider the subset of real numbers of the form

$$S = \{x + y\sqrt{2} \mid x, y \in \mathbb{Z}\}.$$

Determine whether or not this is a subring of $\mathbb{R}$.
SOLUTION: Two typical elements of S are $a = x_1 + y_1\sqrt{2}$ and $b = x_2 + y_2\sqrt{2}$. Then

$$a - b = (x_1 - x_2) + (y_1 - y_2)\sqrt{2},$$

and

$$a \cdot b = (x_1x_2 + 2y_1y_2) + (x_1y_2 + x_2y_1)\sqrt{2}.$$

Since all expressions in parenthesis are integers, these elements are in S. Thus, by Proposition 10.1, S is a subring of $\mathbb{R}$. ∎

Computational Example 10.2
Here is the ring of order 8 we defined by Tables 9.3 and 9.4:

```
InitRing("a", "b")
DefineRing([4, 2],[[a, 0], [0, b]])
R = ListRing(); R
    {0*a, a, 2*a, -a, b, a+b, 2*a+b, -a+b}
```

The set

```
S = [0*a, a, 2*a, -a]; S
    [0*a, a, 2*a, -a]
```

can be seen to be a subring from the addition and multiplication tables in Table 10.1.

One can see that S is closed with respect to both addition and multiplication. Furthermore, additive inverses exist for all elements, so S is also closed with respect to subtraction. Thus, by Proposition 10.1, this is a subring. ∎

Ironically, the subring S has a unity element,

TABLE 10.1: Tables for a particular subring S

$+$	$0a$	a	$2a$	$-a$
$0a$	$0a$	a	$2a$	$-a$
a	a	$2a$	$-a$	$0a$
$2a$	$2a$	$-a$	$0a$	a
$-a$	$-a$	$0a$	a	$2a$

$\cdot$	$0a$	a	$2a$	$-a$
$0a$	$0a$	$0a$	$0a$	$0a$
a	$0a$	a	$2a$	$-a$
$2a$	$0a$	$2a$	$0a$	$2a$
$-a$	0	$-a$	$2a$	a

```
FindUnity(S)
    a
```

which is different than the unity of R. In general the existence of a subring's unity is totally independent of the unity of R.

Recall that the intersection of a number of subgroups was again a subgroup. We could ask whether the same is true for subrings.

PROPOSITION 10.2

Given any non-empty collection of subrings of the group R, denoted by L, then the intersection of all of the subrings in the collection

$$H^* = \bigcap_{H \in L} H$$

is a subring of R.

PROOF: First of all, note that H^* is not the empty set, since 0 is in each H in the collection. We now can apply Proposition 10.1. Let x and y be two elements in H^*. Then, for every $H \in L$, we have $x, y \in H$.

Since each H is a subring of R, we have $x - y \in H$ and $x \cdot y \in H$ for all $H \in L$. Therefore, $x - y$ and $x \cdot y$ are in H^*, and so H^* is a subring of R. ∎

As with subgroups, we now have a general method of producing subrings of a ring R. Let S be any subset of R. We can consider the collection L of all subrings of R that contain the set S. This collection is non-empty since it contains the subring R itself. So by Proposition 10.2,

$$[S] = H^* = \bigcap_{H \in L} H$$

is a subring of R. By the way that the collection was defined, $[S]$ contains S. Actually, $[S]$ is the *smallest* subring of R containing the subset S. For if H is a subring of R that contains S, then $H \in L$, so that $[S] \subseteq H$.

DEFINITION 10.2 We call $[S]$ the subring of R *generated* by the set S.

Example 10.3

Find the subring of T_8 from Table 9.7 generated by the element g.

SOLUTION: Clearly 0 is in the subring, and since $g+g=0$, the set $\{0,g\}$ is closed under subtraction. But $g^2=e$, so this element is in the subring. This causes $g+e=c$ to be in the subring as well. The set $\{0,c,e,g\}$ can be seen to be closed under addition, multiplication, and additive inverses. So $[g]=\{0,c,e,g\}$. □

Just as in the case for the **Group** command, the command **Ring** finds $[S]$ for any set S in *Sage*. For example, we can find some subrings for the non-commutative group of order 8,

```
InitRing("a", "b")
DefineRing([4, 2], [[a, b], [2*a, 0]])
R = ListRing(); R
    {0*a, a, 2*a, -a, b, a+b, 2*a+b, -a+b}
```

with the commands

```
Ring(0*a)
    {0*a}
Ring(a)
    {0*a, a, 2*a, -a}
Ring(2*a)
    {0*a, 2*a}
Ring(2*a, b)
    {0*a, b, 2*a, 2*a+b}
```

In this way, we can find all subrings of the ring R. It turns out that there are six nontrivial subrings for this ring, corresponding to the six nontrivial subgroups of Z_{15}^*.

We can also find all of the subrings for the infinite ring $\mathbb{Z}$.

PROPOSITION 10.3

A subring of the ring of integers $\mathbb{Z}$ consists of all multiples of some non-negative number n. This subring is denoted $n\mathbb{Z}$.

PROOF: First of all, the trivial subring $\{0\}$ can be considered the set of all multiples of 0. Also, the entire ring $\mathbb{Z}$ could be considered all of the multiples of 1. Let S be a nontrivial subring, and let $x \neq 0$ be in S. Then $-x$ is also in S, so S must contain some positive integers. Let n be the smallest positive integer contained in S. Certainly all multiples of n would be in S, but suppose that some element m in S is not a multiple of n. Then by the greatest common divisor theorem (0.4), there exist two integers u and v such that

$$un + vm = \gcd(n, m).$$

Since S is closed under addition, this implies that $\gcd(n, m)$ is in S. But m is not a multiple of n, so $\gcd(n, m) < n$. But this contradicts the fact that n is the *smallest* positive integer in S. Thus, S consists exactly of all of the multiples of n, and so $S = n\mathbb{Z}$. □

Although the subrings of $\mathbb{Z}$ are easily classified, this is not the case with the ring of real numbers. Example 10.1 gives just one of countless subrings of $\mathbb{R}$:

$$S = \{x + y\sqrt{2} \mid x, y \in \mathbb{Z}\}.$$

It is actually possible to define this subring in *Sage*. We can let e represent 1, and a represent $\sqrt{2}$. These two elements are both of infinite additive order. We can convey this to *Sage* by entering "0" for the order of each of the elements. Then $a^2 = 2e$, so the ring can be entered by the commands

```
InitRing("e", "a")
DefineRing([0, 0], [[e, a], [a, 2*e]])
ListRing()
    'Ring is infinite.'
```

Of course we cannot list the elements, since there are an infinite number of elements. But we can still do operations in this ring.

```
(e + 2*a) * (4*e - 3*a)
    -8*e+5*a
```

This last statement demonstrates that

$$(1 + 2\sqrt{2}) \cdot (4 - 3\sqrt{2}) = -8 + 5\sqrt{2}.$$

Clearly, the subrings of the real numbers can be much more complicated than the subrings of integers.

Problems for §10.1

For Problems **1** through **10**: Use Proposition 10.1 to determine if the following subsets are subrings of $\mathbb{R}$.

1 $\{x + y\sqrt{5} \mid x, y \in \mathbb{Z}\}$
2 $\{x + y\sqrt{2} \mid x, y \in \mathbb{Q}\}$
3 $\{x \mid x \in \mathbb{R}, x > 0\}$
4 $\{x/y \mid x \text{ is an even integer}, y \text{ is an odd integer}\}$
5 $\{x/(2^y) \mid x, y \in \mathbb{Z}, y \geq 0\}$
6 $\{x + y\sqrt[3]{2} \mid x, y \in \mathbb{Z}\}$
7 $\{x + y\sqrt[3]{2} + z\sqrt[3]{4} \mid x, y, z \in \mathbb{Z}\}$
8 $\{x + y\sqrt{2} \mid y \in \mathbb{Z}, x \text{ is an even integer}\}$
9 $\{x + y\sqrt{2} \mid x, y \in \mathbb{Z}, x + y \text{ is even}\}$
10 $\{x + y\sqrt{3} \mid x, y \in \mathbb{Z}, x + y \text{ is even}\}$

11 Let y be an element of a ring R. Let

$$A = \{x \in R \mid x \cdot y = 0\}.$$

Show that A is a subring of R.

12 Let y be an element of a ring R. Let

$$B = \{x \cdot y \mid x \in R\}.$$

Show that B is a subring of R.

13 Let R be a ring, and let

$$Z = \{x \in R \mid x \cdot y = y \cdot x \text{ for all } y \in R\}.$$

Show that Z is a subring of R. This subring is called the *center* of R.

14 An element x of a ring R is called *nilpotent* if $x^n = 0$ for some positive number n. Show that the set of all nilpotent elements in a commutative ring R forms an subring of R.

Hint: See Problem 12 of §9.3.

15 Show that $2\mathbb{Z} \cup 3\mathbb{Z}$ is not a subring of $\mathbb{Z}$. (The symbol $\cup$ denotes the *union* of the two sets.)

16 Find all of the subrings of the commutative ring of order 8 defined by Tables 9.3 and 9.4 in Chapter 9.

Hint: There are eight subgroups of the additive group Z_{15}^*. Find the eight subgroups, and determine which subgroups are in fact subrings.

17 Find all of the subrings of T_4 in Table 9.6.

18 Find all of the subrings of T_8 in Table 9.7.

Hint: First find all 16 subgroups of the additive group, Z_{24}^*.

Interactive Problems

19 Find all of the subrings of the ring of order 8:

```
InitRing("a", "b")
DefineRing([4, 2], [[a, b], [0, 0]])
R = ListRing(); R
    {0*a, a, 2*a, -a, b, a+b, 2*a+b, -a+b}
```

20 Find all of the subrings of the ring of order 8:

```
InitRing("a", "b")
DefineRing([4, 2], [[2*a, 0], [2*a, 2*a]])
R = ListRing(); R
    {0*a, a, 2*a, -a, b, a+b, 2*a+b, -a+b}
```

10.2 Quotient Rings and Ideals

When we studied group theory, one of the most important concepts we discovered was being able to form a quotient group out of the cosets of certain subgroups—namely the normal subgroups. A natural question is whether it is possible to form quotient rings out of the cosets of a subring. In this section we will explore the possibility of forming a quotient ring, and in the process we will define the *ideal*, which roughly corresponds to a normal subgroup. We begin by looking at some examples.

Motivating Example 10.4

Here is the non-commutative ring of order 8 from the last section.

```
InitRing("a", "b")
DefineRing([4, 2], [[a, b], [2*a, 0]])
R = ListRing(); R
    {0*a, a, 2*a, -a, b, a+b, 2*a+b, -a+b}
```

Can we form a quotient ring out of this ring, the way that we constructed a quotient group?

SOLUTION: We found this ring has six nontrivial subrings.

$$S_1 = \{0, a, 2a, 3a\}, \qquad S_2 = \{0, 2a\}, \qquad S_3 = \{0, b\},$$
$$S_4 = \{0, a+b, 2a, 3a+b\}, \quad S_5 = \{0, 2a+b\}, \quad S_6 = \{0, 2a, b, 2a+b\}.$$

We would expect the additive structure of the quotient ring to be the additive quotient group R/S. We can use *Sage* to find the cosets of S under the operation of addition. Since left and right cosets are the same when working with rings, we will simply use the **Coset** command.

```
S1 = Ring(a); S1
    {0*a, a, 2*a, -a}
Q = Coset(R, S1); Q
    {{0*a, a, 2*a, -a}, {b, a+b, 2*a+b, -a+b}}
```

We can *add* two cosets together using the following definition:

$$X + Y = \{x + y \mid x \in X \quad \text{and} \quad y \in Y\}.$$

This gives us a natural way to add the elements of the quotient Q, which is shown in Table 10.2, produced by the command **AddTable(Q)**.

The natural way to define the product of two sets is the way we defined such a product for groups:

$$X \cdot Y = \{x \cdot y \mid x \in X \quad \text{and} \quad y \in Y\}.$$

TABLE 10.2: Addition table for the quotient ring Q

$+$	$\{0, a, 2a, 3a\}$	$\{b, a+b, 2a+b, 3a+b\}$
$\{0, a, 2a, 3a\}$	$\{0, a, 2a, 3a\}$	$\{b, a+b, 2a+b, 3a+b\}$
$\{b, a+b, 2a+b, 3a+b\}$	$\{b, a+b, 2a+b, 3a+b\}$	$\{0, a, 2a, 3a\}$

Will such a product of two cosets in Q yield another coset?

Unfortunately no! The multiplication tables in *Sage* reveal black squares—which indicate that the product of two cosets is not a coset. The problem lies in the following two cosets:

```
Q1 = S1; Q1
    {0*a, a, 2*a, -a}
Q2 = b + S1; Q2
    {b, a+b, 2*a+b, -a+b}
Q1 * Q2
    {0*a, b, a+b, 2*a, 2*a+b, -a+b}
```

which produces extra elements. To ensure that S acts as the zero element in the product of cosets, we need to have S times any element of R to produce only elements in S.

Suppose we found a subring S for which $S \cdot x$ always was a subset of S. By the same argument we would also require that $x \cdot S$ be a subset of S. Using *Sage*, we can test the other subrings.

```
S2 = Ring(2*a); S2
    {0*a, 2*a}
S2 * R
    {0*a, 2*a}
R * S2
    {0*a, 2*a}
```

We see that both $R \cdot S_2$ and $S_2 \cdot R$ are subsets of S_2, so this ensures that the additive identity of the quotient group $\{0, 2a\}$ will behave as the zero element in the product of cosets. The multiplication table for the quotient group is as given by the commands

```
Q = Coset(R, S2); Q
    {{0*a, 2*a}, {a, -a}, {b, 2*a+b}, {a+b, -a+b}}
MultTable(Q)
```

which produce Table 10.3. □

This multiplication table is non-commutative, even though all of the subrings of R are commutative. So this quotient is unlike any of the subrings of R.

TABLE 10.3: Multiplying cosets of the subring S_2

$\cdot$	$\{0a, 2a\}$	$\{a, -a\}$	$\{b, 2a+b\}$	$\{a+b, -a+b\}$
$\{0a, 2a\}$	$\{0a\}$	$\{0a, 2a\}$	$\{0a\}$	$\{0a, 2a\}$
$\{a, -a\}$	$\{0a, 2a\}$	$\{a, -a\}$	$\{b, 2a+b\}$	$\{a+b, -a+b\}$
$\{b, 2a+b\}$	$\{0a\}$	$\{0a, 2a\}$	$\{0a\}$	$\{0, 2a\}$
$\{a+b, -a+b\}$	$\{0a, 2a\}$	$\{a, -a\}$	$\{b, 2a+b\}$	$\{a+b, -a+b\}$

However, not every product yields a coset—sometimes it yields only a *subset* of a coset. One way to rectify this slight blemish in our multiplication table is to add the identity coset to each entry in the table. That is, instead of defining the product of the cosets X and Y to be $X \cdot Y$, we define the product of two cosets to be

$$X * Y = X \cdot Y + S.$$

The command

```
QuotientRing = true
```

creates a multiplication table using this new definition of the product of two cosets. Thus, **MultTable(Q)** produces a table similar to Table 10.3, only every $\{0a\}$ is replaced by $\{0a, 2a\}$.

The key to getting the quotient ring to work lies in the fact that $S_2 \cdot R$ and $R \cdot S_2$ were subsets of S_2. Let us first define the special type of subring that will allow quotient rings.

DEFINITION 10.3 A subring I of a ring R is called an *ideal* of R if both $I \cdot R$ and $R \cdot I$ are contained in the subring I.

We already observed that if a subring is not an ideal, then the quotient ring cannot be defined. Let us now show that a quotient ring can be defined provided that I is an ideal.

PROPOSITION 10.4

Let R be a ring, and let I be an ideal of R. Then the additive quotient group R/I forms a ring, with the product of two cosets X and Y being $X * Y = X \cdot Y + I$. This ring is called the quotient ring R/I.

PROOF: The quotient group R/I is an abelian group, so we need only to check that the multiplication is closed, and that the associativity and two distributive laws hold.

Let X and Y be two cosets of R/I. Let x be an element in X, and y an element in Y. Then the product of the cosets X and Y is

$$X * Y = X \cdot Y + I = (x + I) \cdot (y + I) + I = x \cdot y + I \cdot y + x \cdot I + I \cdot I + I.$$

Historical Diversion

Richard Dedekind (1831–1916)

Dedekind was born Julius Wilhelm Richard Dedekind in Braunschweig, Germany, but he never used his first two names as an adult. He attending Collegium Carolinum in 1848, and then moved to the University of Göttingen in 1950. He attended lectures under Carl Gauss, but he was teaching mainly elementary-level mathematics at the time. Dedekind is considered to be Gauss' last student. Dedekind received his doctorate in 1852.

Since the University of Berlin was considered the leading center for mathematics, Dedekind went to Berlin for two years. There he met a contemporary, Bernhard Riemann, and together in 1854 they were awarded the habilitation, which is the highest academic award a scholar could achieve. Dedekind returned to Göttingen to teach as a Privatdozent, and was the first at Göttingen to lecture on Galois theory. Dedekind understood the importance of group theory for algebra and arithmetic.

In 1858, he began to teach at the Polytechnic in Zürich. While teaching calculus for the first time, he came up with the idea we now call the Dedekind cut. He associated every real number a with a set of rational numbers less than a. Limits can then be expressed in terms of set theory. With this idea Dedekind could show that there were no gaps, or discontinuities, on the number line. This put the real number system on a firm foundation.

Dedekind also worked with infinite sets, defining two sets as "similar" if there is a one-to-one and onto mapping between the two sets. This led to the first precise definition of an infinite set. In 1872, he met Georg Cantor while on holiday in Interlaken. Dedekind became a close ally of Cantor during his philosophical battles with Kronecker. (See Historical Diversion on page 33.)

In 1879, Dedekind gereralized Kummer's *ideal numbers* to formulate a definition of an ideal. (See Historical Diversion on page 432.) His definition was a subset of a set of numbers, all of which were *algebraic integers*, that is, they satisfied a polynomial equation with integer coefficients, and a leading coefficient of 1. Dedekind's definition of an ideal would later be generalized by Emmy Noether. (See Historical Diversion on page 306.)

Dedekind is also known for the Dedekind domain, which is an integral domain for which every non-trivial ideal factors into a product of prime ideals. Kummer showed that $\mathbb{Z}[\omega_n]$ has this property for all n, but Dedekind generalized this for all domains of algebraic integers.

Image source: Wikimedia Commons

Because I is an ideal, $I \cdot y$, $x \cdot I$, and $I \cdot I$ are all subsets of I. Hence, the sum $I \cdot y + x \cdot I + I \cdot I + I$ will be a subset of I. But since the last term of this expression is I, $I \cdot y + x \cdot I + I \cdot I + I$ contains the ideal I, so this sum equals I. Thus,

$$(x + I) * (y + I) = X * Y = X \cdot Y + I = x \cdot y + I,$$

which is a coset of R/I.

Now suppose that X, Y, and Z are three cosets of R/I with x, y, and z being representative elements, respectively. Then

$$\begin{aligned}(X * Y) * Z &= ((x + I) * (y + I)) * (z + I) \\ &= (x \cdot y + I) * (z + I) \\ &= ((x \cdot y) \cdot z + I) \\ &= (x \cdot (y \cdot z) + I) \\ &= (x + I) * (y \cdot z + I) \\ &= (x + I) * ((y + I) * (z + I)) \\ &= X * (Y * Z).\end{aligned}$$

So multiplication is associative. Also,

$$\begin{aligned}X * (Y + Z) &= (x + I) * (y + z + I) \\ &= (x \cdot (y + z) + I) \\ &= x \cdot y + x \cdot z + I \\ &= (x \cdot y + I) + (x \cdot z + I) \\ &= X * Y + X * Z,\end{aligned}$$

and

$$\begin{aligned}(X + Y) * Z &= (x + y + I) * (z + I) \\ &= ((x + y) \cdot z + I) \\ &= x \cdot z + y \cdot z + I \\ &= (x \cdot z + I) + (y \cdot z + I) \\ &= X * Z + Y * Z.\end{aligned}$$

Thus, the two distributive laws hold, so R/I is a ring. ∎

This shows that the ideals play the same role for rings that normal subgroups did for groups, namely that subsets with an additional property allow for quotients to be defined.

Example 10.5

Find the ideals of the ring $\mathbb{Z}$, and determine the quotient rings.

SOLUTION: By Proposition 10.3, all subrings are of the form $S = n\mathbb{Z}$ for some n. Yet any multiple of n times an integer yields a multiple of n, so $S \cdot \mathbb{Z} = \mathbb{Z} \cdot S = S$. Therefore, every subring of $\mathbb{Z}$ is an ideal.

The cosets of the quotient ring $\mathbb{Z}/(n\mathbb{Z})$ can be expressed in the form

$$a + n\mathbb{Z},$$

where $a = 0, 1, 2, \ldots n-1$. Clearly the quotient ring behaves exactly like the ring Z_n. We say that the quotient ring is *isomorphic* to Z_n. ▯

In contrast, let us consider a ring like the rational numbers $\mathbb{Q}$. Even though there are a host of subrings of $\mathbb{Q}$, the only ideals are the trivial subrings. This can be generalized by the following proposition.

PROPOSITION 10.5
Any field or skew field can only have trivial ideals.

PROOF: Let K be a field or skew field, and suppose that there is a nontrivial ideal I of K. Then there is a nonzero element x in I, and hence x^{-1} exists in K. Thus

$$e = x \cdot x^{-1} \in I \cdot K \subseteq I.$$

So the multiplicative identity e is contained in I. But then,

$$K = e \cdot K \subseteq I \cdot K \subseteq I.$$

Hence, $I = K$, so the only ideals of K are the trivial ideals. ▯

We have already observed that the intersection of two subrings is again a subring. The natural question is whether the intersection of two ideals gives an ideal. This will help us to find all of the ideals of a given ring.

PROPOSITION 10.6
If L is a non-empty collection of ideals of a ring R, then the intersection of all of these ideals

$$I^* = \bigcap_{I \in L} I$$

is an ideal of R.

PROOF: Since I^* is an intersection of subrings of R, by Proposition 10.2 I^* is a subring of R. Thus, we only need to check that $I^* \cdot R$ and $R \cdot I^*$ are contained in I^*.

Suppose that x is an element of I^*. Then x is in each $I \in L$, and so $x \cdot R$ and $R \cdot x$ are subsets of each I in the collection. Thus, $x \cdot R$ and $R \cdot x$ will both be subsets of I^*. Since this result is true for every x in I^*, we have that $I^* \cdot R$ and $R \cdot I^*$ are both subsets of I^*. Therefore, I^* is an ideal. ▯

We can now define the smallest ideal of R that contains a subset S. We proceed as we did for subrings, and consider the collection L of all ideals of R containing S. Then the smallest ideal of R containing S would be

$$\langle S \rangle = \bigcap_{I \in L} I.$$

We call $\langle S \rangle$ the *ideal generated by* S. Notice the distinction between this notation and the notation $[S]$ of the subring generated by S. If S contains only one element, say a, we will use the notation $\langle a \rangle$ rather than the cumbersome $\langle \{a\} \rangle$ to denote the ideal generated by a.

This proposition allows us to quickly find all ideals of a ring.

Computational Example 10.6

Find the ideals in the non-commutative ring R of order 8,

```
InitRing("a", "b")
DefineRing([4, 2], [[a, b], [2*a, 0]])
R = ListRing(); R
    {0*a, a, 2*a, -a, b, a+b, 2*a+b, -a+b}
```

SOLUTION: We can have *Sage* find $\langle S \rangle$ using the command **Ideal(R, S)** for different subsets S. For example, when $S = \{a\}$,

```
Ideal(R, a)
    {0*a, a, 2*a, -a, b, a+b, 2*a+b, -a+b}
```

we find that this command produces the whole ring, so a cannot be contained in any nontrivial ideal. Likewise, $-a$, $a + b$, and $-a + b$ cannot be in a nontrivial ideal. The three remaining nonzero elements, $2a$, b, and $2a + b$, generate different ideals.

```
Ideal(R, 2*a)
    {0*a, 2*a}
Ideal(R, b)
    {0*a, b, 2*a, 2*a+b}
Ideal(R, 2*a+b)
    {0*a, 2*a+b}
```

These three ideals will be denoted by $\langle 2a \rangle$, $\langle b \rangle$, and $\langle 2a + b \rangle$. It is clear that any ideal containing two out of three of these elements must contain b, and therefore must be $\langle b \rangle$. Hence, there are exactly five ideals in this ring: the two trivial ideals that can be denoted $\langle 0 \rangle$ and $\langle a \rangle$, and the three ideals $\langle 2a \rangle$, $\langle b \rangle$, and $\langle 2a + b \rangle$. ∎

Notice that all five ideals can be generated with only one element. We will give a special name for these ideals.

DEFINITION 10.4 An ideal of R that is generated by only one element of R is called a *principal ideal*. If all of the ideals of R are principal ideals, then the ring is called a *principal ideal ring*.

The ring of integers $\mathbb{Z}$ is a principal ideal ring, since all ideals (in fact all subrings) are of the form $n\mathbb{Z}$, which is generated by the single element n. Since $\mathbb{Z}$ is also an integral domain, we will combine the two terms and call $\mathbb{Z}$ a *principal ideal domain*, or *PID*. We will talk more about PIDs in §12.3.

Problems for §10.2

1 If X and Y are ideals of a ring, show that the *sum* of X and Y,

$$X + Y = \{x + y \mid x \in X \text{ and } y \in Y\}$$

is an ideal.

2 In the ring of integers, find a positive integer n such that

$$\langle n \rangle = \langle 12 \rangle + \langle 16 \rangle.$$

(See Problem 1.)

3 If X and Y are ideals of a ring, show that the *product* of X and Y,

$$X \cdot Y = \{x_1 \cdot y_1 + x_2 \cdot y_2 + \cdots + x_n \cdot y_n \mid x_i \in X \text{ and } y_i \in Y,\ n > 0\},$$

is an ideal.

4 In the ring of integers, find a positive integer n such that

$$\langle n \rangle = \langle 12 \rangle \cdot \langle 16 \rangle.$$

(See Problem 3.)

5 Let X and Y be ideals of a ring. Prove that $X \cdot Y \subseteq X \cap Y$. (See Problem 3.)

6 Let R be a ring and let p be a fixed prime. Define I_p to be the set of elements for which the additive order of the element is a power of p. Show that I_p is an ideal.

7 Find all of the ideals of the commutative ring of order 8 defined by Tables 9.3 and 9.4 in Chapter 9. (See Problem 16.)

8 Find all of the ideals of T_4 in Table 9.6.

9 Find all of the ideals of T_8 in Table 9.7. (See Problem 18 from §10.1.)

10 Verify that $\{0, c\}$ is an ideal of the ring T_4 in Table 9.6. Construct addition and multiplication tables for the quotient ring $T_4/\{0, c\}$.

11 Verify that $\{0, 2a\}$ is an ideal of the commutative ring R of order 8 that is defined by Tables 9.3 and 9.4 in Chapter 9. Construct addition and multiplication tables for the quotient ring $R/\{0, 2a\}$.

12 Verify that $\{0, b\}$ is an ideal of the commutative ring R of order 8 that is defined by Tables 9.3 and 9.4 in Chapter 9. Construct addition and multiplication tables for the quotient ring $R/\{0, b\}$.

13 A *left ideal* I of a ring R is a subring for which $r \cdot x \in I$ when $r \in R$, and $x \in I$. Find a left ideal of T_8 that is not a standard ideal.

14 Verify that $\{0, c\}$ is an ideal of the ring T_8 in Table 9.7. Construct addition and multiplication tables for the quotient ring $T_8/\{0, c\}$.

15 Let $A = \langle 6 \rangle$ be an ideal of the ring $\mathbb{Z}$. Construct addition and multiplication tables of the quotient ring $\mathbb{Z}/\langle 6 \rangle$. What does this ring remind you of?

16 Let $A = \langle 2 \rangle$ and $B = \langle 6 \rangle$ be two ideals of the ring $\mathbb{Z}$. Construct addition and multiplication tables of the quotient ring A/B.

17 If R is a commutative ring and y is a fixed element of R, prove that the set

$$A = \{x \in R \mid x \cdot y = 0\}$$

is an ideal in R. (See Problem 11 in §10.1.)

18 If R is a commutative ring and y is a fixed element of R, prove that the set

$$B = \{x \cdot y \mid x \in R\}$$

is an ideal of R.

Hint: Note that if there is no multiplicative identity, y may not be in I.

19 An element x of a ring R is called *nilpotent* if $x^n = 0$ for some positive number n. Show that the set of all nilpotent elements in a commutative ring R forms an ideal of R. See Problem 14 of §10.1.

20 Let R be a unity ring, and I an ideal of R. Show that R/I is a unity ring.

Interactive Problems

21 Which of the subrings of the ring of order 8 found in Problem 19 of §10.1 are ideals? The ring is given as follows:

```
InitRing("a", "b")
DefineRing([4, 2], [[a, b], [0, 0]])
R = ListRing(); R
    {0*a, a, 2*a, -a, b, a+b, 2*a+b, -a+b}
```

22 Which of the subrings of the ring of order 8 found in Problem 20 of §10.1 are ideals? The ring is given as follows:

```
InitRing("a", "b")
DefineRing([4, 2], [[2*a, 0], [2*a, 2*a]])
R = ListRing(); R
    {0*a, a, 2*a, -a, b, a+b, 2*a+b, -a+b}
```

10.3 Ring Isomorphisms

As we work with different rings, it is natural to ask whether we can consider two rings to be "equivalent" if the elements of one ring can be renamed to form the other ring. We have already seen that the quotient ring $\mathbb{Z}/(n\mathbb{Z})$ was essentially the same ring as Z_n. We will proceed the same way we defined isomorphisms with groups.

DEFINITION 10.5 Let A and B be two rings. A *ring isomorphism* from A to B is a one-to-one mapping $f : A \to B$ such that

$$f(x + y) = f(x) + f(y) \qquad \text{and}$$
$$f(x \cdot y) = f(x) \cdot f(y)$$

for all $x, y, \in A$. If there exists a ring isomorphism from A to B that is surjective, then we say that the rings A and B are *isomorphic,* denoted by $A \approx B$.

Example 10.7
Find an isomorphism from the quotient ring $\mathbb{Z}/(n\mathbb{Z})$ to Z_n.
SOLUTION: The natural mapping would be as follows:

$$f(a + n\mathbb{Z}) = a \textbf{ mod } n,$$

which we can verify is well defined by noting that if $a + n\mathbb{Z} = b + n\mathbb{Z}$, then $a - b$ is a multiple of n, so a **mod** $n = b$ **mod** n. Also, f is an injective and surjective function from $\mathbb{Z}/(n\mathbb{Z})$ to Z_n. Furthermore, $f(a + b) = f(a) + f(b)$, and $f(a \cdot b) = f(a) \cdot f(b)$. So we have that $\mathbb{Z}/(n\mathbb{Z}) \approx Z_n$. ∎

Computational Example 10.8
Two very similar-looking rings of order 10 can be defined in *Sage* as follows:

```
InitRing("a")
DefineRing([10], [[2*a]])
CheckRing()
    This is a ring.
A = ListRing(); A
    {0*a, a, 2*a, 3*a, 4*a, 5*a, 6*a, 7*a, 8*a, -a}
InitRing("b")
DefineRing([10], [[6*b]])
CheckRing()
    This is a ring.
B = ListRing(); B
    {0*b, a, 2*b, 3*b, 4*b, 5*b, 6*b, 7*b, 8*b, -b}
```

Show that these rings are isomorphic.

SOLUTION: The addition and multiplication tables of A are shown in Table 10.4. Note that the multiplicative structure is different than Z_{10}, since there is no multiplicative identity. The addition table for B is similar, but the multiplication table is shown in Table 10.5.

In spite of the similarities between the two tables, they are not the same "color pattern". If they are isomorphic, it is not immediately clear what the isomorphism should be.

Since a is an additive generator of A, we know that it should map to one of the additive generators of B, $\{b, 3b, 7b, 9b\}$. In *Sage*, the command **RingHomo** defines a ring homomorphism, similar to the way that **Homomorph** defined a group homomorphism. So let us see if we can create an isomorphism.

TABLE 10.4: Addition and multiplication in the ring A of order 10

$+$	$0a$	a	$2a$	$3a$	$4a$	$5a$	$6a$	$7a$	$8a$	$-a$
$0a$	$0a$	a	$2a$	$3a$	$4a$	$5a$	$6a$	$7a$	$8a$	$-a$
a	a	$2a$	$3a$	$4a$	$5a$	$6a$	$7a$	$8a$	$-a$	$0a$
$2a$	$2a$	$3a$	$4a$	$5a$	$6a$	$7a$	$8a$	$-a$	$0a$	a
$3a$	$3a$	$4a$	$5a$	$6a$	$7a$	$8a$	$-a$	$0a$	a	$2a$
$4a$	$4a$	$5a$	$6a$	$7a$	$8a$	$-a$	$0a$	a	$2a$	$3a$
$5a$	$5a$	$6a$	$7a$	$8a$	$-a$	$0a$	a	$2a$	$3a$	$4a$
$6a$	$6a$	$7a$	$8a$	$-a$	$0a$	a	$2a$	$3a$	$4a$	$5a$
$7a$	$7a$	$8a$	$-a$	$0a$	a	$2a$	$3a$	$4a$	$5a$	$6a$
$8a$	$8a$	$-a$	$0a$	a	$2a$	$3a$	$4a$	$5a$	$6a$	$7a$
$-a$	$-a$	$0a$	a	$2a$	$3a$	$4a$	$5a$	$6a$	$7a$	$8a$

$\cdot$	$0a$	a	$2a$	$3a$	$4a$	$5a$	$6a$	$7a$	$8a$	$-a$
$0a$	$0a$	$0a$	0	$0a$	$0a$	$0a$	$0a$	$0a$	$0a$	$0a$
a	$0a$	$2a$	$4a$	$6a$	$8a$	$0a$	$2a$	$4a$	$6a$	$8a$
$2a$	$0a$	$4a$	$8a$	$2a$	$6a$	$0a$	$4a$	$8a$	$2a$	$6a$
$3a$	$0a$	$6a$	$2a$	$8a$	$4a$	$0a$	$6a$	$2a$	$8a$	$4a$
$4a$	$0a$	$8a$	$6a$	$4a$	$2a$	$0a$	$8a$	$6a$	$4a$	$2a$
$5a$	$0a$	$0a$	$0a$	$0a$	$0a$	$0a$	$0a$	$0a$	$0a$	$0a$
$6a$	$0a$	$2a$	$4a$	$6a$	$8a$	$0a$	$2a$	$4a$	$6a$	$8a$
$7a$	$0a$	$4a$	$8a$	$2a$	$6a$	$0a$	$4a$	$8a$	$2a$	$6a$
$8a$	$0a$	$6a$	$2a$	$8a$	$4a$	$0a$	$6a$	$2a$	$8a$	$4a$
$-a$	$0a$	$8a$	$6a$	$4a$	$2a$	$0a$	$8a$	$6a$	$4a$	$2a$

TABLE 10.5: The ring B

$\cdot$	$0b$	b	$2b$	$3b$	$4b$	$5b$	$6b$	$7b$	$8b$	$-b$
$0b$	$0b$	$0b$	0	$0b$	$0b$	$0b$	$0b$	$0b$	$0b$	$0b$
b	$0b$	$6b$	$2b$	$8b$	$4b$	$0b$	$6b$	$2b$	$8b$	$4b$
$2b$	$0b$	$2b$	$4b$	$6b$	$8b$	$0b$	$2b$	$4b$	$6b$	$8b$
$3b$	$0b$	$8b$	$6b$	$4b$	$2b$	$0b$	$8b$	$6b$	$4b$	$2b$
$4b$	$0b$	$4b$	$8b$	$2b$	$6b$	$0b$	$4b$	$8b$	$2b$	$6b$
$5b$	$0b$	$0b$	$0b$	$0b$	$0b$	$0b$	$0b$	$0b$	$0b$	$0b$
$6b$	$0b$	$6b$	$2b$	$8b$	$4b$	$0b$	$6b$	$2b$	$8b$	$4b$
$7b$	$0b$	$2b$	$4b$	$6b$	$8b$	$0b$	$2b$	$4b$	$6b$	$8b$
$8b$	$0b$	$8b$	$6b$	$4b$	$2b$	$0b$	$8b$	$6b$	$4b$	$2b$
$-b$	$0b$	$4b$	$8b$	$2b$	$6b$	$0b$	$4b$	$8b$	$2b$	$6b$

```
F = RingHomo(A, B)
HomoDef(F, a, b)
FinishHomo(F)
    b + b is not 6*b
    'Homomorphism failed'
F = RingHomo(A, B)
HomoDef(F, a, 3*b)
FinishHomo(F)
    3*b + 3*b is not 4*b
    'Homomorphism failed'
F = RingHomo(A, B)
HomoDef(F, a, 7*b)
FinishHomo(F)
    'Homomorphism defined'
Kernel(F)
    {0*a}
```

Because the last mapping has a kernel of the additive identity, we know from group homomorphisms that this mapping must be one-to-one. So we have found an isomorphism from A to B, but it was far from obvious. ∎

In this example, we found an isomorphism, but we had to use trial and error. It is not at all clear why we would have to have a map to $7b$.

We would like a way to generalize this example so we can determine if two similar rings are isomorphic.

One way to help find an isomorphism between A and B is to show that both of these are isomorphic to a subring of the Z_n for some n. For example, consider $2Z_{20}$, the even elements of Z_{20}.

TABLE 10.6: Multiplication for the ring $2Z_{20}$

$\cdot$	0	2	4	6	8	10	12	14	16	18
0	0	0	0	0	0	0	0	0	0	0
2	0	4	8	12	16	0	4	8	12	16
4	0	8	16	4	12	0	8	16	4	12
6	0	12	4	16	8	0	12	4	16	8
8	0	16	12	8	4	0	16	12	8	4
10	0	0	0	0	0	0	0	0	0	0
12	0	4	8	12	16	0	4	8	12	16
14	0	8	16	4	12	0	8	16	4	12
16	0	12	4	16	8	0	12	4	16	8
18	0	16	12	8	4	0	16	12	8	4

```
Z20 = ZRing(20); Z20
    {0, 1, 2, 3, 4, 5, 6, 7, 8, 9, 10, 11, 12, 13, 14, 15, 16, 17,
    18, 19}
R = Ring(Z20[2]); R
    {0, 2, 4, 6, 8, 10, 12, 14, 16, 18}
MultTable(R)
```

which produces Table 10.6. One can see that the color patterns for A and R are the same, so that $A \approx 2Z_{20}$.

We can now generalize this example as follows.

PROPOSITION 10.7

Let R be a finite ring whose additive structure is a cyclic group of order n. Let x be a generator of the additive group. Then $x^2 = kx$ for some positive integer $k \le n$, and

$$A \approx kZ_{kn}.$$

PROOF: If $x^2 = 0$, we can let $k = n$, so that k will be positive and $kx = 0 = x^2$. If x^2 is not zero, then since x generates the additive group, there is a k such that $x^2 = kx$ with $0 < k < n$.

Now the natural mapping is one that sends $f(a \cdot x) = k \cdot a$ **mod** (kn). This is obviously one-to-one and onto, since the value of a ranges from 0 to $n - 1$. To check that this is an isomorphism, note that

$$\begin{aligned} f(a \cdot x + b \cdot x) &= f((a+b) \cdot x) = k \cdot (a+b) \ \textbf{mod}\ (kn) \\ &= (k \cdot a \ \textbf{mod}\ (kn) + k \cdot b \ \textbf{mod}\ (kn)) \ \textbf{mod}\ (kn) \\ &= f(a \cdot x) + f(b \cdot x). \end{aligned}$$

Also,

$$\begin{aligned} f((a\cdot x)\cdot(b\cdot x)) &= f(a\cdot b\cdot x^2)\\ &= f(a\cdot b\cdot k\cdot x)\\ &= k\cdot a\cdot b\cdot k \textbf{ mod } (kn)\\ &= ((k\cdot a \textbf{ mod } (kn))\cdot(k\cdot b \textbf{ mod } (kn))) \textbf{ mod } (kn)\\ &= f(a\cdot x)\cdot f(b\cdot x). \end{aligned}$$

Therefore, f is an isomorphism, and $R \approx kZ_{kn}$. □

This proposition shows not only that $A \approx 2Z_{20}$, but also that $B \approx 6A_{60}$, since $b^2 = 6b$ in this ring.

DEFINITION 10.6 A *cyclic ring* is a ring whose additive group is cyclic.

Note that this definition of cyclic rings also includes the infinite rings $\mathbb{Z}$ and its subrings $k\mathbb{Z}$.

In order to prove that in fact $A \approx B$, we will need a few lemmas about number theory. Once these are proven, we will be able to determine *all* non-isomorphic rings of order 10.

LEMMA 10.1

Let d be a positive divisor of n, and let f be the largest divisor of d that is coprime to (n/d). Then if q is coprime to both f and (n/d), then q is coprime to n.

PROOF: Suppose that $\gcd(q, n)$ is not 1. Then there is a prime number p that divides neither f nor (n/d), yet divides n. Thus, p must divide d.

Now $f \cdot p$ will be coprime to (n/d) since both f and p are. Also, since f is not a multiple of p while d is, $f \cdot p$ will be a divisor of d. But we defined f to be the *largest* factor of d coprime to (n/d). This contradiction shows that $\gcd(q, n) = 1$. □

LEMMA 10.2

Given two positive numbers x and y, there exist u and v in $\mathbb{Z}$ such that

$$ux + vy = \gcd(x, y),$$

where u is coprime to y.

PROOF: The greatest common divisor theorem (0.4) would give us values for u and v, but there would be no way to guarantee that u would be coprime to y.

Let $k = \gcd(x, y)$. Then (x/k) and (y/k) are coprime, so (x/k) has an multiplicative inverse in $Z_{(y/k)}$, say n. That is,

$$\frac{x}{k} \cdot n \equiv 1 \left(\text{mod } \frac{y}{k}\right).$$

Let f be the largest divisor of k that is coprime to (y/k). By the Chinese remainder theorem (0.7), there is a number u such that

$$u \equiv n \left(\text{mod } \frac{y}{k}\right)$$

and

$$u \equiv 1 \; (\text{mod } f).$$

Since n is coprime to (y/k), u is coprime to (y/k). Also, u is coprime to f, so by Lemma 10.1 u is coprime to y. Also,

$$u \cdot \frac{x}{k} \equiv 1 \left(\text{mod } \frac{y}{k}\right)$$

so there is a v such that $u \cdot \frac{x}{k} + v \cdot \frac{y}{k} = 1$. Multiplying both sides by k gives us

$$u \cdot x + v \cdot y = k = \gcd(x, y).$$

∎

THEOREM 10.1: The Cyclic Ring Theorem

If x and n are positive integers, then

$$xZ_{x \cdot n} \approx kZ_{k \cdot n},$$

where $k = \gcd(x, n)$.

PROOF: Since $k = \gcd(x, n)$ by Lemma 10.2 we can find integers u and v such that $u \cdot x + v \cdot n = k$, where u is coprime to n. We now define a mapping f from kZ_{kn} to xZ_{xn} as follows:

$$f(k \cdot w \textbf{ mod } (kn)) = uxw \textbf{ mod } (xn).$$

Note that this is well defined, since if $k \cdot w$ is equivalent to $k \cdot p$ (mod kn) then

$$w \equiv p \; (\text{mod } n) \Longrightarrow xw \equiv xp \; (\text{mod } xn)$$
$$\Longrightarrow uxw \equiv uxp \; (\text{mod } xn).$$

Next we need to show that f is a homomorphism from kZ_{kn} to xZ_{xn}. If $a \equiv k \cdot w$ (mod kn) and $b \equiv k \cdot z$ (mod kn), then

$$f(a + b) = f((k \cdot w + k \cdot z) \textbf{ mod } (kn)) = u \cdot (x \cdot w + x \cdot z) \textbf{ mod } (xn)$$
$$= (u \cdot x \cdot w + u \cdot x \cdot z) \textbf{ mod } (xn) = f(a) + f(b).$$

$$\begin{aligned} f(a \cdot b) &= f((k \cdot w \cdot k \cdot z) \ \textbf{mod}\ (kn)) = (u \cdot x \cdot w \cdot k \cdot z) \ \textbf{mod}\ (xn) \\ &= (u \cdot x \cdot w \cdot (u \cdot x + v \cdot n) \cdot z) \ \textbf{mod}\ (xn) \\ &= (u \cdot x \cdot w \cdot u \cdot x \cdot z + u \cdot x \cdot w \cdot v \cdot n \cdot z) \ \textbf{mod}\ (xn) \\ &= ((u \cdot x \cdot w) \cdot (u \cdot x \cdot z)) \ \textbf{mod}\ (xn) = f(a) \cdot f(b). \end{aligned}$$

So f is indeed a homomorphism from kZ_{kn} to xZ_{xn}.

Since u is coprime to n, u has an inverse, u^{-1} (mod n). Then we see that f is onto, since any element $x \cdot a$ (mod xn) in xZ_{xn} can be obtained by taking

$$f(k \cdot a \cdot u^{-1} \ \textbf{mod}\ (kn)) = (u \cdot x \cdot a \cdot u^{-1}) \ \textbf{mod}\ (xn) = x \cdot a \ \textbf{mod}\ (xn).$$

Finally, both xZ_{xn} and kZ_{kn} contain n elements, so by the pigeonhole principle f must be a one-to-one function. Thus, f is an isomorphism, and $xZ_{xn} \approx kZ_{kn}$. ∎

Because $2 = \gcd(6, 10)$, we see that $A \approx 2Z_{20}$ is isomorphic to $B \approx 6Z_{60}$.

In fact, since the only rings of order 10 are cyclic rings, there are four possible non-isomorphic rings of order 10:

$$Z_{10}, \qquad 2Z_{20}, \qquad 5Z_{50}, \qquad \text{and} \qquad 10Z_{100}.$$

It is easy to see that these rings are all distinct by looking at the multiplication tables.

COROLLARY 10.1

The number of non-isomorphic cyclic rings of order n is precisely the number of divisors of n (including 1 and n).

PROOF: By Proposition 10.7 every cyclic ring of order n is isomorphic to kZ_{kn} for some value of k. By the cyclic ring theorem, we see that this is isomorphic to dZ_{dn}, where $d = \gcd(k, n)$. Hence d is a divisor of n. We need to show that two different rings of this form are non-isomorphic. Consider the rings $A = dZ_{dn}$ and $B = fZ_{fn}$, where d and f are different divisors of n. Perhaps the easiest way to show that these are different is to count the number of elements in A and B that can appear in the multiplication tables. The elements that can appear in the table for A are

$$d^2, 2d^2, 3d^2, \ldots, nd = 0$$

while the elements appearing in the multiplication table of B are

$$f^2, 2f^2, 3f^2, \ldots, nf = 0.$$

Thus, there are n/d such elements of A, and n/f elements of B. Since d and f are different, we see that the rings A and B are not isomorphic. Therefore, there is a one-to-one correspondence between the factors of n and the cyclic rings of order n. ∎

Although this corollary seems to be a big help in finding *all* finite rings, there are, in fact, many non-cyclic rings. For example, there are 8 non-cyclic rings of order 4, which when combined with the 3 cyclic rings from Corollary 10.1 gives a total of 11 rings of order 4. There are 52 rings of order 8 (4 cyclic, 20 with additive group Z_{15}^*, and 28 with an additive group Z_{24}^*).

Table 10.7 shows the number of rings of a given order. There are at least 18,590 known rings of order 32, but it has not been proven that these are all of them.

In *Sage*, we can load any of the rings of order 15 or less. The command **NumberSmallRings** will produce the number of rings of a given order, up to order 15.

```
NumberSmallRings(8)
    52
```

Now we can load any of these 52 rings.

```
R = SmallRing(8, 51); R
    {0*a, a, b, a+b, c, a+c, b+c, a+b+c}
MultTable(R)
```

The multiplication table for this ring is shown in Table 10.8.

Problems for §10.3

1 Suppose ϕ is an isomorphism between R and S. Show that if S is commutative, then so is R.

2 Suppose ϕ is a surjective isomorphism between R and S. Show that if S has a unity element, then so does R.

3 Suppose ϕ is an isomorphism between R and S. Show that if R has a zero divisor, then so does S.

TABLE 10.7: Number of rings of order n

n	rings	n	rings	n	rings	n	rings
1	1	9	11	17	2	25	11
2	2	10	4	18	22	26	4
3	2	11	2	19	2	27	59
4	11	12	22	20	22	28	22
5	2	13	2	21	4	29	2
6	4	14	4	22	4	30	8
7	2	15	4	23	2	31	2
8	52	16	390	24	104	32	???

TABLE 10.8: Ring number 51 of order 8

$\cdot$	$0\cdot a$	a	b	$a+b$	c	$a+c$	$b+c$	$a+b+c$
$0\cdot a$	$0\cdot a$	$0\cdot a$	$0\cdot a$	$0\cdot a$	$0\cdot a$	$0\cdot a$	$0\cdot a$	$0\cdot a$
a	$0\cdot a$	a	b	$a+b$	c	$a+c$	$b+c$	$a+b+c$
b	$0\cdot a$	b	$b+c$	c	b	$0\cdot a$	c	$b+c$
$a+b$	$0\cdot a$	$a+b$	c	$a+b+c$	$b+c$	$a+c$	b	a
c	$0\cdot a$	c	b	$b+c$	c	$0\cdot a$	$b+c$	b
$a+c$	$0\cdot a$	$a+c$	$0\cdot a$	$a+c$	$0\cdot a$	$a+c$	$0\cdot a$	$a+c$
$b+c$	$0\cdot a$	$b+c$	c	b	$b+c$	$0\cdot a$	b	c
$a+b+c$	$0\cdot a$	$a+b+c$	$b+c$	a	b	$a+c$	c	$a+b$

4 Suppose ϕ is an isomorphism between R and S. Show that if R has a non-zero idempotent element, then so does S. See Problem 9 of §9.3.

5 Find a subring of the ring T_8 in Table 9.7 that is isomorphic to the ring T_4 in Table 9.6.

6 Let R be a non-commutative ring. Define the operation $x * y = y \cdot x$. Show that the set R forms a ring using the operations $*$ and $+$ instead of $\cdot$ and $+$. This new ring is called the *opposite ring* of R, and is denoted R^{op}.

7 Show that the ring T_4 in Table 9.6 is not isomorphic to its opposite. (See Problem 6.)

8 Show that the quotient ring R/S_2 in Table 10.3 is isomorphic to T_4^{op}. (See Problem 6.)

9 Show that the ring T_8 in Table 9.7 *is* isomorphic to its opposite. (See Problem 6.)

Hint: First construct the multiplication table for T_8^{op}, then determine how to rearrange the elements of T_8 so that the patterns match.

10 Prove that a non-commutative ring of order 4 or less must be isomorphic to either T_4 from Table 9.6 or T_4^{op}. (See Problem 6.)

Hint: Use Problem 12 from §9.2.

11 Is the ring $2\mathbb{Z}$ isomorphic to the ring $3\mathbb{Z}$? Why or why not?

12 Let $A = \langle 2 \rangle$ and $B = \langle 8 \rangle$ be two ideals of the ring $\mathbb{Z}$. Show that the group A/B is isomorphic to Z_4, but the ring A/B is not isomorphic to the ring Z_4.

13 Is the ring $\mathbb{R}$ isomorphic the the ring of complex numbers $\mathbb{C}$?

For Problems **14** through **17**, find all non-isomorphic rings of the following order.

14 6 **15** 21 **16** 30 **17** 210

18 Let R be a ring with unity e, and let S be the subring $[e]$ generated from the unity element. Show that S is isomorphic to either $\mathbb{Z}$ or Z_n for some n.

Interactive Problems

19 Load the rings Z_{12} and Z_6 into *Sage* simultaneously with the commands:

```
Z12 = ZRing(12)
Z6 = ZRing(6)
```

Show that $I = \{0, 6\}$ is an ideal of Z_{12}, and display addition and multiplication tables of the quotient ring Z_{12}/I, showing that Z_{12}/I is isomorphic to Z_6.

20 Use *Sage* to find the eight non-isomorphic non-cyclic rings of order 4.
Hint: The additive group must be isomorphic to Z_8^*, so the ring is defined by:

```
InitRing(" a", " b")
DefineRing([2, 2], [[???, ???], [???, ???]])
CheckRing()
```

Fill in each ??? with a member of $\{0, a, b, a + b\}$ to see whether a ring is formed. Is there a faster way than trying all $4^4 = 256$ combinations?

21 Use *Sage* to display the multiplication tables of all rings of order 6.

10.4 Homomorphisms and Kernels

We found in Chapter 4 that mappings between two groups proved to be an invaluable tool. Group homomorphisms allowed us to prove the three isomorphism theorems, and eventually led us to the group of automorphisms of a group. In this section we will carry over most of these results to rings. We will find that all three of the isomorphism theorems have a corresponding version for rings and ideals. The automorphisms of a ring or field will later lead to Galois theory, which we will study in a later chapter.

Since we defined a ring isomorphism in a similar fashion as group isomorphisms, we naturally will define ring homomorphisms by mimicking group homomorphisms.

DEFINITION 10.7 If A and B are two rings, then a mapping $f : A \to B$ such that

$$f(x+y) = f(x) + f(y), \qquad \text{and}$$
$$f(x \cdot y) = f(x) \cdot f(y),$$

for all x and y in A is called a *ring homomorphism.*

Notice that a ring homomorphism preserves both of the ring operations. In particular, a ring homomorphism will also be a group homomorphism from the additive group of A to the additive group of B. Thus, we can immediately apply the results of group homomorphisms to see two properties of ring homomorphisms.

If f is a ring homomorphism from A to B, then

$$f(0) = 0 \qquad \text{and}$$

$$f(-x) = -f(x) \qquad \text{for all} \qquad x \in A.$$

Any isomorphism is certainly a homomorphism. But let us see how to define a homomorphism between two non-isomorphic rings.

Example 10.9

Let n be a positive integer. Find a homomorphism between $\mathbb{Z}$ and Z_n.

SOLUTION: The natural mapping is

$$f(x) = x \text{ **mod** } n.$$

Proposition 0.2 can be restated as $f(x+y) = f(x) + f(y)$, and $f(x \cdot y) = f(x) \cdot f(y)$. Thus, this is a homomorphism. □

Computational Example 10.10

Use *Sage* to find a homomorphism from Z_3 to Z_6.

SOLUTION: First we define Z_3 and Z_6 simultaneously.

```
Z3 = ZRing(3); Z3
    {0, 1, 2}
Z6 = ZRing(6); Z6
    {0, 1, 2, 3, 4, 5}
a = Z3[1]
b = Z6[1]
```

Here, we defined a and b to be the additive generators of Z_3 and Z_6. The homomorphism is determined completely by the value of $f(1)$. A natural choice would be to let $f(1) = 2$ **mod** 6. To define a ring homomorphism, we use the command **RingHomo** instead of **Homomorph**.

```
F = RingHomo(Z3, Z6)
HomoDef(F, 1, 2)
```

Even though 1 and 2 are technically elements of $\mathbb{Z}$, not Z_3 or Z_6, *Sage* makes the natural translations, knowing the arguments are expected to be in the rings Z_3 and Z_6. We can now use the command **FinishHomo** to check if F is a ring homomorphism.

```
FinishHomo(F)
    2 * 2 is not 2
    'Homomorphism failed'
```

Sage shows that this would not produce a homomorphism. One way to correct this problem would be to send $f(a)$ to the zero element of Z_6.

```
F = RingHomo(Z3, Z6)
HomoDef(F, 1, 0)
FinishHomo(F)
    'Homomorphism defined'
```

Although this works, this is a rather trivial example, since it sends *all* elements to 0. After some experimenting, we can find a more interesting example.

```
F = RingHomo(Z3, Z6)
HomoDef(F, 1, 4)
FinishHomo(F)
    'Homomorphism defined'
```

Thus, $f(1) = 4$, so $f(2) = 2$, and of course $f(0) = 0$. ◻

There will always be at least one homomorphism between two rings, the one which sends all elements to zero.

DEFINITION 10.8 If A and B are any two rings, then the mapping $f : A \to B$

$$f(x) = 0 \quad \text{for all } x \in A$$

is called the *zero homomorphism from A to B*.

As with groups, we define $f(S)$, where S is a *set* of elements in the domain of f, to be the set of all values $f(x)$, where x is in S. We can also define the inverse image of an element y to be $f^{-1}(y)$, the set of elements such that $f(x) = y$. In fact, we can define the inverse image of a set of elements in the same way: $f^{-1}(T)$ is the set of elements such that $f(x)$ is in T. We can find images and inverse images of ring homomorphisms the same way we did for group homomorphisms. Here is a new homomorphism going from Z_6 to Z_3.

```
G = RingHomo(Z6, Z3)
HomoDef(G, 1, 1)
FinishHomo(G)
    'Homomorphism defined'
G(4)
    1
Image(G, Z6)
    {0, 1, 2}
HomoInv(G, 2)
    {2, 5}
HomoInv(G, [0, 1])
    {0, 1, 3, 4}
```

We can ask whether the image or inverse image of a subring will again be a subring. This is actually very easy to prove, as seen in the next proposition.

PROPOSITION 10.8

Suppose f is a homomorphism from the ring A to the ring B. Then if S is a subring of A, $f(S)$ is a subring of B. Likewise, if T is a subring of B, then $f^{-1}(T)$ will be a subring of A.

PROOF: Suppose S is a subring of A. We will use Proposition 10.1 to show that $f(S)$ is a subring of B. The element $f(0) = 0$ is in $f(S)$, so $f(S)$ is non-empty. If u and v are two elements of $f(S)$, then there exist elements x and y in S such that

$$f(x) = u$$

and

$$f(y) = v.$$

But $x \cdot y$ and $x - y$ are also in S, and so

$$f(x \cdot y) = f(x) \cdot f(y) = u \cdot v$$

and

$$f(x - y) = f(x) - f(y) = u - v$$

must be in $f(S)$. Thus, by Proposition 10.1, $f(S)$ is a subring of B.

Now suppose that T is a subring of B. Since 0 is contained in $f^{-1}(T)$, we have that $f^{-1}(T)$ is non-empty. If x and y are two elements of $f^{-1}(T)$, then $f(x)$ and $f(y)$ will be two elements of T. Thus,

$$f(x \cdot y) = f(x) \cdot f(y)$$

and

$$f(x - y) = f(x) - f(y)$$

would be elements of T. Hence, $x \cdot y$ and $x - y$ are in $f^{-1}(T)$. Thus, by Proposition 10.1, $f^{-1}(T)$ is a subring of A. ∎

We can define the kernel and the image of a homomorphism in the same way that we did for group homomorphisms.

DEFINITION 10.9 Given a homomorphism f from the ring A to the ring B, the *kernel* of f is $f^{-1}(0)$, denoted Ker(f). The *image* of f is $f(A)$, denoted Im(f).

In *Sage*, we can use the **HomoInv** command to find the kernel of a homomorphism, or we can use the command

```
Kernel(G)
    {0, 3}
```

as we did for group homomorphisms.

When we have a homomorphism from A to B, we have by Proposition 10.8 that the image will be a subring of B. Likewise, the kernel of a homomorphism will be a subring of A. However, we can say even more about the kernel.

PROPOSITION 10.9

If f is a homomorphism from the ring A to the ring B, then the kernel of f is an ideal of A. Furthermore, f is injective if, and only if, Ker(f)$=\{0\}$.

PROOF: Suppose that x is in the kernel of f, and y is any other element of A. Then

$$f(x \cdot y) = f(x) \cdot f(y) = 0 \cdot f(y) = 0,$$

and

$$f(y \cdot x) = f(y) \cdot f(x) = f(y) \cdot 0 = 0.$$

Hence, $x \cdot y$ and $y \cdot x$ are in the kernel of f, so the kernel is an ideal of A.

If f is injective, then $f^{-1}(0)$ can only contain one element, which must be 0. On the other hand, if $f^{-1}(0) = \{0\}$, then

$$\begin{aligned} f(x) = f(y) &\Longrightarrow f(x) - f(y) = 0 \\ &\Longrightarrow f(x - y) = 0 \\ &\Longrightarrow x - y = 0 \\ &\Longrightarrow x = y. \end{aligned}$$

Therefore, f is injective if, and only if, Ker(f) $= \{0\}$. ∎

Motivational Example 10.11

Find a non-zero homomorphism from the non-commutative ring R of order 8 used throughout §10.2, to some other ring.

SOLUTION: The kernel would have to be an ideal of R. But R has only three nontrivial ideals:

```
InitRing("a", "b")
DefineRing([4, 2], [[a, b], [2*a, 0]])
R = ListRing(); R
    {0*a, a, 2*a, -a, b, a+b, 2*a+b, -a+b}

I1 = Ideal(R, 2*a); I1
    {0*a, 2*a}
I2 = Ideal(R, 2*a + b); I2
    {0*a, 2*a+b}
I3 = Ideal(R, b); I3
    {0*a, b, 2*a, 2*a+b}
```

To produce an interesting homomorphism, we would use one of these ideals as the kernel. To which ring should we map R?

The natural answer would be the quotient ring. Since there is a natural *group* homomorphism from R to R/I, we can ask whether this group homomorphism extends to become a ring homomorphism.

Let us define $Q = R/I_1$.

```
Q = Coset(R, I1); Q
    {{0*a, 2*a}, {a, -a}, {b, 2*a+b}, {a+b, -a+b}}
```

We wish to define a homomorphism $i(x)$ that maps an element in R to the coset of Q containing that element. We only need to define $i(a)$ and $i(b)$ to complete the definition.

```
i = RingHomo(R, Q)
HomoDef(i, a, a + I1)
HomoDef(i, b, b + I1)
FinishHomo(i)
    'Homomorphism defined'
```

The kernel of this homomorphism,

```
Kernel(i)
    {0*a, 2*a}
```

which is of course I_1. ∎

In general, we can form a homomorphism from a ring R to a quotient ring R/I using the same technique. We will state this as a lemma:

LEMMA 10.3
If I is an ideal of the ring R, then the natural mapping $i : R \to R/I$ defined by $i(x) = x + I$ is a surjective ring homomorphism from R to R/I with the kernel being I.

PROOF: It is clear that the rule $i(x) = x + I$ defines a surjective mapping i from R to R/I, and that $\text{Ker}(i) = I$. We need only to check that $i(x)$ is a homomorphism.

Since

$$\begin{aligned} i(x + y) &= (x + y) + I \\ &= (x + I) + (y + I) \\ &= i(x) + i(y) \end{aligned}$$

and

$$\begin{aligned} i(x \cdot y) &= x \cdot y + I \\ &= (x + I) \cdot (y + I) \\ &= i(x) \cdot i(y), \end{aligned}$$

we see that $i(x)$ is indeed a surjective homomorphism. $\square$

In the homomorphisms produced by Lemma 10.3, the image of the homomorphism is isomorphic to $R/\text{Ker}(f)$. The first isomorphism theorem studied in Chapter 4 shows that the additive group on $\text{Im}(f)$ would be group isomorphic to the additive structure of $R/\text{Ker}(f)$. It is easy to show that the ring $\text{Im}(f)$ is isomorphic to the ring $R/\text{Ker}(f)$ as well, giving us an isomorphism theorem for rings.

THEOREM 10.2: The First Ring Isomorphism Theorem
Let f be a ring homomorphism from a ring R to a ring S, whose image is H. If the kernel of f is I, then there is a natural surjective isomorphism $\phi : R/I \to H$ that causes the diagram in Figure 10.1 to commute. (Here, $i(x)$ is the homomorphism defined in Lemma 10.3.) Thus, $H \approx R/I$.

PROOF: Figure 10.1 actually helps us determine how ϕ must be defined. For each coset $(x + I)$ in R/I, we need to have

$$\phi(x + I) = f(x)$$

in order for the diagram to commute. To prove that this rule defines a mapping, we need to show that this is well defined. That is, if $x + I = y + I$ it needs to be true that $f(x) = f(y)$, or else there would be a contradiction in

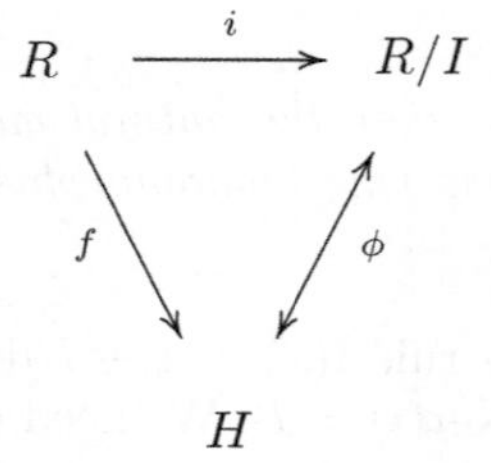

$$\begin{array}{ccc} R & \xrightarrow{i} & R/I \\ & f \searrow \quad \swarrow \phi & \\ & H & \end{array}$$

FIGURE 10.1: Commuting diagram for the first ring isomorphism theorem

the definition of ϕ. But

$$\begin{aligned} x + I = y + I &\iff x - y \in I \\ &\iff f(x - y) = 0 \\ &\iff f(x) = f(y) \\ &\iff \phi(x + I) = \phi(y + I). \end{aligned}$$

So we see that the definition of ϕ will not produce any such contradictions.
To show that ϕ is a homomorphism, we have that

$$\begin{aligned} \phi((x + I) + (y + I)) &= \phi(x + y + I) \\ &= f(x + y) \\ &= f(x) + f(y) \\ &= \phi(x + I) + \phi(y + I), \end{aligned}$$

and

$$\begin{aligned} \phi((x + I) \cdot (y + I)) &= \phi(x \cdot y + I) \\ &= f(x \cdot y) \\ &= f(x) \cdot f(y) \\ &= \phi(x + I) \cdot \phi(y + I). \end{aligned}$$

So ϕ is a homomorphism from R/I to H. It is apparent that this homomorphism is onto, and

$$\begin{aligned} \phi(x + I) = 0 &\iff f(x) = 0 \\ &\iff x \in I \\ &\iff x + I = I. \end{aligned}$$

So the kernel of ϕ is $\{I\}$, the zero element of R/I. Thus, ϕ is an isomorphism from R/I onto H, so $R/I \approx H$. Since the mapping ϕ was defined so that the diagram in Figure 10.1 commutes, the theorem is proved. □

It should be noted that there are second and third ring isomorphism theorems. These are considered in Problems 15 and 16.

Problems for §10.4

1 Find all ring homomorphisms from Z_6 to Z_6.

2 Find all ring homomorphisms from Z_{10} to Z_{10}.

3 Show that if $\phi(x) = 2x$, then ϕ is *not* a ring homomorphism from $\mathbb{R}$ to $\mathbb{R}$.

4 Is the mapping ϕ from Z_5 to Z_{30} given by $\phi(x) = 6x$ a ring homomorphism?

5 Is the mapping ϕ from Z_5 to Z_{20} given by $\phi(x) = 4x$ a ring homomorphism?

6 Is the mapping ϕ from Z_{30} to Z_5 given by $\phi(x) = x$ **mod** 5 a ring homomorphism?

7 Is the mapping ϕ from Z_{20} to Z_5 given by $\phi(x) = x$ **mod** 5 a ring homomorphism?

8 Is the mapping ϕ from Z_{20} to Z_{10} given by $\phi(x) = 6x$ **mod** 10 a ring homomorphism?

9 Is the mapping ϕ from Z_2 to Z_4 given by $\phi(x) = x$ a ring homomorphism?

10 Determine all ring homomorphisms from the rationals $\mathbb{Q}$ to $\mathbb{Q}$.
Hint: What are the possible kernels? If $\phi(1) = 1$, show that $\phi(x) = x$.

11 Let $\mathbb{C}$ denote the set of numbers of the form $a + bi$, where $i = \sqrt{-1}$ and a and b are real. ($\mathbb{C}$ is in fact a subring of the quaternions $\mathbb{H}$.) Let $\phi(a + bi) = a - bi$. Show that ϕ is a ring homomorphism from the ring $\mathbb{C}$ to itself.
Hint: Let $x = a + bi$, and $y = c + di$.

12 Show that if ϕ is a homomorphism from a ring R to a ring S, then an idempotent element of R must be sent to an idempotent element of S. See Problem 9 of §9.3.

13 Show that if ϕ is a homomorphism from a ring R to a ring S, then a nilpotent element of R must be sent to a nilpotent element of S. See Problem 14 of §10.1.

14 Show that if ϕ is a homomorphism from a ring R to a ring S, and R is a principle ideal ring, then $\text{Im}(\phi)$ is also a principle ideal ring.

15 Prove the second ring isomorphism theorem: If K and I are two ideals of a ring R, then

$$K/(K \cap I) \approx (K + I)/I.$$

(See Problem 1 of §10.2 for the definition of $K + I$.)

16 Prove the third ring isomorphism theorem: If K and I are two ideals of a ring R, where $K \subseteq I$, then K is an ideal of I, I/K is an ideal of R/K, and

$$(R/K)/(I/K) \approx R/I.$$

17 Find all the non-trivial homomorphisms from T_8 to T_4.
Hint: Consider Problems 9 and 20 from §10.2.

Interactive Problems

18 The ring of Example 10.11 also has an ideal $I_2 = \{0, 2a + b\}$. Define a homomorphism from the ring R to R/I_2.

19 The ring of Example 10.11 also has an ideal $I_3 = \{0, b, 2a, 2a+b\}$. Define a homomorphism from the ring R to R/I_3.

20 Use *Sage* to find a non-trivial homomorphism from the ring of Example 10.11 to itself, which is not an automorphism.

Chapter 11

Integral Domains and Fields

Although we have already defined integral domains and fields, this chapter focuses on particular cases of integral domains and fields. For example, one can construct a larger integral domain from a field or integral domain by considering polynomials over the original ring. Likewise, one can expand any integral domain into a field by forcing division to be possible. These provide us with useful examples for experimentation in hopes of finding properties of general integral domains and fields. We will also study what may be the most important field of all, the field of complex numbers.

11.1 Polynomial Rings

The study of polynomials is the oldest topic of algebra. The Babylonians were able to solve the quadratic equation around 1600 B.C., and the cubic equations were being solved in Arabia in 825 A.D., even before the modern algebraic notation. (Polynomials were written out with words.) In 1535, Tartaglia demonstrated how to solve the general cubic equation, and shortly thereafter Ferrari found the solution to the general fourth-degree equation. This led to a great surge of interest in the *theory of equations*, as mathematicians raced to find a general formula for the quintic, or fifth-degree equation. Finally, Abel and Galois independently proved in the 1820's that it was in fact impossible to find such a formula for the quintic equation, utilizing group theory.

The reader is obviously familiar with polynomials for which the coeffients are real numbers. However, we can construct polynomials from any ring, and the set of all such polynomials will be a new ring, called a *polynomial ring*. But only the polynomial rings formed either from fields or integral domains will have the properties that we are used to.

DEFINITION 11.1 Let K be a commutative ring. We define the set of polynomials in x over K, denoted $K[x]$, to be the set of all expressions of the form

$$k_0 + k_1x + k_2x^2 + k_3x^3 + \cdots$$

where the coefficients k_n are elements of K, and only a *finite* number of the coefficients are nonzero. If k_d is the last nonzero coefficient, then d is called the *degree* of the polynomial.

Notice that if $d = 0$, we essentially obtain the nonzero elements of K. These polynomials are referred to as *constant polynomials.* The degree for the zero polynomial

$$0 + 0x + 0x^2 + 0x^3 + 0x^4 + \cdots$$

is not defined.

By convention, the terms with zero coefficients are omitted when writing polynomials. Thus, the second degree polynomial in $\mathbb{Z}[x]$

$$1 + 0x + 3x^2 + 0x^3 + \cdots$$

would be written $1 + 3x^2$. The one exception to this convention is the zero polynomial, which is written as 0.

We can define the sum and product of polynomials in the familiar way. If

$$\begin{aligned} A &= a_0 + a_1x + a_2x^2 + a_3x^3 + \cdots \qquad \text{and} \\ B &= b_0 + b_1x + b_2x^2 + b_3x^3 + \cdots \end{aligned}$$

then

$$A + B = (a_0 + b_0) + (a_1 + b_1)x + (a_2 + b_2)x^2 + (a_3 + b_3)x^3 + \cdots$$

and

$$A \cdot B = \sum_{i=0}^{\infty} \sum_{j=0}^{\infty} (a_i \cdot b_j)x^{i+j}.$$

Although this looks like a double infinite sum, only a finite number of the terms will be nonzero. In fact, this product could be written as

$$\begin{aligned} A \cdot B = a_0 \cdot b_0 & \\ &+ (a_0 \cdot b_1 + a_1 \cdot b_0)x \\ &+ (a_0 \cdot b_2 + a_1 \cdot b_1 + a_2 \cdot b_0)x^2 \\ &+ (a_0 \cdot b_3 + a_1 \cdot b_2 + a_2 \cdot b_1 + a_3 \cdot b_0)x^3 + \cdots \end{aligned}$$

so each coefficient is determined by a finite sum.

LEMMA 11.1

Let A and B be two nonzero polynomials in x over K of degree m and n respectively, where K has no zero divisors. Then $A \cdot B$ is a polynomial of degree $m + n$, and $A + B$ is a polynomial of degree no greater than the larger of m or n.

PROOF: Let A be a polynomial of degree m,

$$A = a_0 + a_1x + a_2x^2 + a_3x^3 + \cdots a_mx^m$$

and B be a polynomial of degree n,

$$B = b_0 + b_1x + b_2x^2 + b_3x^3 + \cdots b_nx^n.$$

Here, a_m and b_n are nonzero elements of K. The product is determined by

$$A \cdot B = \sum_{i=0}^{\infty}\sum_{j=0}^{\infty}(a_i \cdot b_j)\, x^{i+j}.$$

Note that a_i and b_j are zero for $i > m$ and $j > n$. If $i + j > m + n$, either $i > m$ or $j > n$, and in either case $a_i \cdot b_j = 0$. Thus, there are no nonzero terms in $A \cdot B$ with coefficients larger than $m+n$. However, if $i+j = m+n$, the only nonzero term would be the one coming from $i = m$ and $j = n$, giving

$$a_m \cdot b_n\, x^{m+n}.$$

Since there are no zero divisors in K, $a_m \cdot b_n$ is nonzero, so $A \cdot B$ is a polynomial of degree $m + n$.

Next we turn our attention to $A + B$. We may assume without loss of generality that m is no more than n. Then the sum of A and B can be expressed as

$$(a_0 + b_0) + (a_1 + b_1)x + (a_2 + b_2)x^2 + \cdots (a_m + b_m)x^m + b_{m+1}x^{m+1} + \cdots b_nx^n.$$

If $m < n$, this clearly is a polynomial with degree n. Even if $m = n$, this still gives a polynomial whose degree cannot be more than n. ∎

We still have to show that $K[x]$ will be a ring. But if K is an integral domain or field, we will be able to say more about $K[x]$.

PROPOSITION 11.1

Let K be an integral domain or a field. Then the set of polynomials in x over K forms an integral domain.

PROOF: We have seen that $K[x]$ is closed under addition and multiplication. By the commutativity of K, addition and multiplication are obviously commutative. It is also clear that the zero polynomial acts as the additive identity in $K[x]$. Also, the additive inverse of

$$A = a_0 + a_1x + a_2x^2 + a_3x^3 + \cdots$$

is given by

$$-A = (-a_0) + (-a_1)x + (-a_2)x^2 + (-a_3)x^3 + \cdots,$$

since the sum of these two polynomials is

$$A + (-A) = 0 + 0x + 0x^2 + 0x^3 + \cdots = 0.$$

The polynomial with $b_0 = 1$, and $b_j = 0$ for all positive j,

$$I = 1 + 0x + 0x^2 + 0x^3 + \cdots,$$

acts as the multiplicative identity, since

$$I \cdot A = A \cdot I = \sum_{i=0}^{\infty}\sum_{j=0}^{\infty} a_i \cdot b_j\, x^{i+j} = \sum_{i=0}^{\infty} a_i \cdot 1\, x^i = A.$$

To check associativity of addition and multiplication, we need three polynomials

$$\begin{aligned} A &= a_0 + a_1x + a_2x^2 + a_3x^3 + \cdots, \\ B &= b_0 + b_1x + b_2x^2 + b_3x^3 + \cdots, \qquad \text{and} \\ C &= c_0 + c_1x + c_2x^2 + c_3x^3 + \cdots. \end{aligned}$$

Then

$$\begin{aligned} (A+B)+C &= (a_0+b_0)+c_0 + ((a_1+b_1)+c_1)x + ((a_2+b_2)+c_2)x^2 + \cdots \\ &= a_0 + (b_0+c_0) + (a_1+(b_1+c_1))x + (a_2+(b_2+c_2))x^2 + \cdots \\ &= A + (B+C). \end{aligned}$$

Also,

$$\begin{aligned} A \cdot (B \cdot C) &= A \cdot \left(\sum_{j=0}^{\infty}\sum_{k=0}^{\infty} b_j \cdot c_k\, x^{j+k} \right) \\ &= \sum_{i=0}^{\infty}\sum_{j=0}^{\infty}\sum_{k=0}^{\infty} a_i \cdot (b_j \cdot c_k) x^{i+j+k} \\ &= \sum_{i=0}^{\infty}\sum_{j=0}^{\infty}\sum_{k=0}^{\infty} (a_i \cdot b_j) \cdot c_k\, x^{i+j+k} = (A \cdot B) \cdot C. \end{aligned}$$

The two distributive laws are also easy to verify using the summation notation.

$$A \cdot (B + C) = A \cdot \left(\sum_{j=0}^{\infty} (b_j + c_j) x^j \right) = \sum_{i=0}^{\infty}\sum_{j=0}^{\infty} a_i \cdot (b_j + c_j) x^{i+j}$$

$$\begin{aligned} &= \sum_{i=0}^{\infty}\sum_{j=0}^{\infty} (a_i \cdot b_j + a_i c_j) x^{i+j} \\ &= \sum_{i=0}^{\infty}\sum_{j=0}^{\infty} a_i \cdot b_j\, x^{i+j} + \sum_{i=0}^{\infty}\sum_{j=0}^{\infty} a_i \cdot c_j\, x^{i+j} = A \cdot B + A \cdot C. \end{aligned}$$

We can use the fact that multiplication is commutative to show that

$$(A + B) \cdot C = A \cdot C + B \cdot C.$$

Thus, $K[x]$ is a commutative ring with identity.

Next, let us show that $K[x]$ has no zero divisors. Suppose that $A \cdot B = 0$, with both A and B being nonzero polynomials. Say that A has degree m and B has degree n. Then by Lemma 11.1, $A \cdot B$ has degree $m + n$, which is impossible if either m or n were positive. But if A and B are constant polynomials, then $a_0 \cdot b_0 = 0$, which would indicate that either a_0 or b_0 is 0, since K has no zero divisors. Thus, either A or B would have to be 0, so we have that $K[x]$ has no zero divisors.

Finally, let us show that $K[x]$ is not a field, by showing that the polynomial $(1 + x)$ is not invertible. Suppose that there was a polynomial A such that $A \cdot (1+x) = 1$. Then $A \neq 0$, so suppose A has degree m. Then by Lemma 11.1, we have $m + 1 = 0$, telling us $m = -1$, which is impossible. Thus, $(1 + x)$ has no inverse in $K[x]$, and therefore $K[x]$ is an integral domain. ∎

Although this proposition holds for polynomials defined over an integral domain, there is no reason why we cannot have *Sage* work with polynomials defined over any commutative ring. However, we will discover that the familiar properties of polynomials radically change.

Let us consider the commutative ring of order 8 from Tables 9.3 and 9.4 in Chapter 9.

```
InitRing("a", "b")
Define([4, 2], [[a, 0], [0, b]])
R = ListRing(); R
    {0*a, a, 2*a, -a, b, a+b, 2*a+b, -a+b}
```

We form a polynomial ring over R by defining a new symbol x.

```
AddRingVar("x")
```

A typical polynomial would be

```
Y = a*x + b; Y
    a*x+b
```

If we consider raising this polynomial to a power,

```
Y^4
    a*x^4+b
```

we find that this polynomial ring has rather bizarre properties. In fact, sometimes the square of a first-degree polynomial is not a second degree polynomial! Consider

```
(2*a*x + a + b)^2
     a+b
```

which yields the identity element in R. Thus, $2ax+a+b$ is its own multiplicative inverse. To further complicate matters, polynomials may be "factored" in more than one way. The two products

```
(2*a*x + b)*(a*x + b)
     2*a*x^2+b
(2*a*x + b)*(a*x + 2*a + b)
     2*a*x^2+b
```

yield the same quadratic polynomial. Because of the bizarre properties of polynomials over general rings, we mainly will focus our attention on polynomial rings $K[x]$, where K is an integral domain or field.

In order to find new integral domains and fields, we will use a simple property that will classify all rings.

DEFINITION 11.2 Let R be a ring. We define the *characteristic of* R to be the smallest positive number n such that $nx = 0$ for all elements x of R. If no such positive number exists, we say the ring has *characteristic 0.*

For integral domains or fields, the characteristic plays an extremely important role, as the next proposition illustrates.

PROPOSITION 11.2

Let R be a nontrivial ring without zero-divisors. If the characteristic is 0, then for n an integer and x a nonzero element of R, $nx = 0$ only if $n = 0$. If the characteristic is positive then it is a prime number p, and for nonzero x, $nx = 0$ if, and only if, n is a multiple of p.

PROOF: Suppose that $nx = 0$ for some nonzero $x \in R$. Then for another nonzero element y of R,

$$0 = (nx) \cdot y = n(x \cdot y) = x \cdot (ny).$$

But x is nonzero, and the ring has no zero divisors, so we have $ny = 0$. This argument works in both ways, so

$$(*) \qquad nx = 0 \Longleftrightarrow ny = 0 \qquad \text{if } x \neq 0 \text{ and } y \neq 0.$$

If n was not zero, then $|n|$ would be a positive number such that $nx = 0$ for all x in the ring. Hence, if the ring has characteristic 0, then $nx = 0$ implies that either $x = 0$ or $n = 0$.

Now suppose that the ring has positive characteristic, and let x be any nonzero element of R. Let p be the smallest positive integer for which $p \cdot x = 0$.

If p is not prime, then $p = ab$ with $0 < a < p$ and $0 < b < p$. But then

$$(ax) \cdot (bx) = (ab)\left(x^2\right) = p(x^2) = 0.$$

Since the ring has no zero divisors, either $ax = 0$ or $bx = 0$. But this contradicts the fact that p was the *smallest* number such that $px = 0$. Thus, p is prime. By $(*)$ we have that $py = 0$ for *every* element in R, and since this cannot be true for any smaller integer, we have that the characteristic of the ring is the prime number p.

It is easy to see that if n is a multiple of p, then $n = cp$ for some integer c. Thus, for any element x in R,

$$nx = (cp)x = c(px) = c0 = 0.$$

Suppose that $nx = 0$ for some n that is not a multiple of p. Then $\gcd(n, p)$ must be 1, and so by the greatest common divisor theorem (0.4), there are integers u and v such that $un + vp = 1$. But then

$$x = 1 \cdot x = (un + vp)x = u(nx) + v(px) = u \cdot 0 + v \cdot 0 = 0.$$

So for nonzero x, $nx = 0$ if, and only if, n is a multiple of p. □

Characteristics are important because they provide a new way of defining integral domains and fields in *Sage*. We begin by telling *Sage* the characteristic p of the ring we want to define with the command **InitDomain(p)**. For example, to define a ring with characteristic 3, we enter

```
InitDomain(3)
```

This actually defines the field Z_3, as we can see with the command

```
Z3 = ListField(); Z3
    {0, 1, 2}
```

We can create polynomials over this new domain by the **AddFieldVar** command.

```
AddFieldVar("i")
```

Now we can do computations in the polynomial ring $Z_3[i]$.

```
2*i + 5*i
    i
(2*i + 1)^2
    i^2 + i + 1
```

Let us try imitating the complex numbers, and tell *Sage* that $i^2 = -1$.

TABLE 11.1: Addition of "complex numbers modulo 3"

$+$	0	1	2	i	$2i$	$1+i$	$2+i$	$1+2i$	$2+2i$
0	0	1	2	i	$2i$	$1+i$	$2+i$	$1+2i$	$2+2i$
1	1	2	0	$1+i$	$1+2i$	$2+i$	i	$2+2i$	$2i$
2	2	0	1	$2+i$	$2+2i$	i	$1+i$	$2i$	$1+2i$
i	i	$1+i$	$2+i$	$2i$	0	$1+2i$	$2+2i$	1	2
$2i$	$2i$	$1+2i$	$2+2i$	0	i	1	2	$1+i$	$2+i$
$1+i$	$1+i$	$2+i$	i	$1+2i$	1	$2+2i$	$2i$	2	0
$2+i$	$2+i$	i	$1+i$	$2+2i$	2	$2i$	$1+2i$	0	1
$1+2i$	$1+2i$	$2+2i$	$2i$	1	$1+i$	2	0	$2+i$	i
$2+2i$	$2+2i$	$2i$	$1+2i$	2	$2+i$	0	1	i	$1+i$

TABLE 11.2: Multiplication for "complex numbers modulo 3"

$\cdot$	0	1	2	i	$2i$	$1+i$	$2+i$	$1+2i$	$2+2i$
0	0	0	0	0	0	0	0	0	0
1	0	1	2	i	$2i$	$1+i$	$2+i$	$1+2i$	$2+2i$
2	0	2	1	$2i$	i	$2+2i$	$1+2i$	$2+i$	$1+i$
i	0	i	$2i$	2	1	$2+i$	$2+2i$	$1+i$	$1+2i$
$2i$	0	$2i$	i	1	2	$1+2i$	$1+i$	$2+2i$	$2+i$
$1+i$	0	$1+i$	$2+2i$	$2+i$	$1+2i$	$2i$	1	2	i
$2+i$	0	$2+i$	$1+2i$	$2+2i$	$1+i$	1	i	$2i$	2
$1+2i$	0	$1+2i$	$2+i$	$1+i$	$2+2i$	2	$2i$	i	1
$2+2i$	0	$2+2i$	$1+i$	$1+2i$	$2+i$	i	2	1	$2i$

```
Define(i^2, -1)
K = ListField(); K
    {0, 1, 2, i, i + 1, i + 2, 2*i, 2*i + 1, 2*i + 2}
AddTable(K)
MultTable(K)
```

This produces Tables 11.1 and 11.2.

We can see that this ring has nine elements and has no zero divisors. By Corollary 9.1, K is a field. We could call K the field of "complex numbers modulo 3."

Sage offers a shortcut for working with polynomials over an integral domain. We can add an additional parameter for the **InitDomain** command that will tell *Sage* the name of the polynomial variable, usually "x". For example, the command

```
InitDomain(3, "x")
```

defines the integral domain $Z_3[x]$ in one step. We can now do operations in $Z_3[x]$.

```
(x + 2)^2
    x^2 + x + 1
factor(x^2 + 2)
    (x + 1) * (x + 2)
factor(x^2 + 1)
    x^2 + 1
```

If we continue to expand the field to the "complex numbers modulo 3,"

```
AddFieldVar("i")
Define(i^2, -1)
```

the variable x automatically promotes to a variable of the larger field. Thus, we can form polynomials like

```
y = (1 + i)*x + 2; y
    (i + 1)*x + 2

z = (2 + i)*x^2 + 2*i*x + 1 + 2*i; z
    (i + 2)*x^2 + 2*i*x + 2*i + 1
y^2
    2*i*x^2 + (i + 1)*x + 1
y*z
    x^3 + (i + 2)*x^2 + (i + 2)*x + i + 2
```

Sage can factor polynomials defined over any finite field. In Chapter 12, we will prove that such factorizations are unique. (We will also see an example of an integral domain, not a field, for which factorizations are *not* unique.) If *Sage* tries to factor $x^2 + 1$ in the standard way (using the ring $\mathbb{Z}$),

```
var("X")
factor(X^2 + 1)
    X^2 + 1
```

it finds the polynomial is irreducible. But if we factor the polynomial over the field of "complex numbers modulo 3,"

```
factor(x^2 + 1)
    (x + i) * (x + 2*i)
```

we find that it does factor. Hence, whether a polynomial factors or not depends largely on which integral domain we are using.

The polynomial rings defined over integral domains give us some good examples of integral domains. In the next chapter we will find other ways of forming integral domains, some of which have some unusual properties. But even these are based on polynomial rings. So polynomials are the basic building blocks that are used for forming new integral domains and fields.

Problems for §11.1

For Problems **1** through **6**: Expand the following polynomials using the ring defined by Tables 9.3 and 9.4.

1 $(2ax + b)^2$

2 $(bx + a)(bx - a)$

3 $(2ax + a + b)(ax + b)$

4 $(2ax^2 + ax + b)(bx + a)$

5 $(ax^2 + (a + b)x + 2a)(2ax + b)$

6 $(bx^2 + (2a + b)x + a)(bx^2 + 2ax - a)$

7 Find the characteristic of the ring defined by Tables 9.3 and 9.4.

8 Find the characteristic of the ring T_8 in Table 9.7.

9 Prove that if $n > 1$, the characteristic of Z_n is n.

10 Let R be a unity ring. If the identity element has a finite order in the additive group, show that this order is the characteristic of the ring.

11 A *Boolean ring* is a nontrivial ring in which all elements x satisfy $x^2 = x$. Prove that every Boolean ring has characteristic 2.

12 Prove that if a ring R has a finite number of elements, then the characteristic of R is a positive integer.

13 Let D be an integral domain with positive characteristic. Prove that all nonzero elements of D have the same additive order.

14 Show an example for which Problem 13 is not true for arbitrary rings.

15 Let $\{0, e, a, b\}$ be a field of order 4, with e as the unity. Construct addition and multiplication tables for the field.

16 Let R be a commutative ring of characteristic 2. Prove that $(x + y)^2 = x^2 + y^2$ for all x and y in R. This property is often referred to as "freshman's dream."

17 Let R be a commutative ring of characteristic 2. Prove that $(x + y)^4 = x^4 + y^4$ for all x and y in R. You can use the result of Problem 16.

18 Find an example of a commutative ring of characteristic 4 for which there are elements x and y such that $(x + y)^4 \neq x^4 + y^4$.

19 Find an example of a non-commutative ring of characteristic 4 for which there are elements x and y such that $(x + y)^4 \neq x^4 + y^4$.

20 List all polynomials in $Z_3[x]$ that have degree 2.

21 Of the second degree polynomials in $Z_3[x]$ listed in Problem 20, which ones cannot be factored?

Hint: A quadratic polynomial in $Z_3[x]$ cannot be factored if neither 0, 1, nor 2 are roots.

22 List all polynomials in $Z_2[x]$ that have degree 3.

23 Of the third-degree polynomials in $Z_2[x]$ listed in Problem 20, which ones cannot be factored?

Hint: A cubic polynomial in $Z_2[x]$ cannot be factored if neither 0 nor 1 are roots.

Interactive Problems

24 In the field of "complex numbers modulo 3":

```
InitDomain(3, "x")
AddFieldVar("i")
Define(i^2, -1)
K = ListField(); K
    {0, 1, 2, i, i + 1, i + 2, 2*i, 2*i + 1, 2*i + 2}
```

Factor the polynomials $x^3 + 1$, $x^3 + 2$, $x^3 + i$, $x^3 + 2i$. What do you notice about the factorizations? Knowing how *real* polynomials factor, explain what is happening.

25 Explain why the ring "complex numbers modulo 5":

```
InitDomain(5)
AddFieldVar("i")
Define(i^2, -1)
```

does not form a field. Can you determine a pattern as to which integers "complex numbers modulo n" form a field?

11.2 The Field of Quotients

In the last section, we found a way to form integral domains by imitating the familiar polynomials from high school algebra. In this section we will show how we can form a field from an integral domain, imitating grade school fractions.

We view a standard fraction as one integer divided by another. We want to extend this idea, and form fractions out of any integral domain. However, even with standard fractions there is a complication, since we consider

$$\frac{2}{4} = \frac{3}{6},$$

even though both the numerators and denominators are different. What we mean to say is that these two fractions are *equivalent*, where we define

$$\frac{x}{y} \equiv \frac{u}{v} \qquad \text{if, and only if,} \qquad x \cdot v = y \cdot u.$$

This forms an equivalence relation on the set of fractions x/y. We have already seen equivalence relations while working with cosets of a group. What we call a rational number is really a set of fractions of the form x/y that are all equivalent.

DEFINITION 11.3 Let K be an integral domain, and let P denote the set of all ordered pairs (x, y) of elements of K, with y nonzero:

$$P = \{(x, y) \mid x, y \in K \quad \text{and} \quad y \neq 0\}.$$

We define a relation on P by

$$(x, y) \equiv (u, v) \qquad \text{if} \qquad x \cdot v = y \cdot u.$$

LEMMA 11.2
The above relation is an equivalence relation on P.

PROOF: We need to show that the relation is reflexive, symmetric, and transitive. Let(x, y), (u, v), and (s, t) be arbitrary elements of P.
Reflexive:

$$(x, y) \equiv (x, y)$$

is equivalent to saying $x \cdot y = x \cdot y$ which is, of course, true. So this relation is reflexive.
Symmetric:

$$(x, y) = (u, v) \Longrightarrow x \cdot v = y \cdot u \Longrightarrow u \cdot y = v \cdot x \Longrightarrow (u, v) \equiv (x, y),$$

so this relation is also symmetric.
Transitive:
If $(x, y) \equiv (u, v)$ and $(u, v) \equiv (s, t)$, then

$$(x, y) \equiv (u, v) \Longrightarrow x \cdot v = y \cdot u \Longrightarrow x \cdot v \cdot t = y \cdot u \cdot t,$$

$$(u, v) \equiv (s, t) \Longrightarrow u \cdot t = v \cdot s \Longrightarrow u \cdot t \cdot y = v \cdot s \cdot y.$$

These two statements imply that $x \cdot v \cdot t = v \cdot s \cdot y$. Notice that in the last step we had to use the commutativity of multiplication. Using commutativity again, we have $x \cdot t \cdot v = y \cdot s \cdot v$, and since K has no zero divisors and v is nonzero, we can use Lemma 9.4 to say that $x \cdot t = y \cdot s$. Then

$$x \cdot t = y \cdot s \Longrightarrow (x, y) \equiv (s, t),$$

so we have the transitive law holding. Therefore, this relation is an equivalence relation. ∎

DEFINITION 11.4 Let K be an integral domain, let P denote the set

$$P = \{(x, y) \mid x, y \in K \quad \text{and} \quad y \neq 0\},$$

and let the equivalence relation on P be

$$(x, y) \equiv (u, v) \qquad \text{if} \qquad x \cdot v = y \cdot u.$$

For each (x, y) in P, let $\left(\frac{x}{y}\right)$ denote the equivalence class of P that contains (x, y). Let Q denote the set of all equivalence classes $\left(\frac{a}{b}\right)$. The set Q is called the *set of quotients* for K.

This definition allows us to replace an equivalence of two expressions with an equality. We now have that

$$\left(\frac{x}{y}\right) = \left(\frac{u}{v}\right) \quad \text{if, and only if,} \quad x \cdot v = u \cdot y.$$

The next step is to define addition and multiplication on our set of quotients Q. Once again, we will use the rational numbers to guide us in the definition.

LEMMA 11.3

Let K be an integral domain, and let Q be the set of quotients for K. The addition and multiplication of two equivalence classes in Q, defined by

$$\left(\frac{x}{y}\right) + \left(\frac{u}{v}\right) = \left(\frac{x \cdot v + u \cdot y}{y \cdot v}\right)$$

and

$$\left(\frac{x}{y}\right) \cdot \left(\frac{u}{v}\right) = \left(\frac{x \cdot u}{y \cdot v}\right),$$

are both well-defined operations on Q. That is, the sum and product do not depend on the choice of the representative elements (x, y) and (u, v) of the equivalence classes.

PROOF: The first observation we need to make is that the formulas for the sum and product both form valid elements of Q, since $y \cdot v$ is nonzero as long as y and v are both nonzero.

Next let us work to show that addition does not depend on the choice of representative elements (x, y) and (u, v). That is, if $\left(\frac{x}{y}\right) = \left(\frac{a}{b}\right)$, and $\left(\frac{u}{v}\right) = \left(\frac{c}{d}\right)$, we need to show that

$$\left(\frac{x}{y}\right) + \left(\frac{u}{v}\right) = \left(\frac{a}{b}\right) + \left(\frac{c}{d}\right).$$

That is, we have to prove that

$$\left(\frac{x \cdot v + u \cdot y}{y \cdot v}\right) = \left(\frac{a \cdot d + c \cdot b}{b \cdot d}\right).$$

Since $\left(\frac{x}{y}\right) = \left(\frac{a}{b}\right)$ and $\left(\frac{u}{v}\right) = \left(\frac{c}{d}\right)$, we have $x \cdot b = a \cdot y$ and $u \cdot d = c \cdot v$. Multiplying the first equation by $v \cdot d$ and the second by $y \cdot b$, we get

$$x \cdot b \cdot v \cdot d = a \cdot y \cdot v \cdot d$$

and

$$u \cdot d \cdot y \cdot b = c \cdot v \cdot y \cdot b.$$

Adding these two equations together and factoring, we get

$$(x \cdot v + u \cdot y) \cdot b \cdot d = (a \cdot d + c \cdot b) \cdot y \cdot v.$$

This gives us

$$\left(\frac{x \cdot v + u \cdot y}{y \cdot v}\right) = \left(\frac{a \cdot d + c \cdot b}{b \cdot d}\right),$$

which is what we wanted.

We also need to show that multiplication is well defined, that is

$$\left(\frac{x}{y}\right) \cdot \left(\frac{u}{v}\right) = \left(\frac{a}{b}\right) \cdot \left(\frac{c}{d}\right).$$

But since $x \cdot b = a \cdot y$ and $u \cdot d = c \cdot v$, we can multiply these two equations together to get

$$x \cdot b \cdot u \cdot d = a \cdot y \cdot c \cdot v,$$

or

$$(x \cdot u) \cdot (b \cdot d) = (a \cdot c) \cdot (y \cdot v).$$

Therefore,

$$\left(\frac{x \cdot u}{y \cdot v}\right) = \left(\frac{a \cdot c}{b \cdot d}\right),$$

so multiplication also is well defined. ∎

THEOREM 11.1: The Field of Quotients Theorem

Let K be an integral domain, and let Q be the set of quotients for K. Then Q forms a field using the above definitions of addition and multiplication. The field Q is called the field of quotients for *K.*

PROOF: We have already noted that addition and multiplication are closed in Q.

We next want to look at the properties of addition. From the definition,

$$\left(\frac{x}{y}\right) + \left(\frac{u}{v}\right) = \left(\frac{x \cdot v + u \cdot y}{y \cdot v}\right) = \left(\frac{u}{v}\right) + \left(\frac{x}{y}\right),$$

we see that addition is commutative. Let z be any nonzero element of K. Then $\left(\frac{0}{z}\right)$ acts as the additive identity:

$$\left(\frac{u}{v}\right)+\left(\frac{0}{z}\right)=\left(\frac{0}{z}\right)+\left(\frac{u}{v}\right)=\left(\frac{0\cdot v+u\cdot z}{z\cdot v}\right)=\left(\frac{u\cdot z}{v\cdot z}\right)=\left(\frac{u}{v}\right).$$

Likewise, $\left(\frac{-u}{v}\right)$ is the additive inverse of $\left(\frac{u}{v}\right)$:

$$\left(\frac{u}{v}\right)+\left(\frac{-u}{v}\right)=\left(\frac{-u}{v}\right)+\left(\frac{u}{v}\right)=\left(\frac{-u\cdot v+u\cdot v}{v\cdot v}\right)=\left(\frac{0}{v\cdot v}\right)=\left(\frac{0}{z}\right).$$

The associativity of addition is straightforward:

$$\begin{aligned}\left(\left(\frac{x}{y}\right)+\left(\frac{u}{v}\right)\right)+\left(\frac{a}{b}\right)&=\left(\frac{x\cdot v+u\cdot y}{y\cdot v}\right)+\left(\frac{a}{b}\right)\\&=\left(\frac{x\cdot v\cdot b+u\cdot y\cdot b+a\cdot y\cdot v}{y\cdot v\cdot b}\right),\end{aligned}$$

while

$$\begin{aligned}\left(\frac{x}{y}\right)+\left(\left(\frac{u}{v}\right)+\left(\frac{a}{b}\right)\right)&=\left(\frac{x}{y}\right)+\left(\frac{u\cdot b+a\cdot v}{v\cdot b}\right)\\&=\left(\frac{x\cdot v\cdot b+u\cdot y\cdot b+a\cdot y\cdot v}{y\cdot v\cdot b}\right).\end{aligned}$$

So Q forms a group with respect to addition.

Next we look at the properties of multiplication. Multiplication is obviously commutative, since

$$\left(\frac{x}{y}\right)\cdot\left(\frac{u}{v}\right)=\left(\frac{x\cdot u}{y\cdot v}\right)=\left(\frac{u\cdot x}{v\cdot y}\right)=\left(\frac{u}{v}\right)\cdot\left(\frac{x}{y}\right).$$

We also have associativity for multiplication:

$$\begin{aligned}\left(\left(\frac{x}{y}\right)\cdot\left(\frac{u}{v}\right)\right)\cdot\left(\frac{a}{b}\right)&=\left(\frac{x\cdot u}{y\cdot v}\right)\cdot\left(\frac{a}{b}\right)\\&=\left(\frac{x\cdot u\cdot a}{y\cdot v\cdot b}\right)=\left(\frac{x}{y}\right)\cdot\left(\frac{u\cdot a}{v\cdot b}\right)=\left(\frac{x}{y}\right)\cdot\left(\left(\frac{u}{v}\right)\cdot\left(\frac{a}{b}\right)\right).\end{aligned}$$

The element $\left(\frac{z}{z}\right)$ acts as the multiplicative identity for any $z\neq 0$.

$$\left(\frac{z}{z}\right)\cdot\left(\frac{x}{y}\right)=\left(\frac{x}{y}\right)\cdot\left(\frac{z}{z}\right)=\left(\frac{x\cdot z}{y\cdot z}\right)=\left(\frac{x}{y}\right).$$

If $x=0$, then $\left(\frac{x}{y}\right)=\left(\frac{0}{z}\right)$. Otherwise, the multiplicative inverse of $\left(\frac{x}{y}\right)$ is $\left(\frac{y}{x}\right)$, since

$$\left(\frac{x}{y}\right)\cdot\left(\frac{y}{x}\right)=\left(\frac{x\cdot y}{y\cdot x}\right)=\left(\frac{z}{z}\right).$$

Thus, every nonzero element of Q has a multiplicative inverse. Finally, we have the two distribution laws. Because of the commutativity of multiplication, we only need to check one. Since

$$\left(\left(\frac{u}{v}\right)+\left(\frac{a}{b}\right)\right)\cdot\left(\frac{x}{y}\right)=\left(\frac{u\cdot b+a\cdot v}{v\cdot b}\right)\cdot\left(\frac{x}{y}\right)=\left(\frac{u\cdot b\cdot x+a\cdot v\cdot x}{v\cdot b\cdot y}\right),$$

while

$$\begin{aligned}\left(\frac{u}{v}\right)\cdot\left(\frac{x}{y}\right)+\left(\frac{a}{b}\right)\cdot\left(\frac{x}{y}\right)&=\left(\frac{u\cdot x}{v\cdot y}\right)+\left(\frac{a\cdot x}{b\cdot y}\right)\\&=\left(\frac{u\cdot x\cdot b\cdot y+a\cdot x\cdot v\cdot y}{v\cdot y\cdot b\cdot y}\right)\\&=\left(\frac{u\cdot x\cdot b+a\cdot x\cdot v}{v\cdot y\cdot b}\right),\end{aligned}$$

we have the distributive laws holding, and therefore Q is a field. ∎

In the construction of the field Q, we never used the identity element of K. Hence, if we started with a commutative ring without zero divisors instead of an integral domain, the construction would still produce a field. We can mention this as a corollary.

COROLLARY 11.1

Let K be any commutative ring without zero divisors. Then the set of quotients Q defined above forms a field.

Although the field of quotients was designed from the way we formed rational numbers from the set of integers, we can apply the field of quotients to any other integral domain. What happens if we form a field of quotients for the polynomial ring $K[x]$?

Let us first consider the most familiar polynomial ring $\mathbb{Z}[x]$—the polynomials with integer coefficients. An element in the field of quotients would be of the form $p(x)/q(x)$, where $p(x)$ and $q(x)$ are polynomials with integer coefficients. But we consider two such fractions $p(x)/q(x)$ and $r(x)/s(x)$ to be equivalent if $p(x)\cdot s(x)=r(x)\cdot q(x)$. For example, the two fractions

```
var("x")
A = (3*x^2 + 5*x - 2)/(2*x^2 + 7*x + 6); A
    (3*x^2 + 5*x - 2)/(2*x^2 + 7*x + 6)
B = (3*x^2 - 4*x + 1)/(2*x^2 + x - 3); B
    (3*x^2 - 4*x + 1)/(2*x^2 + x - 3)
```

can be seen to be equivalent, since

```
expand((3*x^2 + 5*x - 2) * (2*x^2 + x - 3))
    6*x^4 + 13*x^3 - 8*x^2 - 17*x + 6
expand((3*x^2 - 4*x + 1) * (2*x^2 + 7*x + 6))
    6*x^4 + 13*x^3 - 8*x^2 - 17*x + 6
```

yield the same result. Other ways of showing that A and B are equivalent is by computing either of these two commands:

```
Together(A - B)
  0
Together(A/B)
  1
```

We call the field of quotients for the polynomials $\mathbb{Z}[x]$ the *field of rational functions in* x, denoted $\mathbb{Z}(x)$.

It should be mentioned that a rational function, in this context, is not a function! The rational functions A and B are merely *elements* of $\mathbb{Z}(x)$, which may in turn be arguments for some homomorphism. To say that "A is undefined when $x = -2$" or "B is undefined at $x = 1$" is meaningless, since x is not a variable for which numbers can be plugged in. Rather, x is merely a symbol that is used as a place holder. This is why we can say that A and B are truly equal, even though their "graphs" would disagree at two points.

We can form rational functions from any integral domain K. This produces the field $K(x)$, the *rational functions in* x *over* K.

Computational Example 11.1

Simplify the rational function

$$\frac{(1+i)x^2 + (2+2i)x + 2}{x^2 + ix + 1}$$

defined over the field of order 9 that was defined by Tables 11.1 and 11.2.

SOLUTION: First we set up the field.

```
InitDomain(3, "x")
AddFieldVar("i")
Define(i^2, -1)
```

Sage will automatically simplify the rational function for us.

```
A = ((1 + i)*x^2 + (2 + 2*i)*x + 2)/(x^2 + i*x + 1); A
    ((i + 1)*x + i + 2)/(x + 2*i + 1)
```

However, if we consider the simpler looking rational function

```
B = (2*x - i)/(x - i*x + i); B
    (2*x + 2*i)/((2*i + 1)*x + i)
```

we find that they are equal.

```
A - B
   0
```

□

As you can see from this example, the definition of the quotient field does not depend on whether elements in the integral domain can be factored uniquely. However, unique factorization is an important property that we will study in depth in Chapter 12. We will learn that the polynomial ring $K[x]$ used in the above example really does have a type of unique factorization, after we have studied the true definition of what a unique factorization is. But before we go into this, let us look closely at some of the more familiar fields: the rational numbers, the real numbers, and the complex numbers. These fields will be the basis for defining many other fields, so it is natural to learn the properties of these fields before going on to study more complicated fields.

Problems for §11.2

1 If Q is the field of quotients of an integral domain, show that $\left(\frac{-a}{b}\right)$ is the additive inverse of $\left(\frac{a}{b}\right)$ in Q.

2 If Q is the field of quotients of an integral domain, show that the left distributive property holds for Q:

$$\left(\frac{u}{v}\right)\cdot\left(\left(\frac{x}{y}\right)+\left(\frac{z}{w}\right)\right)=\left(\frac{u}{v}\right)\cdot\left(\frac{x}{y}\right)+\left(\frac{u}{v}\right)\cdot\left(\frac{z}{w}\right).$$

3 If Q is the field of quotients of an integral domain, show that the multiplication in Q is associative.

4 Investigate what happens if we compute the field of quotients of a ring that is already a field. Let $K = Z_3$, and let P be the set of ordered pairs

$$P = \{(x, y) \quad | \quad x, y \in Z_3 \quad \text{and} \quad y \neq 0\}.$$

Write a list of all ordered pairs in P, and determine which pairs are equivalent under the relation

$$(x, y) \equiv (u, v) \quad \text{if} \quad x \cdot v \equiv y \cdot u \pmod 3.$$

If Q is the set of equivalence classes, construct addition and multiplication tables for Q and show that Q is isomorphic to Z_3.

5 Repeat Problem 4, using Z_5 instead of Z_3.

6 Prove that if K is a field, then the field of quotients of K is isomorphic to K.

7 Show that if we apply Corollary 11.1 to the ring of even integers, we obtain a field isomorphic to $\mathbb{Q}$.

8 What is the quotient field for the ring given by

$$\{x + y\sqrt{2} \mid x, y \in \mathbb{Z}\}?$$

9 Show by cross multiplying that the two rational functions A and B from Example 11.1 are indeed equal.

For Problems **10** through **17**: Perform the following operations in $Z_2(x)$, the rational functions over Z_2.

10 $\dfrac{x^2+x+1}{x+1} + \dfrac{x+1}{x}$

11 $\dfrac{x+1}{x^2+x+1} + \dfrac{1}{x^2+x}$

12 $\dfrac{x^2+1}{x} + \dfrac{x^2+x+1}{x+1}$

13 $\dfrac{x^2+x}{x^2+x+1} + \dfrac{x}{x+1}$

14 $\dfrac{x^2+x+1}{x+1} \cdot \dfrac{x}{x+1}$

15 $\dfrac{x^2+1}{x^2+x+1} \cdot \dfrac{x^2}{x+1}$

16 $\dfrac{x^2+x+1}{x^2+x} \cdot \dfrac{x^2+1}{x}$

17 $\dfrac{x^2}{x^2+x+1} \cdot \dfrac{x+1}{x^2+x+1}$

Interactive Problems

18 Have *Sage* simplify the rational function over $Z_2(x)$:

$$\frac{x^4+x^3+x+1}{x^3+x^2+x+1}.$$

19 Try squaring different elements of $Z_2(x)$. What do you observe? Any explanations?

20 Have *Sage* compute the following operation in the rational function field of Example 11.1.

$$\frac{(1+i)x+2}{x^2+2ix+2+i} + \frac{2x+1+i}{x^2+(2+i)x+2}.$$

21 It was mentioned that the definition of the quotient field does not depend on whether elements in the integral domain have unique factorization. An example of such a domain is $\mathbb{Z}[\sqrt{-5}]$, which we can enter in *Sage* as follows:

```
InitDomain(0, "x")
AddFieldVar("a")
Define(a^2, -5)
```

Show that the two fractions

$$\frac{3x+3a}{(1+a)x} \quad \text{and} \quad \frac{(1-a)x+5+a}{2x}$$

are in fact equal, even though neither can simplify.

11.3 Complex Numbers

The field of complex numbers may be the most important field of mathematics. Although the real numbers are used more often, the set of real numbers has deficiencies in that not every polynomial equation can be solved with real numbers; for example, $x^2 + 1 = 0$. On the other hand, every polynomial equation can be solved using complex numbers, even if the coefficients are complex. This result is called the *Fundemental theorem of algebra*, even though Carl Gauss's first proof of the theorem was geometric.

Because of the importance of complex numbers, we give a brief summary of complex number theory in this section, since many of these results, in particular DeMoivre's theorem, are needed in later sections. Readers who have had a course in complex numbers will find this section to be mainly review.

We have already seen some examples of complex numbers in the form $a+bi$, where i represents the "square root of negative one." *Sage* uses a capital **I** to enter and display the imaginary number. This allows us to perform standard arithmetic on complex numbers.

```
(2 + 3*I) + (4 - I)
    2*I + 6
(2 + 3*I)*(4 - I)
    10*I + 11
(2 + 3*I)/(4 - I)
    14/17*I + 5/17
```

You may have noticed that *Sage* puts the complex part of the number first. In this presentation, it is not at all clear where the "**I**" came from. This gives the complex numbers a rather mysterious quality that is compounded by their common misnomer, "imaginary numbers."

We would like to show how complex numbers are a natural extension of the real numbers. Instead of considering quantities of the form $a + bi$, we will consider ordered pairs (a, b). We will declare the following properties for ordered pairs of real numbers:

1. $(a, b) = (c, d)$ if, and only if, $a = c$ and $b = d$.

2. $(a, b) + (c, d) = (a + c,\ b + d)$.

3. $(a, b) \cdot (c, d) = (a \cdot c - b \cdot d,\ a \cdot d + b \cdot c)$.

We define $\mathbb{C}$ to be the set of all ordered pairs of real numbers.

PROPOSITION 11.3
The set $\mathbb{C}$ forms a field called the field of complex numbers. *This field contains a subfield isomorphic to the real numbers.*

PROOF: Because the real numbers are closed with respect to both addition and multiplication, it is clear that both $(a+c, b+d)$ and $(a\cdot c-b\cdot d, a\cdot d+b\cdot c)$ would be defined for all real numbers a, b, c, and d. Thus, $\mathbb{C}$ is closed with respect to both addition and multiplication. Furthermore, since

$$(c,d)+(a,b)=(c+a,d+b)=(a+c,b+d)=(a,b)+(c,d)$$

and

$$(c,d)\cdot(a,b)=(c\cdot a-d\cdot b,c\cdot b+d\cdot a)=(a\cdot c-b\cdot d,a\cdot d+b\cdot c)=(a,b)\cdot(c,d),$$

we see that both addition and multiplication are commutative. The element $(0,0)$ acts as the zero element, since

$$(0,0)+(a,b)=(a,b).$$

The addition inverse of (a,b) is $(-a,-b)$, since

$$(a,b)+(-a,-b)=(0,0).$$

Note that the order on the last two sums is irrelevant, since addition has already been shown to be commutative.

To show that addition is associative, we note that

$$\begin{aligned}(a,b)+\big[(c,d)+(e,f)\big]&=(a,b)+(c+e,d+f)=(a+c+e,b+d+f)\\&=(a+c,b+d)+(e,f)=\big[(a,b)+(c,d)\big]+(e,f).\end{aligned}$$

To show that multiplication is associative is a little more complicated. We have

$$\begin{aligned}&(a,b)\cdot\big[(c,d)\cdot(e,f)\big]=(a,b)\cdot(c\cdot e-d\cdot f,c\cdot f+d\cdot e)=\\&(a\cdot c\cdot e-a\cdot d\cdot f-b\cdot c\cdot f-b\cdot d\cdot e,a\cdot c\cdot f+a\cdot d\cdot e+b\cdot c\cdot e-b\cdot d\cdot f),\end{aligned}$$

and

$$\begin{aligned}&\big[(a,b)\cdot(c,d)\big]\cdot(e,f)=(a\cdot c-b\cdot d,a\cdot d+b\cdot c)\cdot(e,f)=\\&(a\cdot c\cdot e-b\cdot d\cdot e-a\cdot d\cdot f-b\cdot c\cdot f,a\cdot c\cdot f-b\cdot d\cdot f+a\cdot d\cdot e+b\cdot c\cdot e).\end{aligned}$$

By comparing these two, we see that they are equal, so multiplication is associative.

We need to test the distributive laws next. The left distributive law we can get by expanding:

$$\begin{aligned}(a,b)\cdot\big[(c,d)+(e,f)\big]&=(a,b)\cdot(c+e,d+f)\\&=(a\cdot c+a\cdot e-b\cdot d-b\cdot f,a\cdot d+a\cdot f+b\cdot c+b\cdot e)\\&=(a\cdot c-b\cdot d,a\cdot d+b\cdot c)+(a\cdot e-b\cdot f,a\cdot f+b\cdot e)\\&=(a,b)\cdot(c,d)+(a,b)\cdot(e,f).\end{aligned}$$

Thus, the left distributive law is satisfied. However, the right distributive law follows from the left distributive law, and using the commutative multiplication:

$$\begin{aligned}\big[(a,b)+(c,d)\big]\cdot(e,f) &= (e,f)\cdot\big[(a,b)+(c,d)\big]\\ &= (e,f)\cdot(a,b)+(e,f)\cdot(c,d)\\ &= (a,b)\cdot(e,f)+(c,d)\cdot(e,f).\end{aligned}$$

We have now shown that the set $\mathbb{C}$ forms a commutative ring. To show that this ring has a multiplicative identity, we consider the element $(1,0)$. Since the ring is commutative, we only need to check

$$(1,0)\cdot(a,b) = (1\cdot a - 0\cdot b, 1\cdot b + 0\cdot a) = (a,b).$$

Finally, we need to show that every nonzero element has an inverse. If (a,b) is nonzero, then a^2+b^2 will be a positive number. Hence

$$\left(\frac{a}{a^2+b^2}, \frac{-b}{a^2+b^2}\right)$$

is an element of $\mathbb{C}$. The product

$$(a,b)\cdot\left(\frac{a}{a^2+b^2}, \frac{-b}{a^2+b^2}\right) = \left(\frac{a^2+b^2}{a^2+b^2}, \frac{-a\cdot b + a\cdot b}{a^2+b^2}\right) = (1,0)$$

verifies that

$$(a,b)^{-1} = \left(\frac{a}{a^2+b^2}, \frac{-b}{a^2+b^2}\right)$$

since multiplication is commutative. Therefore, the set $\mathbb{C}$ forms a field.

The second part of this proposition is to show that $\mathbb{C}$ contains a copy of the real numbers as a subfield. Consider the mapping f, which maps real numbers to $\mathbb{C}$, given by

$$f(x) = (x,0).$$

To check that f is a homomorphism, we check that

$$f(x)+f(y) = (x,0)+(y,0) = (x+y,0) = f(x+y)$$

and

$$f(x)\cdot f(y) = (x,0)\cdot(y,0) = (x\cdot y+0, 0+0) = (x\cdot y,0) = f(x\cdot y).$$

Thus, f is a homomorphism from the reals to $\mathbb{C}$. It is clear that f is one-to-one, since $(x,0)=(y,0)$ if, and only if, $x=y$. Thus the image of f:

$$\{(x,0) \mid x \in \mathbb{R}\}$$

is isomorphic to the real numbers. Hence, we have found a subring of $\mathbb{C}$ isomorphic to $\mathbb{R}$. ∎

The purpose of constructing the complex numbers was to produce a field for which we can take the square root of negative one. We can now show that we have succeeded in doing this.

LEMMA 11.4

There are exactly two solutions to the equation $x^2 = (-1, 0)$ *in the field* $\mathbb{C}$, *given by* $(0, \pm 1)$.

PROOF: If (a, b) solves the equation $x^2 = (-1, 0)$, we have that

$$(a, b)^2 = (a^2 - b^2, 2ab) = (-1, 0).$$

Thus, a and b must satisfy the two equations

$$a^2 - b^2 = -1$$

and

$$2ab = 0.$$

The second equation implies that either a or b must be 0. But if $b = 0$, then the first equation becomes $a^2 = -1$, which has no real solutions. Thus, $a = 0$, and $-b^2 = -1$. There are two real solutions for b, ± 1. Thus, $(0, 1)$ and $(0, -1)$ both solve the equations for a and b, and so

$$(0, 1)^2 = (0, -1)^2 = (-1, 0). \qquad \square$$

By defining the complex numbers as ordered pairs, we have taken some of the mystery out of the complex numbers. Lemma 11.4 shows that the square root of negative one comes as a natural consequence of the way we defined the product.

We can now convert ordered pairs to the customary notation by defining $i = (0, 1)$, and identifying the identity element $(1, 0)$ with 1. Then any complex number (a, b) can be written

$$(a, b) = (a, 0) + (0, b) = a \cdot (1, 0) + b \cdot (0, 1) = a + bi.$$

We can rewrite the rules for addition and multiplication in $\mathbb{C}$ as follows:

$$(a + bi) + (c + di) = (a + c) + (b + d)i.$$

$$(a + bi) \cdot (c + di) = (a \cdot c - b \cdot d) + (b \cdot c + a \cdot d)i.$$

In working with groups, we found that the group automorphisms revealed many of the important properties of the group. This will also be true for rings. Let us extend the group automorphisms to apply to rings.

DEFINITION 11.5 A *ring automorphism* is a one-to-one and onto ring homomorphism that maps a ring to itself.

LEMMA 11.5
The set of all ring automorphisms of a given ring forms a group.

PROOF: We first note that if $f(x)$ is an automorphism of a ring R, then $f^{-1}(x)$ is well defined, since $f(x)$ is both one-to-one and onto. We see that

$$f(f^{-1}(x) + f^{-1}(y)) = f(f^{-1}(x)) + f(f^{-1}(y)) = x + y,$$

so $f^{-1}(x + y) = f^{-1}(x) + f^{-1}(y)$. Also,

$$f(f^{-1}(x) \cdot f^{-1}(y)) = f(f^{-1}(x)) \cdot f(f^{-1}(y)) = x \cdot y,$$

so $f^{-1}(x \cdot y) = f^{-1}(x) \cdot f^{-1}(y)$. Thus, f^{-1} is a ring homomorphism. Since f was both one-to-one and onto, f^{-1} is both one-to-one and onto. Therefore, f^{-1} is a ring automorphism.

If f and ϕ are two ring automorphisms, then

$$f(\phi(x + y)) = f(\phi(x) + \phi(y)) = f(\phi(x)) + f(\phi(y))$$

and

$$f(\phi(x \cdot y)) = f(\phi(x) \cdot \phi(y)) = f(\phi(x)) \cdot f(\phi(y)).$$

The combination $f(\phi(x))$ is also one-to-one and onto, so this product, which we can denote $f \cdot \phi$, is a ring automorphism. Since the set of all ring automorphisms is closed with respect to multiplication and inverses, and the set of all ring automorphisms is a subgroup of the set of all *group* automorphisms with respect to addition, we see that this set is a group. ∎

The natural question that arises is determining the group of ring automorphisms of $\mathbb{C}$. This is in fact a difficult question to answer, but if we only consider the automorphisms that send each real number to itself, the question becomes easy to answer.

PROPOSITION 11.4
Besides the identity automorphism, there is another ring automorphism on $\mathbb{C}$, given by

$$\phi(a + bi) = a - bi.$$

In fact, these are the only automorphisms for which $\phi(x) = x$ for all real numbers x.

PROOF: We check that

$$\begin{aligned}\phi(a + bi) + \phi(c + di) &= (a - bi) + (c - di) = a + c - (b + d)i \\ &= \phi(a + c + (b + d)i) = \phi\big((a + bi) + (c + di)\big).\end{aligned}$$

$$\begin{aligned}\phi(a + bi) \cdot \phi(c + di) &= (a - bi) \cdot (c - di) = (a \cdot c - b \cdot d) - (a \cdot d + b \cdot c)i \\ &= \phi\big((a \cdot c - b \cdot d) + (a \cdot d + b \cdot c)i\big) = \phi\big((a + bi) \cdot (c + di)\big).\end{aligned}$$

Thus, ϕ is a homomorphism. Since $a - bi = 0$ if, and only if, a and b are both 0, the kernel of ϕ is just $\{0\}$, and so ϕ is one-to-one. Also, ϕ is onto, since $\phi(a - bi) = a + bi$. Therefore, ϕ is an automorphism.

To show that there are exactly two such automorphisms, suppose that $f(x)$ is an automorphism of $\mathbb{C}$ for which $f(x) = x$ for all real numbers x. Then $f(i)^2 = f(i^2) = f(-1) = -1$, so by Lemma 11.4, $f(i) = \pm i$. If $f(i) = i$, then $f(x) = x$ for all $x \in \mathbb{C}$, and if $f(i) = -i$, then $f(x) = \phi(x)$ for all x. ∎

The ring automorphism found in Proposition 11.4 is called the *conjugate.* The conjugate of z is generally denoted by $\overline{z}$. That is, if $z = a + bi$, then $\overline{z} = \phi(z) = a - bi$. The conjugate automorphism is defined in *Sage* as

```
conjugate(3 + 4*I)
    -4*I + 3
```

It is an easy computation to see that

$$z \cdot \overline{z} = (a + bi) \cdot (a - bi) = a^2 + b^2.$$

Thus, $z \cdot \overline{z}$ is always a non-negative real number.

DEFINITION 11.6 We say the *absolute value* of a complex number $z = a + bi$ is

$$|z| = \sqrt{z \cdot \overline{z}}.$$

The geometric interpretation of $|z|$ is the distance from (a, b) to the origin. In *Sage*, the function **abs(z)** gives the absolute value for both real and complex numbers.

```
abs(3 + 4*I)
    5
```

The familiar property for the absolute value of real numbers holds for all complex numbers as well.

PROPOSITION 11.5

For any two elements x and y in $\mathbb{C}$,

$$|x \cdot y| = |x| \cdot |y|.$$

PROOF: We have

$$|x \cdot y| = \sqrt{x \cdot y \cdot \overline{x \cdot y}} = \sqrt{x \cdot y \cdot \overline{x} \cdot \overline{y}} = \sqrt{x \cdot \overline{x} \cdot y \cdot \overline{y}} = \sqrt{x \cdot \overline{x}} \cdot \sqrt{y \cdot \overline{y}} = |x| \cdot |y|.$$

Thus, $|x \cdot y| = |x| \cdot |y|$. ∎

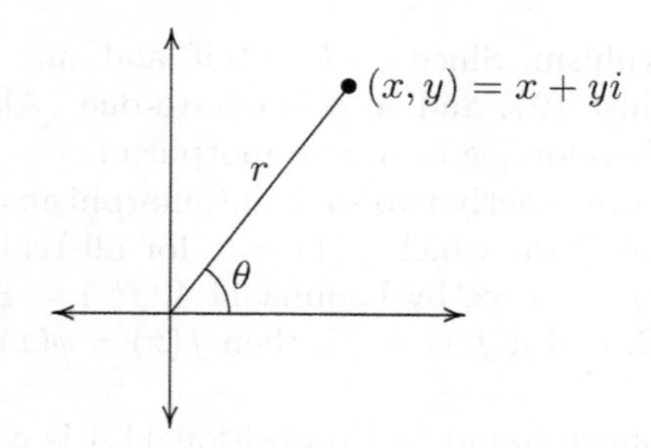

FIGURE 11.1: Polar coordinates for a complex number

Since there is a geometric interpretation of the absolute value, this proposition suggests that there is also a geometric interpretation for the product of two complex numbers.

From polar coordinates it is known that any point in the plane can be located by knowing its distance r from the origin, and its angle θ from the positive x-axis.

Since r is the absolute value of $(x+yi)$, perhaps the angle θ is also significant with respect to the complex number. By using trigonometry in Figure 11.1, we have that

$$x + yi = r(\cos\theta + i\sin\theta).$$

This form is called the *polar form* of the complex number $x + yi$. The angle θ is called the *argument* of $x + yi$. We can find the approximate argument of a complex number (in radians) with the *Sage* command

```
N(arg(3 + 4*I))
    0.927295218001612
```

Sage always finds an angle θ between $-\pi$ and π, but we can also consider the angles

$$\ldots,\ \theta - 6\pi,\ \theta - 4\pi,\ \theta - 2\pi,\ \theta,\ \theta + 2\pi,\ \theta + 4\pi,\ \theta + 6\pi,\ \ldots\ .$$

All of these angles have the same sine and cosine, and hence are interchangeable in the polar coordinate system. We call these angles *co-terminal*. The set of angles co-terminal to θ can be written

$$\{\theta + 2\pi n \mid n \in \mathbb{Z}\}.$$

For example, the polar form of $-\sqrt{3} - i$ is given by

$$2\left(\cos\left(\frac{-5\pi}{6}\right) + i\sin\left(\frac{-5\pi}{6}\right)\right),$$

as seen from the commands

```
simplify(abs(-sqrt(3) - I))
    2
simplify(arg( -sqrt(3) - I))
    -5/6*pi
```

However, we could have used any co-terminal angle instead of the one *Sage* gave us. Thus,

$$2\left(\cos\left(\frac{7\pi}{6}\right)+i\sin\left(\frac{7\pi}{6}\right)\right),\quad 2\left(\cos\left(\frac{19\pi}{6}\right)+i\sin\left(\frac{19\pi}{6}\right)\right),\ \ldots$$

are also polar forms of $-\sqrt{3}-i$. The usefulness of the polar form of a complex number is hinted at by the next lemma, which makes use of the trigonometric identities

$$\begin{aligned}\cos(A+B) &= \cos(A)\cos(B)-\sin(A)\sin(B), \qquad \text{and}\\ \sin(A+B) &= \sin(A)\cos(B)+\cos(A)\sin(B).\end{aligned}$$

LEMMA 11.6
If $z_1 = r_1(\cos\theta_1 + i\sin\theta_1)$ and $z_2 = r_2(\cos\theta_2 + i\sin\theta_2)$, then

$$z_1\cdot z_2 = r_1\cdot r_2\big(\cos(\theta_1+\theta_2)+i\sin(\theta_1+\theta_2)\big).$$

So the argument of the product is the sum of the arguments.

PROOF: We note that

$$z_1\cdot z_2 = r_1(\cos\theta_1 + i\sin\theta_1)\cdot r_2(\cos\theta_2+i\sin\theta_2) =$$
$$r_1\cdot r_2((\cos\theta_1\cdot\cos\theta_2 - \sin\theta_1\cdot\sin\theta_2) + i\cdot(\cos\theta_1\cdot\sin\theta_2+\sin\theta_1\cdot\cos\theta_2)).$$

Using the trigonometric identities, this simplifies to

$$z_1\cdot z_2 = r_1\cdot r_2\big(\cos(\theta_1+\theta_2)+i\sin(\theta_1+\theta_2)\big).$$

∎

We can now use induction to prove the following important theorem:

THEOREM 11.2: De Moivre's Theorem
If n is an integer, and $z = r(\cos\theta + i\sin\theta)$ is a nonzero complex number in polar form, then

$$z^n = r^n\big(\cos(n\theta)+i\sin(n\theta)\big).$$

PROOF: Let us first prove the theorem for positive values of n. For $n = 1$, the statement is obvious. Let us assume that the statement is true for the previous case. That is,

$$z^{n-1} = r^{n-1}\big(\cos((n-1)\theta)+i\sin((n-1)\theta)\big).$$

We want to prove that the theorem holds for n as well. Using Lemma 11.6, we have

$$\begin{aligned} z^n &= z^{n-1} \cdot z \\ &= r^{n-1}\big(\cos((n-1)\theta) + i\sin((n-1)\theta)\big) \cdot \big(r(\cos\theta + i\sin\theta)\big) \\ &= r^n(\cos((n-1)\theta + \theta) + i\sin((n-1)\theta + \theta)) \\ &= r^n(\cos(n\theta) + i\sin(n\theta)). \end{aligned}$$

Thus, the theorem is true for n, and hence by induction it is true whenever n is positive.

If z is nonzero, then letting $n = 0$ gives

$$r^0(\cos(0\,\theta) + i\sin(0\,\theta)) = 1(1 + i \cdot 0) = 1 = z^0.$$

So the theorem holds for $n = 0$. If z is nonzero, then $r > 0$, and so

$$\big(r^{-n}\big(\cos(-n\theta) + i\sin(-n\theta)\big)\big) \cdot \big(r^n\big(\cos(n\theta) + i\sin(n\theta)\big)\big) =$$
$$r^{-n+n}\big(\cos(-n\theta + n\theta) + i\sin(-n\theta + n\theta)\big) = r^0(\cos 0 + i\sin 0) = 1.$$

Now, if $n < 0$, then the theorem holds for $-n$, and so

$$z^{-n}\big(r^n(\cos(n\theta) + i\sin(n\theta))\big) = 1,$$

hence

$$r^n(\cos(n\theta) + i\sin(n\theta)) = z^n$$

even when $n < 0$. ☐

De Moivre's theorem (11.2) allows us to quickly raise a complex number to an integer power.

Example 11.2
Compute $(-\sqrt{3} - i)^5$.

SOLUTION: Since $r = \sqrt{(-\sqrt{3})^2 + (-1)^2} = 2$, and

$$\theta = \tan^{-1}\left(\frac{-1}{-\sqrt{3}}\right) - \pi = -5\pi/6,$$

then $(-\sqrt{3} - i)^5$ is

$$2^5\left(\cos\left(\frac{-25\pi}{6}\right) + i\sin\left(\frac{-25\pi}{6}\right)\right) = 32\left(\frac{\sqrt{3}}{2} - \frac{i}{2}\right) = 16\sqrt{3} - 16i.$$

☐

We can also use De Moivre's theorem (11.2) to find the nth root of 1. We first define

$$\omega_n = \cos\left(\frac{2\pi}{n}\right) + i\sin\left(\frac{2\pi}{n}\right).$$

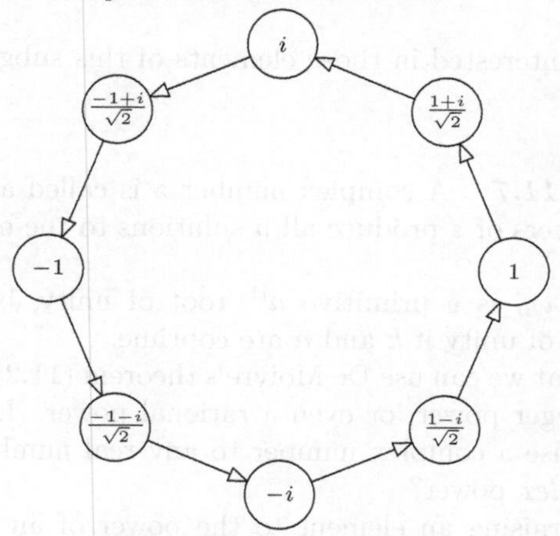

FIGURE 11.2: The eight roots of unity

For example, $\omega_1 = 1$, $\omega_2 = -1$, $\omega_3 = (-1 + i\sqrt{3})/2$, and $\omega_4 = i$, etc. Then

$$(\omega_n)^n = \cos(2\pi) + i\sin(2\pi) = 1,$$

so ω_n is indeed one n^{th} root of unity. In fact, all n^{th} roots of 1 are given by the numbers $\omega_n, \omega_n^2, \omega_n^3, \ldots$ up to $(\omega_n)^n = 1$.

Computational Example 11.3

The eighth root of unity, ω_8, can be entered into *Sage* using the commands

```
w8 = (1/2 + I/2)*sqrt(2); w8
    (1/2*I + 1/2)*sqrt(2)
```

This allows us to consider the group generated by ω_8:

```
G = Group(w8); G
    {(1/2*I + 1/2)*sqrt(2), I, (1/2*I - 1/2)*sqrt(2), -1,
     -(1/2*I + 1/2)*sqrt(2), -I, -(1/2*I - 1/2)*sqrt(2), 1}
```

This gives the eight roots of unity, and shows that these elements form a group. In fact, the nth roots of unity will form a cyclic group isomorphic to Z_n. □

By rearranging the elements of G, we can create a circle graph as in Figure 11.2, with the elements in the proper positions in the complex plane.

```
G = [I, (1/2+I/2)*sqrt(2), 1 ,(1/2-I/2)*sqrt(2), -I,
(-1/2-I/2)*sqrt(2), -1, (-1/2+I/2)*sqrt(2)]
CircleGraph(G, Mult(w8))
```

We are mainly interested in those elements of this subgroup that are generators.

DEFINITION 11.7 A complex number z is called a *primitive* n^{th} *root of unity* if the powers of z produce all n solutions to the equation $x^n = 1$.

It is clear that ω_n is a primitive n^{th} root of unity, but also $(\omega_n)^k$ is a primitive n^{th} root of unity if k and n are coprime.

We have seen that we can use De Moivre's theorem (11.2) to raise a complex number to an integer power, or even a rational power. Is it possible to use this formula to raise a complex number to any real number, or even raise a number to a *complex* power?

In most fields, raising an element to the power of an *element* is absurd. Even in the real number system we will discover that we must utilize the exponential function e^x to compute quantities such as $2^{\sqrt{2}}$. We use that fact that $2 = e^{\ln 2}$, and so

$$2^{\sqrt{2}} = \left(e^{\ln 2}\right)^{\sqrt{2}} = e^{((\ln 2)\sqrt{2})}.$$

The key algebraic property of the exponential function is that

$$e^{x+y} = e^x \cdot e^y \qquad \text{for all} \quad x, y \in \mathbb{R}.$$

This indicates that the exponential function is a *group homomorphism* mapping the additive group of real numbers to the multiplicative group of real numbers. This homomorphism enables us to consider raising an element of the real numbers to the power of an *element.*

Can we extend the exponential function into a group homomorphism from the additive structure of $\mathbb{C}$ (denoted $\mathbb{C}^+$), to the multiplicative structure $\mathbb{C}^*$? If such a group homomorphism exists, then

$$e^{a+bi} = e^a \cdot e^{bi} = e^a \cdot (e^i)^b.$$

Sage indicates that the value of e^i is $(\cos 1 + i \sin 1)$. Problems 1 through 3 show three ways of proving this, all involving calculus. There is in fact no way to prove that $e^i = \cos 1 + i \sin 1$ without calculus. But given that this is true, we then have by De Moivre's theorem (11.2) that

$$e^{a+bi} = e^a \cdot (e^i)^b = e^a \cdot (\cos b + i \sin b)$$

whenever b is an integer. We will define this as the exponential function for all complex numbers. Notice that radian measure must be used in this formula.

PROPOSITION 11.6
For $z = a + bi$, the function

$$f(z) = e^a \cdot (\cos b + i \sin b)$$

defines a group homomorphism from $\mathbb{C}^+$ *to* $\mathbb{C}^*$*, which is an extension of the standard exponential function. This function is called the* complex exponential *function, and is also denoted* e^z*.*

PROOF: If $z_1 = a_1 + b_1 i$, and $z_2 = a_2 + b_2 i$, we observe that

$$f(z_1 + z_2) = e^{a_1+a_2}(\cos(b_1 + b_2) + i \sin(b_1 + b_2)).$$

By Lemma 11.6, this equals

$$e^{a_1}(\cos(b_1) + i \sin(b_1)) \cdot e^{a_2}(\cos(b_2) + i \sin(b_2)) = f(z_1) \cdot f(z_2).$$

Thus, f is a group homomorphism from $\mathbb{C}^+$ to $\mathbb{C}^*$. ☐

This allows us another way of expressing ω_n. Notice that

$$e^{2\pi i/n} = \cos\left(\frac{2\pi}{n}\right) + i \sin\left(\frac{2\pi}{n}\right) = \omega_n.$$

So we now have a more succinct way of defining the nth root of 1.

The real exponential function is one-to-one, but it is not onto since there is no number for which $e^x = -1$. However, the complex exponential function *is* onto, since for every nonzero complex number in polar form, $z = r(\cos\theta + i\sin\theta)$, there is a complex number whose exponential is z, namely $\ln(r) + i\theta$. The drawback of the complex exponential function is that it is *not* one-to-one! The kernel of this homomorphism is the set

$$N = f^{-1}(1) = \{2k\pi i \mid k \in \mathbb{Z}\}.$$

DEFINITION 11.8 For any nonzero complex number z, we define the *complex logarithm* of z, denoted $\log(z)$, to be the set of elements x such that $e^x = z$.

Notice that we use the function $\ln(x)$ to denote the *real* logarithm, while we use $\log(z)$ to denote the complex logarithm. We have already observed that when z is written in polar form, $z = r(\cos\theta + i\sin\theta)$, that one value of x that satisfies the equation is $x = \ln(r) + \theta i$. We also know that $f^{-1}(z)$ will be a coset of the kernel of f. Thus, we have $\log(z) = \ln(r) + \theta i + N$.

For example, $\log(-1)$ is the set

$$\{\pi i + 2k\pi i \mid k \in \mathbb{Z}\} = \{\ldots, -5\pi i, -3\pi i, -\pi i, \pi i, 3\pi i, 5\pi i, \ldots\}.$$

The *Sage* **log** function works for complex numbers, but only gives one element of the set. Thus, we must add the kernel N to this result to obtain the set given by $\log(z)$.

To help visualize the complex logarithm, we can graph the complex part of $\log(x + iy)$, but since this gives multiple values for each input value, we get a surface that resembles a parking garage or a spiral staircase. See Figure 11.3.

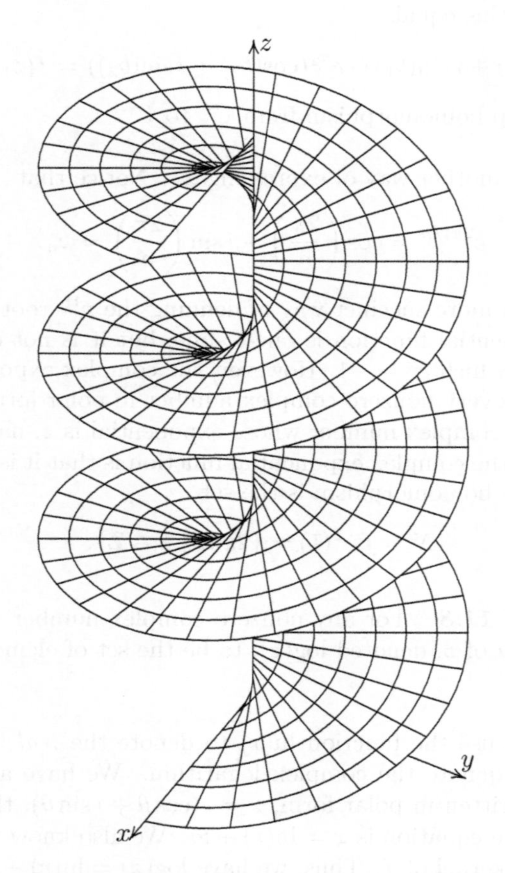

FIGURE 11.3: Imaginary portion of the complex logarithm function

We can now define a complex number raised to a complex power, by saying

$$x^z = (e^{\log(x)})^z = e^{z \cdot \log(x)}.$$

Notice that this gives a *set* of numbers, not just a single number. Although there will at times be an infinite number of elements in the set x^z, this will not always be the case.

PROPOSITION 11.7

For each integer $n > 0$, and any nonzero complex number z, there are exactly n values for $z^{(1/n)}$. Thus, there are exactly n solutions for x to the equation $x^n = z$.

PROOF: Let z have the polar form

$$z = r(\cos\theta + i\sin\theta).$$

Then $\log(z)$ is the set

$$\{\ln(r) + \theta i + 2k\pi i \mid k \in \mathbb{Z}\}.$$

Thus, $\log(z)/n$ is given by the set

$$\left\{ \frac{\ln(r)}{n} + \frac{(\theta + 2k\pi)i}{n} \,\middle|\, k \in \mathbb{Z} \right\}.$$

Thus, the exponential function of the elements of this set is given by

$$\left\{ e^{(\ln(r)/n} \cdot \left(\cos\left(\frac{(\theta + 2k\pi)}{n}\right) + i\sin\left(\frac{(\theta + 2k\pi)}{n}\right)\right) \,\middle|\, k \in \mathbb{Z} \right\}$$

$$= \left\{ r^{(1/n)} \cdot \left(\cos\left(\frac{(\theta + 2k\pi)}{n}\right) + i\sin\left(\frac{(\theta + 2k\pi)}{n}\right)\right) \,\middle|\, k \in \mathbb{Z} \right\}.$$

Notice that for two different values of k that differ by n, the arguments of the cosine and sine will differ by 2π. Hence, we only have to consider the values of k from 0 to $(n-1)$. This gives us the set

$$\left\{ r^{(1/n)} \cdot \left(\cos\left(\frac{(\theta + 2k\pi)}{n}\right) + i\sin\left(\frac{(\theta + 2k\pi)}{n}\right)\right) \,\middle|\, k = 0, 1, 2, \ldots, n-1 \right\}.$$

However, these n solutions will have arguments that differ by less than 2π, so these n solutions are distinct.

Finally, we must show that x is an element of $z^{(1/n)}$ if, and only if, x solves the equation $x^n = z$. But for any element in the above expression, we have that

$$x^n = r^{n(1/n)} \cdot \left(\cos\left(\frac{n(\theta + 2k\pi)}{n}\right) + i\sin\left(\frac{n(\theta + 2k\pi)}{n}\right)\right)$$

$$= r(\cos\theta + i\sin\theta) = z.$$

Likewise, if $x^n = z$, we can raise both sides to the $(1/n)^{\text{th}}$ power to get that the two sets $(x^n)^{(1/n)}$ and $z^{(1/n)}$ are equal. Since the element x is certainly in the first set, it must also be in the set $z^{(1/n)}$ that we have just computed. □

This last proposition is very useful for finding square roots and cube roots of complex numbers. This turns out to have some important applications in finding the roots of real polynomials! In fact, complex numbers and the functions we have defined in this section have many applications in the real world. The complex exponential function was fundamental to the invention of the shortwave radio. The complex logarithm can be used in solving real valued differential equations. So even though these numbers are labeled "imaginary," they are by no means just a figment of someone's imagination.

Problems for §11.3

1 Assume that the Taylor series for the exponential function

$$e^x = 1 + \frac{x}{1!} + \frac{x^2}{2!} + \frac{x^3}{3!} + \cdots + \frac{x^n}{n!} + \cdots$$

is valid for complex numbers as well as for real numbers. Prove that $e^i = (\cos 1 + i \sin 1)$.

Hint: Recall the Taylor series for $\sin(x)$ and $\cos(x)$.

2 Suppose we can write $e^{ix} = u(x) + iv(x)$, where $u(x)$ and $v(x)$ are real functions of a real variable x. If we assume that

$$\frac{d}{dx}e^{ix} = u'(x) + iv'(x) = ie^{ix},$$

use differential equations to prove that $u(x) = \cos(x)$ and $v(x) = \sin(x)$.

Hint: Since $e^0 = 1$, we know that $u(0) = 1$ and $v(0) = 0$.

3 Assume that the limit from calculus

$$e^x = \lim_{n\to\infty} \left(1 + \frac{x}{n}\right)^n$$

is valid for complex values of x as well as real values. Prove that $e^i = (\cos 1 + i \sin 1)$.

Hint: Convert $(1 + i/n)$ into polar form using an arctangent.

4 Find all possible values of $\log(-1)$.

5 Find all possible values of $\log(\sqrt{3} - i)$.

6 Find all possible values of $1^{1/6}$.

7 Find all complex solutions to the equation $z^4 + 1 = 0$.

8 Find all complex solutions to the equation $z^3 + 8 = 0$.

9 Find all possible values of $(8i)^{1/3}$.

10 Find five values of the expression i^i.

11 Find five values of the expression $(-i)^{(i/2)}$.

12 Show that when x and y are both complex, the set of all values of the expression x^y forms a geometric sequence:

$$\{\ldots, a \cdot r^{-3}, a \cdot r^{-2}, a \cdot r^{-1}, a, a \cdot r, a \cdot r^2, a \cdot r^3, \ldots\}.$$

13 Find complex numbers x and y such that the set of values for x^y are the powers of 2:

$$\{\ldots, \frac{1}{16}, \frac{1}{8}, \frac{1}{4}, \frac{1}{2}, 1, 2, 4, 8, 16, \ldots\}.$$

(See Problem 12. There will be more than one solution to this problem.)

14 Show that for a fixed n, the set of all n^{th} roots of 1 forms a group with respect to multiplication.

15 Prove that the group in exercise 14 is cyclic, with

$$\omega_n = \cos\left(\frac{2\pi}{n}\right) + i \sin\left(\frac{2\pi}{n}\right)$$

as a generator. Show that any generator of this group is a primitive n^{th} root of unity.

16 Prove or disprove: For all complex numbers x, y, and z,

$$(x^z) \cdot (y^z) = (x \cdot y)^z.$$

Note: x^z and y^z may both represent *sets* of complex numbers, so the left-hand side of this equation is the set of all possible products formed.

17 Prove or disprove: For all complex numbers x, y, and z,

$$(z^x)^y = z^{(x \cdot y)}.$$

(See the note on Problem 16.)

18 Prove or disprove: For all complex numbers x, y, and z,

$$(z^x) \cdot (z^y) = z^{(x+y)}.$$

(See the note on Problem 16.)

Interactive Problems

19 Find the twelfth roots of unity, and arrange them in such a way that the circle graph puts the elements in the correct place in the complex plane, as was done in Example 11.3.

20 Use *Sage* to plot the real part of $\log(x+iy)$, the companion of Figure 11.3. Would this surface be multi-valued, as was Figure 11.3?

11.4 Ordered Commutative Rings

The integers, the rational numbers, and the real numbers all have one property that most rings do not have. Given two different elements in the ring, we can say that one of them is greater than the other. Most rings do not have such an ordering, but we will find that some rings can be ordered in more than one way! The orderings of a ring can give us new insight into the structure of the ring.

Although this section is not referred to elsewhere in the book, it does introduce the concept of an automorphism of a field. These automorphisms will become central to the study of Galois theory in Chapter 15.

We begin by making a formal definition of an ordered ring R. If there is a way to tell whether one element is greater than another, we should be able to distinguish those elements that are greater than zero, called the *positive elements* P.

DEFINITION 11.9 A commutative ring R is *ordered* if there exists a set P such that the three properties hold:

1. P is closed under addition.
2. P is closed under multiplication.
3. For each x in R, one and only one of the following statements is true:

$$x \in P, \qquad x = 0, \qquad -x \in P.$$

The third property is sometimes called the *law of trichotomy*. With this law, we can define what it means for one element to be greater than another.

DEFINITION 11.10 We say that x is greater than y, denoted $x > y$, if $x - y \in P$. Likewise, we say that x is smaller than y, denoted $x < y$, if $y - x \in P$. By the law of trichotomy, either

$$x > y, \qquad x < y, \qquad \text{or} \qquad x = y.$$

This notation keeps us from having to constantly refer to the set P. Instead of writing $x \in P$, we can merely write $x > 0$.

We begin by proving some simple properties of the "greater than" sign.

LEMMA 11.7
If x, y, and z are elements in an ordered ring, then we have the following three properties:

1. *If $x > y$, then $x + z > y + z$.*
2. *If $x > y$ and $z > 0$, then $x \cdot z > y \cdot z$.*
3. *If $x > y$ and $y > z$, then $x > z$.*

PROOF: To prove the first statement, note that since $x > y$, we have that

$$x - y \in P.$$

But then

$$(x + z) - (y + z) \in P$$

and so $x + z > y + z$.

For the second statement, we have that $x > y$ and $z > 0$, and so $(x - y) \in P$ and $z \in P$. Since P is closed under multiplication, we have that

$$(x - y) \cdot z = x \cdot z - y \cdot z \in P,$$

and so $x \cdot z > y \cdot z$.

Finally, if $x > y$ and $y > z$, then both $x - y \in P$ and $y - z \in P$. Since P is closed under addition, we have that

$$(x - y) + (y - z) = x - z \in P,$$

and so $x > z$. ▯

Given a ring that has an ordering, one of the great challenges is determining the set of positive elements P. There are at least some elements that must be in P.

PROPOSITION 11.8

For any nonzero element x in an ordered ring, x^2 is in P.

PROOF: Since $x \neq 0$, by the law of trichotomy either $x > 0$, or $-x > 0$. If $x > 0$ then

$$x^2 = x \cdot x > 0.$$

On the other hand, if $-x > 0$, then

$$x^2 = (-x) \cdot (-x) > 0.$$

Thus, in either case x^2 is in P. ▯

An immediate consequence of this is that if the ring has an identity e, then $e > 0$, since $e = e^2$. An additional statement can be proved if the ring is an integral domain.

COROLLARY 11.2

If R is an ordered integral domain with multiplicative identity 1, and n is any positive integer, then $n \cdot 1$ is in P. In particular, the characteristic of R must be 0.

PROOF: Since $1^2 = 1$ we have from Proposition 11.8 that $1 > 0$. Proceeding by induction, let us assume that $(n-1) \cdot 1 > 0$, and show that $n \cdot 1 > 0$. But this is easy, since

$$n \cdot 1 = (n-1) \cdot 1 + 1 \cdot 1 = (n-1) \cdot 1 + 1 > 0.$$

Thus, we have that $n \cdot 1 > 0$ for every positive number n. This immediately implies that the characteristic is zero, for if R had a positive characteristic p, then $p \cdot 1 = 0$, and we would have $0 > 0$, a contradiction. ▯

The standard examples of ordered rings are the integers, the rationals, and the real numbers. It should be noted that the complex numbers do *not* form an ordered ring, since $i^2 = -1 < 0$, and by Proposition 11.8, any square must be positive.

Example 11.4

Consider the subring of $\mathbb{R}$ from Example 10.1.

$$S = \{x + y\sqrt{2} \mid x, y \in \mathbb{Z}\}.$$

We will call this ring $\mathbb{Z}[\sqrt{2}]$, the ring formed by adjoining $\sqrt{2}$ to $\mathbb{Z}$. Find a non-standard ordering on this ring.

SOLUTION: By Proposition 0.5, this ring has no zero divisors, so this is an integral domain.

The standard ordering of $\mathbb{Z}[\sqrt{2}]$ would be to let P consist of all numbers that are positive when viewed as a real number. By Corollary 11.2, the positive integers must be in P, but there is no way of proving that $\sqrt{2}$ is in P. Thus, we can consider an ordering where $-\sqrt{2} \in P$. We can determine whether any other element was in P or not in P. For example, $1 + \sqrt{2}$ would be negative, since

$$(1 + \sqrt{2}) \cdot (1 - \sqrt{2}) = -1 < 0,$$

and $1 - \sqrt{2}$ is the sum of two numbers in P, so this term is in P. ▯

To see what is really going on in this example, it is helpful to look at the ring automorphisms, which were introduced in the last section. The automorphism of particular interest is as follows:

$$f : \mathbb{Z}[\sqrt{2}] \to \mathbb{Z}[\sqrt{2}],$$

$$f(x + y\sqrt{2}) = x - y\sqrt{2}.$$

This automorphism can be defined in *Sage*. Since *Sage* already knows that **sqrt(2)** · **sqrt(2)** is 2, we do not need to tell *Sage* anything to define the ring $\mathbb{Z}[\sqrt{2}]$. We now can define the homomorphism. For a homomorphism on an infinite set of objects, the format is slightly different.

```
InitDomain(0)
F = FieldHomo()
HomoDef(F, sqrt(2), -sqrt(2))
CheckHomo(F)
    True
F(2 + 3*sqrt(2))
    -3*sqrt(2) + 2
```

We do not have to indicate the domain and target, since the domain will be the currently defined field, which in this case is a subset of the real numbers. The command **InitDomain(0)** merely clears out any previous fields that have been defined, such as the "complex numbers modulo 3."

If we let P denote the set of positive elements using the "standard" ordering, and let P' be the set of positive elements under the unusual ordering we saw above, then $P' = f(P)$. In fact, for any automorphism ϕ on an ordered ring, we can construct an alternative way to order the ring by using $\phi(P)$ instead of P for the set of positive elements.

While we are working with the integral domain $\mathbb{Z}[\sqrt{2}]$ we might mention what happens if we consider the field of quotients of this ring. Certainly this must include all numbers of the form

$$x + y\sqrt{2}, \qquad x, y \in \mathbb{Q},$$

but could there be other elements? We need to check that all non-zero elements of this form have a multiplicative inverse. For example,

```
Together(1/(2 + sqrt(2)))
    -1/2*sqrt(2) + 1
```

produces a number in the correct form. In fact, we can learn how to take the inverse of $a + b\sqrt{2}$ by multiplying by $a - b\sqrt{2}$.

```
var("a b")
expand((a + b*sqrt(2))*(a - b*sqrt(2)))
    a^2 - 2*b^2
```

Since this will be rational whenever a and b are rational, we see that

$$\frac{1}{a + b\sqrt{2}} = \frac{a - b\sqrt{2}}{a^2 - 2b^2}.$$

By Proposition 0.5, the denominator will not be zero for rational a and b, so this is a field. We will call this field $\mathbb{Q}[\sqrt{2}]$.

As one might guess from the **FieldHomo** command, we were really defining a homomorphism on $\mathbb{Q}(\sqrt{2})$. Hence, the automorphism f that we discovered earlier on $\mathbb{Z}[\sqrt{2}]$ extends to an automorphism on $\mathbb{Q}[\sqrt{2}]$. Thus, the unusual ordering that we gave to $\mathbb{Z}[\sqrt{2}]$ extends to the field of quotients. In fact this generally happens, as seen in the next proposition.

PROPOSITION 11.9
Let R be an ordered integral domain, with P the set of positive elements. Then if Q is the field of quotients on R, then the ordering on R can be extended in a unique way to an ordering on Q. That is, there is a unique set P' that forms an ordering on Q, with

$$p \in P \Rightarrow \left(\frac{p}{1}\right) \in P'.$$

PROOF: We will begin by showing that the ordering is uniquely determined. Since for any p in P, we have

$$\left(\frac{1}{p}\right) \cdot \left(\frac{p}{1}\right) = \left(\frac{p}{p}\right) = \left(\frac{1}{1}\right) = 1 \in P,$$

so $\left(\frac{1}{p}\right)$ must be considered to be positive in the new ordering. But then $\left(\frac{n}{p}\right)$ must be positive whenever n and p are in P. Thus P' contains at least those elements of the form $\left(\frac{n}{p}\right)$, where n and p are in P. Note that every nonzero element in the field of quotients Q must be of one of the four forms

$$\left(\frac{n}{p}\right), \left(\frac{-n}{p}\right), \left(\frac{n}{-p}\right), \left(\frac{-n}{-p}\right),$$

where n and p are in P. But the first and the last expressions are equivalent, and the middle two are also equivalent. Thus, for every nonzero element of Q, either that element or its negative is of the form $\left(\frac{n}{p}\right)$, with n and p in P. Thus, P' cannot contain any more elements besides those of the form $\left(\frac{n}{p}\right)$, and hence P' is uniquely determined.

Now, suppose we consider the set of elements P' that can be expressed in the form $\left(\frac{n}{p}\right)$, where n and p are in P. Does this form an ordering on Q? We have already seen that the law of trichotomy has already been demonstrated. All we need to show is that P' is closed under addition and multiplication. But this is clear by looking at the formulas

$$\left(\frac{x}{y}\right) + \left(\frac{u}{v}\right) = \left(\frac{x \cdot v + u \cdot y}{y \cdot v}\right)$$

and

$$\left(\frac{x}{y}\right) \cdot \left(\frac{u}{v}\right) = \left(\frac{x \cdot u}{y \cdot v}\right).$$

Thus, P' forms an ordering on Q, and is an extension of the ordering P. $\square$

Example 11.5
What possible orderings can be put on the field

$$\{x + y\sqrt[3]{2} + z\sqrt[3]{4} \mid x, y, z \in \mathbb{Q}\}?$$

SOLUTION: First, we need to verify that this is indeed a field. We can try some calculations in *Sage.*

```
Together(1/(1 + 2^(1/3))
    1/3*2^(2/3) - 1/3*2^(1/3) + 1/3
```

It seems that *Sage* is always able to rationalize the denominator, but it is harder to prove that this is always possible. However, if we compute

```
var("a b c")
expand((a + b*2^(1/3) + c*2^(2/3)) *
  (a^2 - 2*b*c + (2*c^2 - a*b)*2^(1/3) + (b^2 - a*c)*2^(2/3)))
    a^3 + 2*b^3 - 6*a*b*c + 4*c^3
```

we find that this product will always produce a rational number. Thus,

$$\frac{1}{a + b\sqrt[3]{2} + c\sqrt[3]{4}} = \frac{a^2 - 2bc + (2c^2 - ab)\sqrt[3]{2} + (b^2 - ac)\sqrt[3]{4}}{a^3 + 2b^3 + 4c^3 - 6abc}.$$

It takes a bit more work to show that the denominator will never be zero when a, b, and c are rational (see Problem 15). Once this has been proven, we see that $\mathbb{Q}[\sqrt[3]{2}]$ is a field.

Does this field have an unusual ordering, as the field $\mathbb{Q}[\sqrt{2}]$ did? In this field,

$$2\sqrt[3]{2} = (\sqrt[3]{4})^2 > 0,$$

so both $\sqrt[3]{2}$ and $\sqrt[3]{4}$ must be positive. Also note that this field does not have a nontrivial automorphism, since the only element in the field for which $x^3 = 2$ is $\sqrt[3]{2}$. Thus, an automorphism f on this field sends $\sqrt[3]{2}$ to itself, and hence $f(x) = x$ for all x in this field. It is not surprising, then, that this field does not have an unusual ordering, as the field $\mathbb{Z}[\sqrt{2}]$ did. ∎

Computational Example 11.6
Find several possible ways of defining an ordering on the field

$$S = \left\{ x + y\cos\left(\frac{\pi}{9}\right) + z\cos\left(\frac{2\pi}{9}\right) \mid x, y, z \in \mathbb{Q} \right\}.$$

SOLUTION: Using trigonometric identities we can multiply two such numbers together to get a number in the same form. This can be verified by the command

```
var("x1 x2 y1 y2 z1 z2")
TrigReduce( (x1 + y1*cos(pi/9) + z1*cos(2*pi/9) ) *
  (x2 + y2*cos(pi/9) + z2*cos(2*pi/9) ) )
    1/2*y1*y2*cos(2/9*pi) + x2*z1*cos(2/9*pi) +
    x1*z2*cos(2/9*pi) - 1/2*z1*z2*cos(2/9*pi) +
    x2*y1*cos(1/9*pi) + x1*y2*cos(1/9*pi) +
    1/2*y2*z1*cos(1/9*pi) + 1/2*y1*z2*cos(1/9*pi) +
    1/2*z1*z2*cos(1/9*pi) + x1*x2 + 1/2*y1*y2 + 1/4*y2*z1 +
    1/4*y1*z2 + 1/2*z1*z2
```

As messy as this is, one can see that it is an element of S, so S is closed under multiplication. In fact, S is a field, but it is very difficult to show this.

Since the elements of this field are all real, there is a natural ordering of the elements of S. Are there other ways to order this field? We want to look for automorphisms on the field S. Consider a mapping sending $\cos(\pi/9)$ to $-\cos(2\pi/9)$, and sending $\cos(2\pi/9)$ to $\cos(4\pi/9) = \cos(\pi/9) - \cos(2\pi/9)$.

```
InitDomain(0)
F = FieldHomo()
HomoDef(F, cos(pi/9), - cos(2*pi/9) )
HomoDef(F, cos(2*pi/9), cos(pi/9) - cos(2*pi/9) )
CheckHomo(F)
    True
```

Apparently, this is an automorphism. Furthermore, we could consider the homomorphism $f^2(x) = f(f(x))$:

```
F(F(cos(pi/9)))
    cos(2/9*pi) - cos(1/9*pi)
F(F(cos(2*pi/9)))
    -cos(1/9*pi)
```

Are there any other automorphisms on the field S? We can show that this is all of them. We will take advantage of the trig identity $\cos(3x) = 4\cos^3 x - 3\cos x$.

Thus,

$$\frac{1}{2} = \cos\left(\frac{3\pi}{9}\right) = 4\cos^3\left(\frac{\pi}{9}\right) - 3\cos\left(\frac{\pi}{9}\right).$$

Thus, $\cos(\pi/9)$ satisfies the polynomial equation $4x^3 - 3x = 1/2$. Because f is an automorphism, we have to have $f(\cos(\pi/9))$ satisfying the same polynomial equation. But there are only three roots to a cubic equation, and so there are only three possible values for $f(\cos(\pi/9))$. Each of these three solutions produces a unique automorphism on S. By Lemma 11.5, we see that the group of automorphisms of this ring is isomorphic to Z_3. In fact, we can compute

```
F(F(F(cos(pi/9))))
    cos(1/9*pi)
F(F(F(cos(2*pi/9))))
    cos(2/9*pi)
```

and see that $f(f(f(x))) = x$ for all x.

The three automorphisms give us three ways to define an ordering on the field S:

1. $a >_1 b$ if a is larger than b as real numbers.
2. $a >_2 b$ if $f(a) >_1 f(b)$.
3. $a >_3 b$ if $f(f(a)) >_1 f(f(b))$. $\square$

We can actually use the automorphisms to prove that S is a field. If we let

```
var("a b c")
x = a + b*cos(pi/9) + c*cos(2*pi/9); x
    c*cos(2/9*pi) + b*cos(1/9*pi) + a
```

we can consider what happens if we multiply the three automorphisms together.

```
TrigReduce(x * F(x) * F(F(x)))
    a^3 - 3/4*a*b^2 + 1/8*b^3 - 3/4*a*b*c + 3/4*b^2*c
    - 3/4*a*c^2 + 3/8*b*c^2 - 1/8*c^3
```

We discover that this product will be a rational number! See Problem 17 for an explanation. With this, we can see that

$$\frac{1}{x} = \frac{8f(x) \cdot f(f(x))}{8a^3 - 6ab^2 + b^3 - 6abc + 6b^2c - 6ac^2 + 3bc^2 - c^3}.$$

Since we can compute

```
TrigReduce(8 * F(x) * F(F(x)))
    4*b^2*cos(2/9*pi) - 8*a*c*cos(2/9*pi) - 4*c^2*cos(2/9*pi) -
    8*a*b*cos(1/9*pi) + 8*b*c*cos(1/9*pi) + 4*c^2*cos(1/9*pi) +
    8*a^2 - 2*b^2 - 2*b*c - 2*c^2
```

we find that

$$\frac{1}{a + b\cos(\pi/9) + c\cos(2\pi/9)} =$$

$$\frac{(8a^2-2b^2-2bc-2c^2)+(8ab+8bc+4c^2)\cos\left(\frac{\pi}{9}\right)+(4b^2-8ac-4c^2)\cos\left(\frac{2\pi}{9}\right)}{8a^3 - 6ab^2 + b^3 - 6abc + 6b^2c - 6ac^2 + 3bc^2 - c^3}.$$

We have seen that some fields may have many ways of assigning an order to the elements, while others have only 1. The key is the number of ring automorphisms. These ring automorphisms will play a major role in the following chapters.

Problems for §11.4

1 Show that the equation $x^2 + e = 0$ has no solutions in an ordered ring.

2 Prove that if a is an element in a nontrivial ordered ring, then there exists an element b such that $b > a$.

3 Show that if $0 < x < y$ in an ordered ring, then $x^2 < y^2$.

4 Show that if $0 < a < b$ and $0 < c < d$ in an ordered ring, then $ac < bd$.

5 Show that if a and b are invertible elements in an ordered ring, and $0 < a < b$, then $a^{-1} > b^{-1}$.

6 Prove that if x and y are two elements in an ordered ring,

$$x^2 + y^2 \geq 2xy.$$

7 Prove that if x and y are two elements in an ordered ring,

$$x^2 + y^2 \geq -2xy.$$

8 In the integral domain $\mathbb{Z}[x]$, let $(\mathbb{Z}[x])^+$ denote the set of all polynomials whose leading coefficient is positive. Prove that $\mathbb{Z}[x]$ is an ordered integral domain by proving that $(\mathbb{Z}[x])^+$ is a set of positive elements for $\mathbb{Z}[x]$.

9 Show that in the integral domain $\mathbb{Z}[x]$, there is a ring automorphism that sends x to $-x$. Hence, there is a second way to order the integral domain $\mathbb{Z}[x]$. Describe the set of positive elements in this new ordering. (See Problem 8.)

10 Show that the ring of real numbers $\mathbb{R}$ does not have a nontrivial ring automorphism.
Hint: First show that there is no nonstandard ordering on $\mathbb{R}$.

11 Although the definition of an ordered ring assumed that the ring R is commutative, there is no reason why we cannot use Definition 11.9 on a noncommutative ring. Consider the set W of all linear differential operators with polynomial coefficients. Each member of this set sends the function $y(x)$ to

$$p_n(x)y^{(n)} + p_{n-1}(x)y^{(n-1)} + \cdots + p_2(x)y'' + p_1(x)y' + p_0(x)y,$$

where each $p_i(x)$ is a polynomial in $\mathbb{Z}[x]$, and $y^{(n)}$ represents the nth derivative of y. Addition can be defined on this set in the standard way, and multiplication is done by composition, using implicit differentiation. For example, if D is the element sending y to y', and X is the element sending y to xy, then $D \cdot X$ sends y to $(xy)' = xy' + y$. Show that W is a ring. This ring is called the *Weyl algebra* (of one variable).

12 In the ring W from Problem 11, let T_1 map y to $xy''+y$, and T_2 map y to $(x^2+1)y'-2xy$. Find $T_1 \cdot T_2$ and $T_2 \cdot T_1$, and show the ring is non-commutative.

13 Let $\phi : W \to \mathbb{Z}[x]$ be the mapping that maps every non-zero element of w from Problem 11 to the coefficient polynomial $p_n(x)$ of the highest derivative of the linear operator. For example, if T maps y to $(3x^2+5x+2)y''+(4x+3)y'-5x^2y$, $\phi(T) = 3x^2+5x+2$. Show that $\phi(T \cdot S) = \phi(T) \cdot \phi(S)$.

14 Show that the Weyl algebra W from Problem 11 has an ordering according to Definition 11.9.

Hint: Let P be the set of non-zero elements for which the leading coefficient of $\phi(T)$ is positive, using ϕ from Problem 13. You can use the result of Problem 8.

15 Prove that $a^3+2b^3+4c^3-6abc$ is never zero for rational a, b, and c, not all 0. This result was used in Example 11.5.

Hint: If there were a rational solution, we could multiply by the common denominator to get an integer solution. Furthermore, we could assume that no prime divides all three numbers.

16 Find the multiplicative inverse of $\sqrt[3]{4}-\sqrt[3]{2}-3$ in $\mathbb{Q}(\sqrt[3]{2})$.

17 Show that for the automorphism from Example 11.6, $y = x \cdot f(x) \cdot f(f(x))$ will always be a rational number.

Hint: What is $f(y)$?

18 Show that in Example 11.6, the denominator

$$8a^3-6ab^2+b^3-6abc+6b^2c-6ac^2+3bc^2-c^3 \neq 0$$

for rational a, b, and c, not all 0. See the hint for Problem 15.

Interactive Problems

19 Follow the example of $\mathbb{Z}[\sqrt[3]{2}]$ to define the integral domain $\mathbb{Z}[\sqrt{5}]$ in *Sage*. Then define F to be a nontrivial ring automorphism for this domain.

20 Use *Sage* to show that numbers of the form

$$x + y\cos(\pi/7) + z\cos(2\pi/7)$$

are closed under multiplication, using **`TrigReduce`**. Assuming that this forms a field, find a non-trivial ring automorphism on this field. This problem will not work in *Mathematica*.

Chapter 12

Unique Factorization

We have already proven the unique factorization for the set of integers, namely, that all numbers greater than one can be factored uniquely into one or more positive primes. In this chapter we will focus on integral domains for which a similar property holds. These will be referred to as unique factorization domains, or UFD's.

The study of unique factorization may well be considered the origin of modern abstract algebra. When Gabriel Lamé announced his proof of Fermat's last theorem, it was quickly pointed out that he assumed certain integral domains were UFD's, which in fact were not. Ernst Kummer partially repaired the proof by introducing *ideal numbers*. (See Historical Diversion on page 432.) Later, Kummer's ideal numbers were developed into *ideals*, and Dedekind would later classify those integral domains for which ideals can be expressed uniquely as a product of prime ideals.

12.1 Factorization of Polynomials

In the last chapter, we defined the integral domain $F[x]$ of all polynomials with coefficients in a field F. In this section we will investigate how such polynomials factor. Most of the statements in this section are very familiar from the properties of polynomials $\mathbb{Q}[x]$ studied in a standard algebra course.

We say that $f(x)$ factors if there are two non-constant polynomials $g(x)$ and $h(x)$ such that $f(x) = g(x) \cdot h(x)$. We also say that both $g(x)$ and $h(x)$ divide the polynomial $f(x)$. But $g(x)$ and $h(x)$ may also factor into non-constant polynomials. We want to show that we can factor $f(x)$ into polynomials that cannot be factored further. We also want to lay down the groundwork for showing that the polynomials produced by this factorization are in some sense uniquely determined.

One of the techniques from a standard algebra course is doing "long division" on polynomials.

Example 12.1
Find the quotient and remainder when the polynomial $x^3 - 3x^2 + 4x - 5$ is

$$
\begin{array}{r|l}
 & x/2 \quad - 3/2 \\
\hline
2x^2 - 5 & x^3 - 3x^2 + 4x \quad - 5 \\
 & x^3 \qquad - 5/2x \\
\hline
 & \quad - 3x^2 + 13/2x - 5 \\
 & \quad - 3x^2 \qquad + 15/2 \\
\hline
 & \qquad 13/2x - 25/2
\end{array}
$$

FIGURE 12.1: Sample long division problem

divided by $2x^2 - 5$.

SOLUTION: The work is done in Figure 12.1. This shows that $x^3 - 3x^2 + 4x - 5$ divided by $2x^2 - 5$ yields $x/2 - 3/2$, with a remainder of $13/2x - 25/2$. We can write this as

$$x^3 - 3x^2 + 4x - 5 = (2x^2 - 5) \cdot (x/2 - 3/2) + (13/2x - 25/2). \quad \square$$

Fortunately, *Sage* can do this tedious long division for you.

```
var("x")
PolynomialQuotient(x^3 - 3*x^2 + 4*x - 5, 2*x^2 - 5)
    1/2*x - 3/2
PolynomialRemainder(x^3 - 3*x^2 + 4*x - 5, 2*x^2 - 5)
    13/2*x - 25/2
```

This "long division" algorithm works for any field, not just the rational numbers $\mathbb{Q}$. We can prove this by induction on the degree of the dividend.

THEOREM 12.1: The Division Algorithm Theorem

Let F be a field, and let $F[x]$ be the set of polynomials in x over F. Let $f(x)$ and $g(x)$ be two elements of $F[x]$, with g nonzero. Then there exist unique polynomials $q(x)$ and $r(x)$ in $F[x]$ such that

$$f(x) = g(x) \cdot q(x) + r(x)$$

and either $r(x) = 0$ or the degree of $r(x)$ is less than the degree of $g(x)$.

PROOF: We begin by showing that $q(x)$ and $r(x)$ exist, and then prove that they are unique. If $f(x) = 0$, or if the degree of $f(x)$ is less than the degree of $g(x)$, we can simply let $q(x) = 0$, and $r(x) = f(x)$. So we may suppose that the degree of $f(x)$ is at least as large as the degree of $g(x)$. Let n be the degree of $f(x)$ and let m be the degree of $g(x)$.

If $n = m = 0$, then $f(x)$ and $g(x)$ are both nonzero constants in the field F, so we may pick $q(x)$ to be the constant polynomial $f \cdot g^{-1}$, and pick $r(x) = 0$. Thus, we can find a suitable $q(x)$ and $r(x)$ when $n = 0$.

Now let us proceed by induction on n. That is, we will assume that we can find a suitable $q(x)$ and $r(x)$ whenever the degree of $f(x)$ is less than n. Let

$$f(x) = a_n x^n + a_{n-1}x^{n-1} + \cdots + a_0,$$

and

$$g(x) = b_m x^m + b_{m-1}x^{m-1} + \cdots + b_0.$$

Since n is at least as large as m, we can consider the polynomial

$$p(x) = a_n b_m^{-1} x^{n-m}$$

of degree $n - m$. By Lemma 11.1, $p(x) \cdot g(x)$ has degree n, and in fact, since

$$p(x) \cdot g(x) = a_n x^n + a_n b_m^{-1} b_{m-1} x^{n-1} + \cdots + a_n b_m^{-1} b_0 x^{n-m},$$

the coefficient of the x^n term would be a_n. Thus, $f(x) - p(x) \cdot g(x)$ is of degree less than n. So by the induction hypothesis, there exist polynomials $z(x)$ and $r(x)$ such that

$$f(x) - p(x) \cdot g(x) = z(x) \cdot g(x) + r(x)$$

with the degree of $r(x)$, less than the degree of $g(x)$. Thus,

$$f(x) = (p(x) + z(x)) \cdot g(x) + r(x).$$

By letting $q(x) = p(x) + z(x)$ we have proved that suitable $q(x)$ and $r(x)$ exist.

Next, let us prove that $q(x)$ and $r(x)$ are unique. Suppose that there is a second pair $\overline{q}(x)$ and $\overline{r}(x)$ such that $f(x) = \overline{q}(x) \cdot g(x) + \overline{r}(x)$. Then

$$\overline{q}(x) \cdot g(x) + \overline{r}(x) = q(x) \cdot g(x) + r(x),$$

or

$$(\overline{q}(x) - q(x)) \cdot g(x) = r(x) - \overline{r}(x).$$

The left-hand side is either 0 (when $\overline{q}(x) = q(x)$), or has degree at least m, since $g(x)$ is of degree m. The right-hand side is either 0, or has a degree less than m. This is a contradiction unless both sides of the equation are 0. Thus, $\overline{q}(x) = q(x)$ and $\overline{r}(x) = r(x)$, and the uniqueness has been proven. □

This theorem not only shows that the quotient $q(x)$ and remainder $r(x)$ are unique, but the proof basically follows the procedure that is used in Figure 12.1. This means that the familiar long division algorithm used for real polynomials will in fact work for polynomials over any field. In many circumstances, we can do this algorithm on polynomials over any integral domain.

COROLLARY 12.1

Let R be an integral domain, and let $f(x)$ and $g(x)$ be two polynomials in $R[x]$. If there is a field F containing R such that $g(x)$ divides $f(x)$ as polynomials *in $F[x]$, and if the leading coefficient of $g(x)$ is 1, then $g(x)$ divides $f(x)$ in $R[x]$.*

PROOF: The only time that we needed to use a division in the proof of the division algorithm theorem (12.1) is when we divided by the leading coefficient of $g(x)$. Thus, if the leading coefficient of $g(x)$ is 1, we can do all of the operations in $R[x]$ instead of $F[x]$. The result is that there are polynomials $q(x)$ and $r(x)$ such that

$$f(x) = g(x) \cdot q(x) + r(x)$$

in $R[x]$. But $g(x)$ divides $f(x)$ in the ring $F[x]$. So there is an $h(x)$ in $F[x]$ such that

$$f(x) = g(x) \cdot h(x).$$

But $q(x)$ and $r(x)$ can also be viewed as polynomials in $F[x]$, and the division algorithm shows that these are uniquely defined, even in $F[x]$. Thus, $q(x) = h(x)$ and $r(x) = 0$. Therefore, $g(x)$ divides $f(x)$ in $R[x]$. □

We are used to thinking of polynomials as functions, rather than as elements in a domain. If we want to "evaluate" a polynomial $f(x)$ at a particular value y, we run into a technical problem, since $f(x)$ is not a function. The division algorithm comes to our rescue on the occasion when we do need to evaluate polynomials at a particular value.

DEFINITION 12.1 Let K be a field or integral domain, and let $K[x]$ be the set of polynomials in x over K. For a fixed element y in K, define the mapping $\phi_y : K[x] \to K$ by

$$\phi_y(f(x)) = \text{ the remainder } r(x) \text{ when } f(x) \text{ is divided by } (x - y).$$

Since the polynomial $(x - y)$ is first degree, either $r(x)$ is 0 or is of degree 0, so $r(x)$ is in fact in K.

PROPOSITION 12.1

The mapping $\phi_y : K[x] \to K$ is a homomorphism, called the evaluation homomorphism at *y.*

PROOF: Let $f_1(x)$ and $f_2(x)$ be two polynomials in $K[x]$. By the division algorithm theorem (12.1) there exists $q_1(x)$, $q_2(x)$, $\phi_y(f_1(x)) = r_1(x)$, and $\phi_y(f_2(x)) = r_2(x)$ such that

$$f_1(x) = (x - y) \cdot q_1(x) + r_1(x),$$

and

$$f_2(x) = (x - y) \cdot q_2(x) + r_2(x).$$

Then

$$f_1(x) + f_2(x) = (x - y)(q_1(x) + q_2(x)) + r_1(x) + r_2(x),$$

and

$$\begin{aligned} f_1(x) \cdot f_2(x) &= ((x - y) \cdot q_1(x) + r_1(x)) \cdot ((x - y) \cdot q_2(x) + r_2(x)) \\ &= (x - y) \cdot ((x - y) \cdot q_1(x)q_2(x) + q_1(x)r_2(x) + q_2(x)r_1(x)) + r_1(x) \cdot r_2(x). \end{aligned}$$

By the uniqueness of the division algorithm, we have that

$$\phi_y(f_1(x) + f_2(x)) = r_1(x) + r_2(x) = \phi_y(f_1(x)) + \phi_y(f_2(x)),$$

and

$$\phi_y(f_1(x) \cdot f_2(x)) = r_1(x) \cdot r_2(x) = \phi_y(f_1(x)) \cdot \phi_y(f_2(x)).$$

Thus, ϕ_y is a homomorphism. □

We will often denote $\phi_y(f(x))$ by the conventional notation, $f(y)$. However, whenever we want to emphasize the homomorphism property, we will use the notation $\phi_y(f(x))$ for the evaluation homomorphism.

In *Sage*, one can use the **.subs** function to find the value of a polynomial in one variable at a particular number. To evaluate the polynomial $x^3 + 5x^2 + 4x - 4$ at $x = 3$, enter

```
var("x")
(x^3 + 5*x^2 + 4*x - 4).subs(x == 3)
    80
```

Notice that we had to use a "double equal sign" inside of the **.subs** command. *Sage* also can use the standard functional notation:

```
f(x) = x^3 + 5*x^2 + 4*x - 4
f(3)
    80
```

The difference here is that *Sage* is actually defining a simple function, and evaluating the function at $x = 3$

The **.subs** commands suggest a way to determine what it means for a polynomial to have a root.

DEFINITION 12.2 Let $f(x)$ be a polynomial over the field or integral domain F. If r is an element of F such that $\phi_r(f(x)) = 0$, then r is called a *zero*, or a *root*, of $f(x)$. Of course this is equivalent to saying that $(x - r)$ is a factor of $f(x)$.

Example 12.2

Consider the polynomial $x^2 + 1$ in $Z_5[x]$. We can visually evaluate this polynomial at $x = 2$ to see that

$$\phi_2(x^2 + 1) = 2^2 + 1 = 0$$

in the field Z_5. Thus, 2 is a root, or zero, of $x^2 + 1$. □

As one can imagine, the factorization of a polynomial over an arbitrary field can be more cumbersome than the customary factorization. For a finite field (such as Z_5), almost the only way to find roots is by trial and error. Fortunately, *Sage* can do this very quickly. However, the good news is that if we have found enough roots to a polynomial, we already have the factorization.

PROPOSITION 12.2

Let $f(x)$ be a polynomial over the field F that has positive degree n and leading coefficient a_n. If $r_1, r_2, r_3, \ldots r_n$ are n distinct zeros of $f(x)$, then

$$f(x) = a_n \cdot (x - r_1) \cdot (x - r_2) \cdot (x - r_3) \cdots (x - r_n).$$

PROOF: Again, we will proceed by induction on the degree of $f(x)$, which we will call n. If $n = 1$, then $f(x) = a_1x + a_0$, and since r_1 is a root, $a_1r_1 + a_0 = 0$. Thus, $a_0 = -a_1r_1$, and hence

$$f(x) = a_1x - a_1r_1 = a_1(x - r_1).$$

So the proposition is true when $n = 1$.

Now we will apply the induction hypothesis on n. Since r_n is a root of $f(x)$, we have that

$$f(x) = (x - r_n)g(x)$$

for some $g(x)$, which by Lemma 11.1 is of degree $n - 1$. Furthermore, $g(x)$ and $f(x)$ have the same leading coefficient, a_n. For $m = 1, 2, \ldots, n - 1$, we have

$$0 = \phi_{r_m}(f(x)) = (r_m - r_n) \cdot \phi_{r_m}(g(x)).$$

Since $(r_m - r_n)$ is not 0, we have that $g(x)$ has $n - 1$ distinct roots, namely r_1, r_2, r_3, $\ldots$, r_{n-1}. Thus, by induction,

$$g(x) = a_n(x - r_1)(x - r_2)(x - r_3) \cdots (x - r_{n-1}).$$

Thus,

$$f(x) = a_n(x - r_1)(x - r_2)(x - r_3) \cdots (x - r_n).$$ □

COROLLARY 12.2

A polynomial of positive degree n over the field F has at most n distinct zeros in F.

PROOF: Suppose that $f(x)$ has at least $n+1$ roots, $r_1, r_2, \ldots, r_n, r_{n+1}$. From Proposition 12.2,

$$f(x) = a_n(x - r_1)(x - r_2)(x - r_3) \cdots (x - r_n).$$

Since r_{n+1} is also a root, we have

$$0 = \phi_{r_{n+1}}(f(x)) = a_n(r_{n+1} - r_1)(r_{n+1} - r_2)(r_{n+1} - r_3) \cdots (r_{n+1} - r_n).$$

But all of the factors on the right-hand side are nonzero, which is a contradiction. Thus, there can be at most n distinct zeros of $f(x)$. ∎

We can use Proposition 12.2 to do some factorizations in different fields.

Example 12.3

Find the factorization of $x^2 + 1$ in $Z_5[x]$.

SOLUTION: We can verify that both 2 and 3 are roots of the polynomial $x^2 + 1$. Thus

$$x^2 + 1 = (x - 2)(x - 3) \quad \text{in} \quad Z_5.$$ ∎

Here is an application of Corollary 12.2 that has many applications even using the real number field.

COROLLARY 12.3

Let F be a field, let $x_0, x_1, x_2, x_3, \ldots x_n$ be $n + 1$ distinct elements of F, and let $y_0, y_1, y_2, y_3, \cdots y_n$ be $n + 1$ values in F (not necessarily distinct). Then there is a unique polynomial $f(x)$ with degree at most n such that

$$f(x_0) = y_0, \quad f(x_1) = y_1, \quad f(x_2) = y_2, \quad \cdots \quad f(x_n) = y_n.$$

PROOF: To prove uniqueness, suppose that $f(x)$ and $g(x)$ are two such polynomials. Then $h(x) = f(x) - g(x)$ will have roots at $x_0, x_1, x_2, x_3, \ldots, x_n$. But $h(x)$ would have degree at most n, which contradicts Corollary 12.2. Thus, the polynomial $f(x)$ is unique.

To show that this polynomial exists, we will first construct the nth degree polynomial

$$f_0(x) = \frac{(x - x_1) \cdot (x - x_2) \cdot (x - x_3) \cdots (x - x_n)}{(x_0 - x_1) \cdot (x_0 - x_2) \cdot (x_0 - x_3) \cdots (x_0 - x_n)}$$

for which $f_0(x_0) = 1$ but $x_1, x_2, x_3, \ldots x_n$ are roots of $f_0(x)$. (Note that since all of the x_i are distinct, the denominator is not 0.)

We can likewise define $f_1(x), f_2(x), f_3(x), \ldots, f_n(x)$ such that

$$f_1(x_1) = f_2(x_2) = f_3(x_3) = \cdots f_n(x_n) = 1,$$

yet the remaining n x_i's are roots for each polynomial. Finally, we construct the polynomial

$$g(x) = y_0 f_0(x) + y_1 f_1(x) + y_2 f_2(x) + y_3 f_3(x) + \cdots + y_n f_n(x).$$

Clearly $g(x)$ will be a polynomial of degree at most n, and also $g(x_0) = y_0$, $g(x_1) = y_1$, $g(x_2) = y_2$, $g(x_3) = y_3, \ldots g(x_n) = y_n$. Thus, we have constructed the required polynomial. ∎

This corollary shows, for example, that knowing just three points of a quadratic function is sufficient to determine the quadratic function. *Sage* has the built-in function **InterpolatingPolynomial** that finds this polynomial from a list of points.

```
InitDomain(0)
InterpolatingPolynomial([[1,2], [2,4], [3,8]], "x")
    x^2 - x + 2
```

This finds the polynomial in x such that $f(1) = 2$, $f(2) = 4$, and $f(3) = 8$. The first command **InitDomain(0)** defines the field to be rational numbers. The **InterpolatingPolynomial** command works for other fields as well, such as the "complex numbers modulo 3." Later, we will see how this interpolating polynomial, applied to a different field, is used to store information on a CD or DVD.

We are now ready to define the polynomials that in many ways act as the prime numbers of number theory.

DEFINITION 12.3 A polynomial $f(x)$ in $F[x]$ is said to be *reducible* over F if $f(x)$ has positive degree, and $f(x)$ can be expressed as a product $f(x) = g(x) \cdot h(x)$ where both $g(x)$ and $h(x)$ have positive degree. If $f(x)$ has positive degree and is not reducible, it is called *irreducible.*

We saw above that $x^2 + 1$ was reducible over Z_5. However, *Sage* will claim that this polynomial is irreducible.

```
var("x")
factor(x^2 + 1)
    x^2 + 1
```

The reason of course is that *Sage* is viewing this polynomial as an element of $\mathbb{Q}[x]$, not $Z_5[x]$. Yet this polynomial *would* have a factorization if we were allowed to work with complex numbers:

```
expand( (x + I)*(x - I) )
    x^2 + 1
```

Thus, $x^2 + 1$ is reducible over $\mathbb{C}$, the field of complex numbers. Thus, whether a polynomial is reducible or irreducible over F greatly depends on the field F.

It should be noted that if $g(x)$ and $h(x)$ both have positive degree, then $g(x) \cdot h(x)$ has degree at least 2. Thus, all polynomials of degree 1 must be irreducible. Constant polynomials, however, are not considered to be irreducible.

Although it can be tricky to decide whether a polynomial is reducible or irreducible, there is a way to test polynomials of low degree.

PROPOSITION 12.3

If $f(x)$ is a polynomial of degree 2 or 3 over the field F, then $f(x)$ is reducible over F if, and only if, $f(x)$ has a zero in F.

PROOF: Suppose that $f(x)$ has a zero in F, say r. Then

$$f(x) = (x - r)q(x)$$

where $q(x)$ has degree one less than $f(x)$. This shows that $f(x)$ is reducible.

Now suppose that $f(x)$ is reducible. Then $f(x) = g(x) \cdot h(x)$, where the degree of $g(x)$ plus the degree of $h(x)$ is 2 or 3. Thus, either $g(x)$ or $h(x)$ has degree 1. We may suppose $g(x)$ has degree 1, and so

$$f(x) = (a_1 x + a_0)h(x).$$

Then $-a_0 a_1^{-1}$ is a root of $f(x)$, and the proof is complete. □

We can use this proposition to determine whether polynomials of degree less than 4 are irreducible over a finite field. Simply plug in all elements of the field, and see if any of them produce 0 in that field.

Example 12.4

Determine whether the polynomial $x^3 + 2x^2 - 3x + 4$ is reducible in Z_5.

SOLUTION: If we let $f(x) = x^3 + 2x^2 - 3x + 4$, we find that $f(0) = 4$, $f(1) = 4$. $f(2) = 14$, $f(3) = 40$, and $f(4) = 88$. One of these, namely when x was replaced by 3, produced a multiple of 5, which is equivalent to 0 in the field Z_5. Thus, this polynomial is reducible. □

PROPOSITION 12.4

If F is a field, then all polynomials in $F[x]$ of positive degree are either irreducible, or can be expressed as a product of irreducible polynomials.

PROOF: If $f(x)$ has degree 1, then we have seen that it is irreducible. Let us proceed by induction on the degree n of $f(x)$. If $f(x)$ is not irreducible, then we can express $f(x) = g(x) \cdot h(x)$, where $g(x)$ and $h(x)$ are polynomials of degree at least 1. But $g(x)$ and $h(x)$ must have degree less than n. Thus, by induction, $g(x)$ and $h(x)$ are either irreducible, or can be written as a product of irreducible polynomials. Thus, $f(x)$ can be written as a product of irreducible polynomials. □

One last tool we have to help us find irreducible polynomials is the Greatest Common Divisor (GCD) of two polynomials. The proof of the next theorem mimics the proof of the greatest common divisor theorem for integers (0.4).

THEOREM 12.2: The Greatest Common Divisor Theorem for Polynomials

Let F be a field, and let $F[x]$ be the polynomials in x over the field F. Given two nonzero polynomials $f(x)$ and $g(x)$ in $F[x]$, there exists a nonzero polynomial $h(x)$ such that

1. *$h(x)$ divides both $f(x)$ and $g(x)$.*

2. *There exist polynomials $s(x)$ and $t(x)$ such that*

$$f(x) \cdot s(x) + g(x) \cdot t(x) = h(x).$$

Furthermore, the polynomial $h(x)$ is unique except for multiplication by a constant.

PROOF: Let us consider the set of all polynomials that can be produced by

$$f(x) \cdot s(x) + g(x) \cdot t(x)$$

where $s(x)$ and $t(x)$ are in $F[x]$. Call this set A. Both $f(x)$ and $g(x)$ are in A, so A contains nonzero polynomials. Consider a nonzero polynomial $h(x)$ in A of the lowest degree. By the division algorithm theorem (12.1), we can find polynomials $q(x)$ and $r(x)$ such that

$$f(x) = q(x) \cdot h(x) + r(x),$$

where $r(x)$ is either 0, or has lower degree than $h(x)$. But then

$$r(x) = f(x) - q(x) \cdot h(x) = (1 - q(x) \cdot s(x)) \cdot f(x) - q(x) \cdot g(x) \cdot t(x),$$

which is in A. But if $r(x)$ is not zero, the degree of $r(x)$ would be less than the degree of $h(x)$, and we picked $h(x)$ to be of the lowest degree. Thus, $r(x) = 0$, and $h(x)$ divides $f(x)$. By a similar argument, $h(x)$ divides $g(x)$.

To prove that $h(x)$ is unique, note that since $h(x)$ divides $f(x)$ and $g(x)$, then $h(x)$ divides all polynomials in A. So if there is another polynomial $d(x)$ in A that divides both $f(x)$ and $g(x)$, then $h(x)$ would divide $d(x)$. But $d(x)$ would also divide $h(x)$. Thus, $h(x)$ and $d(x)$ would have to have the same degree, and

$$d(x) = u \cdot h(x)$$

where u is a constant polynomial. Thus, $h(x)$ is unique up to multiplication by a constant. ∎

DEFINITION 12.4 Given two polynomials in $F[x]$, the *greatest common divisor* is the polynomial given in the above theorem whose leading coefficient is 1.

The *Sage* command **gcd** will find the greatest common divisor of two polynomials as well as integers. For example, $\gcd(x^4 - 1, x^3 - 1)$ is found by the commands

```
var("x")
gcd(x^3 - 1, x^4 - 1)
    x - 1
```

Thus, there are two polynomials $s(x)$ and $t(x)$ such that

$$(x^3 - 1) \cdot s(x) + (x^4 - 1) \cdot t(x) = x - 1.$$

See Problem 20.

COROLLARY 12.4

Let F be a field, and let $f(x)$, $g(x)$, and $h(x)$ be polynomials in $F[x]$. If $f(x)$ is an irreducible divisor of $g(x) \cdot h(x)$, then either $g(x)$ or $h(x)$ is a multiple of $f(x)$.

PROOF: Suppose that $f(x)$ divides neither $g(x)$ nor $h(x)$. Then the greatest common divisor of $f(x)$ and $g(x)$ must have degree less than the degree of $f(x)$. But $\gcd(f(x), g(x))$ must divide $f(x)$, and $f(x)$ is irreducible. Thus, the greatest common divisor of $f(x)$ and $g(x)$ must be 1. Likewise $\gcd(f(x), h(x))$ must also be 1. By the greatest common divisor theorem (12.2), there exist polynomials $r(x)$, $s(x)$, $t(x)$, and $u(x)$ such that

$$f(x) \cdot r(x) + g(x) \cdot s(x) = 1,$$

and

$$f(x) \cdot t(x) + h(x) \cdot u(x) = 1.$$

By multiplying these two together, we have

$$\begin{aligned} 1 &= (f(x) \cdot r(x) + g(x) \cdot s(x)) \cdot (f(x) \cdot t(x) + h(x) \cdot u(x)) \\ &= f(x)^2 \cdot r(x) \cdot t(x) + f(x) \cdot r(x) \cdot h(x) \cdot u(x) \\ &\qquad + f(x) \cdot g(x) \cdot s(x) \cdot t(x) + g(x) \cdot h(x) \cdot s(x) \cdot u(x). \end{aligned}$$

Note that all of the terms on the right-hand side are multiples of $f(x)$ (including the last term, since $g(x) \cdot h(x)$ is a multiple of $f(x)$). But the left-hand side is 1, which cannot be a multiple of $f(x)$. Thus, we have a contradiction, and so either $g(x)$ or $h(x)$ is a multiple of $f(x)$. $\square$

The irreducible polynomials will play the same role in the domain $F[x]$ as prime numbers play in the domain $\mathbb{Z}$. The key property of integer factorizations is that every positive number greater than one can be factored *uniquely* into a product of primes. We would like to prove something similar for polynomials in $F[x]$, but find we will have to modify our definition of unique factorization. In the next section, we will explain what it means for a general ring to have a unique factorization, and apply this to both polynomial rings and integers.

Problems for §12.1

For Problems **1** through **6**: Use the division algorithm to determine polynomials $q(x)$ and $r(x)$ in $F[x]$ such that $f(x) = q(x) \cdot g(x) + r(x)$, where the degree of $r(x)$ is less than the degree of $g(x)$.

1	$f(x) = 2x^3 + 3x^2 - 5x + 4,$	$g(x) = 2x^2 - x + 1,$	$F = \mathbb{Q}.$
2	$f(x) = x^3 + 4x^2 + 2x - 1,$	$g(x) = 2x^2 + x - 1,$	$F = \mathbb{Q}.$
3	$f(x) = x^5 + x^3 + x^2 + x,$	$g(x) = x^3 + x^2 + 1,$	$F = Z_2.$
4	$f(x) = x^6 + x^5 + x^2 + 1,$	$g(x) = x^3 + x + 1,$	$F = Z_2$
5	$f(x) = x^4 + 2x^2 + x + 1,$	$g(x) = x^2 + x + 2,$	$F = Z_3$
6	$f(x) = x^3 + 4x^2 + 3x + 2,$	$g(x) = 2x^2 + 3x + 1,$	$F = Z_5$

7 Find a quadratic polynomial $f(x)$ such that $f(-1) = 6$, $f(1) = 2$, and $f(2) = 9$.

Hint: Either solve three equations for three unknowns, or use the proof of Corollary 12.3.

8 Find a quadratic polynomial in $Z_3[x]$ such that $f(0) = f(1) = 2$, and $f(2) = 0$.

9 Prove that $x^2 + 5$ is irreducible over the field $\mathbb{R}$ of real numbers.

10 Determine whether $x^3 + 2x^2 + 3x + 2$ is irreducible over Z_5.

11 Determine whether $x^3 + 2x^2 + 3x + 5$ is irreducible over Z_7.

12 Show that $x^3 - 9$ is irreducible over the field Z_{13}.

13 Let F be a field that is contained in a larger field K. Let $f(x)$ and $g(x)$ be two polynomials in $F[x]$ that are coprime in $F[x]$. Show that $f(x)$ and $g(x)$ are also coprime in $K[x]$.

For Problems **14** through **19**: Find the greatest common divisors of the polynomials over the field F. Hint: Mimic the Euclidean algorithm.

14	$f(x) = x^3 + 2x^2 - 2x - 3,$	$g(x) = x^3 - x^2 - 5x + 6,$	$F = \mathbb{Q}.$
15	$f(x) = 2x^4 - 3x^3 + 7x^2 + 7x - 5,$	$g(x) = x^3 - 3x^2 + 7x - 5,$	$F = \mathbb{Q}.$
16	$f(x) = x^5 + x^4 + 1,$	$g(x) = x^5 + x + 1,$	$F = Z_2.$
17	$f(x) = x^3 + 2,$	$g(x) = x^2 + 1,$	$F = Z_5.$
18	$f(x) = x^4 + 2x^2 + 2x + 2,$	$g(x) = x^4 + 2x^3 + x^2 + 2,$	$F = Z_3.$
19	$f(x) = x^3 + 2x^2 + 4x + 1,$	$g(x) = x^3 + 3x^2 + 5x + 3,$	$F = Z_7.$

20 Find $s(x)$ and $t(x)$ such that

$$(x^3 - 1) \cdot s(x) + (x^4 - 1) \cdot t(x) = x - 1.$$

Interactive Problems

21 Use the *Sage* command **`InterpolatingPolynomial`** to find a third-degree polynomial such that $f(n) = n!$ for $n = 1, 2, 3$, and 4. How close is $f(5)$ to 120?

22 Use *Sage* to find polynomials $s(x)$ and $t(x)$ in $\mathbb{Q}[x]$ such that

$$(x^2 + 2x - 3) \cdot s(x) + (x^2 - x + 4) \cdot t(x) = 1.$$

Hint: **`xgcd`** does not work for polynomials. Imitate the Euclidean algorithm with **`PolynomialQuotient`** and **`PolynomialRemainder`**.

12.2 Unique Factorization Domains

In this section we wish to determine a general definition of unique factorization that would apply not only to $F[x]$, but for *any* ring. We will mainly be interested in integral domains for which factorizations are unique.

DEFINITION 12.5 Let R be a commutative ring. We say that an element x in R is a *unit* if x has a multiplicative inverse.

In Proposition 9.4 we defined the set of invertible elements of R as R^*, and showed that they formed a group under multiplication. The units of R will play the same role as the constant polynomials do in the ring $F[x]$. In fact, we can model the definition of reducible and irreducible elements of a ring on the definition of irreducible polynomials in $F[x]$.

DEFINITION 12.6 Let R be a commutative ring. If a nonzero element x in R is not a unit, and can be expressed as a product $x = y \cdot z$, where neither y nor z are units, then we say that x is *reducible*. If a nonzero element is neither a unit nor reducible, we say it is *irreducible*.

Although this definition is mainly applied to integral domains, we can apply the definition to any ring with an identity.

Example 12.5
Find an irreducible element in the ring defined by Tables 9.3 and 9.4 in Chapter 9.

SOLUTION: We see from Table 9.4 that the units of this ring are $a + b$ and $-a + b$. We observe that a, $2a$, $3a$, and b can be expressed as a product of two non-units, so these would be reducible. But the only way that $2a + b$ can be expressed as a product is if $a + b$ or $-a + b$ is one of the factors, so this element is irreducible. ∎

Let us consider the more familiar ring, $\mathbb{Z}$. The only two elements with multiplicative inverses are ± 1. The irreducible elements are of course the prime numbers 2, 3, 5, 7, 11, 13, ... But by this definition, the *negative* of a prime number is also irreducible. But by introducing negative primes, we find that numbers can be written as a product of primes in more than one way:

$$12 = 2 \cdot 2 \cdot 3 = 2 \cdot (-2) \cdot (-3) = (-2) \cdot (-2) \cdot 3.$$

Because we now are including negative primes, we also have to redefine what is meant by unique factorization. The first step is to understand the relationship between these different factorizations.

DEFINITION 12.7 Let R be a commutative unity ring. We say that the element x is an *associate* of an element y if there is a unit z such that $y = x \cdot z$.

Note that if x is an associate of y, then $x = y \cdot z^{-1}$, so that y is an associate of x. Even though we saw three different factorizations of 12, note that these are related via associates. We now can explain what unique factorization means for a general ring.

DEFINITION 12.8 A ring R has *unique factorization* if the following two conditions are satisfied:

1. If x is nonzero, and is not a unit of R, then x can be written as a product of irreducible elements of R.

2. If
$$x = y_1 \cdot y_2 \cdot y_3 \cdots y_m = z_1 \cdot z_2 \cdot z_3 \cdots z_n$$
are two expressions of x as a product of irreducible elements, then $m = n$ and there is a permutation $\sigma \in S_n$ such that y_i and $z_{\sigma(i)}$ are associates. In other words, each y_i is an associate of some z_j, and vice versa.

Furthermore, if R is an integral domain, then R is a *unique factorization domain*, abbreviated as *UFD*.

We would like to find a quick way to determine whether an integral domain is a UFD. The needed tool will be the definition of the *prime* elements. Although we have already defined a prime element in the integers $\mathbb{Z}$, for a general ring we wish to define a prime element as one that satisfies a different property.

DEFINITION 12.9 A nonzero element x of a commutative ring is *prime* if x is not a unit, and whenever $y \cdot z$ is a multiple of x, then either y or z must be a multiple of x.

Although primes and irreducible elements are the same in $\mathbb{Z}$, for many other rings they are totally different.

Example 12.6

Show that a is a prime element of the ring of Example 12.5.

SOLUTION: We need to show that whenever $y \cdot z$ is a multiple of a, then either y or z is a multiple of a. Another way of saying this is that whenever y and z are not multiples of a, then $y \cdot z$ is not a multiple of a. The multiples of a are $\{0, a, 2a, 3a\}$, so the non-multiples of a are $\{b, a+b, 2a+b, -a+b\}$. Table 9.4 shows that multiplication is closed under this latter set, so a is prime, even though it is not irreducible. □

Although this ring has prime elements that are not irreducible, we can show that this can only happen when the ring has zero divisors.

LEMMA 12.1

If K is an integral domain, and x is a prime element of K, then x is irreducible.

PROOF: Since x is prime, it is neither 0 nor a unit. Suppose that $x = y \cdot z$, where neither y nor z are units. Since x is prime, we have that either y or z is a multiple of x. Suppose that y is a multiple of x. Then $y = x \cdot w$ for some number w. Then

$$x = y \cdot z = x \cdot w \cdot z.$$

Since K is an integral domain, we know that x is not a zero divisor, so we can use Lemma 9.4 and say that

$$1 = w \cdot z.$$

But this indicates that z is a unit, which contradicts the original assumption that neither y nor z were units. Thus, x is irreducible. □

Even though a prime element is irreducible in an integral domain, it is *not* true that an irreducible element is prime! Consider for example the ring $\mathbb{Z}[\sqrt{-5}]$, whose elements are the numbers of the form $x + y\sqrt{-5}$, where x and y are integers. To determine the irreducible elements of this ring, let us define the following function on $\mathbb{Z}[\sqrt{-5}]$:

$$N(x + y\sqrt{-5}) = (x + y\sqrt{-5})(x - y\sqrt{-5}) = x^2 + 5y^2.$$

Notice that $N(z)$ is the product of the number z with its complex conjugate. We can observe that if a and b are in $\mathbb{Z}[\sqrt{-5}]$, $N(a \cdot b) = N(a) \cdot N(b)$. This function will help us to determine the irreducible elements of $\mathbb{Z}[\sqrt{-5}]$.

Let us begin by finding the units of $\mathbb{Z}[\sqrt{-5}]$. If $a = x + y\sqrt{-5}$ is invertible, then $N(a)$ must be invertible. Hence $x^2 + 5y^2 = 1$. The only integer solution to this equation is when $y = 0$ and $x = \pm 1$. Thus, ± 1 are the two units of this ring.

Next, let us find an irreducible element. Since $N(2) = 4$, the only way a product of non-units a and b could equal 2 is if $N(a) = N(b) = 2$. But the equation $x^2 + 5y^2 = 2$ clearly has no integer solutions. Thus, 2 is an irreducible element in this ring. By the same reasoning, 3 is also irreducible.

However, neither 2 nor 3 is a prime element of this ring! Consider the product

$$(1 + \sqrt{-5})(1 - \sqrt{-5}) = 1 + 5 = 6.$$

This product is a multiple of 2 and 3, but neither factor is a multiple of 2 or 3. Thus, 2 and 3 are not prime in this ring.

This example shows a domain that is *not* a unique factorization domain. We have seen two ways of factoring the number 6 that are not equivalent in terms of associates.

The ring $\mathbb{Z}[\sqrt{-5}]$ can be generalized to produce similar rings, some of which are UFD's, and some are not.

DEFINITION 12.10 Let n be an integer that is not divisible by the square of any integer other than 1. Then the ring $\mathbb{Z}[\sqrt{n}]$ is called a *quadratic domain.*

We have already worked with some examples of quadratic domains. For example, we found two possible ways to order the ring $\mathbb{Z}[\sqrt{2}]$, using ring homomorphisms. In §12.4, we will generalize the $N(z)$ function to determine whether or not many of these quadratic domains are UFD's.

The fact that neither 2 nor 3 was prime in the quadratic domain $\mathbb{Z}[\sqrt{-5}]$ is a clue as to why this ring is not a UFD.

PROPOSITION 12.5

An integral domain is a UFD if, and only if, all nonzero non-units can be written as a product of primes.

PROOF: We begin by showing that if K is a UFD, then all irreducible elements are prime. Suppose w is irreducible, and $x \cdot w = y \cdot z$ is a multiple of w. Then x, y, and z have factorizations into irreducible elements:

$$x = x_1 \cdot x_2 \cdots x_n,$$

$$y = y_1 \cdot y_2 \cdots y_m,$$
$$z = z_1 \cdot z_2 \cdots z_k.$$

Thus,

$$x_1 \cdot x_2 \cdots x_n \cdot w = y_1 \cdot y_2 \cdots y_m \cdot z_1 \cdot z_2 \cdots z_k.$$

Since a factorization is unique, and all terms in this product are irreducible, we have that w is an associate to one of the terms on the right-hand side. Thus, either y or z is a multiple of w, and hence w is prime.

Since a nonzero element that is not a unit in a UFD can be expressed as a product of irreducible elements, we have shown that all such elements can be expressed as a product of primes.

Now let us suppose that all nonzero, non-unit elements in an integral domain can be expressed as a product of primes. The first part of the definition of a UFD is obviously fulfilled since the prime elements are irreducible. Suppose we have another factorization in terms of irreducible elements.

$$p_1 \cdot p_2 \cdot p_3 \cdots p_n = z_1 \cdot z_2 \cdot z_3 \cdots z_m.$$

Here, the p_i are prime elements, while the z_j are merely irreducible elements. We need to prove that $n = m$, and that, after a rearrangement of the z_j's, we have that p_i and z_i are associates. We will proceed by induction on n, the number of primes in the factorization. If $n = 1$, then $m = 1$; otherwise we would have a prime number (which is irreducible) expressed as a product of two or more irreducible elements. Also, $p_1 = z_1$, and so trivially the p's are associates of the z's.

Next, we will consider the general case. Since the right-hand side of

$$p_1 \cdot p_2 \cdot p_3 \cdots p_n = z_1 \cdot z_2 \cdot z_3 \cdots z_m$$

is a multiple of p_n, one of the z's must be a multiple of p_n. Suppose that

$$z_k = p_n \cdot u.$$

Since z_k is irreducible, we find that u is a unit, hence z_k and p_n are associates. We now can write

$$p_1 \cdot p_2 \cdot p_3 \cdots p_{n-1} \cdot p_n = z_1 \cdot z_2 \cdots z_{k-1} \cdot p_n \cdot u \cdot z_{k+1} \cdots z_m.$$

Since the ring is an integral domain, we can use Lemma 9.4 and cancel out the p_n.

$$p_1 \cdot p_2 \cdot p_3 \cdots p_{n-1} = z_1 \cdot z_2 \cdots z_{k-1} \cdot (u \cdot z_{k+1}) \cdots z_m.$$

The unit u may be multiplied by any of the irreducible elements z to produce another irreducible element. We now can apply the induction hypothesis, which says that there are $n-1$ z's left, and that a rearrangement of the z's would make p_i and z_i associates. Therefore, $m = n$, and some rearrangement of the z's in

$$p_1 \cdot p_2 \cdot p_3 \cdots p_n = z_1 \cdot z_2 \cdot z_3 \cdots z_m$$

will allow p_i and z_i to be associates, proving that the ring is a UFD. ∎

This proposition will help us greatly in determining whether an integral domain is a UFD. We usually will proceed in two steps: proving that any element can be written as a product of irreducible elements, and then proving that any irreducible element is prime.

COROLLARY 12.5
If F is a field, then the ring $F[x]$ is a UFD.

PROOF: From Proposition 12.4, every polynomial of positive degree is either irreducible, or can be expressed as a product of irreducible polynomials. By Corollary 12.4, all irreducible polynomials are prime. Thus, by Proposition 12.5, $F[x]$ is a UFD. ∎

Although this corollary proves that polynomials over the rational numbers have a unique factorization, we still have not proven that $\mathbb{Z}[x]$, the polynomials over the integers, is a unique factorization domain. Corollary 12.5 will not help us, since $\mathbb{Z}$ is not a field. Yet it seems plausible that we could prove that $\mathbb{Z}[x]$ is a UFD, merely by using the fact that $\mathbb{Q}[x]$ is a UFD. In the process, let us prove that $R[x]$ is a UFD whenever R is a UFD. First, we will need to prove a few lemmas. This next lemma, commonly referred to as Gauss's lemma, uses the formula for the product of two polynomials.

LEMMA 12.2: Gauss's Lemma
If R is an integral domain, then a prime element of R is also a prime element of $R[x]$.

PROOF: We need to show that if p is a prime of R that divides $h(x) = f(x) \cdot g(x)$, then p must divide either $f(x)$ or $g(x)$. Suppose that p does not divide all of the coefficients of $f(x)$ nor does p divide all of the coefficients of $g(x)$. Let

$$f(x) = a_0 + a_1x + a_2x^2 + a_3x^3 + \cdots,$$
$$g(x) = b_0 + b_1x + b_2x^2 + b_3x^3 + \cdots,$$
$$h(x) = f(x) \cdot g(x) = c_0 + c_1x + c_2x^2 + c_3x^3 + \cdots.$$

Let a_i be the first coefficient of $f(x)$ that is not divisible by p, and let b_j be the first coefficient of $g(x)$ that is not divisible by p.

Since $h(x)$ is divisible by p, we know that the coefficient c_{i+j} must be divisible by p. But

$$c_{i+j} = a_0 b_{i+j} + a_1 b_{i+j-1} + \cdots + a_{i-1} b_{j+1} + a_i b_j + a_{i+1} b_{j-1} + \cdots a_{i+j} b_0.$$

Note that all terms on the right-hand side except $a_i b_j$ are divisible by p (since $a_0, a_1, \ldots a_{i-1}$ and $b_0, b_1, \ldots b_{j-1}$ are all multiples of p). So $a_i b_j$ is also a multiple of p. But this contradicts the fact that p is a prime element of R, and neither a_i nor b_j is a multiple of p. Thus, p is prime in $R[x]$. ∎

With Gauss's lemma (12.2), we can see that whenever a product of several polynomials in $R[x]$ is divisible by a p, a prime number of R, then one of those polynomials must have been divisible by p. We can use induction to extend this argument to any element of R.

LEMMA 12.3

Let R be a unique factorization domain, and let

$$g_1(x),\ g_2(x),\ g_3(x),\ \ldots,\ g_n(x)$$

be polynomials in $R[x]$ that are not divisible by any prime element of R. Let $f(x)$ be a polynomial in $R[x]$, and let c and d be two elements in R such that

$$c \cdot f(x) = d \cdot g_1(x) \cdot g_2(x) \cdot g_3(x) \cdots g_n(x).$$

Then d is divisible by c in R.

PROOF: If c is a unit in R, then obviously d is a multiple of c. We will now use induction on the number of prime factors of c in the ring R. If c contains a prime p, then by Lemma 12.2, one of the terms on the right-hand side must be a multiple of p. But none of the $g_i(x)$ are divisible by a prime, so we find that d is a multiple of p. Then we have

$$\frac{c}{p} \cdot f(x) = \frac{d}{p} \cdot g_1(x) \cdot g_2(x) \cdot g_3(x) \cdots g_n(x),$$

where c/p and d/p are both in R. Since c/p contains one less prime factor than c, we can use induction to say that d/p is a multiple of c/p. Then d would be divisible by c in R. ∎

The next step in proving that $R[x]$ is a UFD is to find the irreducible elements of $R[x]$. If there is a field F that contains R, we can use the irreducible elements of $F[x]$ to find the irreducible elements of $R[x]$.

LEMMA 12.4

Let R be a unique factorization domain, and let F be a field containing R. Then if $f(x)$ is a polynomial in $R[x]$ that is irreducible in $F[x]$, then $f(x)$ can *be written*

$$f(x) = c \cdot g(x),$$

where c is an element of R, and $g(x)$ is irreducible in $R[x]$.

PROOF: We want to first show that we can express

$$f(x) = c \cdot g(x),$$

where the only elements of R that divide $g(x)$ are units. Let a_n be the leading coefficient of $f(x)$. Notice that if an element of R divides $f(x)$, then that element must divide a_n. Since R is a UFD, there are only a finite number of primes in the factorization of a_n. Let us proceed by induction on the number of primes in this factorization.

If there are no prime elements of R that divide $f(x)$ we can let $c = 1$ and $g(x) = f(x)$. If there *is* a prime element of R that divides $f(x)$, we can write

$$f(x) = p \cdot h(x),$$

where p is a prime in R, and $h(x)$ is in $R[x]$. But then the leading coefficient of $h(x)$ will contain one less prime in its prime factorization, so by induction we have

$$h(x) = d \cdot g(x),$$

where the only elements of R that divide $g(x)$ are units. Then we let $c = p \cdot d$, and

$$f(x) = c \cdot g(x).$$

All that is left to show is that $g(x)$ is irreducible in $R[x]$. Suppose that

$$g(x) = r(x) \cdot s(x),$$

where $r(x)$ and $s(x)$ are in $R[x]$. We then have

$$f(x) = c \cdot r(x) \cdot s(x).$$

But there is a field F containing R such that $f(x)$ is irreducible in $F[x]$. Thus, either $r(x)$ or $s(x)$ are units in $F[x]$, which are constant polynomials. But we designed $g(x)$ so that the only constants in $R[x]$ that divide $g(x)$ are units of R. Thus, $g(x)$ is irreducible in $R[x]$. □

Although this lemma refers to some field F that contains R, there is a natural field to use—the field of quotients in R. We can use this field to show that, in fact, the irreducible elements of R that we found in Lemma 12.4 are in fact prime elements of $R[x]$.

LEMMA 12.5

Let R be a unique factorization domain, and let F be the field of quotients for R. Then if $g(x)$ is irreducible over $R[x]$ and $F[x]$, then $g(x)$ is prime in $R[x]$.

PROOF: Suppose that $r(x) \cdot s(x)$ is divisible by $g(x)$ in $R[x]$. We need to show that either $r(x)$ or $s(x)$ is divisible by $g(x)$ in $R[x]$. Yet $g(x)$ is irreducible in $F[x]$, which is a UFD since F is a field. Thus, either $r(x)$ or $s(x)$ is divisible by $g(x)$ in $F[x]$. Suppose that $r(x)$ is divisible. Then we have

$$r(x) = g(x) \cdot k(x),$$

where $k(x)$ is in $F[x]$. The coefficients of $k(x)$ are in the quotient field of R, so we may write

$$k(x) = \frac{a_0}{b_0} + \frac{a_1}{b_1}x + \frac{a_2}{b_2}x^2 + \frac{a_3}{b_3}x^3 + \cdots \frac{a_n}{b_n}x^n.$$

Let c be the product of $b_0 \cdot b_1 \cdot b_2 \cdot b_3 \cdots b_n$. Then $j(x) = c \cdot k(x)$ is a polynomial in $R[x]$. Thus we have

$$c \cdot r(x) = g(x) \cdot (c \cdot k(x)) = g(x) \cdot j(x),$$

where $g(x)$ and $j(x)$ are in $R[x]$. As in Lemma 12.4, there will only be a finite number of primes in R that divide all of the coefficients of $j(x)$, so we can factor out these primes and write

$$j(x) = d \cdot q(x),$$

where $q(x)$ is not divisible by any prime in R. Then

$$c \cdot r(x) = d \cdot g(x) \cdot q(x),$$

so we can apply Lemma 12.3, since neither $g(x)$ nor $q(x)$ is divisible by a prime of R. Hence, d is divisible by c, and

$$r(x) = \frac{d}{c} \cdot g(x) \cdot q(x).$$

Therefore, $r(x)$ is divisible by $g(x)$, and hence $g(x)$ is prime in $R[x]$. ▯

At this point all of the major battles have been fought. All that is left to do is put the pieces together to show that $R[x]$ is UFD.

THEOREM 12.3: The Unique Factorization Domain Theorem

$R[x]$ is a unique factorization domain if, and only if, R is a unique factorization domain.

PROOF: First of all, if R is not a UFD, then there is some element c of R that is not expressible as a product of primes. But then c cannot be expressed as a product of primes in $R[x]$, since such a product must consist of constant polynomials, and this would contradict the fact that c cannot be expressed as a product of primes in R. Thus, $R[x]$ would not be a UFD.

Now suppose that R is a UFD. We need to show that any nonzero polynomial $f(x)$ in $R[x]$ is either a unit, or is expressible as a product of prime polynomials. If $f(x)$ has degree 0, and is not a unit of R, then since R is a UFD, the constant $f(x)$ can be expressed as a product of primes in R. By Lemma 12.2, any prime in R is also a prime in $R[x]$. Thus, if the degree of $f(x)$ is zero, $f(x)$ is either a unit, or can be expressed as a product of primes in $R[x]$.

Now suppose $f(x)$ has positive degree. Let F be the field of quotients over R. Then $F[x]$ is a unique factorization domain by Corollary 12.5. Thus, we can write

$$f(x) = g_1(x) \cdot g_2(x) \cdot g_3(x) \cdot \cdots \cdot g_n(x),$$

where each $g_i(x)$ is irreducible in $F[x]$. For each $g_i(x)$, let c_i be the product of the denominators of all of the coefficients. Then $h_i(x) = c_i \cdot g_i(x)$ will be in $R[x]$, and we have

$$\begin{aligned} c_1 \cdot c_2 \cdot c_3 \cdot \cdots \cdot c_n \cdot f(x) &= c_1 g_1(x) \cdot c_2 g_2(x) \cdot c_3 g_3(x) \cdot \cdots \cdot c_n g_n(x) \\ &= h_1(x) \cdot h_2(x) \cdot h_3(x) \cdot \cdots \cdot h_n(x). \end{aligned}$$

Since c_i is a unit in $F[x]$, the $h_i(x)$ will still all be irreducible in $F[x]$. We can now apply Lemma 12.4 on each of the $h_i(x)$ and find an element d_i in R such that

$$h_i(x) = d_i \cdot j_i(x),$$

where the $j_i(x)$ are irreducible in $R[x]$. By Lemma 12.5, the $j_i(x)$ are prime in $R[x]$. We now can express

$$c_1 \cdot c_2 \cdot c_3 \cdot \cdots \cdot c_n \cdot f(x) = d_1 j_1(x) \cdot d_2 j_2(x) \cdot d_3 j_3(x) \cdot \cdots \cdot d_n j_n(x).$$

Let $C = c_1 \cdot c_2 \cdot c_3 \cdots c_n$ and $D = d_1 \cdot d_2 \cdot d_3 \cdots d_n$. We can then write

$$C \cdot f(x) = D \cdot j_1(x) \cdot j_2(x) \cdot j_3(x) \cdots j_n(x),$$

where C and D are in R, and the $j_i(x)$ are prime polynomials in $R[x]$. We can now apply Lemma 12.3, which states that D must be a multiple of C in R. Thus

$$f(x) = \frac{D}{C} \cdot j_1(x) \cdot j_2(x) \cdot j_3(x) \cdots j_n(x),$$

where D/C is in R. Since R is a UFD, D/C can be expressed as a product of primes in R, which by Lemma 12.2 are primes in $R[x]$. Thus, $f(x)$ can be expressed as a product of primes in $R[x]$ and so by Proposition 12.5, $R[x]$ is a UFD. □

Not only does this theorem determine when we can consider polynomial factorization to be unique, but this theorem also applies to factoring polynomials in more than one variable.

Since $R[x]$ is an integral domain, we can consider another variable y, and consider the polynomial ring $R[x][y]$. A typical element of $R[x][y]$ would be

$$c_0(x) + c_1(x)y + c_2(x)y^2 + c_3(x)y^3 + \cdots c_n(x)y^n,$$

where each $c_i(x)$ is a polynomial in $R[x]$. If each $c_i(x)$ is written

$$c_i(x) = d_0 + d_1x + d_2x^2 + d_3x^3 + \cdots$$

we find that the polynomial in $R[x][y]$ could be written

$$d_{0\,0} + d_{1\,0}x + d_{0\,1}y + d_{2\,0}x^2 + d_{1\,1}x \cdot y + d_{0\,2}y^2 + \cdots.$$

If we make the convention that $x \cdot y = y \cdot x$, we see that $R[x][y] = R[y][x]$.

DEFINITION 12.11 We will denote the polynomial ring of two variables by $R[x, y] = R[x][y]$. The variables x and y are called *indeterminates.* Likewise, we denote the polynomial ring of n indeterminates by

$$R[x_1, x_2, x_3, \ldots, x_n].$$

COROLLARY 12.6

Let R be a unique factorization domain and let $x_1, x_2, x_3, \ldots x_n$ be indeterminates over R. Then $R[x_1, x_2, x_3, \ldots x_n]$ is a unique factorization domain.

PROOF: We will use induction on n. If $n = 1$, the unique factorization domain theorem (12.3) shows that $R[x]$ is a UFD. Otherwise, we write

$$R[x_1, x_2, x_3, \ldots, x_n] = R[x_1, x_2, x_3, \ldots, x_{n-1}][x_n].$$

By the induction hypothesis, $R[x_1, x_2, x_3, \ldots, x_{n-1}]$ is a UFD. So by the unique factorization domain theorem (12.3), $R[x_1, x_2, x_3, \ldots, x_n]$ is a UFD. ▯

Polynomials in several variables are of considerable importance in geometry, since curves and surfaces are described by equations in several variables. Although *Sage*'s **factor** command will be able to factor polynomials in many variables, its ability is limited to when R is either $\mathbb{Z}$ or $\mathbb{Q}$. For example, *Sage* can factor

```
var("x y")
factor(x^3*y^2 + x^2*y - x*y^2 - 2*x + y)
    (x^2*y + 2*x - y)*(x*y - 1)
```

over the integers, but cannot factor this over any other ring, even a finite field. Fortunately, we will not have a need for factoring polynomials in two variables over any other field.

Problems for §12.2

1 Prove that $x^3 - 3x + 3$ is irreducible over the field $\mathbb{Q}$ of rational numbers.
Hint: Assume p/q is a root, and show both p and q are multiples of 3.

2 Find the factorization of $x^3 + 2x^2 + 2$ over the field Z_3.

3 Find the factorization of $x^3 + 2x^2 + 2$ over the field Z_5.

4 Find the factorization of $x^3 + 2x^2 + 2$ over the field Z_7.

5 Find the factorization of $x^4 + 2x^2 + 2$ over the field Z_5.

6 Find the factorization of $x^4 + 2x^3 + 2$ over the field Z_5.

7 Find the factorization of $x^5 + x + 1$ over the field Z_2.
Hint: If there were an irreducible quadratic factor, what would it be?

8 Find the factorization of $x^4 + 2x^3 + 2x + 2$ over the field Z_3.
Hint: First do Problem 21 of §11.1.

9 Find all of the irreducible elements of Z_{12}.
Hint: First find all of the units. Construct a multiplication table of the non-units. Which elements do not appear in the interior of the table?

10 Find all of the irreducible elements of Z_{18}. (See the hint for Problem 9.)

11 Show that the ring Z_8 has unique factorization, even though it is not an integral domain.

12 Is the ring Z_9 a unique factorization ring?

13 Let $f(x)$ be a polynomial in $\mathbb{Z}[x]$, and let p be a prime that does not divide the leading term. If the polynomial $f(x)$ **mod** $p \in Z_p[x]$ is irreducible in $Z_p[x]$, show that $f(x)$ is irreducible in $\mathbb{Z}[x]$.

14 Show that $x^3 - x + 2$ is irreducible in $\mathbb{Z}[x]$.
Hint: Use Problem 13.

15 Show that if $f(x) = g(x) \cdot h(x)$ in $\mathbb{Q}[x]$, with $g(x)$ and $h(x)$ non-constant, and $f(x) \in \mathbb{Z}[x]$, then $f(x)$ is also reducible over $\mathbb{Z}[x]$.
Hint: Find the common denominator of $g(x)$ and $h(x)$, and use Gauss's Lemma (12.2).

16 Let $f(x) \in \mathbb{Z}[x]$ be of degree d, and suppose there are $2d + 1$ integers a_i for which $f(a_i)$ is prime. Show that $f(x)$ is irreducible in $\mathbb{Z}[x]$.
Hint: If $f(x) = g(x) \cdot h(x)$, how many times can $g(x)$ or $h(x)$ be ± 1?

17 Use Problem 16 to show that $x^4 + x^2 + 41$ is irreducible in $\mathbb{Z}[x]$.

18 Show that $x^4 + 10x^2 + 2$ is irreducible in $\mathbb{Q}[x]$.
Hint: Use Problems 15 and 16.

19 Consider the subring of the elements of $\mathbb{Q}[x]$ for which the constant term is an integer. Show that this subring is not a UFD.
Hint: Show that the only units are ± 1, and that 2 is irreducible. Consider the sequence $x, x/2, x/4, x/8, \ldots x/(2^n), \ldots$

Interactive Problems

20 Define the domain $\mathbb{Z}[\sqrt{6}]$ in *Sage* as follows:

```
InitDomain(0)
AddFieldVar("a")
Define(a^2, 6)
```

Show that the element $u = 5 + 2a$ is a unit by finding its inverse. Use the element u to find yet another unit of $\mathbb{Z}[\sqrt{6}]$.

21 We can have *Sage* explore the domain $\mathbb{Z}[\sqrt[3]{2}]$:

```
InitDomain(0)
AddFieldVar("a")
Define(a^3, 2)
factor(2 + 0*a)
    a^3
```

Try factoring other standard prime numbers, to see if they factor in the new domain. Which standard primes are still primes in this new domain? Try factoring other elements in the domain to see if you always get a factorization. Note that trying this experiment on a non-UFD produces an error message. This exercise is not available in *Mathematica.*

12.3 Principal Ideal Domains

Although we have found that polynomial rings created from unique factorization domains produce more unique factorization domains, there still is the question of how to tell whether a given ring is a unique factorization domain. The answer lies in the ideals of the ring. In this section we will explore the interconnection between the ideals of a ring, and the prime and irreducible elements of the ring.

We begin by recalling that many ideals can be generated with only one element. In fact, many rings, such as the integers $\mathbb{Z}$, are such that every ideal

Historical Diversion

Ernst Kummer (1810–1893)

Kummer was a German mathematician, although he was born in what was then Prussia. He started out teaching for 10 years at a *gymnasium*, which is the German equivalent to high school. During these years, he inspired the future mathematician Leopold Kronecker.

Kummer made significant contributions to several areas of mathematics. He worked with Gauss's hypergeometric functions, and used the Maclaurin series of these functions to prove that any three such functions, whose parameters differ by integers, are linearly related. This is known as the *contiguous relations* of the hypergeometric series.

Kummer's greatest accomplishment came in an attempt to prove Fermat's last theorem. (See the Historical Diversion on page 96.) Several years earlier, Gabrial Lamé had a flawed proof of the theorem, based on the assumption that $\mathbb{Z}[\omega_n]$ had unique factorization. In the cases where $\mathbb{Z}[\omega_n]$ is a UFD, such as $n = 3$ and $n = 4$, one can prove Fermat's last theorem from

$$z^n = x^n + y^n = (x + y)(x + \omega_n y)(x + \omega_n^2 y) \cdots (x + \omega_n^{n-1} y).$$

However, Kummer had shown three years before Lamé's proof that $\mathbb{Z}[\omega_n]$ is not a UFD for $n = 23$. (It is now known that there are only a finite set of integers for which $\mathbb{Z}[\omega_n]$ is a UFD.)

Kummer had an idea of replacing elements in a domain with "ideal integers," which represented a special subring of the domain. This would later lead to the terminology of "ideals" of a ring. Kummer's plan, expressed in modern terminology, was to first prove that every non-trivial ideal can be uniquely expressed as a product of prime ideals, even if the domain was not a UFD. Since some of the ideals were not principle ideals, some of the prime ideals did not correspond to an element in the original domain. By using this "extension" of the domain, Kummer was able to prove Fermat's last theorem for most prime numbers, in particular for all primes less than 100 except 37, 59, and 67.

Because of Kummer's attempt, and partial success, in proving Fermat's last theorem, he paved the way for modern ring theory. Richard Dedekind and Emmy Noether would later use Kummer's ideal numbers to formulate the definition of the "ideal" and "prime ideal" that we use today. (See Historical Diversions on pages 336 and 306.)

Image source: Wikimedia Commons

is generated by only one element. We called such rings *principal ideal rings,* or PIRs. When the ring is also a domain, we call it a *principal ideal domain,* or PID. In fact, PIDs are so common that it is somewhat tricky to find an example of a UFD that is not a PID.

Example 12.7

Show that the ring $R = \mathbb{Z}[x, y]$ is not a PID, even though it is a UFD.

SOLUTION: By Corollary 12.6 this is a UFD. Consider the ideal of elements without a constant term. This ideal can be expressed as $\langle\{x, y\}\rangle$, that is, the ideal generated by x and y. But since both x and y are in this ideal, we cannot express this ideal as the multiples of a single polynomial. Thus, it requires at least two elements to generate this ideal in $\mathbb{Z}[x, y]$. Thus, this ideal is not a principal ideal, so $\mathbb{Z}[x, y]$ is not a PID, even though it is a UFD. ▯

DEFINITION 12.12 Let R be a commutative ring, and let P be a nontrivial ideal of R. (Thus, P is neither $\{0\}$ nor R.) We say that P is a *prime ideal* if, whenever x and y are in R, and $x \cdot y$ is in P, then either x or y is in P.

When we first defined a prime element of a ring, we were careful to mention that the ring did not have to be an integral domain. By defining prime elements for all commutative rings, we open the door to showing a connection between prime ideals and prime elements.

PROPOSITION 12.6

Let R be a commutative unity ring. Then p is a prime element of R if, and only if, the principal ideal $\langle p\rangle$ is a prime ideal.

PROOF: Suppose that p is prime. Then p is neither 0 nor a unit, so $\langle p\rangle$ cannot be the zero ring. If $\langle p\rangle = R$, then there must be some element x of R that makes $p \cdot x = 1$. But this is impossible, since p is not a unit. Thus, $\langle p\rangle$ would be a nontrivial ideal of R. Now suppose that $x \cdot y$ is in $\langle p\rangle$. Then there must be some z such that $x \cdot y = p \cdot z$. Since p is prime, either x or y is a multiple of p. So either x or y is in $\langle p\rangle$, making $\langle p\rangle$ a prime ideal.

Now suppose that $\langle p\rangle$ is a prime ideal. Then $\langle p\rangle$ is neither $\{0\}$ nor R, so p is neither 0 nor a unit. If $x \cdot y$ is a multiple of p, then $x \cdot y$ would be in $\langle p\rangle$. Since $\langle p\rangle$ is a prime ideal, either x or y would then be in $\langle p\rangle$. But this would indicate that x or y is a multiple of p. Thus, p is a prime element of R. ▯

Although this proposition refers to principal ideals, it is certainly possible for an ideal to be a prime ideal without being a principal ideal.

Example 12.8
Show that the ideal of Example 12.7 is a prime ideal.
SOLUTION: Note that we can characterize the ideal $\langle\{x, y\}\rangle$ as

$$\langle\{x, y\}\rangle = \{f(x, y) \in \mathbb{Z}[x, y] \mid f(0, 0) = 0\}.$$

Thus, if $f(x, y) \cdot g(x, y)$ is in $\langle\{x, y\}\rangle$, we have $f(0, 0) \cdot g(0, 0) = 0$, so either $f(0, 0) = 0$ or $g(0, 0) = 0$. So $\langle\{x, y\}\rangle$ is a prime ideal, even though it is not a principal ideal. ∎

Although Proposition 12.6 gives us a test for determining whether an element is prime, to implement this we need a way to see whether an ideal is a prime ideal.

PROPOSITION 12.7
Let R be a commutative unity ring, and let P be a nontrivial ideal of R. Then P is a prime ideal if, and only if, the quotient ring R/P has no zero divisors.

PROOF: Assume that P is a prime ideal. Let us suppose that the product of two elements of R/P, $a + P$ and $b + P$, is the zero element. That is,

$$(a + P) \cdot (b + P) = a \cdot b + P = 0 + P.$$

This implies that $a \cdot b$ is in P. Since P is a prime ideal, either a or b is in P. Thus, either

$$a + P = 0 + P \quad \text{or} \quad b + P = 0 + P.$$

Thus, we have shown that R/P has no zero divisors.

Now suppose that R/P has no zero divisors. If $a \cdot b$ is in P, then we have the following holding in R/P:

$$(a + P) \cdot (b + P) = a \cdot b + P = 0 + P.$$

Since R/P has no zero divisors, either $a + P$ or $b + P$ must be equal to $0 + P$. Thus, either a or b is in P, and since P is a nontrivial ideal, P is a prime ideal. ∎

Let us try to use this proposition to find the prime elements of a ring.

Computational Example 12.9
Determine whether $2a + b$ is a prime element of the ring from Example 12.5.
SOLUTION: We begin by loading the ring.

```
InitRing("a", "b")
DefineRing([4, 2], [[a, 0], [0, b]])
R = ListRing(); R
    {0*a, a, 2*a, -a, b, a+b, 2*a+b, -a+b}
```

To determine whether $2a + b$ is prime, we compute the quotient $R/\langle 2a + b\rangle$. First, we find the principal ideal generated by $2a + b$:

```
S = Ideal(R, 2*a + b); S
    {0*a, b, 2*a, 2*a+b}
```

This forms a non-trivial ideal, so we can now consider the quotient ring.

```
Q = Coset(R, S); Q
    {{0*a, b, 2*a, 2*a+b}, {a, a+b, -a, -a+b}}
Q[1]*Q[1]
    {a, a+b, -a, -a+b}
```

The quotient ring has only two elements, and in fact is isomorphic to Z_2. So $2a + b$ is a prime element of R. ∎

We are mainly interested in finding the prime elements of an *infinite* ring. *Sage* can still often help us out, since the quotient ring $R/\langle p\rangle$ will usually be finite.

Computational Example 12.10

Determine whether 3 is prime in the ring $\mathbb{Z}[\sqrt{-5}]$.

SOLUTION:

We saw in the last section that 3 was an irreducible element of this ring. First we define the ring $\mathbb{Z}[\sqrt{-5}]$ by letting $a = \sqrt{-5}$.

```
InitDomain(0)
AddFieldVar("a")
Define(a^2, -5)
```

We can now try to factor in this new ring.

```
factor(3 + 2*a)
    2*a + 3
```

This shows that $3 + 2\sqrt{-5}$ is a prime number in this domain. But when we try to factor 3,

```
factor(3 + 0*a)
    ValueError: Non-principal ideal in factorization
```

we get an error message, complaining about a non-principal ideal. Thus, we see that 3 is *not* prime in $\mathbb{Z}[\sqrt{-5}]$, even though it is irreducible.

To really see what is going on, let us construct the ring $\mathbb{Z}[\sqrt{-5}]/\langle 3\rangle$. We have to plan ahead to see that this ring has nine elements. We can let e represent the identity element $1 + \langle 3\rangle$, and a represent $\sqrt{-5} + \langle 3\rangle$. Both of these will have additive order of 3, so we can define the ring by

```
InitRing("e", "a")
DefineRing([3, 3], [[e, a], [a, -5*e]])
R = ListRing(); R
    {0*e, e, -e, a, e+a, -e+a, -a, e-a, -e-a}
(e + a)*(e - a)
    0*e
```

We find that this ring has zero divisors, so by Proposition 12.7, the element 3 is not prime in this domain. ∎

We have seen that Proposition 12.7 is a useful way of determining whether an element is prime. Let us use this proposition to show that in a principal ideal domain, irreducible elements are also prime elements. This amounts to showing that $R/\langle p\rangle$ has no zero divisors whenever p is irreducible. However, we can actually prove more, which will be very useful later on.

LEMMA 12.6

Let R be a principal ideal domain, and let p be an irreducible element of R. Then the quotient ring $R/\langle p\rangle$ is a field.

PROOF: Since R is an integral domain, it is clear that $R/\langle p\rangle$ is a commutative ring, and contains the identity element $1 + \langle p\rangle$. Thus, we have to show that all nonzero elements of $R/\langle p\rangle$ have an inverse. Let $x + \langle p\rangle$ be a nonzero element of $R/\langle p\rangle$. We immediately have that x is not a multiple of p. Thus, we can consider the ideal generated by both x and p, that is, $\langle\{x, p\}\rangle$.

Since R is a PID, there is some element d in R such that $\langle\{x, p\}\rangle = \langle d\rangle$. Then both x and p would be multiples of d. But we already observed that x is not a multiple of p, so d cannot be a multiple of p. But p is irreducible, so d must be a unit. Then $\langle d\rangle = R$, and so $\langle\{x, p\}\rangle = R$. This means that there are elements u and v in R such that

$$x \cdot u + p \cdot v = 1.$$

We now claim that $u + \langle p\rangle$ is our sought-after inverse. Note that

$$(x + \langle p\rangle) \cdot (u + \langle p\rangle) = x \cdot u + \langle p\rangle = x \cdot u + p \cdot v + \langle p\rangle = 1 + \langle p\rangle.$$

Since every nonzero element of $R/\langle p\rangle$ is invertible, we have that $R/\langle p\rangle$ is a field. ∎

From this lemma, it is easy to see that an irreducible element of a PID must also be a prime element. Thus, we are on our way to showing that a PID is a unique factorization domain. By Proposition 12.5, we only need to show that every non-invertible element can be expressed as a product of irreducible factors. In order to eliminate the possibility of an "infinite chain" of irreducible elements, each one dividing the previous element, we will use the following lemma.

LEMMA 12.7
Let R be a principal ideal ring. If there is an infinite sequence of larger and larger ideals of R satisfying

$$I_1 \subseteq I_2 \subseteq I_3 \subseteq \cdots \subseteq I_n \subseteq I_{n+1} \subseteq \cdots,$$

then there exists an integer m such that $I_n = I_m$ for all $n > m$.

PROOF: Since we have an infinite sequence of ideals, we can consider taking the union of all of them:

$$I = \bigcup_{n=1}^{\infty} I_n.$$

Let us show that I is an ideal of R. Note that any element of I is in I_k for some integer k. In fact, if x and y are two elements of I, we can pick the larger of the two values of k to show that x and y are both in I_k. Then $x \pm y$ is in I_k, since I_k is an ideal. Thus $x \pm y$ is in I. This shows that I is a subgroup of R under addition. Now let z be in R. Then $x \cdot z$ and $z \cdot x$ are both in I_k, so $x \cdot z$ and $z \cdot x$ are in I. Therefore, $I \cdot R = R \cdot I = I$. This shows that I is an ideal.

Since R is a principal ideal ring, there is some element a in I such that $I = \langle a \rangle$. Then a is in I_m for some m. But I_m is contained in I, so we must have that $I = I_m$. Thus, $I_n = I_m$ for all $n > m$. ∎

We now have all we need to show that a PID is in fact a UFD.

THEOREM 12.4: The Principal Ideal Domain Theorem
Every principal ideal domain is a unique factorization domain.

PROOF: Our strategy is to first show that an irreducible element is a prime element, and then show that every element is a finite product of irreducible elements. Let p be an irreducible element of R, which is a PID. By Lemma 12.6 $R/\langle p \rangle$ is a field, so it certainly has no zero divisors. Thus, by Proposition 12.7, $\langle p \rangle$ is a prime ideal, so by Proposition 12.6, p is prime. Let us now show that every non-invertible element of R can be written as a product of irreducible elements. Suppose this is not true for some element x_0. Then x_0 is not irreducible, so we can find elements x_1 and y_1 in R such that $x_1 \cdot y_1 = x_0$. But x_1 and y_1 cannot both be irreducible, so we can assume x_1 is reducible. By induction we can continue this process to form a sequence

$$\{x_0, x_1, x_2, x_3, \cdots\}$$

for which each term in the sequence divides the previous term. Then we have an infinite chain of strictly increasing ideals,

$$\langle x_0 \rangle \subset \langle x_1 \rangle \subset \langle x_2 \rangle \subset \langle x_3 \rangle \subset \cdots.$$

By Lemma 12.7, such an infinite chain of ideals is impossible in a PID. This contradiction shows that every element of R can be expressed as a product of irreducible elements, which in turn are prime. By Proposition 12.5, R is a unique factorization domain. ∎

This theorem reveals the most important use of principal ideal domains—it enables us to find unique factorization domains. For example, $\mathbb{Z}$ was proven to be a PID from Proposition 10.3, so we now can see that $\mathbb{Z}$ is a UFD, a result that was proven in §0.1.

It should be noted that not all unique factorization domains are PIDs—in fact we discovered that $\mathbb{Z}[x, y]$ is not a PID, even though it is a UFD. However, many of the important unique factorization domains are also principal ideal domains.

Of course, there still is the problem of how to determine whether an integral domain is a PID. In the next section, we will find the main way of determining whether a certain domain is in fact a PID, which would then prove that it is a UFD.

Problems for §12.3

1 Show that $\mathbb{Z}[\sqrt{-5}]$ is not a principal ideal domain by finding an ideal of this ring that is not a principal ideal.

Hint: Consider the ideal $\langle\{2, 1 + \sqrt{-5}\}\rangle$.

2 Find all of the prime ideals of Z_{12}. (Note that this ring has zero divisors.)

3 Find all of the prime elements of Z_{12}. (Note that this ring has zero divisors. See Problem 2.)

4 Find all of the prime ideals of Z_{18}.

5 Find all of the prime elements of Z_{18}. (See Problem 4.)

6 Can a field have irreducible or prime elements? Explain.

7 We say that an ideal I of a ring R is a *maximal ideal* if $I \neq R$, and the only ideals containing I are I and R. Show that the prime ideals of $\mathbb{Z}$ are also maximal ideals.

8 Let R be a commutative unity ring. Show that if I is a maximal ideal, then R/I is a field. See Problem 7.

Hint: For $b \in R$, consider the ideal $\{br + a \mid r \in R,\ a \in I\}$.

9 Let R be a commutative unity ring, and I an ideal of R. Show that if R/I is a field, then I is a maximal ideal. See Problem 7.

10 Let R be a commutative ring with identity. Show that if I is a maximal ideal, then I is a prime ideal. See Problems 7 and 8.

11 Let $R = \mathbb{Z}[x]$, and let I be the ideal of polynomials for which $f(0)$ is even. Show that I is a maximal ideal. See Problems 7 and 9.

12 Let $R = \mathbb{Z}[x]$, and let I be the ideal of polynomials for which $f(0) = 0$. Show that I is a prime ideal, but not a maximal ideal. See Problems 7 and 11.

13 Show that for a *finite* commutative unity ring, every prime ideal is a maximal ideal. See Problems 7 and 9.

14 Show that for a PID, every prime ideal is a maximal ideal. See Problems 7 and 9.

15 Find a ring with exactly two maximal ideals. See Problem 7.

16 Let R be a non-trivial commutative unity ring with no zero divisors. Prove that if every nontrivial ideal of R is a prime ideal, then R is a field.
Hint: If x is an element of R, show that x is contained in x^2R.

17 Let R be a commutative ring, and let I be an ideal of R. If P is a prime ideal of I, prove that P is an ideal of R.

18 Let R be a PID. Prove that every element that is neither 0 nor a unit is divisible by some prime element.

Interactive Problems

19 Use *Sage* to show that the ring $\mathbb{Z}[\sqrt{6}]/\langle 11\rangle$ has no zero divisors. Use this to prove that 11 is a prime element of $\mathbb{Z}[\sqrt{6}]$.

20 We saw in Example 12.10 that 3 is not a prime number in $\mathbb{Z}[\sqrt{-5}]$. Find a prime in the ordinary sense that is also prime in this ring, showing that there are prime elements. What other primes can you find?

12.4 Euclidean Domains

We have already seen the importance of principal ideal domains to determine whether a ring is a unique factorization domain. However, we still have the problem of determining whether a given integral domain is a principal ideal domain. In this section we will develop the standard method for proving that a ring is a principal ideal domain, using the idea of a *division algorithm*.

Example 12.11

Show that $F[x]$ is a PID for any field F.

SOLUTION: We examine what the ideals could be. If I is a nontrivial ideal of $F[x]$, we can find a nonzero element $f(x)$ in I with the lowest degree. If $g(x)$ is also in I, then by the division algorithm

$$g(x) = f(x) \cdot q(x) + r(x),$$

with the degree of $r(x)$ less than $f(x)$. But $r(x)$ would also be in I, and since $f(x)$ has least degree of all the nonzero elements in I, we must have $r(x) = 0$. Therefore all elements of I are multiples of $f(x)$, so $I = \langle f(x) \rangle$. □

Rather than making this a formal proposition, we want to study this example, since we can prove that *many* different domains are PIDs the same way. There were two keys to the proof that $F[x]$ was a PID: the fact that every polynomial had a degree, and the division algorithm. Whenever we have an integral domain that has a property like a division algorithm, there is a good chance that we can use this division algorithm to prove that the ring is a PID. Let us formulate what we mean by a "division algorithm."

DEFINITION 12.13 An integral domain R is called a *Euclidean domain* if there is a function $\mu(x)$ defined on the nonzero elements of R such that the following three properties hold:

1. $\mu(x)$ is a non-negative integer for every nonzero x in R.
2. Whenever both x and y are nonzero, $\mu(x \cdot y) \geq \mu(x)$.
3. For any x and y in R, with y nonzero, there exist elements q and r in R such that
$$x = q \cdot y + r,$$
where either $r = 0$ or $\mu(r) < \mu(y)$.

The function $\mu(x)$ is called the *Euclidean valuation* on R.

Let us first look at some examples of Euclidean domains.

Example 12.12

Since this definition was modeled after the ring $F[x]$, it is expected that $F[x]$ is a Euclidean domain. The function $\mu(f(x))$ would be the degree of the polynomial $f(x)$. Properties 1 and 2 come from the definition of the degree, and Lemma 11.1. Property 3 we observed in the division algorithm theorem (12.1). Thus, $F[x]$ is a Euclidean domain whenever F is a field. ∎

However, there are many other examples of Euclidean domains.

Example 12.13

Consider the set of integers, $\mathbb{Z}$. We can use the absolute value for the valuation: $\mu(x) = |x|$. Clearly properties 1 and 2 hold, and the third property comes from Theorem 0.1. Thus, $\mathbb{Z}$ is also a Euclidean domain. ∎

Whenever we have a Euclidean domain, we can prove that the domain is a PID, using the exact same argument as we did for $F[x]$.

THEOREM 12.5: The Euclidean Domain Theorem

Every Euclidean domain is a principal ideal domain.

PROOF: Let R be a Euclidean domain, and let $\mu(x)$ be the valuation. If I is an ideal, we consider the set

$$P = \{\mu(x) \mid x \in I, x \neq 0\}.$$

The set P consists of non-negative integers, so there is a smallest number in P. Pick an element y in I so that $\mu(y)$ is the minimal number in P. Then for any other x in I, we have

$$x = y \cdot q + r$$

for some q and r in R, with $\mu(r) < \mu(y)$ or $r = 0$. Then r is in I, but if r were nonzero, then this would contradict the minimality of $\mu(y)$. Thus, $r = 0$, and so x is a multiple of y. Since this is true for all x in I, we see that $I = \langle y \rangle$. Thus, every ideal of R is a principal ideal, so R is a PID. ∎

We started this section by showing that $F[x]$ is a principal ideal ring whenever F is a field, but let us formally make this a corollary of the Euclidean domain theorem.

COROLLARY 12.7

Let F be a field. Then the ring of polynomials $F[x]$ is a principal ideal domain.

PROOF: We have already seen in Example 12.12 that $F[x]$ is a Euclidean domain whenever F is a field. By the Euclidean domain theorem (12.5), $F[x]$ is a PID. ∎

The only problem with this definition of the Euclidean domain is that it gives no help in determining what the valuation function $\mu(x)$ should be. In fact, there may be many possible valuation functions for a given integral domain. See Problem 1 for an alternative definition of a Euclidean domain that does not involve a valuation function.

For the remainder of this chapter, we will consider the problem of determining whether some quadratic domains are Euclidean domains. This class of domains will help us to see some general techniques for finding a valuation function for a domain. We have already seen that $\mathbb{Z}[\sqrt{-5}]$ is not a UFD, so this clearly is not a Euclidean domain.

We saw before that $\mathbb{Z}[\sqrt{2}]$ had two automorphisms, and in general the quadratic domain $\mathbb{Z}[\sqrt{n}]$ will have two automorphisms, the identity mapping, and the automorphism

$$f(x + y\sqrt{n}) = x - y\sqrt{n}.$$

We define the function N as the product of the two automorphisms:

$$N(x + y\sqrt{n}) = (x + y\sqrt{n}) \cdot (x - y\sqrt{n}) = x^2 - y^2 n.$$

Note that $N(a)$ will always be an integer.

At first glance it may be difficult to see what the $N(a)$ has to do with the Euclidean domains. Our goal is to construct a valuation function from $N(a)$. We first need to verify some elementary properties of this function. In the process, we will notice that these properties are still valid if we extend $N(a)$ to be defined on $\mathbb{Q}[\sqrt{n}]$.

LEMMA 12.8

Let $\mathbb{Z}[\sqrt{n}]$ be a quadratic domain, and let $N(x + y\sqrt{n}) = x^2 - y^2 n$. Then for the rings $\mathbb{Z}[\sqrt{n}]$ and $\mathbb{Q}[\sqrt{n}]$,

1. *$N(a) = 0$ if, and only if, $a = 0$.*
2. *$N(a \cdot b) = N(a) \cdot N(b)$.*
3. *$N(\pm 1) = 1$.*

PROOF:

1. It is easy to see that $N(0) = 0$ by definition. If $N(x + y\sqrt{n}) = 0$, then

$$(x + y\sqrt{n}) \cdot (x - y\sqrt{n}) = x^2 - y^2 n = 0.$$

If y is nonzero, then we find that $\sqrt{n} = |\frac{x}{y}|$, which is ridiculous since n is not a perfect square, and so $\sqrt{n}$ is irrational. Thus, $y = 0$, and hence x is also 0. So $N(a) = 0$ if, and only if, $a = 0$.

2. A quick computation shows that if $a = x_1 + y_1\sqrt{n}$, and $b = x_2 + y_2\sqrt{n}$, then

$$a \cdot b = (x_1 + y_1\sqrt{n}) \cdot (x_2 + y_2\sqrt{n}) = (x_1 \cdot x_2 + y_1 \cdot y_2 \cdot n) + (x_1 \cdot y_2 + y_1 \cdot x_2)\sqrt{n}.$$

So

$$\begin{aligned} N(a \cdot b) &= (x_1 \cdot x_2 + y_1 \cdot y_2 \cdot n)^2 - (x_1 \cdot y_2 + y_1 \cdot x_2)^2 \cdot n \\ &= x_1^2 x_2^2 + 2x_1 x_2 y_1 y_2 n + y_1^2 y_2^2 n^2 - x_1^2 y_2^2 n - 2x_1 x_2 y_1 y_2 n - y_1^2 x_2^2 n \\ &= x_1^2 x_2^2 + y_1^2 y_2^2 n^2 - x_1^2 y_2^2 n - y_1^2 x_2^2 n \\ &= (x_1^2 - y_1^2 n) \cdot (x_2^2 - y_2^2 n) = N(a) \cdot N(b). \end{aligned}$$

3. This is easy, since $\pm 1 = \pm 1 + 0\sqrt{n}$. So $N(\pm 1) = (\pm 1)^2 - 0 \cdot n = 1$. ∎

We can use the $N(a)$ function to prove that $\mathbb{Q}[\sqrt{n}]$ is a field.

COROLLARY 12.8

Let n be an integer that is not divisible by the square of any integer greater than 1. Then the ring $\mathbb{Q}[\sqrt{n}]$ is a field.

PROOF: Since $\mathbb{Q}[\sqrt{n}]$ is obviously a commutative ring with an identity, all we need to show is that every nonzero element has an inverse. Let $b = x + y\sqrt{n}$ be a nonzero element. Then $N(b)$ is nonzero by Lemma 12.8. Consider the element

$$c = (x - y\sqrt{n})/N(b).$$

Then

$$b \cdot c = (x + y\sqrt{n}) \cdot (x - y\sqrt{n})/N(b) = N(b)/N(b) = 1.$$

So every nonzero element has an inverse. Thus, $\mathbb{Q}[\sqrt{n}]$ is a field. ∎

Using the three properties of the *norm function* $N(a)$, we are able to determine at least some of the irreducible elements of the ring $\mathbb{Z}[\sqrt{n}]$.

PROPOSITION 12.8

Let $\mathbb{Z}[\sqrt{n}]$ be a quadratic domain, and let $N(x + y\sqrt{n}) = x^2 - y^2 n$. Then

1. *$N(a) = \pm 1$ if, and only if, a is a unit in $\mathbb{Z}[\sqrt{n}]$.*
2. *If $N(a)$ is a prime number in $\mathbb{Z}$, then a is an irreducible element of $\mathbb{Z}[\sqrt{n}]$.*

PROOF: Suppose that $N(a) = N(x + y\sqrt{n}) = \pm 1$. Consider the element

$$b = (x - y\sqrt{n})/N(a).$$

Then

$$a \cdot b = (x + y\sqrt{n}) \cdot (x - y\sqrt{n})/N(a) = N(a)/N(a) = 1.$$

So a has an inverse, and therefore is a unit in $\mathbb{Z}[\sqrt{n}]$.

Now suppose that a is a unit in $\mathbb{Z}[\sqrt{n}]$. Then a has an inverse, a^{-1}. Then

$$1 = N(1) = N(a \cdot a^{-1}) = N(a) \cdot N(a^{-1}),$$

which shows that $N(a)$ must be ± 1.

Now suppose that $N(a) = p$, a prime number in $\mathbb{Z}$, and that $a = b \cdot c$. Then

$$p = N(a) = N(b \cdot c) = N(b) \cdot N(c).$$

Since p is prime, either $N(b)$ or $N(c)$ is ± 1. So either b or c must be a unit in $\mathbb{Z}[\sqrt{n}]$, so a is irreducible in $\mathbb{Z}[\sqrt{n}]$. □

We can now use the Euclidean valuation function $\mu(x) = |N(x)|$ to prove the following.

PROPOSITION 12.9

The integral domains $\mathbb{Z}[\sqrt{-2}]$, $\mathbb{Z}[\sqrt{-1}]$, $\mathbb{Z}[\sqrt{2}]$, and $\mathbb{Z}[\sqrt{3}]$ are Euclidean domains.

PROOF: Let us work with all four domains at the same time by considering $\mathbb{Z}[\sqrt{n}]$, where $n = -2, -1, 2$, or 3.

If we let $\mu(x) = |N(x)|$, then clearly $\mu(x)$ is a non-negative integer. Furthermore, $\mu(x) = 0$ only when $x = 0$. Thus, if u and v are two non-zero elements of $\mathbb{Z}[\sqrt{n}]$, then

$$\mu(u \cdot v) = |N(u \cdot v)| = |N(u)| \cdot |N(v)| = \mu(u) \cdot \mu(v) \geq \mu(u) \cdot 1 = \mu(u).$$

So the first two conditions for the valuation function are easily satisfied. The last condition is harder to prove. We need to show that for any x and y in $\mathbb{Z}[\sqrt{n}]$, with y nonzero, there are elements q and r such that

$$x = q \cdot y + r,$$

with either $r = 0$, or $\mu(r) < \mu(y)$. We can consider x and y to be in $\mathbb{Q}[\sqrt{n}]$, which is a field from Corollary 12.8, so we can compute

$$t = x \cdot y^{-1} = u + v\sqrt{n}.$$

Of course, t will be in $\mathbb{Q}[\sqrt{n}]$ instead of $\mathbb{Z}[\sqrt{n}]$, so we cannot use this for our q. However, we can find an element "closest" to t in $\mathbb{Z}[\sqrt{n}]$ by finding the integers p and k nearest to u and v. That is, we will select integers p and k such that

$$(*) \qquad |p - u| \leq \frac{1}{2} \qquad \text{and} \qquad |k - v| \leq \frac{1}{2}.$$

We now let $q = p + k\sqrt{n}$, which is in $\mathbb{Z}[\sqrt{n}]$. The remainder r would be given by $q \cdot y - x$. All we need to do is show that $r = 0$, or $\mu(r) < \mu(y)$.

Now, the norm $N(x)$ is valid on $\mathbb{Q}[\sqrt{n}]$, so we can compute

$$N(q-t) = N\left((p-u) + (k-v)\sqrt{n}\right) = (p-u)^2 - n(k-v)^2.$$

By $(*)$ we see that if $n > 0$,

$$-n/4 \le (p-u)^2 - n(k-v)^2 \le 1/4.$$

On the other hand, if $n < 0$, then

$$0 \le (p-u)^2 - n(k-v)^2 \le (1-n)/4.$$

Thus, as long as $-2 \le n \le 3$ we have that

$$|N(q-t)| = |(p-u)^2 - n(k-v)^2| \le 3/4 < 1.$$

Thus,

$$\begin{aligned}\mu(r) &= |N(r)| = |N(q\cdot y - x)| \\ &= |N((q - x\cdot y^{-1})\cdot y)| \\ &= |N(q-t)|\cdot|N(y)| \\ &< |N(y)| = \mu(y).\end{aligned}$$

Therefore, the function $\mu(x)$ serves as a valuation function on $\mathbb{Z}[\sqrt{n}]$, and so $\mathbb{Z}[\sqrt{n}]$ is a Euclidean domain for $n = -2, -1, 2$, or 3. ∎

One of these four domains has special applications. The ring $\mathbb{Z}[\sqrt{-1}] = \mathbb{Z}[i]$ is called the domain of *Gaussian integers. Sage*'s **factor** command can find the prime factorization over the Gaussian integers by first defining the field of complex rational numbers.

```
InitDomain(0)
AddFieldVar("i")
Define(i^2, -1)
```

For example, we can factor the number 5 as follows:

```
factor(5 + 0*i)
    (-i - 2) * (i - 2)
```

Notice that we added $0 \cdot i$ to the integer so that *Sage* will interpret it as an element of $\mathbb{Z}[i]$, and not just an integer. The two factors $-i-2$ and $i-2$ are both prime in $\mathbb{Z}[i]$. By investigating further the divisibility properties of $\mathbb{Z}[i]$, one can prove the classic "two squares theorem" of Fermat: Every prime number of the form $4n+1$ is the sum of two squares. (See Problem 10 of §13.2.) It is interesting that the study of domains other than the familiar integers yields new information about the integers.

We can also explore factorizations in the other Euclidean domains found in Proposition 12.9. For example, the domain $\mathbb{Z}[\sqrt{-2}]$ can be set up with the commands:

```
InitDomain(0)
AddFieldVar("a")
Define(a^2, -2)
```

Then we can factor numbers such as

```
factor(3 + 0*a)
    (-1) * (a - 1) * (a + 1)
```

So 3 is not prime in $\mathbb{Z}[\sqrt{-2}]$. Note that this time, a unit was included in the factorization.

The domain $\mathbb{Z}[\sqrt{2}]$ is even stranger, for there is an infinite number of units.

```
InitDomain(0)
AddFieldVar("a")
Define(a^2, 2)
```

Although a is not a unit (it is prime), we find that $1 + a$ is.

```
1/(1+a)
    a - 1
```

Thus, the sequence

```
(1+a)^2
    2*a + 3
(1+a)^3
    5*a + 7
(1+a)^4
    12*a + 17
```

produces an infinite number of units. In this domain, 3 is prime, but 2 is not, since $2 = a^2$.

Since every Euclidean domain is a PID, the natural question to ask is whether there is a PID that is *not* a Euclidean domain. There actually are such domains, although known examples are rare. The simplest example is $\mathbb{Z}[(1 + \sqrt{-19})/2]$, but it is tricky to prove that this example works, for two reasons. First of all, to show that this ring is *not* a Euclidean domain, we must show that no valuation function $\mu(x)$ can be defined whatsoever. Problem 1 gives an alternative way to define a Euclidean domain that does not depend on a valuation function, and hence helps in showing that $\mathbb{Z}[(1 + \sqrt{-19})/2]$ is not a Euclidean domain. But then we must show that this ring is still a PID, which is especially hard since the main tool for proving that a domain is a PID is the Euclidean domain theorem (12.5). For a sketch of how this is proven, see Problems 15 to 22. A similar proof can be used to show that $\mathbb{Z}[(1 + \sqrt{-43})/2]$, $\mathbb{Z}[(1 + \sqrt{-67})/2]$, and $\mathbb{Z}[(1 + \sqrt{-163})/2]$ are PIDs, but not Euclidean domains.

Problems for §12.4

1 Suppose that R is an integral domain. Let S_0 be the set containing all units of R, along with the zero element. Let S_1 be the set of all elements x such that either $x = 0$ or

$$\langle x \rangle + S_0 = R.$$

(That is, every element of R can be written as a multiple of x plus an element of S_0.) Define S_i inductively as the set of elements x such that either $x = 0$ or

$$\langle x \rangle + S_{i-1} = R.$$

Prove that R is a Euclidean domain if, and only if, every element of R is in S_n for some n. This result is sometimes referred to as Motzkin's lemma.
Hint: Let $\mu(x)$ be the smallest value of n for which x is in S_n.

2 Show that the elements q and r in part 3 of the definition of a Euclidean domain are not necessarily unique.
Hint: In $\mathbb{Z}[i]$, let $x = -4 + i$, $y = 5 + 3i$. Consider $q = -1 + i$ and $q = -1$.

3 Let D be a Euclidean domain, and let μ be the valuation function. Show that $u \neq 0$ is a unit in D if, and only if, $\mu(u) = \mu(1)$.

4 Let D be a Euclidean domain, and let μ be the valuation function. Show that if a and b are associates, then $\mu(a) = \mu(b)$.

5 Show that $\mathbb{Z}[\sqrt{-6}]$ is not a unique factorization domain.
Hint: Factor 10 in two ways.

6 Prove that 7 is prime in $\mathbb{Z}[\sqrt{6}]$.
Hint: First show that $x^2 - 6y^2 \equiv 0 \pmod 7$ only when x and y are both $0 \pmod 7$.

7 Show that if $n \equiv 3 \pmod 4$, then n cannot be expressed as the sum of two square integers.

8 If $a^2 + b^2$ is a prime number in the ordinary sense, prove that $a + bi$ is a prime number in the domain $\mathbb{Z}[i]$.
Hint: Use Proposition 12.8.

9 If $p = a^2 + b^2$ is a prime number in the ordinary sense, find the prime factorization of p in the domain $\mathbb{Z}[i]$. (See Problem 8.)

10 Let $p > 0$ be a prime number in the ordinary sense. Show that p factors in the larger domain $\mathbb{Z}[i]$ if, and only if, there are two integers a and b for which $p = a^2 + b^2$. (See Problem 9.)

11 Suppose that n is an integer for which $\sqrt{4n+1}$ is irrational. Let

$$q = \frac{1+\sqrt{4n+1}}{2},$$

and consider the domain $\mathbb{Z}[q] = \{x + yq \mid x, y \in \mathbb{Z}\}$. Define the function $N(a)$ on $\mathbb{Z}[q]$ by

$$\begin{aligned} N(x+yq) &= \left(x + y\left(\frac{1+\sqrt{4n+1}}{2}\right)\right) \cdot \left(x + y\left(\frac{1-\sqrt{4n+1}}{2}\right)\right) \\ &= x^2 + xy - ny^2. \end{aligned}$$

Show that $N(x)$ satisfies the properties of Lemma 12.8, that is, $N(a) = 0$ if, and only if, $a = 0$, $N(a \cdot b) = N(a) \cdot N(b)$, and $N(\pm 1) = 1$. These domains are called *semi-quadratic domains.*

12 Prove Proposition 12.8 for the semi-quadratic domains $\mathbb{Z}[q]$ of Problem 11.

13 Show that $\mathbb{Z}[(1+\sqrt{-3})/2]$ is a Euclidean domain. This is the ring of *Eulerian integers.* (See Problems 11 and 12.)
Hint: Use the same trick used in Proposition 12.9. Since $\mathbb{Q}[q] = \mathbb{Q}[\sqrt{-3}]$ is a field by Corollary 12.8, we can find $t = x \cdot y^{-1} = u + vq$ in $\mathbb{Q}[q]$, and then round u and v to the nearest integer to find an element in $\mathbb{Z}[q]$.

14 Show that $\mathbb{Z}[(1+\sqrt{5})/2]$ is a Euclidean domain. This ring is called the *Golden ratio domain.* (See the hint for Problem 13.)

15 Show that the only units of $\mathbb{Z}[(1+\sqrt{-19})/2]$ are ± 1.
Hint: Use Problems 11 and 12 with $n = -5$.

16 Show that 2 and 3 are prime numbers in $\mathbb{Z}[(1+\sqrt{-19})/2]$.
Hint: Use Problems 11 and 12. When can $x^2 + xy + 5y^2$ be even or a multiple of 3?

17 Use Problem 1 to show that $\mathbb{Z}[(1+\sqrt{-19})/2]$ is *not* a Euclidean domain.
Hint: Use Problems 15 and 16 to show that $S_1 = S_0$, and hence $S_i = S_0$ for all i.

18 For every complex number z, show that there is a $x \in \mathbb{Z}[(1+\sqrt{-19})/2]$ such that $|\mathrm{Re}(z-x)| \le 1/2$ and $0 \le \mathrm{Im}(z-x) \le \sqrt{19}/2$.
Hint: First find an x for which $0 \le \mathrm{Im}(z-x) \le \sqrt{19}/2$, then add an integer to x to get $|\mathrm{Re}(z-x)| \le 1/2$.

19 For every complex number z, show that there is a $y \in \mathbb{Z}[(1+\sqrt{-19})/2]$ such that either $|z-y| < 1$ or $|2z-y| < 1$.
Hint: First pick a y using Problem 18, and draw a picture in the complex plane to show where y could be. Show that three circles of radius 1 centered at $(1 \pm \sqrt{-19})/2$ and 0, and two circles of radius 1/2 centered at $(1 \pm \sqrt{-19})/4$ cover this region.

20 Let I be an ideal of $R = \mathbb{Z}[(1+\sqrt{-19})/2]$, and let m be a nonzero element of I for which $N(m)$ is as small as possible. (See Problems 11 and 12 for the definition of $N(m)$.) Show that if $x \in I$, then there is a $y \in R$ such that $2x = my$.

Hint: Let $z = m^{-1}x \in \mathbb{Q}[\sqrt{-19}]$. We can extend the $N(x)$ function to $\mathbb{Q}[\sqrt{-19}]$, so Problem 19 shows that there is a $y \in R$ for which $N(m^{-1}x - y) < 1$ or $N(2m^{-1}x - y) < 1$.

21 Let I be an ideal of $R = \mathbb{Z}[(1+\sqrt{-19})/2]$, and let $m \in I$, $m \neq 0$ have minimum $N(m)$ as in Problem 20. Show that if $x \in I$, but $x \notin \langle m \rangle$, then m is a multiple of 2, and that $x = (m/2)y$ for some $y \in R$ that is not a multiple of 2.

Hint: Problem 16 shows that 2 is prime in R.

22 Show that $\mathbb{Z}[(1+\sqrt{-19})/2]$ is a PID.

Hint: Use Problem 21 to show that if I is an ideal that is not a principal ideal, and m is the element of I with the least nonzero $N(m)$, then $(m/2)y\bar{y} \in I$, and hence $m/2 \in I$, but $N(m/2) < N(m)$.

Interactive Problems

23 Use the *Sage* command

```
InitDomain(0)
AddFieldVar("i")
Define(i^2, -1)
factor(2 + 0*i)
```

to determine whether 2 is prime in the domain $\mathbb{Z}[i]$. Try this using the numbers 3, 5, 7, 11, 13, 17, 19, 23, 29, and 31 in place of 2. Which of these numbers are prime in the domain $\mathbb{Z}[i]$? Can you find a pattern?

24 We saw that $\mathbb{Z}[\sqrt{2}]$ is a Euclidean domain, and that 3 is prime. What other primes in the ordinary sense are prime in this ring? Can you find a pattern?

Chapter 13

Finite Division Rings

As we begin to study the properties of fields, it is reasonable to start by looking at *finite* fields. Finite fields are much easier to visualize, since we can display the addition and multiplication tables to find patterns. Also, finite fields are completely understood. We can classify all finite fields in terms of their size. Finally, finite fields have many applications, playing a key role in the classification of finite simple groups, and also in error corrections codes such as the ones used for compact disks.

13.1 Entering Finite Fields in *Sage*

In order to experiment with finite fields to discover patterns, we need to understand how to describe a finite field in terms that a computer program could understand. This process will later be generalized to infinite fields, as we explore *field extensions* in the next chapter. In fact, fields have special properties that allow for shortcuts in the process of entering them into *Sage*.

We have already seen several examples of finite fields. The first example was the discovery that whenever p is prime, the ring Z_p forms a field with p elements. In §11.1 we found another example of a finite field—the "complex numbers modulo 3." This ring was defined in *Sage* with the commands

```
InitDomain(3)
AddFieldVar("i")
Define(i^2, -1)
K = ListField(); K
    {0, 1, 2, i, i + 1, i + 2, 2*i, 2*i + 1, 2*i + 2}
```

Motivational Example 13.1

Find a connection between the field K and the polynomials in $Z_3[x]$.

SOLUTION: Notice that the polynomial $x^2 + 1$ is irreducible in $Z_3[x]$.

```
InitDomain(3, "x")
factor(x^2 + 1)
    x^2 + 1
```

Each element of the field K can be thought of as evaluating some polynomial in $Z_3[x]$ at $x = i$. Even though i is not an element of Z_3, we can consider any polynomial in $Z_3[x]$ as being also a polynomial in $K[x]$. This suggests that we should use the evaluation homomorphism

$$\phi_i : K[x] \to K.$$

However, we can restrict this homomorphism to apply only to polynomials in $Z_3[x]$.

$$\phi'_i : Z_3[x] \to K.$$

The image will still be all of K, since ϕ'_i sends the polynomial x to i. The kernel of this homomorphism will consist of all polynomials in $Z_3[x]$ that yield 0 when evaluated at $x = i$. For example, $x^2 + 1$ is in the kernel, as are all multiples of $x^2 + 1$. In fact, if $f(x)$ is an element of the kernel, then $\gcd(f(x), x^2 + 1)$ must be in the kernel, and $x^2 + 1$ is irreducible in $Z_3[x]$. Thus, the kernel must be precisely the multiples of $x^2 + 1$. This ideal can be described as $\langle x^2 + 1\rangle$, the ideal generated by $x^2 + 1$.

By the first ring isomorphism theorem (10.2), we now have that

$$K \approx Z_3[x]/\langle x^2 + 1\rangle$$

since the field K is the image of the homomorphism ϕ'_i. □

We can try a similar process to produce other fields.

Computational Example 13.2

Find a field of order 25.

SOLUTION: Recall that we tried to form a field by extending Z_5 by an element i, where $i^2 = -1$. However, we failed to produce a field, since the ring had zero divisors. We succeeded in producing the ring

$$K \approx Z_5[x]/\langle x^2 + 1\rangle$$

but $x^2 + 1$ factors in Z_5: $(x + 2)(x + 3)$. This factorization apparently causes the zero divisors to appear in the quotient ring. Perhaps we should try using a polynomial that is irreducible in Z_5.

We first define Z_5 in *Sage*:

```
InitDomain(5, "x")
Z5 = ListField(); Z5
    {0, 1, 2, 3, 4}
```

Next, we find a polynomial that is irreducible in Z_5.

```
factor(x^2 + 2*x + 3)
    x^2 + 2*x + 3
```

So x^2+2x+3 is irreducible over Z_5. To find a new field for which x^2+2x+3 has a zero, we will denote one of the zeros by the letter w. Then it is clear that $w^2 = -2w - 3$, so we can enter this into *Sage*.

```
AddFieldVar("w")
Define(w^2, -2*w - 3)
```

Sage can now generate the ring containing w.

```
H = ListField(); H
    {0, 1, 2, 3, 4, w, w + 1, w + 2, w + 3, w + 4, 2*w, 2*w + 1,
    2*w + 2, 2*w + 3, 2*w + 4, 3*w, 3*w + 1, 3*w + 2, 3*w + 3,
    3*w + 4, 4*w, 4*w + 1, 4*w + 2, 4*w + 3, 4*w + 4}
```

The ring formed has 25 elements, and the fact that *Sage* was able to form the ring this way proves that it is a field. ∎

As in the case of $Z_3[x]/\langle x^2+1\rangle$, we can describe this field as

$$Z_5[x]/\langle x^2+2x+3\rangle.$$

Thus we have found a way to form fields out of polynomial rings.

PROPOSITION 13.1
Let F be a field, and let $f(x)$ be an irreducible polynomial of $F[x]$. Then $F[x]/\langle f(x)\rangle$ is a field that contains F as a subfield.

PROOF: Since F is a field, by Corollary 12.7, $F[x]$ is a principal ideal domain. Since $f(x)$ is an irreducible element of $F[x]$, we have by Lemma 12.6 that the quotient $H = F[x]/\langle f(x)\rangle$ is a field.

Finally, we need to show that the field H contains F as a subfield. Consider the mapping $\phi : F \to H$ given by

$$\phi(y) = y + \langle f(x)\rangle.$$

This is certainly a homomorphism, since it is a restriction of the natural homomorphism from $F[x]$ to $F[x]/\langle f(x)\rangle$. The kernel of ϕ is just 0, so the image is isomorphic to F. Thus, $F[x]/\langle f(x)\rangle$ contains F as a subfield. ∎

DEFINITION 13.1 The field formed in Proposition 13.1 is called the *extension field of K through the irreducible polynomial $f(x)$.*

The next step is to determine the size of this new field.

PROPOSITION 13.2
Let p be a prime number, and let $A(x)$ be an irreducible polynomial in $Z_p[x]$ of degree d. Then the field $K = Z_p[x]/\langle A(x)\rangle$ has order p^d.

PROOF: By the division algorithm theorem (12.1), every element $f(x)$ of $Z_p[x]$ can be written

$$f(x) = q(x) \cdot A(x) + r(x),$$

where either $r(x)$ is 0, or the degree of $r(x)$ is less than d. Thus, the typical element of K,

$$f(x) + \langle A(x) \rangle,$$

could be written as $r(x) + \langle A(x) \rangle$. Furthermore, the $r(x)$ is uniquely determined from the division algorithm. Thus, there are as many elements in K as there are polynomials in $Z_p[x]$ with degree less than d, counting the zero polynomial. All such polynomials can be written

$$a_0 + a_1 x + a_2 x^2 + a_3 x^3 + \cdots + a_{d-1} x^{d-1},$$

with each a_i between 0 and $p-1$, inclusively. Since there are d coefficients, each of which can be p different numbers, there are exactly p^d possible polynomials of degree less than d. Thus, $|K| = p^d$. □

Whenever a finite field is defined by an extension through an irreducible polynomial, the order of the field will be a power of a prime. We would like to show that all finite fields are produced in this way. So naturally we begin by showing that all finite fields have an order that is a power of a prime number.

PROPOSITION 13.3

Suppose K is a finite division ring. Then $|K| = p^n$ for some prime p and some integer n.

PROOF: Let q be the order of K. From the additive structure of the ring, we see that $q \cdot x = 0$ for all x in K. Thus, the characteristic is positive, and by Proposition 11.2, the characteristic is a prime number, p.

Suppose that q has a prime factor r other than p. Then the additive group of K must have a subgroup of order r, according to Lemma 6.2. Hence $r \cdot x = 0$ for some element x in K. But this contradicts Proposition 11.2, since r is not divisible by p. Therefore, q has no prime factors other than p, so $q = p^n$ for some integer n. □

According to this proposition, it is impossible to find a field of order 6. However, it is still possible to find a field of order 4. An irreducible polynomial of degree 2 in $Z_2[x]$ is $x^2 + x + 1$. Thus the commands

```
InitDomain(2)
AddFieldVar("a")
Define(a^2, -a - 1)
F = ListField(); F
    {0, 1, a, a + 1}
```

TABLE 13.1: Tables for a field of order 4

+	0	1	a	$1+a$
0	0	1	a	$1+a$
1	1	0	$1+a$	a
a	a	$1+a$	0	1
$1+a$	$1+a$	a	1	0

$\cdot$	0	1	a	$1+a$
0	0	0	0	0
1	0	1	a	$1+a$
a	0	a	$1+a$	1
$1+a$	0	$1+a$	1	a

find a field of order 4 shown in Table 13.1.

As we see from this example, it is fairly easy to enter finite fields into *Sage*, as long as they can be expressed as an extension field of Z_p through some irreducible polynomial of $Z_p[x]$. In the next section, we will show that all finite fields can be obtained in this way. In fact, our goal will be to classify *all* finite fields, which will give us a more natural way of defining the fields in *Sage*.

Problems for §13.1

For Problems 1 through 9: Perform the following computations in the field of order 25 from Example 13.2.

1 $(w+2)\cdot(w+4)$
2 $(2w+3)\cdot(3w+2)$
3 $(4w+2)\cdot(3w+2)$
4 $(3w+1)^2$
5 $(w+2)^3$
6 $(2w+3)^3$
7 $(3w+2)^4$
8 $(w+3)^{-1}$
9 $(3w+4)^{-1}$

10 The polynomial x^2+x+1 is irreducible in the field Z_2. Write out by hand the addition and multiplication tables of the field $Z_2[x]/\langle x^2+x+1\rangle$.

11 The polynomial x^3+x+1 is irreducible in the field Z_2. Write out by hand the addition and multiplication tables of the field $Z_2[x]/\langle x^3+x+1\rangle$.

12 The polynomial x^2+x+2 is irreducible in the field Z_3. Write out by hand the addition and multiplication tables of the field $Z_3[x]/\langle x^2+x+2\rangle$.

13 Construct addition and multiplication tables for a field with 16 elements.

14 Find a field with 27 elements.

15 Show that the field $\mathbb{C}$ is isomorphic to $\mathbb{R}[x]/\langle x^2+1\rangle$.

16 Show that the field $\mathbb{Q}(\sqrt{2})$ is isomorphic to $\mathbb{Q}[x]/\langle x^2-2\rangle$.

17 Prove that every element in a finite field can be written as the sum of two squares.

Interactive Problems

18 The polynomial $x^4 + x + 1$ is irreducible in the field Z_2. Use this polynomial to define a field of order 16 in *Sage*. Show that there is a subfield of order 4 in this field. Is there a subfield of order 8 in this field?

19 The polynomial $x^6 + x + 1$ is irreducible in the field Z_2. Use this polynomial to define a field of order 64 in *Sage*. Show that there is a subfield of order 4 in this field. Is there a subfield of order 8 in this field?

20 The polynomial $x^4 + x + 2$ is irreducible in the field Z_3. Use this polynomial to define a field of order 81 in *Sage*. Show that there is a subfield of order 9 in this field. Is there a subfield of order 27 in this field?

13.2 Properties of Finite Fields

In the last example we starting looking at examples of finite fields. In this section we want to explore the properties that all finite fields have in common. In the process, we will begin to classify all finite fields.

We begin by observing that if F is a finite field, then the multiplicative group F^* must be a finite abelian group. If the field is of order p^n, the group F^* has order $p^n - 1$. For example, the field of order 4 has a multiplicative group of order 3, so this group must be isomorphic to Z_3.

Example 13.3

Determine the group F^* for the "complex numbers modulo 3."

SOLUTION: Since this group has 8 elements, there are several possibilities for an abelian group of order 8. However, observing Table 11.2 shows that $1 + i$ is a generator. Thus, $F^* \approx Z_8$. □

This example is not a coincidence. Let us show that in general, there is a generator of the multiplicative group.

PROPOSITION 13.4

If F is a finite field, then the multiplicative group F^* is a cyclic group.

PROOF: F^* is abelian, and so by the fundamental theorem of abelian groups (6.2),

$$F^* \approx Z_{d_1} \times Z_{d_2} \times Z_{d_3} \times \cdots \times Z_{d_n},$$

where the d_i are all powers of prime numbers. Let d be the least common multiple of the set $\{d_1, d_2, d_3, \ldots, d_n\}$. Then for all x in F^*, we have that

$x^d = 1$. Thus, the polynomial $x^d - 1$ has $|F^*|$ solutions. By Corollary 12.2, d must be at least $|F^*|$. But we also have

$$|F^*| = d_1 \cdot d_2 \cdot d_3 \cdots d_n,$$

so d is at most $|F^*|$. Thus, $d = |F^*|$, and so $d_1, d_2, d_3, \cdots, d_n$ are coprime. Therefore, the group F^* is cyclic. ∎

Now that the multiplicative group is completely understood for a finite field, let us turn our attention to the group of automorphisms on the field. We have previously seen examples where the group of automorphisms gave us insight into the structure of a ring, and finite fields are no exception. We begin by proving some basic lemmas in number theory.

LEMMA 13.1
If p is a prime, then

$$n^p \equiv n \pmod{p}$$

for all integers n.

PROOF: Since Z_p^* is of order $p - 1$, we have by Corollary 3.2 that

$$n^{p-1} = 1$$

for all elements n in Z_p^*. (This result is commonly called Fermat's little theorem.) If we multiply both sides by n,

$$n^p = n,$$

we have a statement that is true for $n = 0$ as well. Thus, $n^p = n$ for all n in the ring Z_p. This statement, when converted into modular notation, becomes

$$n^p \equiv n \pmod{p} \qquad \text{for all integers } n.$$ ∎

LEMMA 13.2
If F is a field of characteristic p, then for all $g \in F$, the polynomial

$$f(x) = (x + g)^p - x^p - g^p$$

is the zero polynomial in $F[x]$.

PROOF: If $g = 0$, $f(x) = x^p - x^p = 0$, so the result is trivial. Let us suppose that g is nonzero.

Note that the leading term of $(x+g)^p$ is x^p, which will cancel in $f(x)$. Thus, $f(x)$ has degree at most $p - 1$. We will show that for every n, $n \cdot g$ is a root. Observe that

$$f(n \cdot g) = (n \cdot g + g)^p - (n \cdot g)^p - g^p = ((n + 1)^p - n^p - 1) \cdot g^p.$$

By Lemma 13.1,

$$(n+1)^p \equiv (n+1) \pmod{p}$$

and

$$n^p \equiv n \pmod{p}.$$

Thus,

$$(n+1)^p - n^p - 1 \equiv (n+1) - n - 1 \equiv 0 \pmod{p}.$$

So because F has characteristic p, we have $f(n \cdot g) = 0$. Since g is nonzero, the values

$$\{0, g, 2g, 3g, \cdots, (p-1)g\}$$

are all distinct in F. Thus, $f(x)$ has p distinct roots. But Corollary 12.2 shows us that if $f(x)$ were nonzero, there would be at most $p-1$ roots. Thus, $f(x)$ must be the zero polynomial. ∎

We are now ready to produce one automorphism on a finite field, which we will use to generate all other automorphisms.

THEOREM 13.1: The Frobenius Automorphism Theorem

If F is a finite field of characteristic p, then the mapping

$$f : x \to x^p$$

forms an automorphism of F to itself. Furthermore, $f(y) = y$ if, and only if, y is in the subfield Z_p. This automorphism is called the Frobenius automorphism on F.

PROOF: We first need to show that f is a homomorphism. If F is a field of characteristic p, then by Lemma 13.2 we have that

$$(x+g)^p - x^p - g^p = 0$$

for all g in F. Thus, we have the identity

$$f(x+y) = (x+y)^p = x^p + y^p = f(x) + f(y).$$

It is also obvious that

$$f(x \cdot y) = (x \cdot y)^p = x^p \cdot y^p = f(x) \cdot f(y).$$

So f is a homomorphism. The kernel of f is obviously just 0, since $x^p = 0$ implies that $x = 0$, since F has no zero divisors. Therefore, the mapping is one-to-one. Since F is a finite field, we can use the pigeonhole principle to show that the mapping is also onto. Therefore, f is an automorphism.

Finally, we need to show that $f(y) = y$ if, and only if, y is in the subfield Z_p. Note that this subfield is generated by the multiplicative identity, 1:

$$Z_p = \{0, 1, 2, 3, \cdots, p-1\}.$$

Historical Diversion

Georg Frobenius (1849–1917)

Georg Frobenius was born in a suburb of Berlin, and went to the Joachimsthal Gymnasium when he was almost 11. After he graduated, he went to the University of Göttingen for a semester, then moved back to Berlin to study under the mathematical giants of Kummer, Weierstrass, and Kronecker. His doctorate was awarded, with distinction, in 1870. After this he started out teaching at the Joachimsthal Gymnasium, and then at the Sophienrealschule.

In 1874 he was appointed as an extraordinary professor at the University of Berlin. One year later, he took up an appointment at Zürich, where he did most of his important work in mathematics. In 1891, Frobenius was appointed chair at the University of Berlin.

Frobenius started his career working with differential equations and elliptic functions. In 1873 he demonstrated a way to find an infinite series solution to a second order differential equation in the vicinity of a regular singular point. He also made many advancements in the theory of elliptic and Jacobi functions.

In the second half of his career, Frobenius concentrated on group theory. He reproved Sylow's theorems using Cayley's abstract definition of a group, which is the proof most often used today. (Previous proofs had only been for permutation groups. See Historical Diversion on page 169.) Frobenius also proved that if $n > 0$ divides the order of a finite group G, then the number of solutions to $x^n = e$, denoted by $R_n(G)$, is also divisible by n. Frobenius conjectured that if $R_n(G) = n$, then the set of solutions to $x^n = e$ forms a subgroup of G. This conjecture was finally proven in 1991, using the classification of finite simple groups.

In 1895, Frobenius called a subgroup N of a group G *characteristic* if $\phi(N) = N$ for all automorphisms ϕ of G. He was able to prove several important properties of characteristic subgroups. More importantly, he created the theory of group characters and group representations. These tools are fundamental in studying finite groups, particularly simple groups.

Frobenius also worked with rings of characteristic p, and is known for the *Frobenius endomorphism* that sends every element x to x^p. Only in certain contexts, such as a finite field, is this mapping an automorphism.

By Lemma 13.1, for any element in this subfield, $f(x) = x^p = x$. On the other hand, by Corollary 12.2, the polynomial $x^p - x$ in $F[x]$ cannot have more than p roots in F. We have already found p solutions, so there cannot be any more. Therefore, $f(y) = y$ if, and only if, y is in Z_p. □

Once we have one automorphism $f(x)$, we can consider creating other automorphisms such as $f(f(x))$ and $f(f(f(x)))$. It is not hard to determine the order of $f(x)$.

COROLLARY 13.1

Let F be a finite field of order p^n. Then the Frobenius automorphism f is of order n in the group of automorphisms.

PROOF: Note that the multiplicative group F^* has order $p^n - 1$. Thus, by Corollary 3.2, for every element x in F^*, we have

$$x^{(p^n-1)} = 1.$$

Multiplying both sides by x gives us $x^{p^n} = x$ for all x in F^*, and also $x = 0$. Thus, this statement is true for all x in F.

We now note that

$$f^n(x) = \underbrace{f(f(f(\cdots(f(x))\cdots)))}_{n \text{ times}} = x^{p^n} = x$$

for all x in F, so f^n yields the identity automorphism.

To show that the order of f is not less than n, suppose that the order was $i < n$. Then $f^i(x) = x^{p^i}$ would be x for all x. But then the polynomial

$$x^{p^i} - x$$

would have p^n solutions. This contradicts Corollary 12.2, since $n > i$. Therefore, the order of the Frobenius automorphism is n. □

We next need to show a simple lemma to indicate how to apply the Frobenius automorphism to the set of polynomials over the field.

LEMMA 13.3

Any isomorphism f that maps an integral domain K to an integral domain M extends to an isomorphism mapping $K[x]$ to $M[x]$, with f sending the polynomial x in $K[x]$ to the polynomial x in $M[x]$.

PROOF: Suppose f is an isomorphism mapping K to M. If $w(x)$ is in $K[x]$, with coefficients a_i, we can define $f(w(x))$ by

$$f(w(x)) = f\left(\sum_{i=0}^{\infty} a_i x^i\right) = \sum_{i=0}^{\infty} f(a_i)x^i.$$

If $v(x)$ is another polynomial in $K[x]$ with coefficients b_i, then

$$f(w(x)+v(x)) = f\left(\sum_{i=0}^{\infty}(a_i+b_i)x^i\right) = \sum_{i=0}^{\infty} f(a_i+b_i)x^i.$$

$$= \sum_{i=0}^{\infty} f(a_i)x^i + \sum_{i=0}^{\infty} f(b_i)x^i = f(w(x)) + f(v(x)).$$

Likewise, we have

$$f(w(x)\cdot v(x)) = f\left(\sum_{i=0}^{\infty}\sum_{j=0}^{\infty}(a_i\cdot b_j)x^{i+j}\right)$$

$$= \sum_{i=0}^{\infty}\sum_{j=0}^{\infty} f(a_i\cdot b_j)x^{i+j} = \sum_{i=0}^{\infty}\sum_{j=0}^{\infty} f(a_i)\cdot f(b_j)x^{i+j}$$

$$= f(w(x))\cdot f(v(x)).$$

Thus, f extends to a homomorphism mapping $K[x]$ to $M[x]$. But the kernel of f is just the identity element, since f preserves the degree of any nonzero polynomial. Thus, f extends to an isomorphism from $K[x]$ to $M[x]$, and f maps x to x.

∎

We can apply Lemma 13.3 to the case where f is an automorphism on $K[x]$, such as the Frobenius automorphism. By extending the Frobenius automorphism to a polynomial, we can generate irreducible polynomials in $Z_p[x]$. These irreducible polynomials are important, since we can define the field in terms of these polynomials.

PROPOSITION 13.5

Let F be a finite field of characteristic p. For any y in F, let n be the smallest number such that $y^{p^n} = y$. If $f(x)$ is the Frobenius automorphism, then

$$g(x) = (x-y)\cdot(x-f(y))\cdot(x-f(f(y)))\cdots(x-f^{n-1}(y))$$

is an irreducible polynomial of degree n in $Z_p[x]$. Here f^{n-1} means f applied to itself $n-1$ times.

PROOF: Consider the extension of the Frobenius automorphism onto $F[x]$, as given in Lemma 13.3. If we apply this mapping to the polynomial $g(x)$, we get

$$f(g(x)) = (x-f(y))\cdot(x-f(f(y)))\cdot(x-f(f(f(y))))\cdot\cdots\cdot(x-f^n(y)).$$

Recall we picked n to be the smallest number such that $f^n(y) = y$. Thus,

$$f(g(x)) = (x-f(y))\cdot(x-f(f(y)))\cdot(x-f(f(f(y))))\cdot\cdots\cdot(x-f^{n-1}(y))\cdot(x-y),$$

which after rearranging the factors gives us $g(x)$ again.

Since $g(x)$ is fixed by the Frobenius automorphism, each coefficient of $g(x)$ must be fixed by $f(x)$. But the only elements fixed by $f(x)$ are those in Z_p. Thus, $g(x)$ must have all of its coefficients in Z_p, and so is a polynomial in $Z_p[x]$.

To show that $g(x)$ is irreducible, suppose that

$$g(x) = h(x) \cdot j(x),$$

where both $h(x)$ and $j(x)$ are polynomials in $Z_p[x]$ of positive degree. Then $f(h(x)) = h(x)$ and $f(j(x)) = j(x)$ since the Frobenius automorphism fixes x and the elements in Z_p. By the unique factorization in $F[x]$, $(x - y)$ has to be a factor of $h(x)$ or $j(x)$, but not both, since $(x - y)$ is a factor of $g(x)$ but $(x-y)^2$ is not. Let us suppose that $h(x)$ has $(x-y)$ as a factor. Any factor of $j(x)$ would have to be a factor of $g(x)$, so such a factor would have the form

$$(x - f^m(y))$$

for some $m > 0$. Thus, $f^m(y)$ is a root of $j(x)$, but y is not. But this is impossible, since $f^m(j(x)) = j(x)$, and so $f^m(j(y)) = j(f^m(y)) = 0$. Therefore, $g(x)$ is an irreducible polynomial in $Z_p[x]$. ☐

DEFINITION 13.2 The polynomial produced by Proposition 13.5 is called the *irreducible polynomial of y over Z_p*. If y is in Z_p, this polynomial is simply $x - y$.

We can now use Proposition 13.5 to show us that every finite field can be produced as an extension of Z_p over an irreducible polynomial. While we are at it, we will prove a statement that is true for all fields, not just finite fields.

PROPOSITION 13.6

Let K be any field, and F be a subfield of K. Suppose there is an element y of K such that there are no proper subfields of K containing both F and y. Suppose that there is a polynomial $f(x)$ in $K[x]$ with coefficients in F such that $f(y) = 0$. Suppose further that $f(x)$ is an irreducible polynomial when treated as a polynomial in $F[x]$. Then K is isomorphic to $F[x]/\langle f(x)\rangle$.

PROOF: Consider the evaluation homomorphism

$$\phi_y : K[x] \to K.$$

We can consider the homomorphism ϕ'_y as the restriction of ϕ_y on $F[x]$. Let us consider the kernel of this homomorphism. Because $f(y) = 0$, $f(x)$ is certainly in the kernel of ϕ'_y. But the kernel cannot be all of $F[x]$, since the constant polynomials are not in the kernel. We know that the kernel is an ideal, and by Corollary 12.7, $F[x]$ is a PID, so the kernel can be written as $\langle g(x)\rangle$ for some $g(x)$ in $F[x]$. Yet $f(x)$ is in the kernel, so $g(x)$ divides $f(x)$. But $f(x)$ is irreducible in $F[x]$, and $g(x)$ cannot be a unit, since we have already observed that $\langle g(x)\rangle$ is not all of $F[x]$. Therefore, the kernel of ϕ'_y is $\langle f(x)\rangle$.

From the first ring isomorphism theorem (10.2), the image of ϕ'_y is isomorphic to

$$F[x]/\langle f(x)\rangle.$$

We have already mentioned that $F[x]$ is a PID, so by Lemma 12.6 the image is a field. But the field must contain F, since this is the image of the constant polynomials, and also must contain y, the image of the polynomial x. The only subfield of K that contains both y and F is K itself, so $F[x]/\langle f(x)\rangle$ is isomorphic to K. ∎

One immediate application of Proposition 13.6 is to show us that every finite field can be produced as an extension of Z_p over an irreducible polynomial. We will use the polynomial derived in Proposition 13.5.

COROLLARY 13.2

For every finite field K of characteristic p, there is an irreducible polynomial $f(x)$ of $Z_p[x]$ such that K is isomorphic to $Z_p[x]/\langle f(x)\rangle$.

PROOF: If K is a finite field, by Proposition 13.4, the multiplicative group of K^* is cyclic. Thus, there must be an element y that generates K^* as a group. Since K must have finite characteristic p, we will let F be the subfield Z_p. Let $f(x)$ be the irreducible polynomial of y over Z_p given by Proposition 13.5.

Even though $f(x)$ is irreducible in $Z_p[x]$, $f(x)$ has $(x - y)$ as a factor when viewed as a polynomial in $K[x]$. Note that since y generates all of K, we see that the conditions for Proposition 13.6 are satisfied. Therefore K is isomorphic to $Z_p[x]/\langle f(x)\rangle$. ∎

We have already seen one field of order 9, produced by the polynomial x^2+1. But there are two other irreducible second degree polynomials in $Z_3[x]$, x^2+x+2 and x^2+2x+2. What if we formed fields using these polynomials? Note that both of these polynomials factor in the field $Z_3[x]/\langle x^2+1\rangle$:

```
InitDomain(3, "x")
AddFieldVar("i")
Define(i^2, -1)
factor[x^2 + x + 2)
    (x + i + 2) * (x + 2*i + 2)
factor[x^2 + 2*x + 2)
    (x + i + 1) * (x + 2*i + 1)
```

Proposition 13.6 hints at what must be happening. The field $Z_3[x]/\langle x^2+1\rangle$ is the smallest field of characteristic 3 for which x^2+1 factors. But this field also happens to be the smallest field of characteristic 3 for which $x^2 + x + 2$ and $x^2 + 2x + 2$ factor. This suggests that $Z_3[x]/\langle x^2 + 1\rangle$, $Z_3[x]/\langle x^2 + x + 2\rangle$, and $Z_3[x]/\langle x^2 + 2x + 2\rangle$ are in fact the same field. Could this be so?

The first step in proving this is to find a large field containing both fields.

LEMMA 13.4

Let F and K be two finite fields with the same characteristic p. Then there is a field that contains isomorphic copies of both F and K.

PROOF: Since F is a finite field, by Corollary 13.2 there is a polynomial $f(x)$ in $Z_p[x]$ such that F is isomorphic to $Z_p[x]/\langle f(x)\rangle$.

Since F and K have the same characteristic, we can consider $f(x)$ to be a polynomial in $K[x]$ as well. Let $g(x)$ be an irreducible factor of $f(x)$ over the domain $K[x]$. Of course, $f(x)$ may already be irreducible in $K[x]$, in which case we let $g(x) = f(x)$.

Now consider $E = K[x]/\langle g(x)\rangle$. Since $K[x]$ is a PID, by Lemma 12.6 E is a field. In fact, E contains an element that is a root of the polynomial $g(x)$, namely

$$y = x + \langle g(x)\rangle,$$

since

$$g(y) = g(x + \langle g(x)\rangle) = g(x) + \langle g(x)\rangle = 0 + \langle g(x)\rangle.$$

We can now consider the evaluation homomorphism

$$\phi_y : E[x] \to E.$$

Let us first consider the restriction of this homomorphism to the ring $Z_p[x]$, which we will call ψ. Thus ψ is the homomorphism

$$\psi : Z_p[x] \to E : \psi(w(x)) = w(y).$$

Since y is a root of $g(x)$ in the field E, and $g(x)$ in turn is a factor of $f(x)$, we see that y is a root of $f(x)$ in the field E. Thus, $f(x)$ is in the kernel of the homomorphism ψ. Since $Z_p[x]$ is a PID, the kernel can be written as $\langle h(x)\rangle$ for some polynomial $h(x)$ in $Z_p[x]$. But since $f(x)$ is in the kernel, $h(x)$ must divide $f(x)$. But $f(x)$ is irreducible, and $h(x)$ cannot be a unit, or else the kernel would be all of $Z_p[x]$, which is impossible since the constant polynomials are not in the kernel. Therefore, the kernel must be $\langle f(x)\rangle$, and so by the first ring isomorphism theorem (10.2), the image of ψ is isomorphic to

$$Z_p[x]/\langle f(x)\rangle,$$

which is in turn isomorphic to F. Thus, there is a subfield of E isomorphic to F.

All we have to do is show that there is a copy of the field K inside of

$$E = K[x]/\langle g(x)\rangle.$$

But we can consider the natural homomorphism

$$i : K[x] \to E$$

given by

$$i(p(x)) = p(x) + \langle g(x) \rangle.$$

If we restrict this homomorphism onto the constant polynomials, we get

$$i' : K \to E.$$

Since $g(x)$ is not a unit, it is clear that the kernel of this homomorphism is just 0. Thus, there is a subfield of E isomorphic to K. Therefore, we have constructed a field that contains isomorphic copies of both F and K as subfields. ∎

We can now use this lemma to show that there is only one non-isomorphic field of a given order.

COROLLARY 13.3

Any two finite fields of the same order are isomorphic to each other.

PROOF: If two fields F and K have the same order, by Proposition 13.3, both must have order p^n for some prime number p, and some positive integer n. Thus, both F and K have characteristic p, so by Lemma 13.4 there exists a field E that contains isomorphic copies of both F and K as subfields. Let F' and K' be the subfields of E isomorphic to F and K, respectively. Consider the polynomial

$$f(x) = x^{p^n} - x$$

in $E[x]$. Since F' is a subfield of E, the Frobenius automorphism is of order n on this subfield. Thus, every element of F' is a root of $f(x)$. Likewise, every element of K' is also a root of $f(x)$. But by Corollary 12.2, $f(x)$ can have at most p^n roots. Thus, the subfields F' and K' must coincide, so certainly they are isomorphic. Hence F and K must be isomorphic. ∎

This proposition explains the strange behavior of fields that we discovered in our experiment. Whenever a finite field F is extended through an irreducible polynomial, all irreducible polynomials in $F[x]$ of the same degree factor completely in the new field. The reason is now clear: The field

$$F[x]/\langle f(x) \rangle$$

only depends on the degree of the irreducible polynomial $f(x)$.

We have already seen fields of order 4, 9, and 27 in this chapter. We in fact can refer to them as *the* fields of order 4, 9, or 27. However, there is one question we have yet to answer. Given a prime number p and an integer n, is there a field of order p^n? It seems like all we would need to construct such a field is an irreducible polynomial $f(x)$ in $Z_p[x]$ of degree n, and then the field

$$Z_p[x]/\langle f(x)\rangle$$

would have order p^n. The only problem with this argument is that we have not shown that there *is* an irreducible polynomial of degree n in $Z_p[x]$. In order to construct such irreducible polynomials, we will need to utilize a special class of polynomials—the cyclotomic polynomials. These polynomials have many different uses that crop up in unexpected places.

Problems for §13.2

1 Using Table 11.2 of the field of "complex numbers modulo 3," find all the generators of the multiplicative group of this field.

For Problems **2** through **5**, by Proposition 13.4, the nonzero elements of Z_p form a cyclic group under multiplication. Any generator of this group is called a *primitive root* of p. Find the primitive roots of the following primes.

2 17 **3** 23 **4** 31 **5** 37

6 For a given prime, determine a formula for the number of primitive roots there will be.

7 Show that if F is a finite field of characteristic p, and x is a generator of the multiplicative group, then x^p is also a generator of the multiplicative group.

8 If p is a prime number of the form $4n + 1$, show that there is a solution to the equation

$$x^2 \equiv -1 \pmod{p}.$$

Hint: By Proposition 13.4, Z_p^* is isomorphic to Z_{p-1}. A solution to the equation would have order 4.

9 Use Problem 8 to show that a prime of the form $4n+1$ is not prime in the domain $Z[i]$.

Hint: What is $(x+i)(x-i)$, if x is the solution to the equation in Problem 8?

10 Use Problem 9 to prove the two-square theorem of Fermat: Every prime number of the form $4n + 1$ can be expressed as the sum of two squares.

Hint: Since p is not prime in the domain $Z[i]$, and $Z[i]$ is a UFD, p is reducible in $Z[i]$. If $a + bi$ is one factor, what is the other factor?

11 Let F be a field of prime characteristic p. Show that the intersection of all of the non-trivial subfields of F is a field of order p.

12 Let F be a finite field of characteristic p. Show that $F(x)$, the field of quotients of the polynomial ring $F[x]$, is an infinite field of characteristic p.

13 Let F be any field. Show that no two finite subfields of F can have the same number of elements.

Hint: See the proof for Corollary 13.3.

14 Let F be a field of order p^n. Show that if K is a subfield of F then K has order p^d for some number d that divides n.

15 Let F be a field of order p^n. Show that if d divides n, then there is a unique subfield of order p^d.

Hint: See Problem 13 for the uniqueness part.

16 Let p be prime and $f(x)$ an irreducible polynomial of degree 2 in $Z_p[x]$. If K is a finite field of order p^3, show that $f(x)$ is also irreducible in $K[x]$.

17 Prove that the group of automorphisms of a field of order p^n is isomorphic to Z_n. That is, prove that there are no other automorphisms other than the ones generated by the Frobenius automorphism.

18 Let p be a prime number. Show that every irreducible polynomial with a leading term of x^n in the field Z_p is found in the factorization of the polynomial $x^{p^n} - x$.

Hint: If $f(x)$ is a degree n irreducible polynomial, then $F = Z_p[x]/\langle f(x)\rangle$ is field of order p^n. Show that every element in this field is a root of the polynomial $x^{p^n} - x$. Therefore, the roots of $f(x)$ in the field F are also roots of $x^{p^n} - x$.

19 Suppose $2^n + 1 = p$ is a prime number. Show that n is a power of 2. Such primes are called *Fermat primes.*

Hint: What is the order of 2 in the field Z_p^*?

Interactive Problems

20 First define $Z_3[x]$ as follows:

```
InitDomain(3, "x")
```

Then find the factorization of the polynomial $x^{3^3} - x$. Show that all irreducible polynomials with leading term of x^3 are in this factorization. For an explanation see Problem 18.

21 First define $Z_2[x]$ as follows:

```
InitDomain(2, "x")
```

Then find the factorization of the polynomial $x^{2^5} - x$. Show that all irreducible polynomials of degree 5 are in this factorization.

13.3 Cyclotomic Polynomials

At the end of the last section, we had *almost* classified all finite fields. We have shown that a finite field must have order p^n for some prime p, and that there is at most one field of that order, up to isomorphism. However, we have yet to prove that there will be a finite field of order p^n for every prime p and every $n > 0$.

In order to demonstrate this, we will need to take a detour. We need to discuss a special class of polynomials in $\mathbb{Z}[x]$. These polynomials occur in the factorizations of the simple polynomial $x^n - 1$. Although these polynomials are constructed easily, they have a tendency to appear in many different applications. Not only will they help us to classify all finite fields, but they will be used in the demonstration that fifth-degree polynomial equations cannot be solved in terms of roots.

To introduce the cyclotomic polynomials, we will begin by noticing a pattern in the following factorizations:

```
var("x")
factor(x - 1)
    x - 1
factor(x^2 - 1)
    (x + 1)*(x - 1)
factor(x^3 - 1)
    (x^2 + x + 1)*(x - 1)
factor(x^4 - 1)
    (x^2 + 1)*(x + 1)*(x - 1)
factor(x^5 - 1)
    (x^4 + x^3 + x^2 + x + 1)*(x - 1)
factor(x^6 - 1)
    (x^2 + x + 1)*(x^2 - x + 1)*(x + 1)*(x - 1)
```

In each factorization there is exactly one new polynomial appearing that has not appeared in any previous factorization. Our plan is to find a formula for the irreducible polynomials produced in these factorizations. A natural starting place would be to find all of the complex roots of the polynomial $x^n - 1$. But we have already seen that the primitive n^{th} roots of unity are of the form ω_n^k, where k is coprime to n, and $\omega_n = e^{2\pi i/n}$.

How are the primitive roots of unity related to the factorizations of $x^n - 1$? It is clear that the primitive roots are precisely the complex zeros of $x^n - 1$ that are not zeros of $x^m - 1$ for $m < n$. Thus, if we wish to find the factor of $x^n - 1$ that does not appear in any previous factorizations, we should look for a polynomial whose only complex roots are the primitive n^{th} roots of unity.

Motivational Example 13.4
Find a factor of $x^8 - 1$ that does not appear in any previous factorizations of $x^n - 1$.
SOLUTION: The primitive eighth roots of unity were found to be

$$\omega_8, \quad {\omega_8}^3, \quad {\omega_8}^5, \quad \text{and} \quad {\omega_8}^7.$$

Thus, the simplest polynomial that has these four complex roots would be:

```
w8 = (1 + I)/sqrt(2); w8
    (1/2*I + 1/2)*sqrt(2)
expand((x - w8)*(x - w8^3)*(x - w8^5)*(x - w8^7))
    x^4 + 1
```

Apparently not only did the imaginary part cancel, but also the square roots simplified. We can check that $x^4 + 1$ is a factor of $x^8 - 1$.

```
factor(x^8 - 1)
    (x^4 + 1)*(x^2 + 1)*(x + 1)*(x - 1)
```

□

We can use this example for our definition.

DEFINITION 13.3 For $n > 0$, we define the nth *cyclotomic polynomial* to be the product

$$\Phi_n(x) = (x - {\omega_n}^{k_1}) \cdot (x - {\omega_n}^{k_2}) \cdot (x - {\omega_n}^{k_3}) \cdots (x - {\omega_n}^{k_i}),$$

where $k_1, k_2, k_3, \ldots, k_i$ are the integers between 0 and n that are coprime to n.

It is sometimes convenient to use a special notation for a product of many factors. Just as the sigma Σ can be used to denote the sum of many terms, a large Π (the upper case π) is used to denote such a product. Thus, we could write

$$\Phi_n(x) = \prod_{\substack{k=1 \\ \gcd(k,n)=1}}^{n} (x - {\omega_n}^k).$$

In this product, the index k ranges from 1 to n, but we only consider the values of k for which $\gcd(k, n) = 1$. It is apparent from the definition that the degree of the nth cyclotomic polynomial is $\phi(n)$, where ϕ is Euler's totient function.

Although this definition uses complex numbers, we observed that the polynomials always produced integer coefficients. The next proposition shows us how to find the cyclotomic polynomials without having to work with complex numbers.

PROPOSITION 13.7
For any positive integer n, we have

$$x^n - 1 = \prod_{k|n} \Phi_k(x).$$

Here, the product is taken over all values of k that divide n.

PROOF: We will first show that each nth root of unity is a primitive k^{th} root of unity for exactly one positive divisor k of n. If $z = {\omega_n}^s$ is an nth root of unity, we can let $k = n/\gcd(n, s)$. Then $k \cdot s = n \cdot (s/\gcd(n, s))$ is a multiple of n, so $z^k = 1$. Yet if $z^m = 1$, then $s \cdot m$ must be a multiple of n, so $(s/\gcd(n, s)) \cdot m$ is a multiple of $n/\gcd(n, s)$. But $(s/\gcd(n,s))$ and $(n/\gcd(n, s))$ are coprime, so m would be a multiple of k. Thus, ${\omega_n}^s$ is a primitive k^{th} root of unity, with $k = n/\gcd(n, s)$.

Since

$$x^n - 1 = (x - \omega_n) \cdot (x - {\omega_n}^2) \cdot (x - {\omega_n}^3) \cdot \cdots \cdot (x - {\omega_n}^n),$$

we can collect those factors $(x - {\omega_n}^s)$ for which ${\omega_n}^s$ is a primitive k^{th} root of unity. The result is the formula

$$x^n - 1 = \prod_{k|n} \Phi_k(x).$$

□

To help understand this notation, let us look at the case where $n = 12$. Then Proposition 13.7 states that

$$x^{12} - 1 = \prod_{k|12} \Phi_k(x) = \Phi_1(x) \cdot \Phi_2(x) \cdot \Phi_3(x) \cdot \Phi_4(x) \cdot \Phi_6(x) \cdot \Phi_{12}(x).$$

We can observe this factorization using *Sage*.

```
var("x")
factor(x^12 - 1)
    (x^4 - x^2 + 1)*(x^2 + x + 1)*(x^2 - x + 1)*(x^2 + 1)*(x + 1)*
    (x - 1)
```

Proposition 13.7 at least explains our observation that the factorization of $x^n - 1$ always produces a new factor. However, we have not proven that the cyclotomic polynomials are irreducible in $\mathbb{Z}[x]$. We have to begin by showing that $\Phi_n(x)$ is indeed in $\mathbb{Z}[x]$.

COROLLARY 13.4
The nth cyclotomic polynomial $\Phi_n(x)$ has integer coefficients for all $n > 0$.

PROOF: We will prove this using induction on n. Obviously the first cyclotomic polynomial is $x - 1$, which has integer coefficients. Let $n > 1$, and suppose the claim is valid for all previous cyclotomic polynomials. By Proposition 13.7, we can find the nth cyclotomic polynomial as

$$\Phi_n(x) = (x^n - 1)/f(x)$$

where

$$f(x) = \prod_{\substack{k|n \\ k<n}} \Phi_k(x).$$

Since all previous cyclotomic polynomials have integer coefficients, we see by induction that $f(x)$ has integer coefficients. Furthermore, from the definition of the cyclotomic polynomials, we see that the leading coefficients must be 1, hence the leading coefficient of $f(x)$ is 1. So by Corollary 12.1 the quotient $(x^n - 1)/f(x)$ must in fact have integer coefficients. Therefore, all cyclotomic polynomials have integer coefficients. ∎

It is actually very easy to generate the nth cyclotomic polynomial in *Sage*. The commands

```
Cyclotomic(3, "x")
    x^2 + x + 1
Cyclotomic(6, "x")
    x^2 - x + 1
```

find the third and sixth cyclotomic polynomial. Notice that the coefficients for these cyclotomic polynomials are either 0 or ± 1. This is the case for $n \leq 100$, but for larger values of n, the coefficients of $\Phi_n(x)$ can be larger. For example, there are two coefficients of -2 in $\Phi_{105}(x)$.

```
Cyclotomic(105, "x")
    x^48 + x^47 + x^46 - x^43 - x^42 - 2*x^41 - x^40 - x^39 +
    x^36 + x^35 + x^34 + x^33 + x^32 + x^31 - x^28 - x^26 - x^24 -
    x^22 - x^20 + x^17 + x^16 + x^15 + x^14 + x^13 + x^12 - x^9 -
    x^8 - 2*x^7 - x^6 - x^5 + x^2 + x + 1
```

The next corollary is another easy consequence of Corollary 13.4.

COROLLARY 13.5

If n is divisible by m, with $n > m$, then the polynomial $x^n - 1$ is divisible by $x^m - 1$ in $\mathbb{Z}[x]$. Furthermore, $\Phi_n(x)$ divides

$$\frac{x^n - 1}{x^m - 1}$$

in $\mathbb{Z}[x]$.

PROOF: Since n is divisible by m, whenever m is divisible by k, then n is divisible by k. Thus, every factor appearing in

$$x^m - 1 = \prod_{k|m} \Phi_k(x)$$

also appears in

$$x^n - 1 = \prod_{k|n} \Phi_k(x).$$

In fact, the quotient would be the product of the cyclotomic polynomials $\Phi_k(x)$ for which k is a divisor of n, but not of m. Since the cyclotomic polynomials have integer coefficients,

$$\frac{x^n - 1}{x^m - 1}$$

would have integer coefficients. Furthermore, $\Phi_n(x)$ is one of the cyclotomic polynomials in the factorization of $x^n - 1$ that is not in $x^m - 1$. Thus, the nth cyclotomic polynomial divides $(x^n - 1)/(x^m - 1)$ in $\mathbb{Z}[x]$. ∎

We now want to find some properties of the cyclotomic polynomials, such as showing that $\Phi_n(x)$ is irreducible in $\mathbb{Z}[x]$. One of the most important properties is that two different cyclotomic polynomials cannot share a root in the complex numbers. (This is obvious from the definition.) However, we will be working with other fields besides the complex numbers, so we could ask whether a cyclotomic polynomial could have multiple roots in *any* field.

DEFINITION 13.4 If r is a root of a polynomial $f(x)$, and $(x - r)^2$ divides $f(x)$, we say r is a *multiple root* of $f(x)$.

We would like to determine when $x^n - 1$ has multiple roots. Our strategy is to discover the form of the quotient

$$\frac{x^n - 1}{x - 1}.$$

For example, $(x^4 - 1)/(x - 1)$ and $(x^5 - 1)/(x - 1)$ is given by

```
var("x")
Together((x^4 - 1)/(x - 1))
    x^3 + x^2 + x + 1
Together((x^5 - 1)/(x - 1))
    x^4 + x^3 + x^2 + x + 1
```

We can start to see the general pattern. Using this pattern, we can prove the following lemma.

LEMMA 13.5
If F is any field, then the polynomial $x^n - 1$ has a multiple root if, and only if, n is a multiple of the characteristic of F.

PROOF: We first will ask whether 1 is a multiple root of $x^n - 1$. Since 1 is clearly a root,

$$x^n - 1 = (x-1) \cdot f(x)$$

for some polynomial $f(x)$. But we can use the division algorithm to produce $f(x)$. We claim that

$$f(x) = \sum_{k=0}^{n-1} x^k = 1 + x + x^2 + x^3 + \cdots + x^{n-2} + x^{n-1}.$$

To see this, note that

$$\begin{aligned}(x-1) \cdot f(x) &= x \cdot f(x) - f(x)\\ &= (x + x^2 + x^3 + \cdots x^{n-1} + x^n)\\ &\quad - (1 + x + x^2 + x^3 + \cdots + x^{n-2} + x^{n-1})\\ &= x^n - 1.\end{aligned}$$

To see whether 1 is a double root, we observe that

$$f(1) = \sum_{k=0}^{n-1} 1^k = 1^0 + 1^1 + 1^2 + 1^3 + \cdots + 1^{n-2} + 1^{n-1} = n.$$

Thus, $f(1)$ is zero if, and only if, n is a multiple of the characteristic of F. Therefore, 1 is a double root of $f(x)$ precisely when the characteristic is positive and divides n.

Now suppose that n is not a multiple of the characteristic, and that r is a double root of $x^n - 1$. Then

$$\frac{x^n - 1}{(x-r)^2}$$

is a polynomial in $F[x]$. If we replace x with $x \cdot r$ we get

$$\frac{(x \cdot r)^n - 1}{(x \cdot r - r)^2} = \frac{x^n r^n - 1}{(x-1)^2 \cdot r^2} = \frac{x^n - 1}{(x-1)^2 \cdot r^2}$$

since $r^n = 1$. However, we have already shown that 1 is not a double root of $x^n - 1$, so the right-hand side of this equation cannot be a polynomial. Thus, r is not a double root whenever n is not a multiple of the characteristic. ∎

We are finally ready to show the irreducibility of $\Phi_n(x)$.

THEOREM 13.2: Gauss's Theorem on Cyclotomic Polynomials
For all $n > 0$, $\Phi_n(x)$ is an irreducible polynomial in $\mathbb{Z}[x]$.

PROOF: We see from Corollary 13.4 that $\Phi_n(x)$ has integer coefficients. Let $f(x)$ be an irreducible factor of $\Phi_n(x)$, with leading coefficient x^n. Our goal is to show that $f(x) = \Phi_n(x)$. Since $\Phi_n(x)$ divides $x^n - 1$, we have $x^n - 1 = f(x) \cdot g(x)$ for some $g(x) \in \mathbb{Z}[x]$. Suppose $y = \omega_n^s$ is a complex root of $f(x)$, which is a primitive n^{th} root of unity since it is also a root of $\Phi_n(x)$, so s will be coprime to n. Let p be a prime that does not divide n. We want to show that y^p is also a root of $f(x)$.

Suppose $y^p = \omega_n^{sp}$ is not a root of $f(x)$. Since y^p is also a primitive n^{th} root of unity, $\Phi_n(y^p) = 0$, so $f(y^p) \cdot g(y^p) = 0$. Since we are assuming that $f(y^p) \neq 0$, we see that $g(y^p) = 0$. In particular, this means that y is a root of $g(x^p)$.

Since y is a root of the irreducible polynomial $f(x)$, and also a root of $g(x^p)$, we see that $f(x)$ is a factor of $g(x^p)$ in $\mathbb{Z}[x]$. Hence, we can write $g(x^p) = f(x) \cdot h(x)$ for some $h(x)$ in $\mathbb{Z}[x]$.

We now consider the polynomials $F(x)$, $G(x)$, and $H(x)$ to be the polynomials $f(x)$, $g(x)$, and $h(x)$ modulo p in $Z_p[x]$. Because of the Frobenius automorphism, $[G(x)]^p = G(x^p) = F(x) \cdot H(x)$ in $Z_p[x]$. Since $Z_p[x]$ is a UFD, $F(x)$ and $G(x)$ have a common irreducible factor, say $m(x)$, in $Z_p[x]$. This would indicate that $m(x)$ is a repeated factor of $x^n - 1$ in $Z_p[x]$. But then $x^n - 1$ would have a multiple root in $Z_p[x]/\langle m(x)\rangle$, which contradicts Lemma 13.5. Thus, we find that $\omega_n^{sp} = y^p$ is a root of $f(x)$.

At this point, we have shown that whenever ω_n^s is a root of $f(x)$, and the prime p is coprime to n, then ω_n^{sp} is a root of $f(x)$. By repeating this process, we see that ω_n^{sk} is a root of $f(x)$ whenever k is coprime to n. But this means that all primitive n^{th} roots of unity are roots of $f(x)$, so $f(x) = \Phi_n(x)$. Hence $\Phi_n(x)$ is irreducible. □

Lemma 13.5 can also be used to generate irreducible polynomials in $Z_p[x]$ of any degree. In fact, these irreducible polynomials are the key to proving that a field of order p^n exists.

PROPOSITION 13.8

Let p be a prime integer, and let $n > 1$. Consider the cyclotomic polynomial

$$\Phi_{(p^n-1)}(x)$$

of order $\phi(p^n - 1)$. Let us consider $g(x)$ to be this polynomial modulo p in $Z_p[x]$. Then $g(x)$ factors into irreducible polynomials in $Z_p[x]$, all of which have degree n.

PROOF: Let $h(x)$ be an irreducible factor of $g(x)$, and let K be the field $Z_p[x]/\langle h(x)\rangle$. We wish to show that the order of K is p^n, since by Proposition 13.2 this would indicate that the degree of $h(x)$ is n. Let y be the element

$$y = x + \langle h(x)\rangle$$

in the field K. Then $h(y) = 0$, and hence $g(y) = 0$ in the field K. In fact, $g(x)$ would be a factor of

$$x^{(p^n-1)} - 1,$$

and so $y^{p^n} = y$. In other words, if $f(x)$ is the Frobenius automorphism on K, then $f^n(y) = y$. In fact, $f^n(1) = 1$, and $Z_p[x]$ is generated by x and 1, so we find that $f^n(x) = x$ for all x in K. Thus, the polynomial

$$x^{p^n} - x$$

has at least $|K|$ roots. By Corollary 12.2, $|K|$ can have at most p^n elements.

To show that $|K| = p^n$, let us suppose that $|K| = p^i$, where $i < n$. Then i is the smallest number for which $f^i(x) = x$ for all x in K. It is clear that i would have to divide n, since $f^n(x)$ is also x for all x in K.

Since $f^i(y) = y$, we see that y is a root of the polynomial

$$x^{(p^i-1)} - 1.$$

By Corollary 13.5, $\Phi_{(p^n-1)}(x)$ divides

$$j(x) = \frac{x^{(p^n-1)} - 1}{x^{(p^i-1)} - 1}$$

in $\mathbb{Z}[x]$, since $(p^i - 1)$ divides $(p^n - 1)$. Thus, in $Z_p[x]$, $g(x)$ divides $j(x)$.

Since $g(y) = 0$, and also $y^{(p^i-1)} = 1$, we see that y would be a multiple root of $x^{(p^n-1)} - 1$. But by Lemma 13.5, this polynomial can only have a multiple root if $(p^n - 1)$ is a multiple of p, which it clearly is not. Thus, $i = n$, and so $|K| = p^n$. By Proposition 13.2, the irreducible factors of $g(x)$ over $Z_p[x]$ all have degree n. □

We can now prove what we had suspected was true from the experiments: that there is precisely one field of order p^n, where $n > 0$ and p is a prime number.

COROLLARY 13.6

If p is a prime number, and n is a positive integer, there exists a unique field (up to isomorphism) of order p^n.

PROOF: We have already shown in Corollary 13.3 that finite fields of the same order are isomorphic, so all we have to show is that there is a field of order p^n. By Proposition 13.8, the cyclotomic polynomial

$$\Phi_{(p^n-1)}(x)$$

factors in the domain $Z_p[x]$ into irreducible factors of degree n. If we let $A(x)$ be one of those irreducible factors, then by Proposition 13.2, the field

$$K = Z_p[x]/\langle A(x)\rangle$$

has order p^n. □

DEFINITION 13.5 If $q = p^n$, where p is prime and $n > 0$, then the *Galois field of order q*, denoted GF(q), is the unique field of order q given in Corollary 13.6.

For example, the *official* name for the "complex numbers modulo 3" we have been working with is GF(9). Whenever p is prime, we can write GF(p) for the field Z_p.

When we first defined GF(9), we used the irreducible polynomial $x^2 + 1$ in $Z_3[x]$ to make the definition. But there are two other second-degree irreducible polynomials with a leading coefficient of 1 in $Z_3[x]$, namely, $x^2 + x + 2$ and $x^2 + 2x + 2$. We could have used either of these to define GF(9):

```
InitDomain(3)
AddFieldVar("a")
Define(a^2, -a - 2)
K = ListField(); K
    {0, 1, 2, a, a + 1, a + 2, 2*a, 2*a + 1, 2*a + 2}
```

or

```
InitDomain(3)
AddFieldVar("b")
Define(b^2, -2*b - 2)
K = ListField(); K
    {0, 1, 2, b, b + 1, b + 2, 2*b, 2*b + 1, 2*b + 2}
```

The addition tables are similar, but the multiplication tables are very different. We know from Corollary 13.3 that these are both isomorphic to GF(9), but this is not obvious from the multiplication tables. This begs the question as to whether there is an *official* way to describe the elements of GF(p^n).

It is clear that we must first pick an irreducible polynomial $f(x)$ of degree n over $Z_p[x]$, and then we let $a = x + \langle f(x)\rangle$ be a root of this polynomial in $Z_p[x]/\langle f(x)\rangle$. This will allow every element of GF(p^n) to be expressible in terms of a.

But we also know from Proposition 13.4 that the group GF(p^n)* is a cyclic group, and so we can choose the polynomial $f(x)$ so that a root a will be a generator of this group. Using that fact that GF(p^n)* is cyclic, we can determine the following definition.

DEFINITION 13.6 The *Conway polynomial* of degree n over Z_p is the polynomial $f(x)$ of degree n in $Z_p[x]$ with the following characteristics:

1. *Primitive*: The polynomial $f(x)$ is irreducible, has a leading coefficient of 1, and $x + \langle f(x)\rangle$ is a multiplicative generator of the finite field $Z_p[x]/\langle f(x)\rangle$. Such polynomials are called *primitive polynomials*.

2. *Compatibility*: The polynomial is compatible with the way that the subfields of $\mathrm{GF}(p^n)$ are defined. To be compatible, for all divisors d of n less than n, the $\left(\frac{p^n-1}{p^d-1}\right)^{\text{th}}$ power of the zeros of the polynomial must be zeros of the Conway polynomial of degree d over Z_p.

3. *Tie breaker*: If two or more primitive polynomials satisfy the compatibility condition, let m be the highest power of x for which the coefficients differ. If $n - m$ is even, pick the one with the smallest coefficient from the set $\{0, 1, \ldots p - 1\}$. If $n - m$ is odd, pick the largest, unless there is one with a coefficient of 0.

The tie-breaker at first seems counter-intuitive. Logically, a zero coefficient is always preferred over a nonzero term, but sometimes we pick the polynomial with the largest coefficient, and sometimes use the one with the smallest. But to understand why this is so, consider the first degree Conway polynomials. Since all of the primitive polynomials are of the form $x + c$, with $c \neq 0$, they differ only in the constant term. Hence $m = 0$, so $n - m$ will be odd, and we should select the primitive polynomial with the largest c. This in turn will make the root of this polynomial be as *small* as possible. So for p prime, the root of the Conway polynomial will represent the smallest generator of the group Z_p^*. In general, the Conway polynomial is designed so that the roots will be minimized.

Example 13.5

Let us use this definition to find the Conway polynomial of degree 2 over Z_3.

SOLUTION: There are three irreducible polynomials of degree 2 in $Z_3[x]$ with a leading coefficient of 1: $x^2 + 1$, $x^2 + x + 2$, and $x^2 + 2x + 2$. The roots of $x^2 + 1$ in $Z_3[x]/\langle x^2 + 1\rangle$ have order 4, not 8. Since the multiplicative group is isomorphic to Z_8, which has 4 generators, we know that there will be 2 primitive polynomials. So $x^2 + x + 2$ and $x^2 + 2x + 2$ pass the first test.

In order to understand the compatibility condition, we must first find the Conway polynomial of degree 1 over Z_3. Since there is only one multiplicative generator of Z_3, namely 2, there is only one primitive polynomial of degree 1, $x - 2 = x + 1$.

Now in order for a primitive polynomial of degree 2 to be compatible, the 4th power of the roots must be a root of $x + 1$ $((3^2 - 1)/(3^1 - 1) = 4)$. But the 4th power of all four generators in GF(9) produces 2, so both $x^2 + x + 2$ and $x^2 + 2x + 2$ satisfy the compatibility condition, but $x^2 + 1$ does not, since $i^4 = 1 \neq 2$.

Of the two possible primitive polynomials remaining, we look for the largest power of x for which these differ, (x^1), and since $n - m = 1$ is odd, and neither x^1 coefficient is 0, we pick the larger of the two possible coefficients. So the Conway polynomial is $x^2 + 2x + 2$. ☐

We can use *Sage* to verify this.

```
ConwayPolynomial(3, 2)
    x^2 + 2*x + 2
```

Hence, the *official* notation for GF(9) is

```
InitDomain(3)
AddFieldVar("a")
Define(a^2, -2*a - 2)
F = ListField(); F
    {0, 1, 2, a, a + 1, a + 2, 2*a, 2*a + 1, 2*a + 2}
```

Computational Example 13.6

Find the Conway polynomial of degree 4 over Z_3.

SOLUTION: We can find all of the primitive polynomials of degree $n = 4$ by factoring $\Phi_{(p^n-1)}(x)$ in $Z_3[x]$. (See Problem 15.)

```
Cyclotomic(80, "x")
    x^32 - x^24 + x^16 - x^8 + 1
InitDomain(3, "x")
factor(x^32 - x^24 + x^16 - x^8 + 1)
    (x^4 + x + 2) * (x^4 + 2*x + 2) * (x^4 + x^3 + 2) *
    (x^4 + x^3 + x^2 + 2*x + 2) *
    (x^4 + x^3 + 2*x^2 + 2*x + 2) * (x^4 + 2*x^3 + 2) *
    (x^4 + 2*x^3 + x^2 + x + 2) * (x^4 + 2*x^3 + 2*x^2 + x + 2)
```

So we have 8 primitive polynomials of degree 4 over Z_3, each having 4 roots that are generators of GF(81). But we need the compatibility condition to be satisfied. That is, for a root r in one of these polynomials, we need the $(3^4 - 1)/(3^1 - 1) = 40^{\text{th}}$ power of r to satisfy $x + 1 = 0$, while the $(3^4 - 1)/(3^2 - 1) = 10^{\text{th}}$ power of r must satisfy $x^2 + 2x + 2 = 0$. In other words, r will satisfy $r^{40} + 1 = 0$ and $r^{20} + 2r^{10} + 2 = 0$.

```
factor(x^40 + 1)
    (x^4 + x + 2) * (x^4 + 2*x + 2) * (x^4 + x^2 + 2) *
    (x^4 + 2*x^2 + 2) * (x^4 + x^3 + 2) *
    (x^4 + x^3 + x^2 + 2*x + 2) * (x^4 + x^3 + 2*x^2 + 2*x + 2) *
    (x^4 + 2*x^3 + 2) * (x^4 + 2*x^3 + x^2 + x + 2) *
    (x^4 + 2*x^3 + 2*x^2 + x + 2)
factor(x^20 + 2*x^10 + 2)
    (x^4 + x^2 + 2) * (x^4 + x^3 + 2) *
    (x^4 + x^3 + 2*x^2 + 2*x + 2) * (x^4 + 2*x^3 + 2) *
    (x^4 + 2*x^3 + 2*x^2 + x + 2)
```

Four polynomials are in common with all three factorizations, so these four pass the compatibility test:

$$x^4+x^3+2 \qquad x^4+x^3+2x^2+2x+2 \qquad x^4+2x^3+2 \qquad x^4+2x^3+2x^2+x+2.$$

Since they differ in the x^3 power, and none of them are missing the x^3 term, we pick the one with the largest coefficient. (Again, $n-m=4-3$ is odd.) Of the 2, we pick the one with the smallest x^2 coefficient ($n-m=4-2$ is even), giving us x^4+2x^3+2. We can verify this with *Sage*.

```
ConwayPolynomial(3, 4)
  x^4 + 2*x^3 + 2
```

□

The Galois fields have many applications. A code very similar to the RSA code studied in Chapter 3 was developed using Galois fields of characteristic 2. For a long time the field of order 2^{127} was used, since the multiplicative group is of order $2^{127}-1$, which happens to be prime. (Primes of this form are called Mersenne primes.) This code had the advantage that the key was much shorter than the RSA key, and multiplication in this field could be quickly implemented in binary hardware. However, due to the special properties of finite fields, this code was recently cracked. In order to ensure safety of the encryption, the size of the field had to be upped to order 2^{2201}, which diminished the advantage over the RSA code.

But there is another type of code based on Galois fields, called the Reed-Solomon code, which is not used for security but rather for the storage or transfer of digital data. All digital information, such as the storage of a file in a computer or a song on a compact disc, is stored as a string of "bits" that are either 0 or 1. We will let K denote a finite field of characteristic 2. For example, if $K=\text{GF}(256)$, then each element of K would correspond to a computer "byte." (Each byte is eight bits.) A string of n bytes $(a_0, a_1, a_2, a_3, \ldots, a_{n-1})$ is encoded as a polynomial in K:

$$f(x)=a_0+a_1x+a_2x^2+a_3x^3+\cdots a_{n-1}x^{n-1}.$$

The encryption of this list of elements is simply the evaluation of this polynomial at the 256 elements of K. That is, if g is a generator of the multiplicative group K^*, then

$$f(0),\, f(g),\, f(g^2),\, f(g^3),\, \ldots,\, f(g^{255})$$

is transmitted in place of the numbers $a_0, a_1, a_2, \ldots a_{n-1}$. We know from Corollary 12.3 that we can reconstruct the original list of elements from any n of the numbers transmitted. Thus, if there are some errors in the transmission, the original list can still be determined. Using combinatorial reasoning, Reed and Solomon showed that as many as $(255-n)/2$ errors could occur, and yet the original list of elements can be decoded.

For example, if $n=251$, then every 251 bytes is converted to a 250-degree polynomial, which is evaluated at the 256 elements of K. Even if two of these bytes are transmitted incorrectly, the 251 original bytes can be correctly reconstructed. This is an example of what is called an *error-correcting code.* This code was used by the *Voyager II* spacecraft to transmit pictures of Uranus and Neptune back to Earth [16]. A version of this code (using a larger field K) is used to store the digital music on a compact disc. Current CD players can cope with errors as long as 4000 consecutive bits on the CD, typically caused by a scratch on the CD surface. The Reed-Solomon code also allows over 500 channels of digital television.

The ironic part of this code is that, when Reed and Solomon first discovered the code in 1960 [15], it was described as "interesting, but probably not practical." It wasn't until hardware technology advanced to the point that the code could be implemented that the real value of this code was evident. As with most mathematics, the usefulness of a particular result is not seen until long after the result is published.

One final application of finite fields arises from the study of simple groups. Almost all of the simple groups besides the alternating groups are the Chevalley groups, which are defined in terms of finite fields. For example, the simple group $\text{Aut}(Z_{24}^*)$ can be expressed as the 3-by-3 matrices in the field Z_2 with determinant 1. This example can be generalized to a group G of m-by-m matrices over any finite field of order p^n. When $p^n > 2$, there may be a nontrivial center Z formed by diagonal matrices. However, we can form the quotient group G/Z. The group generated, denoted $L_m(p^n)$, will be simple if $m > 2$, or if $m = 2$ and $p^n > 3$ [9, p. 223].

There are several other ways of forming simple groups using finite fields. In fact, besides the alternating groups, there are only 26 finite simple groups that are not expressed using finite fields. Thus, finite fields are of key importance in the classification of all finite simple groups.

Problems for §13.3

For Problems **1** through **4**: Find the cyclotomic polynomial.

1 $\Phi_6(x)$ **2** $\Phi_9(x)$ **3** $\Phi_{10}(x)$ **4** $\Phi_{13}(x)$

5 For $n > 1$, prove that the sum of all the nth roots of unity is 0.
Hint: Look at the proof of Lemma 13.5.

6 For $n > 1$, prove that the product of all the nth roots of unity is $(-1)^{n+1}$.

7 Find the smallest field of characteristic 2 that has an element with a multiplicative order of 11.

8 Find the smallest field of characteristic 3 that has an element with a multiplicative order of 11.

9 Prove that the constant coefficient of the nth cyclotomic polynomial $\Phi_n(x)$ is equal to -1 when $n = 1$, and is 1 when $n > 1$.
Hint: Use induction along with Proposition 13.7.

10 Prove that the nth cyclotomic polynomial $\Phi_n(x)$ is a "palindrome polynomial" when $n > 1$. That is, the list of coefficients read the same going forward or backward.

Hint: Whenever x is a primitive n^{th} root of unity, x^{-1} will also be a primitive n^{th} root. What happens if we replace x with $1/y$ in the polynomial? You may use the result of Problem 9.

11 Prove that if n is odd, and $n > 1$, then $\Phi_{2n}(x) = \Phi_n(-x)$.

12 Prove that if p is a prime, and $n > 0$, then

$$\Phi_{p^n}(x) = \Phi_p(x^{p^{n-1}}).$$

13 Use Problems 11 and 12 to find $\Phi_{54}(x)$.

14 Prove that $\phi(p^n - 1)$ is divisible by n, where ϕ is Euler's totient function.

Hint: See Proposition 13.8.

15 Prove that the primitive polynomials of degree n over Z_p are precisely the factors of $\Phi_{p^n-1}(x)$ over the field Z_p.

Interactive Problems

16 First define $Z_2[x]$ in *Sage*,

```
InitDomain(2, "x")
```

and then show that the cyclotomic polynomial $\Phi_{(2^3-1)}(x)$ factors in the field Z_2 into irreducible polynomials of degree 3. Show by process of elimination that the only irreducible polynomials of degree 3 are the ones given in this factorization.

17 First define $Z_2[x]$ in *Sage* as in Problem 16. Then show that the cyclotomic polynomial $\Phi_{(2^4-1)}(x)$ factors in the field Z_2 into irreducible polynomials of degree 4. Find one more irreducible polynomial of degree 4 besides the ones given in this factorization.

Hint: Factor the polynomial $x^{2^4} - x$.

18 First define $Z_2[x]$ in *Sage* as in Problem 16. Then show that the cyclotomic polynomial $\Phi_{2^5-1}(x)$ factors in the field Z_2 into irreducible polynomials of degree 5. Does this factorization give all of the irreducible polynomials of degree 5 over Z_2?

19 First define $Z_3[x]$ in *Sage*:

```
InitDomain(3, "x")
```

and then show that the cyclotomic polynomial $\Phi_{3^2-1}(x)$ factors in the field Z_3 into irreducible polynomials of degree 2. What irreducible quadratic polynomial in Z_3 have we seen that is not in the list of factors?

20 Use *Sage* to find the Conway polynomial of degree 6 over Z_2. Show that raising a root of this polynomial to the 9th power produces a zero of the Conway polynomial of degree 3 over Z_2, and raising this root to the 21st power produces a zero of the Conway polynomial of degree 2 over Z_2. Hence, the compatibility condition is satisfied.

13.4 Finite Skew Fields

Since we have completely classified all finite fields, a natural question is whether we can classify all finite *skew* fields, and whether these can be easily entered into *Sage*. At first this seems like it would be a harder problem, since there are many non-abelian groups, and many non-commutative rings. However, a surprising result is that there are *no* finite skew fields. In this section we will prove this remarkable result, known as Wedderburn's theorem.

We begin by carrying over some ideas from group theory. One of the ways we studied non-abelian groups was to find the center of the group, since this was always a normal subgroup. We can ask whether the set of elements of a skew field that commute with all of the elements forms a special set.

DEFINITION 13.7 Let K be a skew field. Then the set of all elements x of K such that $x \cdot y = y \cdot x$ for all $y \in K$ is called the *center* of K.

Example 13.7
Find the center of the quaternions, $\mathbb{H}$. Although we might guess at the answer, we can prove what the center is with *Sage*'s help.

```
InitQuaternions()
    {1,i,j,k,-1,-i,-j,-k}
```

To find the center, let us first define two typical elements in $\mathbb{H}$.

```
var("u0 u1 u2 u3 v0 v1 v2 v3")
A = u0 + u1*i + u2*j + u3*k; A
    u0 + u1*i + u2*j + u3*k
B = v0 + v1*i + v2*j + v3*k; B
    v0 + v1*i + v2*j + v3*k
```

These will commute as long as $A \cdot B - B \cdot A = 0$.

```
D = A*B - B*A
Together(D)
    (-2*u3*v2 + 2*u2*v3)*i + (2*u3*v1 - 2*u1*v3)*j +
    (-2*u2*v1 + 2*u1*v2)*k
```

The only way that this could be zero for all v_1, v_2, and v_3 is for $u_1 = u_2 = u_3 = 0$. Thus, the center of $\mathbb{H}$ is basically the field of real numbers. ∎

LEMMA 13.6

The center of a skew field forms a field.

PROOF: Let K be a skew field, and let Z be its center. We first will show that Z is a subring. If x and y are two elements in Z, and k is any element in K, then

$$(x - y) \cdot k = x \cdot k - y \cdot k = k \cdot x - k \cdot y = k \cdot (x - y)$$

and

$$(x \cdot y) \cdot k = x \cdot (y \cdot k) = x \cdot (k \cdot y) = (x \cdot k) \cdot y = (k \cdot x) \cdot y = k \cdot (x \cdot y).$$

Thus, both $x - y$ and $x \cdot y$ are in Z. By Proposition 10.1, Z is a subring of K.

Both 0 and the identity element are obviously in Z, so Z is nontrivial. Since Z is commutative, all we have left to prove is that every nonzero element of Z is invertible. If $x \neq 0$ is an element in Z and k is in K, then $x \cdot k = k \cdot x$. The inverse of x exists in K, so we can multiply both sides of the equation on both the left and the right by x^{-1}:

$$x^{-1} \cdot (x \cdot k) \cdot x^{-1} = x^{-1} \cdot (k \cdot x) \cdot x^{-1}.$$

Thus,

$$k \cdot x^{-1} = x^{-1} \cdot k$$

for all k in K, and so x^{-1} is in the center Z. Thus, Z is a field. ∎

Another concept from group theory that carries over into the study of fields is the normalizer. Recall the definition of a normalizer of a subset S of a group G. We defined

$$N_G(S) = \{g \in G \mid g \cdot S \cdot g^{-1} = S\}.$$

We would like to apply the normalizer to the multiplicative group of a field. In particular, we would like to consider the normalizer of a particular element, that is, when $S = \{y\}$.

Example 13.8

Let us find the normalizer of the element i in the nonzero quaternions. This consists of all elements A such that $A \cdot i \cdot A^{-1} = i$. The *Sage* command

```
Together(A*i*A^-1 - i)
    (-2*(u2^2 + u3^2)/(u0^2 + u1^2 + u2^2 + u3^2))*i +
    (2*(u1*u2 + u0*u3)/(u0^2 + u1^2 + u2^2 + u3^2))*j +
    (-2*(u0*u2 - u1*u3)/(u0^2 + u1^2 + u2^2 + u3^2))*k
```

shows that $A \cdot i \cdot A^{-1} = i$ whenever

$$\frac{2((u_1u_2 + u_0u_3)j + (-u_0u_2 + u_1u_3)k - i(u_2^2 + u_3^2))}{u_0^2 + u_1^2 + u_2^2 + u_3^2}$$

is zero, which can only happen if $u_2 = u_3 = 0$. In fact, if A is nonzero, this is sufficient, so we see that the normalizer of i is the set of nonzero elements of the form $u_0 + u_1 i$. □

The normalizer does not quite form a field, since it does not include the zero element. Yet if we added the zero element to $N_{\mathbb{H}^*}(\{i\})$, we get a field equivalent to the complex numbers. It is not hard to show that for any skew field, whenever we add the zero element to the normalizer, we will either get a field or a skew field.

LEMMA 13.7

Let K be a skew field, and let k be an element of K. Then if we let

$$Y_k = \{0\} \cup N_{K^*}(\{k\}),$$

then Y_k is a division ring containing the center of K.

PROOF: Let us begin by rewriting the set Y_k. Because

$$N_{K^*}(\{k\}) = \{x \in K^* \mid x \cdot k \cdot x^{-1} = k\},$$

we can simply say $N_{K^*}(\{k\})$ consists of all elements of K^* such that $x \cdot k = k \cdot x$. Of course 0 satisfies this equation as well, so we can write

$$Y_k = \{x \in K \mid x \cdot k = k \cdot x\}.$$

When written in this form, it is obvious that the center is in Y_k. Furthermore, if x and y are in Y_k, then

$$(x - y) \cdot k = x \cdot k - y \cdot k = k \cdot x - k \cdot y = k \cdot (x - y)$$

and

$$(x \cdot y) \cdot k = x \cdot (y \cdot k) = x \cdot (k \cdot y) = (x \cdot k) \cdot y = (k \cdot x) \cdot y = k \cdot (x \cdot y).$$

Thus, by Proposition 10.1, Y_k is a subring of K.

Finally, if x is a nonzero element in Y_k, then $x \cdot k = k \cdot x$. Thus,

$$x^{-1} \cdot (x \cdot k) \cdot x^{-1} = x^{-1} \cdot (k \cdot x) \cdot x^{-1},$$

so

$$k \cdot x^{-1} = x^{-1} \cdot k.$$

Thus, every nonzero element of Y_k has its inverse in Y_k, so Y_k is a division ring. □

We now can apply the center and normalizer to *finite* division rings. We first need a lemma that will help us out regarding the divisibility of the orders of finite fields.

LEMMA 13.8

Let y, n, and m be positive integers, with $y > 1$. Then

$$\frac{y^n - 1}{y^m - 1}$$

is an integer if, and only if, n is divisible by m. Furthermore, if n is divisible by m, with $n > m$, then

$$\frac{y^n - 1}{y^m - 1}$$

is divisible by the number $\Phi_n(y)$.

PROOF: First suppose that n is divisible by m. Then by Corollary 13.5, $x^m - 1$ divides $x^n - 1$, and in fact $\Phi_n(x)$ divides

$$\frac{x^n - 1}{x^m - 1}.$$

Note that since $y > 1$, $y^m > 1$, so $y^m - 1 > 0$. Thus, y is not a root of $x^m - 1$, so we can apply the evaluation homomorphism ϕ_y and find that

$$\frac{y^n - 1}{y^m - 1}$$

is divisible by $\Phi_n(y)$.

Now suppose that n is not divisible by m. Then $n = m \cdot k + p$ for some $0 < p < m$. But note that

$$y^n - 1 = y^{(m \cdot k + p)} - 1 = y^{m \cdot k} \cdot y^p - 1 = y^p(y^{m \cdot k} - 1) + y^p - 1.$$

Thus,

$$\frac{y^n - 1}{y^m - 1} = y^p \cdot \frac{y^{m \cdot k} - 1}{y^m - 1} + \frac{y^p - 1}{y^m - 1}.$$

We have already seen that $y^{(m\cdot k} - 1)/(y^m - 1)$ is an integer, but $y^p < y^m$, so the last term cannot possibly be an integer. Therefore, $(y^n - 1)/(y^m - 1)$ is an integer if, and only if, n is a multiple of m. ∎

This lemma reveals the possible orders of division rings within a finite division ring.

COROLLARY 13.7

Let K be a finite division ring of order p^n, and let F be a subring that is a division ring of order p^m. Then n is a multiple of m.

PROOF: Consider the multiplicative groups K^* and F^*. Certainly F^* is a subgroup of K^*, since F is a subring of K. Notice that K^* contains $p^n - 1$ elements, while $|F^*| = p^m - 1$. By Lagrange's theorem (3.1), $p^m - 1$ must be a factor of $p^n - 1$. So by Lemma 13.8, n must be a multiple of m. ∎

Note that this corollary has applications in finite fields. For example, it shows that the field of order 16 cannot have a subfield of order 8.

There is one more tool that we need from group theory, which stems from the normalizer. We discovered in §7.4 that the class equation was a powerful tool in analyzing groups. In fact, all three Sylow theorems hinge on the class equation. So let us observe how this tool applies to skew fields. Recall that the class equation theorem (7.2) stated that when G is a finite group, then

$$|G| = \sum_g \frac{|G|}{|N_G(\{g\})|}$$

where the sum runs over one g from each conjugacy class.

If K is a finite skew field, we can apply the class equation theorem to the multiplicative group K^*, and find that

$$|K^*| = \sum_k \frac{|K^*|}{|N_{K^*}(\{k\})|}.$$

We can make the obvious substitutions $|K^*| = |K| - 1$, and $|N_{K^*}(\{k\})| = |Y_k| - 1$. The equation now looks like

$$|K| - 1 = \sum_k \frac{|K| - 1}{|Y_k| - 1}$$

where the sum runs from one k from each conjugacy class of K^*.

We are almost ready to use the class equation to prove that finite skew fields cannot exist. But first we need to prove a simple inequality about the evaluation of a cyclotomic polynomial at a positive integer.

LEMMA 13.9
If $n > 1$, then the cyclotomic polynomial evaluated at $y \geq 2$, $\Phi_n(y)$, is greater than $y - 1$.

PROOF: From the definition,

$$\Phi_n(x) = \prod_{\substack{k=1 \\ \gcd(k,n)=1}}^{n} (x - {\omega_n}^k).$$

Plugging in $x = y$, and taking the absolute value of both sides, we get

$$|\Phi_n(y)| = \prod_{\substack{k=1 \\ \gcd(k,n)=1}}^{n} |y - {\omega_n}^k|$$
$$> \prod_{\substack{k=1 \\ \gcd(k,n)=1}}^{n} (y-1) \geq y - 1.$$

Here, the inequality $|y - (\omega_n)^k| > (y-1)$ comes from the fact that the real part of ${\omega_n}^k$ is less than 1 when $n > 1$. ∎

The final step is to use Lemma 13.9 to prove a contradiction in the class equation for finite skew fields.

THEOREM 13.3: Wedderburn's Theorem
There are no finite skew fields.

PROOF: Suppose that K is a finite skew field. By Proposition 13.3 K is of order p^m for some prime p and some $m > 0$. Let Z be the center of K. Since Z is a subring of K, which is a field by Corollary 13.7, Z is of order $y = p^a$, where $m = n \cdot a$ for some $n > 0$. Thus, $|K| = p^{n \cdot a} = y^n$. Note that since K is a skew field, n must be greater than 1. We have from the class equation theorem (7.2)

$$|K| - 1 = \sum_k \frac{|K| - 1}{|Y_k| - 1},$$

where the sum runs from one k from each conjugacy class of K^*. Note that when k is in Z^*, k is in its own conjugacy class, and $Y_k = K$. Thus, the terms in the sum corresponding to elements in Z^* are equal to 1. There are of course $|Z^*| = y - 1$ such terms. For the other terms in the sum, Y_k is a proper subring of K that contains Z. By Lemma 13.7, Y_k is a division ring, and so by Corollary 13.7, $|Y_k| = y^r$ for some r which is a factor of n. If we let

$w = \Phi_n(y)$ we see by Lemma 13.8 that w divides the term

$$\frac{|K| - 1}{|Y_k| - 1} = \frac{y^n - 1}{y^r - 1}.$$

Furthermore, w divides the left-hand side of the class equation, $|K| - 1$. In fact, the only terms in the class equation that are not divisible by w are the $y - 1$ terms that are equal to 1, coming from the non-zero elements of the center Z. Thus, $y - 1$ must be divisible by w. But this is impossible, since $y - 1 < w$ by Lemma 13.9, for $n > 1$. This contradiction proves that finite skew fields cannot exist. ▯

In a sense, the non-existence of finite skew fields is sad, since there would have been plenty of applications for finite skew fields in cryptography and group theory had they existed. On the other hand, this result, when combined with the classification of all finite fields, means that we have found all finite division rings.

Problems for §13.4

1 Use Wedderburn's theorem (13.3) to show that for any prime p, there exist integers $0 \le a, b, c, d < p$ such that

$$a^2 + b^2 + c^2 + d^2 = mp$$

for some positive integer m.

Hint: Consider the ring of "integer quaternions modulo p".

2 Suppose that for p prime, there exist integers $0 \le a, b, c, d < p$ such that

$$a^2 + b^2 + c^2 + d^2 = mp$$

for some even integer m. Show that $mp/2$ can be expressed as the sum of 4 square integers.

Hint: What is

$$\left(\frac{a+b}{2}\right)^2 + \left(\frac{a-b}{2}\right)^2 \left(\frac{c+d}{2}\right)^2 + \left(\frac{c-d}{2}\right)^2?$$

3 Use quaternions to prove Euler's four-square identity:

$$\begin{aligned}(a_1^2 + a_2^2 + a_3^3 + a_4^4)(b_1^2 + b_2^2 + b_3^2 + b_4^4) = {} & (a_1b_1 - a_2b_2 - a_3b_3 - a_4b_4)^2 + \\ & (a_1b_2 + a_2b_1 + a_3b_4 - a_4b_3)^2 + \\ & (a_1b_3 - a_2b_4 + a_3b_1 + a_4b_2)^2 + \\ & (a_1b_4 + a_2b_3 - a_3b_2 + a_4b_1)^2.\end{aligned}$$

Hint: Use the result of Problem 11 of §9.1.

4 Let x be an integer quaternion such that $|x|^2 = mp$ for some prime p. Show that if $y = x + mz$ for an integer quaternion z, then $x \cdot \overline{y}$ is a multiple of m. See Problem 8 of §9.1 for the definition of $\overline{y}$.

5 Let x be an integer quaternion such that $|x|^2 = mp$ for some prime p and some odd number m. Show that we can find an $y = x + mz$, where z is an integer quaternion, such that $|y|^2 < m^2$.

Hint: Get each component of y to be smaller than $m/2$ in absolute value.

6 Show that every prime number p can be expressed as the sum of four squares, $a^2 + b^2 + c^2 + d^2 = p$.

Hint: start with the result of Problem 1, and consider the smallest value of m for which $a^2 + b^2 + c^2 + d^2 = mp$. Using Problems 2 through 5, show that m must be 1.

7 Use Problems 3 and 6 to prove Lagrange's four-square theorem: Every positive integer can be expressed as the sum of four squares.

8 Define a Hurwitz integer as a quaternion $x = a + bi + cj + dk$ for which a, b, c, and d are either all integers, or are all half-integers (integer + 1/2). Show that the Hurwitz integers form a subring of $\mathbb{H}$.

Hint: Let $z = (1 + i + j + k)/2$. Then either x or $x - z$ is an integer quaternion.

9 Define the *norm* on the quaternions $\mathbb{H}$ by $N(x) = x \cdot \overline{x} = |x|^2$. Show that the norm of a Hurwitz integer is always a non-negative integer. See Problem 8. Note that Problem 11 of §9.1 shows that $N(x \cdot y) = N(x)N(y)$.

10 Show that the Hurwitz integers form a *skew Euclidean domain*, using $N(x)$ as a valuation function. That is, for x and y with $y \neq 0$, there are q and r with $N(r) < N(y)$ such that

$$x = q \cdot y + r.$$

See Problem 9.

Hint: Use the same strategy as Proposition 12.9. Let $t = a + bi + cj + dk = x \cdot y^{-1}$, and pick $q = e + fi + gj + hk$ so that $|a - e| \leq 1/4$, $|b - f| \leq 1/2$, $|c - g| \leq 1/2$, and $|d - h| \leq 1/2$.

11 Show that u is a unit in the ring of Hurwitz integers if, and only if, $N(u) = 1$. Find all of the units in the ring of Hurwitz integers. See Problem 9.

12 Use Problem 11 to show that if $N(x) = p$, which is a prime in the ordinary sense, then x is an irreducible Hurwitz integer. See Problem 9. (This turns out to be an if and only if condition.)

13 Use Problem 6 to show that if p is a prime in the ordinary sense, then p factors in the Hurwitz integers into two irreducible elements. See Problem 12.

14 We say that two Hurwitz integers x and y are *associates* if there exist units u_1 and u_2 such that $x = u_1 \cdot y \cdot u_2$. Show that associates form an equivalence relation on the ring of Hurwitz integers.

15 A *right ideal* I of a ring R is a subset such that $I \cdot R \subseteq I$. Show that every right ideal of the Hurwitz integers is a principal right ideal. That is, $I = x \cdot R$ for some x in I.

Hint: Follow the proof of the Euclidean domain theorem (12.5). It should be noted that the Hurwitz integers do not have unique factorization in the usual sense, because of the non-commutative multiplication.

Interactive Problems

16 *Sage* can be used to explore skew fields besides $\mathbb{H}$. Consider the following ring of characteristic 0:

```
InitSkew9()
     {a, b}
```

The ring is defined in terms of 2 generators a and b, such that $a^3 = 3a+1$, $b^3 = 2$, and $b \cdot a = (2-a^2) \cdot b$. This produces a ring that is a 9-dimensional extension of $\mathbb{Q}$. A basis for this ring would be $\{1, a, a^2, b, a \cdot b, a^2 \cdot b, b^2, a \cdot b^2, a^2 \cdot b^2\}$. If

```
var("c1 c2 c3 c4 c5 c6 c7 c8 c9")
w1 = c1 + c2*a + c3*a^2
w2 = c4 + c5*a + c6*a^2
w3 = c7 + c8*a + c9*a^2
w = w1 + w2*b + w3*b^2; w
     c1 + c2*a + c3*a^2 + c4*b + c5*a*b + c6*a^2*b + c7*b^2 +
     c8*a*b^2 + c9*a^2*b^2
```

then w is the general element of this ring. To show that this ring is in fact a skew field for rational values of $c_1, c_2, \ldots c_9$, perform the following operations:

```
v1 = b*w1*b*w1*b - 2*b*w2*b*w3*b
v2 = 2*w3*b^2*w3*b - w2*b^2*w1*b
v3 = w2*b*w2*b^2 - w3*b*w1*b^2
v = v1 + v2*b + v3*b^2
R = v*w
```

Notice that R does not depend on a or b, hence is an element of $\mathbb{Q}$. To simplify it, evaluate

```
expand(R.vector()[0])
```

Using this value of R, find a formula for w^{-1}. Can you prove that R is never zero if $c_1, c_2, c_3, \ldots c_9$ are rational?

Hint: If $R = 0$ for rational values of $c_1, \ldots c_9$, we can multiply by the common denominator to find a solution to $R = 0$ for integer values. In fact, we may assume that $c_1, c_2, c_3, \ldots c_9$ have no common factors. Show that the first three constants must be even. After a substitution, show that c_4, c_5, c_6 must be even. After yet another substitution, show that the remaining constants are even, leading to a contradiction.

17 Find the center of the skew field from Problem 16.

18 Find the normalizer of the element $a \cdot b$ in the skew field from Problem 16. Hint: Use the simplified form of the normalizer from Lemma 13.7.

19 Load the unit Hurwitz integers into *Sage* as follows:

```
InitQuaternions()
    {1, i, j, k, -1, -i, -j, -k}
U = Group(i, (1 + i + j + k)/2)
```

What is the center of this group? What group is $U/Z(U)$ isomorphic to?

20 Since 13 can be expressed as the sum of 4 squares in 2 different ways, show that there are two ways of factoring 13 in Hurwitz integers. See Problem 13. Show that these factorizations are not related by associates in the sense of Problem 14. The easiest way to show that the primes **p** and **q** are not associates is with a nested **for** loop.

```
for x in U:
    for y in U:
        if(x*p*y == q):
            print x, y
```

Chapter 14

The Theory of Fields

As we learn the laws of arithmetic in school, we start out with the simple counting numbers $1, 2, 3, \ldots$ in kindergarten, and expand into more complicated number systems as the need arises. When we learn about subtraction, we find there are problems that can't be solved with counting numbers, so we learn about 0 and negative numbers. As we master division, we find that most problems require rational numbers to solve. When important applications required numbers like π and $\sqrt{2}$, we considered the set of real numbers. Finally, as we studied quadratic equations, we found that some polynomials required complex numbers to be solved.

The pattern is clear. Each time we had a number system that had a problem that couldn't be solved, we expanded the number system to allow for a solution. This idea of expanding a number system plays a key role in field theory. We have already seen that every finite field can be thought of as an extension of a cyclic prime order field Z_p for which a polynomial admits a solution. In this chapter we apply this process to infinite fields, creating larger fields for which a given polynomial will have a root.

14.1 Vector Spaces

In order to study fields in depth, we will first need a few results from a first-year linear algebra course about vector spaces. However, most linear algebra courses work with vectors and matrices with real numbers for entries, whereas we will generalize the notations to allow arbitrary fields. Nonetheless, most of the proofs will follow the same way for arbitrary fields as for real numbers.

DEFINITION 14.1 Let F be a field. We say that V is a *vector space* over F if V is an abelian group under addition $(+)$, and for which there is defined a multiplication $a \cdot v$ for all $a \in F$ and $v \in V$ such that:

1. Whenever $a \in F$ and $v \in V$, $a \cdot v \in V$.
2. When $a \in F$, and $v, w \in V$, then $a \cdot (v + w) = a \cdot v + a \cdot w$.

3. When $a, b \in F$, and $v \in V$, then $(a + b) \cdot v = a \cdot v + b \cdot v$.

4. When $a, b \in F$, and $v \in V$, then $(a \cdot b) \cdot v = a \cdot (b \cdot v)$.

5. If e is the identity of F, then $e \cdot v = v$ for all $v \in V$.

The members of V are called *vectors.* The best way to understand vector spaces is to give some examples.

Example 14.1
Consider the set of 3-tuples $\langle u_1, u_2, u_3 \rangle$ where u_1, u_2, and $u_3 \in \mathbb{R}$. Addition of two vectors is done component-wise, and $k \cdot \langle u_1, u_2, u_3 \rangle = \langle ku_1, ku_2, ku_3 \rangle$ when $k \in \mathbb{R}$. This is a vector space over $\mathbb{R}$, and can be denoted by $\mathbb{R}^3$. □

Example 14.2
We can generalize the previous example using any field F in place of $\mathbb{R}$, and consider n-tuples $\langle u_1, u_2, \ldots, u_n \rangle$. Addition is still defined component-wise, and $k \cdot \langle u_1, u_2, \ldots, u_n \rangle = \langle k \cdot u_1, k \cdot u_2, \ldots, k \cdot u_n \rangle$. This will give us a vector space over F, which we can denote by F^n. □

Example 14.3
Let K be a field, and F any subfield of K. Then K is a vector space over F, defining $a \cdot v$ as a product in the field K. Property 5 follows from the fact that the identity of F must also be the identity of K. The other properties follow from the distributive and associative properties of K. □

This last example demonstrates the usefulness in studying vector spaces over a field F. In fact, this is the example that we will concentrate on for the remainder of the chapter.

The next definition is the key to understanding the properties of a vector space.

DEFINITION 14.2 Let V be a vector space over a field F. We say that a finite set $B = \{x_1, x_2, x_3, \ldots x_n\}$ of vectors in V are *linearly dependent* if there are elements $a_1, a_2, \ldots a_n \in F$, not all zero, for which

$$a_1x_1 + a_2x_2 + \cdots + a_nx_n = 0.$$

We say that the vectors are *linearly independent* if they are not linearly dependent, that is, if the only way for $c_1x_1 + c_2x_2 + \cdots + c_nx_n = 0$ is for $c_1 = c_2 = \cdots = c_n = 0$.

Example 14.4

The vectors $\langle 1,4,-1\rangle$, $\langle 2,-3,1\rangle$, $\langle 4,5,-1\rangle$ are linearly dependent, since there is a nonzero solution to $c_1\langle 1,4,-1\rangle + c_2\langle 2,-3,1\rangle + c_3\langle 4,5,-1\rangle = 0$, namely $c_1 = 2$, $c_2 = 1$, and $c_3 = -1$. On the other hand, $\langle 2,0,1\rangle$, $\langle 0,0,3\rangle$, and $\langle 1,4,0\rangle$ are linearly independent, since in order to get $c_1\langle 2,0,1\rangle + c_2\langle 0,0,3\rangle + c_3\langle 1,4,0\rangle = 0$, we need $4c_3 = 0$, $2c_1 + c_3 = 0$, and $c_1 + 3c_2 = 0$. This forces $c_3 = 0$, $c_1 = 0$, and $c_2 = 0$, so there are no nonzero solutions. ∎

DEFINITION 14.3 Let V be a vector space over a field F. A finite set of vectors $\{x_1, x_2, x_3, \ldots x_n\}$ in V is called a *basis of* V *over* F if the set is linearly independent, and every element of V can be expressed in the form

$$a_1x_1 + a_2x_2 + a_3x_3 + \cdots + a_nx_n$$

with $a_1, a_2, a_3, \ldots, a_n$ in F.

Here are some examples, all of which are fairly routine to check:

1. The complex numbers $\mathbb{C}$ have a basis $\{1, i\}$ over the real numbers $\mathbb{R}$.
2. The quaternions $\mathbb{H}$ have a basis $\{1, i, j, k\}$ over $\mathbb{R}$.
3. The field $\mathbb{Q}[\sqrt{2}]$ has a basis $\{1, \sqrt{2}\}$ over the rational numbers $\mathbb{Q}$.
4. From Example 14.3, the set of real numbers $\mathbb{R}$ is a vector space over the rationals. However, there can be no finite basis $\{x_1, x_2, x_3, \ldots x_n\}$ in $\mathbb{R}$ for which every real number could be expressed as $a_1x_1 + a_2x_2 + a_3x_3 + \cdots + a_nx_n$, with $a_1, a_2, \ldots a_n \in \mathbb{Q}$, lest the set of reals be countable, which contradicts Cantor's diagonalization theorem (0.8). See Problem 19.

There is an easy way to determine if a particular set of vectors is a basis.

LEMMA 14.1

$B = \{x_1, x_2, x_3, \ldots x_n\}$ *is a basis of a vector space* V *over* F *if, and only if, every element of* V *can be expressed* uniquely *in the form*

$$v = c_1x_1 + c_2x_2 + c_3x_3 + \cdots + c_nx_n.$$

The ordered n*-tuple* $\langle c_1, c_2, c_3, \ldots, c_n\rangle$ *is called the* coefficients *of* v *with respect to* B.

PROOF: If B is a basis, then every element $v \in V$ can be expressed in the form $c_1x_1 + c_2x_2 + c_3x_3 + \cdots + c_nx_n$. Suppose that $v = a_1x_1 + a_2x_2 + a_3x_3 + \cdots + a_nx_n$ is another such expression. Then

$$(a_1 - c_1)x_1 + (a_2 - c_2)x_2 + (a_3 - c_3)x_3 + \cdots + (a_n - c_n)x_n = v - v = 0.$$

But the vectors in B are linearly independent, so the only way that the combination of vectors could be 0 is for $a_i - c_i = 0$ for all $1 \le i \le n$. Hence, $a_i = c_i$ for all i, and the representation is unique.

On the other hand, if every $v \in V$ can be uniquely represented as $c_1x_1 + c_2x_2 + c_3x_3 + \cdots + c_nx_n$, then in particular 0 has only one representation, namely $0 = 0x_1 + 0x_2 + 0x_3 + \cdots + 0x_n$. Thus, the vectors in B are linearly independent, and so B is a basis. □

We can define a basis in *Sage* with the command **ToBasis**.

```
B = ToBasis([[1, 4, -1], [2, -3, 1], [4, 5, -1]])
    Error: linearly dependent.
```

Notice that we entered a list of lists, which *Sage* interprets as a list of vectors. This failed because, as we saw before, the set of vectors $\langle 1, 4, -1\rangle$, $\langle 2, -3, 1\rangle$, and $\langle 4, 5, -1\rangle$ were linearly dependent.

```
B = ToBasis([[2, 0, 1], [0, 0, 3], [1, 4, 0]])
    Successful mapping constructed.
```

Once we have defined the basis, we can find the coefficients $c_1, c_2, \ldots c_n$ for any element of the vector space.

```
Coefficients(B, [2, 3, 4])
    [5/8, 9/8, 3/4]
```

This shows that

$$\langle 2, 3, 4\rangle = \frac{5}{8}\langle 2, 0, 1\rangle + \frac{9}{8}\langle 0, 0, 3\rangle + \frac{3}{4}\langle 1, 4, 0\rangle.$$

LEMMA 14.2

Suppose that V is a vector space over F, and $B = \{x_1, x_2, x_3, \ldots x_n\}$ is a basis of V over F. Then any set $\{y_1, y_2, y_3, \cdots y_n, y_{n+1}\}$ of $n+1$ elements of V is linearly dependent.

PROOF: Suppose that $Y = \{y_1, y_2, y_3, \ldots, y_n, y_{n+1}\}$ are linearly independent, so that all of these vectors are nonzero.

Our goal is to show, with a suitable rearrangement of the vectors in B, that $\{y_1, y_2, \cdots y_{k-1}, y_k, x_{k+1}, \ldots, x_n\}$ is a basis for every $0 \le k \le n$. If $k = 0$, then this set is the original set B, which is a basis. So let us use induction to assume that it is true for the previous case, that is, that $\{y_1, y_2, \cdots y_{k-1}, x_k, x_{k+1}, \ldots, x_n\}$ is a basis.

We then can express

$$y_k = a_1y_1 + a_2y_2 + \cdots a_{k-1}y_{k-1} + a_kx_k + a_{k+1}x_{k+1} + \cdots + a_nx_n.$$

Since the vectors in Y are linearly independent, we see that at least one of a_k, $a_{k+1} \dots a_n$ is nonzero. By rearranging the remaining elements of B, we can suppose that $a_k \neq 0$. Then

$$x_k = a_k{}^{-1}(y_k - a_1 y_1 - a_2 y_2 - \cdots - a_{k-1} y_{k-1} - a_{k+1} x_{k+1} - \cdots - a_n x_n).$$

Any element $v \in V$ can be expressed as $v = c_1 y_1 + c_2 y_2 + \cdots + c_{k-1} y_{k-1} + c_k x_k + \cdots + c_n x_n$. By substituting for the value of x_k, we see that v can be expressed as a linear combination of $\{y_1, y_2, \ldots, y_{k-1}, y_k, x_{k+1}, \ldots, x_n\}$. If this set were linearly dependent, there would be a nonzero solution to

$$c_1 y_1 + c_2 y_2 + \cdots + c_{k-1} y_{k-1} + c_k y_k + \cdots + c_n x_n = 0.$$

Then $c_k \neq 0$, lest there also be a nonzero solution to

$$c_1 y_1 + c_2 y_2 + \cdots + c_{k-1} y_{k-1} + c_k x_k + \cdots + c_n x_n = 0,$$

but we are assuming that $\{y_1, y_2, \ldots y_{k-1}, x_k, x_{k+1}, \ldots, x_n\}$ is a basis. But substituting the value for y_k gives

$$\begin{aligned} &c_k\ (a_1 y_1 + a_2 y_2 + \cdots a_{k-1} y_{k-1} + a_k x_k + \cdots a_n x_n) \\ &+ c_1 y_1 + c_2 y_2 + \cdots c_{k-1} y_{k-1} + c_{k+1} x_{k+1} + \cdots c_n x_n = 0. \end{aligned}$$

This is a nonzero solution to

$$b_1 y_1 + b_2 y_2 + \cdots + b_{k-1} y_{k-1} + b_k x_k + \cdots b_n x_n = 0,$$

since $b_k = c_k a_k \neq 0$. Thus, the set $\{y_1, y_2, \ldots y_{k-1}, y_k, x_{k+1}, \ldots, x_n\}$ is linearly independent, and hence is a basis of V.

Now we can use the induction to say that $\{y_1, y_2, \ldots, y_n\}$ is a basis of V, but then y_{n+1} can be expressed in terms of $\{y_1, y_2, \ldots, y_n\}$, which shows that Y is in fact linearly dependent. □

We can now use this lemma to show that any two bases must have the same number of elements.

PROPOSITION 14.1

Let V be a vector space over F. If the sets $X = \{x_1, x_2, x_3, \ldots x_n\}$ and $Y = \{y_1, y_2, y_3, \ldots y_m\}$ are both bases of V over F, then $n = m$.

PROOF: Suppose that n is not equal to m. By exchanging the roles of X and Y if necessary, we can assume that $n < m$. Then we can use Lemma 14.2 to show that $\{y_1, y_2, y_3, \ldots y_{n+1}\}$ is linearly dependent, hence Y is not a basis of V. So we must have $n = m$. □

This proposition allows us to make the following definition.

DEFINITION 14.4 Let V be a vector space over F. If there is a basis $\{x_1, x_2, x_3, \ldots x_n\}$ of V over F, we define the *dimension of V over F* to be the size n of the basis. If there does not exist a finite basis, we say the dimension of V over F is *infinite.*

Looking back at our examples, we see that $\mathbb{R}^3$ is a 3-dimensional vector space over $\mathbb{R}$, $\mathbb{C}$ is a 2-dimensional vector space over $\mathbb{R}$, $\mathbb{H}$ is a 4-dimensional vector space over $\mathbb{R}$, and $\mathbb{R}$ is an infinite-dimensional vector space over $\mathbb{Q}$.

Computational Example 14.5

Since Z_3 is a subfield of GF(9), we can view GF(9) as a vector space over Z_3. Determine a basis for this vector space.

SOLUTION: In *Sage* we need to know that the Conway polynomial of degree 2 over Z_3 is $x^2 + 2x + 2$, so if we let a be a root of this polynomial, we can define GF(9) with the commands:

```
InitDomain(3)
AddFieldVar("a")
Define(a^2, a + 1)
ListField()
    {0, 1, 2, a, a + 1, a + 2, 2*a, 2*a + 1, 2*a + 2}
```

To find a basis, we can observe that every element of GF(9) can be written as $x + ya$, where x and y are in Z_3. Thus, one possible basis would be the set $\{1, a\}$.

```
B = ToBasis([1, a])
    Successful mapping constructed.
Coefficients(B, a^3)
    [1, 2]
```

This shows that indeed $\{1, a\}$ is a basis of GF(9) over Z_3. It is logical that GF(9) will be a 2-dimensional vector space over Z_3, since there are 3^2 elements. □

Computational Example 14.6

It is apparent that GF(81) is a 4-dimensional vector space over Z_3. But can we also can consider GF(81) as a 2-dimensional vector space over GF(9)? If so, find a basis.

SOLUTION: The Conway polynomial of degree 4 over Z_3 is

```
ConwayPolynomial(3, 4)
    x^4 + 2*x^3 + 2
```

so we could define GF(81) by telling *Sage* that the generator q raised to the 4th power is $q^3 + 1$. However, we ought to be able to define GF(81) as an extension of GF(9). This means that we need an irreducible polynomial in GF(9) for which the generator of GF(81) will satisfy. Let us factor the above Conway polynomial in the field GF(9).

```
AddFieldVar("x")
factor(x^4 + 2*x^3 + 2)
    (x^2 + (a + 2)*x + a) * (x^2 + 2*a*x + 2*a + 1)
```

This gives us two polynomials. But we also have the compatibility condition, which says that the $(3^4 - 1)/(3^2 - 1) = 10$th power of the new generator b will be the standard generator a of GF(9). Hence, we can also factor

```
factor(x^10 - a)
    (x^2 + a) * (x^2 + x + a) * (x^2 + 2*x + a) *
    (x^2 + (a + 2)*x + a) * (x^2 + (2*a + 1)*x + a)
```

The only polynomial that is in both factorizations is $x^2 + (2+a)x + a$. So we know that b satisfies this polynomial, so we can define b^2 to be $(1+2a)b+2a$.

```
AddFieldVar("b")
Define(b^2, (1 + 2*a)*b + 2*a)
```

We can now define a basis of GF(81) over GF(9).

```
B = ToBasis([1, a], [b, 2])
    Successful mapping constructed.
```

Note that we included two lists for arguments of the **ToBasis** command. The first list gives a basis for the root field F, which in this case is GF(9). We can now find components for elements in GF(81).

```
Coefficients(B, b^2)
    [2*a + 1, a]
```

This tells us that $b^2 = (2a+1)b + 2a$, which is indeed how we define b^2. ▯

This last example shows that it is possible to have a vector space over a vector space, if the latter vector space happens to be a field. What can we say about the dimension of a vector space over a vector space?

PROPOSITION 14.2

If E is a vector space over F of dimension m, which also happens to be a field, and V is a vector space over E of dimension n, then V is a vector space over

F of dimension mn. Furthermore, if $\{x_1, x_2, x_3, \ldots x_m\}$ *is a basis of E over F, and* $\{y_1, y_2, y_3, \ldots y_n\}$ *is a basis of V over E, then the set*

$$S = \{ x_1y_1, x_2y_1, x_3y_1, \ldots x_my_1, \\ x_1y_2, x_2y_2, x_3y_2, \ldots x_my_2, \\ x_1y_3, x_2y_3, x_3y_3, \ldots x_my_3, \\ \ldots\ldots\ldots \\ x_1y_n, x_2y_n, x_3y_n, \ldots x_my_n\}$$

is a basis of V over F.

PROOF: Since $\{y_1, y_2, y_3, \ldots, y_n\}$ is a basis for V over E, we can write any element of V in the form

$$c_1y_1 + c_2y_2 + c_3y_3 + \cdots + c_ny_n,$$

where $c_1, c_2, c_3, \ldots, c_n$ are in E.

Since $\{x_1, x_2, x_3, \ldots x_m\}$ is a basis of E over F, we can in turn write

$$c_1 = a_{1,1}x_1 + a_{2,1}x_2 + a_{3,1}x_3 + \cdots a_{m,1}x_m, \\ c_2 = a_{1,2}x_1 + a_{2,2}x_2 + a_{3,2}x_3 + \cdots a_{m,2}x_m, \\ c_3 = a_{1,3}x_1 + a_{2,3}x_2 + a_{3,3}x_3 + \cdots a_{m,3}x_m, \\ \ldots\ldots\ldots \\ c_n = a_{1,n}x_1 + a_{2,n}x_2 + a_{3,n}x_3 + \cdots a_{m,n}x_m,$$

where each $a_{i,j}$ is in F. Combining these, we see that every element of V can be expressed in the form

$$a_{1,1}x_1y_1 + a_{2,1}x_2y_1 + a_{3,1}x_3y_1 + \cdots + a_{m,1}x_my_1 \\ + a_{1,2}x_1y_2 + a_{2,2}x_2y_2 + a_{3,2}x_3y_2 + \cdots + a_{m,2}x_my_2 \\ + a_{1,3}x_1y_3 + a_{2,3}x_2y_3 + a_{3,3}x_3y_3 + \cdots + a_{m,3}x_my_3 \\ \ldots\ldots\ldots \\ + a_{1,n}x_1y_n + a_{2,n}x_2y_n + a_{3,n}x_3y_n + \cdots + a_{m,n}x_my_n.$$

Thus, to show that the set S is a basis of V over F, we merely have to show that these vectors are linearly independent. Let us switch to a summation notation for the remainder of the proof. Suppose that there is a nonzero linear combination of these vectors that produces 0, that is

$$\sum_{i=1}^{m}\sum_{j=1}^{n} a_{i,j}x_iy_j = 0$$

for $a_{i,j}$ in F. Then we have

$$0 = \sum_{i=1}^{m}\sum_{j=1}^{n} a_{i,j}x_iy_j = \sum_{j=1}^{n}\left(\sum_{i=1}^{m} a_{i,j}x_i\right) y_j.$$

Since $\{y_1, y_2, y_3, \ldots, y_n\}$ is a basis of V over E, the only way that the right-hand expression could be zero is if

$$\sum_{i=1}^{m} a_{i,j} x_i = 0$$

for all $j = 1, 2, 3, \ldots n$. Now $\{x_1, x_2, x_3, \ldots x_m\}$ is a basis of E over F, so the only way that each of these sums could be 0 is if $a_{i,j} = 0$ for all values of i and j. Since all of the coefficients must be 0, the vectors in S are linearly independent, and therefore the S is a basis of V over F of dimension mn. ▯

The main use of vector spaces in abstract algebra is in the case where the vector space happens to be a field. We will explore this possibility in the next section.

Problems for §14.1

For Problems **1** through **9**: Find a basis for the following fields over $\mathbb{Q}$.

1 $\mathbb{Q}(\sqrt{2})$
2 $\mathbb{Q}(\sqrt{5})$
3 $\mathbb{Q}(\sqrt{2}, \sqrt{3})$
4 $\mathbb{Q}(\sqrt{2}, \sqrt{3}, \sqrt{5})$
5 $\mathbb{Q}(\sqrt[3]{2})$
6 $\mathbb{Q}(\omega_9)$
7 $\mathbb{Q}(\omega_8)$
8 $\mathbb{Q}(\omega_9, \omega_6)$
9 $\mathbb{Q}(\sqrt[3]{2}, \omega_3)$

10 Find a basis for the field $\mathbb{Q}(\sqrt{2}, \sqrt{3})$ over the field $\mathbb{Q}(\sqrt{2})$.

11 Let $\{x_1, x_2, x_3, \ldots x_n\}$ be a set of vectors in a vector space V over F, such that every element of V can be expressed in the form

$$a_1 x_1 + a_2 x_2 + a_3 x_3 + \cdots a_n x_n$$

with $a_1, a_2, a_3, \ldots a_n$ in F. Show there is a subset of the x_i's that is a basis for V.

12 Let $\{x_1, x_2, x_3, \ldots x_n\}$ be a set of linearly independent vectors in a finite dimensional vector space V over F. Show that we can add vectors $\{z_1, z_2, \ldots z_m\}$ so that $\{x_1, x_2, x_3, \ldots x_n, z_1, z_2, \ldots z_m\}$ is a basis for V.

13 We can generalize Definition 14.1 by replacing the field F with a unity ring R. The resulting space M is called a *left R-module.* Show that any unity ring R can be considered to be a left R-module.

14 Show that any ring can be considered a left $\mathbb{Z}$-module. See Problem 13.

15 Let M be a left R-module. A subset N is an *R-submodule* of M if the N is a subgroup under addition, and rn is in N whenever $r \in R$ and $n \in N$. Show that the R-submodules of the module of Problem 13 are precisely the left ideals of R. See Problem 13 from §10.2.

16 Let R be a unity ring, and I an ideal of R. Show that R/I is a left R-module, where we define $r(a+I) = r \cdot a + I$.

17 Given two R-submodules A and B of a left R-module M, we define

$$A + B = \{a + b \mid a \in A, b \in B\}.$$

Show that $A + B$ is a R-submodule of M.

18 Let M be a left R-module, and let A, B, and C be three R-submodules, such that $A \subseteq C$. Prove the *modular law*: $(A + B) \cap C = A + (B \cap C)$. See Problem 17.

19 Show that any finite dimensional vector space of $\mathbb{Q}$ is countable.
Hint: First show that the direct product of two countable sets is countable.

Interactive Problems

20 Use *Sage* to find the coefficients of the vector $\langle 3, -2, 5\rangle$ in $\mathbb{R}^3$ using the basis $\{\langle 2, -1, 4\rangle, \langle 5, 2, 1\rangle, \langle 4, -3, 2\rangle\}$.

21 Use *Sage* to find the coefficients of the element a^5 in GF(27) over Z_3 using the basis $\{1, a, a^2\}$. Here, a is a root of the Conway polynomial of degree 3 over Z_3, which is $x^3 + 2x + 1$.

14.2 Extension Fields

In the last chapter, we saw that given an irreducible polynomial in $Z_p[x]$, we could find a larger finite field for which that polynomial has a root. In this section we imitate this process for infinite fields. The result will be a new field that will contain the old field as a subfield. As a result, the new field will also be a vector space over the old field. We will give a special name to this situation.

DEFINITION 14.5 If F is a nontrivial subfield of the field K, and K is a finite-dimensional vector space over F, we say that K is a *finite extension* of F. We say the degree of the extension, or *dimension* of the extension, is the size of a basis $\{x_1, x_2, x_3, \ldots x_n\}$ of K over F.

For example, the complex numbers $\mathbb{C}$ are a 2-dimensional extension of $\mathbb{R}$. The field GF(27) is a 3-dimensional extension of Z_3, regardless of which basis we use.

It seems as though isomorphic fields should have the same dimension over some field F contained in both of the fields. Yet this is only true if the isomorphism ϕ maps the base field F to itself.

PROPOSITION 14.3
If K and E are two finite extensions of F, and supposing that there is an isomorphism ϕ from K onto E such that $\phi(x) = x$ for all x in F, then K and E have the same dimension over F.

PROOF: Suppose that $\{x_1, x_2, x_3, \ldots x_n\}$ is a basis of K over F. We want to show that $\{\phi(x_1), \phi(x_2), \phi(x_3), \ldots, \phi(x_n)\}$ is a basis of E over F. If v is in E, then $\phi(u) = v$ for some u in K. Since K is generated by the elements in the basis, we have

$$u = c_1x_1 + c_2x_2 + c_3x_3 + \cdots c_nx_n$$

for some $c_1, c_2, c_3, \ldots, c_n$ in F. Then

$$\begin{aligned} v = \phi(u) &= \phi(c_1)\phi(x_1) + \phi(c_2)\phi(x_2) + \phi(c_3)\phi(x_3) + \cdots + \phi(c_n)\phi(x_n) \\ &= c_1\phi(x_1) + c_2\phi(x_2) + c_3\phi(x_3) + \cdots + c_n\phi(x_n). \end{aligned}$$

Thus, $\{\phi(x_1), \phi(x_2), \phi(x_3), \ldots, \phi(x_n)\}$ generates the field E. Also, if

$$c_1\phi(x_1) + c_2\phi(x_2) + c_3\phi(x_3) + \cdots + c_n\phi(x_n) = 0,$$

then $\phi(c_1x_1 + c_2x_2 + c_3x_3 + \cdots c_nx_n) = 0$, which implies that

$$c_1x_1 + c_2x_2 + c_3x_3 + \cdots c_nx_n = 0$$

since K and E are isomorphic. But since $\{x_1, x_2, x_3, \ldots x_n\}$ is a basis for K, this can only happen if $c_1 = c_2 = c_3 = \cdots c_n = 0$. So

$$\{\phi(x_1), \phi(x_2), \phi(x_3), \ldots, \phi(x_n)\}$$

is a basis for E over F, and hence K and E have the same dimension over the field F. ∎

If K is a finite extension of a field F, then F is a subfield of K. Of course there will probably be many other subfields of K, and we need a way to identify these subfields. We have already seen how to find the smallest subgroup or a subring that contains certain elements, and we can follow the same logic for subfields.

DEFINITION 14.6 Let K be a field, and let E be a field containing the field K. Let S be a set of elements in E. Let L denote the collection of all subfields of E that contain the field K, along with the set S. Then we define

$$K(S) = \bigcap_{H \in L} H.$$

That is, $K(S)$ is the intersection of all subfields of E that contain both K and S. If $S = \{a_1, a_2, a_3, \ldots a_n\}$, we will write $K(a_1, a_2, a_3, \ldots a_n)$ for $K(S)$. Thus, if S consists of a single element a, we can write $K(a)$ for $K(S)$.

LEMMA 14.3

Let K be a subfield of E, and let S be a collection of elements of E. Then $K(S)$ is the smallest field that contains both K and the elements S.

PROOF: First, we must show that $K(S)$ is a subfield of E. If x and y are in $K(S)$, $y \neq 0$, then x and y are in each of the subfields in the collection L. Then $x - y$ and $x \cdot y^{-1}$ are also in each of the subfields in this collection. Thus, $x - y$ and $x \cdot y^{-1}$ are in $K(S)$, and so $K(S)$ is a subfield of E.

To show that $K(S)$ is the smallest field containing both K and the elements S, note that $K(S)$ is one of the subfields in the collection L. Thus, any subfield containing both K and the elements of S must also contain $K(S)$. □

For example, if K is the real numbers, and $i = \sqrt{-1}$, then $\mathbb{R}(i)$ gives us the complex numbers $\mathbb{C}$. The field $\mathbb{Q}(\sqrt{2})$ is the smallest field containing $\mathbb{Q}$ and $\sqrt{2}$, which happens to be the same as the ring $\mathbb{Q}[\sqrt{2}]$.

The strategy for defining a field extension in *Sage* is very similar to that of defining a finite field. We begin by finding an irreducible polynomial $f(x)$ in the field F, and creating the field $K = F[x]/\langle f(x)\rangle$.

PROPOSITION 14.4

Let F be a field, and let $f(x)$ be an irreducible polynomial in $F[x]$ of degree n. Then the field $K = F[x]/\langle f(x)\rangle$ is a finite extension of F of dimension n.

PROOF: From Proposition 13.1, $K = F[x]/\langle f(x)\rangle$ is a field that contains F as a subfield. Let $y = x + \langle f(x)\rangle$ in K. If we treat $f(x)$ as a polynomial in $K[x]$, we find that $f(y) = 0$. Consider the set $\{1, y, y^2, y^3, \cdots y^{n-1}\}$. We wish to show that this set is a basis for K. That is, we wish to show that every element of K can be expressed uniquely as

$$k = a_1 1 + a_2 y + a_3 y^2 + \cdots + a_n y^{n-1},$$

where the $a_1, a_2, a_3, \ldots, a_n$ are in F. Any element $k \in K$ can be expressed as $k = g(x) + \langle f(x)\rangle$ for some polynomial $g(x)$ in $F[x]$. By the division algorithm theorem (12.1), there exist unique polynomials $q(x)$ and $r(x)$ such that

$$g(x) = f(x) \cdot q(x) + r(x),$$

where either $r(x) = 0$, or the degree of $r(x)$ is less than n. Then

$$r(x) = a_1 + a_2 x + a_3 x^2 + \cdots + a_n x^{n-1}$$

for some $a_1, a_2, a_3, \ldots, a_n$ in F. Note that we can now write

$$k = g(x) + \langle f(x)\rangle = r(x) + \langle f(x)\rangle = a_1 + a_2y + a_3y^2 + \cdots + a_ny^{n-1}.$$

Since $r(x)$ is unique, k is uniquely determined as a linear combination of $\{1, y, y^2, \ldots, y^{n-1}\}$. Thus, by Lemma 14.1, $\{1, y, y^2, \ldots, y^{n-1}\}$ is a basis. ☐

Computational Example 14.7

Let F be the field of rational numbers, and let $f(x) = x^3 - 2$. Form the field extension $\mathbb{Q}[x]/\langle x^3 - 2\rangle$ in *Sage*.

SOLUTION: Since the characteristic of $\mathbb{Q}$ is 0, we begin the definition by the command

```
InitDomain(0)
```

Next, we let a be a root to the equation $x^3 - 2$. That is, we define a^3 to be 2.

```
AddFieldVar("a")
Define(a^3, 2)
```

That's all there is to it! The basis of this extension field is $\{1, a, a^2\}$. We can verify this with *Sage*.

```
B = ToBasis([1, a, a^2])
    Successful mapping constructed.
```

We can also do divisions in this field.

```
1/(a + a^2)
    1/6*a^2 + 1/3*a - 1/3
```

This shows that $1/(\sqrt[3]{2} + \sqrt[3]{4}) = (2\sqrt[3]{2} + \sqrt[3]{4} - 2)/6$. ☐

Although this example demonstrates that any extension field of the form $F[x]/\langle f(x)\rangle$ can be entered into *Sage*, we would like to show that *any* extension field can be entered into *Sage* in the same way. That is, we must show that any finite extension of F is isomorphic to $F[x]/\langle f(x)\rangle$ for some polynomial $f(x)$.

PROPOSITION 14.5

Suppose a field K is a finite extension of F of dimension n. Let y be an element of K. Then there is an irreducible polynomial $f(x)$ in $F[x]$ of degree at most n such that $f(y) = 0$. That is, when $f(x)$ is treated as a polynomial in $K[x]$, y is a root of $f(x)$. Furthermore, there is a unique polynomial of lowest degree that satisfies these conditions and for which the leading coefficient is equal to 1.

PROOF: Consider the set $\{1, y, y^2, y^3, \ldots, y^n\}$. Since there are $n+1$ elements in this set, and K has dimension n over F, by Lemma 14.2 these are linearly dependent, so there is a nonzero solution to

$$a_0 + a_1 y + a_2 y^2 + a_3 y^3 + \cdots + a_n y^n = 0$$

with $a_0, a_1, a_2, \cdots, a_n$ in F. Thus, there is a nonzero polynomial

$$a_0 + a_1 x + a_2 x^2 + a_3 x^3 + \cdots + a_n x^n$$

in $F[x]$ for which y is a root when treated as a polynomial in $K[x]$.

Let us now show uniqueness. Let $f(x)$ be a polynomial of lowest possible degree in $F[x]$ such that $f(y) = 0$. Since F is a field, we can divide this polynomial by its leading coefficient to obtain a polynomial with a leading coefficient of 1. Now, if there were two such polynomials, $f(x)$ and $g(x)$, then by the division algorithm theorem (12.1), there exist polynomials $q(x)$ and $r(x)$ such that $f(x) = g(x) \cdot q(x) + r(x)$, where either $r(x) = 0$ or the degree of $r(x)$ is strictly less than the degree of $g(x)$. But note that

$$0 = f(y) = g(y) \cdot q(y) + r(y) = 0 + r(y) = 0.$$

Thus, y is a root of the polynomial $r(x)$. But the degree of $f(x)$ and $g(x)$ was chosen to be minimal. So $r(x) = 0$, and $f(x)$ is a multiple of $g(x)$. Finally, since both $f(x)$ and $g(x)$ have the same degree and have the same leading term of 1, we have $f(x) = g(x)$. Therefore, there is a unique polynomial in $F[x]$ of minimal degree and leading coefficient of 1 such that $f(y) = 0$. □

The unique polynomial in Proposition 14.5 will be given a special name.

DEFINITION 14.7 If a field K is a finite extension of F, and a is an element of K, we define the polynomial $f(x)$ given by Proposition 14.5 that has a leading coefficient of 1 to be the *irreducible polynomial of a over F*, denoted $\mathrm{Irr}_F(a, x)$.

For example, $\mathrm{Irr}_{\mathbb{Q}}(\sqrt{2}, x) = x^2 - 2$, since $x^2 - 2$ is the simplest polynomial with rational coefficients for which $\sqrt{2}$ is a root. Note that if we were to allow real coefficients, we could come up with a simpler polynomial: $\mathrm{Irr}_{\mathbb{R}}(\sqrt{2}, x) = x - \sqrt{2}$. Finally, consider the number $\cos(\pi/9)$. We found in §11.4 that this number is a root of the polynomial $4x^3 - 3x - \frac{1}{2}$. However, we want the leading coefficient of the polynomial to be 1, so we write

$$\mathrm{Irr}_{\mathbb{Q}}(\cos(\pi/9), x) = x^3 - \frac{3x}{4} - \frac{1}{8}.$$

Once we find the irreducible polynomial for an element a, it is not hard to program *Sage* to mimic the field $\mathbb{Q}(a)$. For example, let us enter the field $\mathbb{Q}(\cos(\pi/9))$ into *Sage*. If we let $a = \cos(\pi/9)$, we can enter the field by the commands

```
InitDomain(0)
AddFieldVar("a")
Define[a^3, 3*a/4 + 1/8]
```

The first command tells *Sage* that we are working with a field of characteristic 0, and the second command introduces the variable a. Finally, the last command identifies a as one solution to the equation $x^3 - 3x/4 - 1/8$. We can now do operations in this field.

```
a^5
    1/8*a^2 + 9/16*a + 3/32
1/a
    8*a^2 - 6
```

Have we really defined the field $\mathbb{Q}(\cos(\pi/9))$? Actually, we have defined the field

$$\mathbb{Q}[x]/\langle x^3 - 3x/4 - 1/8\rangle$$

in *Sage*, but we can prove that these two fields are isomorphic.

PROPOSITION 14.6

Let F be a subfield of K, and suppose $f(x)$ is an irreducible polynomial in $F[x]$ that has a root w in the larger field K. Then

$$F(w) \approx F[x]/\langle f(x)\rangle.$$

PROOF: Let us consider the evaluation homomorphism ϕ_w that maps polynomials in $F[x]$ to elements in $F(w)$:

$$\phi_w(g(x)) = g(w).$$

By Proposition 12.1, ϕ_w is a ring homomorphism. The image of this homomorphism contains both F and w, and since $F(w)$ is the smallest field containing both F and w, the image is all of $F(w)$. The kernel of ϕ_w is the set of polynomials in $F[x]$ that have w as a root. But $f(x)$ is an irreducible polynomial in $F[x]$ containing w as a root. Thus, any polynomial in the kernel is a multiple of $f(x)$. Hence, the kernel of ϕ_w is $\langle f(x)\rangle$. Finally, by the first ring isomorphism theorem (10.2), we have that $F(w) \approx F[x]/\langle f(x)\rangle$. ∎

It is now easy to see that the dimension of the field extension $F(u)$ will be the degree of the irreducible polynomial $f(x) = \text{Irr}_F(u, x)$.

COROLLARY 14.1

Let K be a finite extension of a field F, and let u be an element in K. If $f(x) = \text{Irr}_F(u, x)$ has degree n, then $F(u)$ has dimension n over F.

PROOF: By Proposition 14.5, $f(x) = \text{Irr}_F(u, x)$ exists. By Proposition 14.6,

$F(u)$ is isomorphic to the field $F[x]/\langle f(x)\rangle$. By Proposition 14.4, $F[x]/\langle f(x)\rangle$ has dimension n over F. Finally, by Proposition 14.3, two isomorphic extensions of F must have the same dimension over F provided that the isomorphism fixes the elements of F, which the isomorphism in Proposition 14.6 clearly does. Thus, the dimension of $F(u)$ over F is n. □

Notice that we never had to tell *Sage* that $a = \cos(\pi/9)$ in our definition of $\mathbb{Q}(\cos(\pi/9))$. Rather, we merely entered the information that a satisfies the equation $a^3 - 3a/4 - 1/8 = 0$.

But there are two *other* solutions to this equation, namely $-\cos(2\pi/9)$ and $\cos(4\pi/9)$. How does *Sage* know that the field is not $\mathbb{Q}(-\cos(2\pi/9))$ or $\mathbb{Q}(\cos(4\pi/9))$?

The answer of course is that these fields are both isomorphic to $\mathbb{Q}(\cos(\pi/9))$, so *Sage* did not need to know the exact value of a. In fact, we can prove that if we start with isomorphic fields, and extend both of them by two elements for which the irreducible polynomials correspond, then the two field extensions will be isomorphic.

PROPOSITION 14.7

Let f be an isomorphism between a field K and a field E. Let M be a finite extension of K, and let u be in M. Let

$$p(x) = c_0 + c_1x + c_2x^2 + c_3x^3 + \cdots + c_nx^n$$

be $\mathrm{Irr}_K(u, x)$. *Define*

$$h(x) = f(c_0) + f(c_1)x + f(c_2)x^2 + f(c_3)x^3 + \cdots + f(c_n)x^n$$

which is in $E[x]$. Suppose there is a finite extension of E for which there is a root of $h(x)$, called v. Then there is an isomorphism μ from $K(u)$ to $E(v)$ for which $\mu(u) = v$, and $\mu(t) = f(t)$ for all t in K.

PROOF: By Lemma 13.3, we can extend f to an isomorphism from $K[x]$ to $E[x]$. By Proposition 12.1, ϕ_v is a ring homomorphism from $E[x]$ to $E(v)$. We can combine these homomorphisms to produce the homomorphism

$$\phi_v \cdot f : K[x] \to E[x] \to E(v).$$

Since the isomorphism in Lemma 13.3 sends x to x, we have that $(\phi_v \cdot f)(x) = \phi_v(f(x)) = \phi_v(x) = v$. So v is in the image of this combination of homomorphisms, as well as the subfield E. Thus, the image of $\phi_v \cdot f$ is $E(v)$. The kernel of ϕ_v is the set of polynomials in $E[x]$ with v as a root. But $h(x)$ is an irreducible polynomial in $E[x]$ for which $h(v) = 0$. Thus, the kernel of ϕ_v is the ideal $\langle h(x)\rangle$. Since $h(x) = f(p(x))$, we have that the kernel of $f \cdot \phi_v$ is $\langle p(x)\rangle$. Thus, by the first ring isomorphism theorem (10.2),

$$K[x]/\langle p(x)\rangle \approx E(v).$$

By Proposition 14.6, we also have

$$K(u) \approx K[x]/\langle p(x)\rangle,$$

and in this isomorphism, u mapped to the coset $x + \langle p(x)\rangle$. If we let μ be the combination of these two isomorphisms,

$$\mu : K(u) \to K[x]/\langle p(x)\rangle \to E(v),$$

then $\mu(u) = \phi_v(f(x)) = v$, and $\mu(t) = f(t)$ for all t in K. ◻

The usual application of this proposition is when K and E are the same field, as in the case $\mathbb{Q}(\cos(\pi/9))$ and $\mathbb{Q}(-\cos(2\pi/9))$, in which case we not only can prove that $\mathbb{Q}(\cos(\pi/9))$ and $\mathbb{Q}(-\cos(2\pi/9))$ are isomorphic, but we can impose further conditions on the isomorphism.

COROLLARY 14.2

If K is a finite extension of a field F, and u and v are two elements in K such that $\mathrm{Irr}_F(u,x) = \mathrm{Irr}_F(v,x)$, then there is an isomorphism μ between $F(u)$ and $F(v)$ such that $\mu(u) = v$, and $\mu(t) = t$ for all t in F.

PROOF: We simply let f be the identity mapping from F to itself, and use Proposition 14.7. Then $p(x)$ and $h(x)$ are both equal to $\mathrm{Irr}_F(u,x)$. Since v is another root of $h(x)$, the conclusion follows from the conclusion of Proposition 14.7. ◻

We discovered in §13.2 that every finite field could be expressed in the form $Z_p[x]/\langle f(x)\rangle$, with $f(x)$ an irreducible polynomial in $Z_p[x]$. It is natural to ask whether any finite extension of a field can be represented in the form $F[x]/\langle f(x)\rangle$ for some polynomial $f(x)$ in $F[x]$. Although there are some fields that are exceptions, $\mathbb{Q}$ and $\mathbb{R}$ are not among them. Once we have proven this, we will be able to enter any finite extension of $\mathbb{Q}$ or $\mathbb{R}$ into *Sage* using the same technique that was used for finite fields.

Problems for §14.2

For Problems **1** through **8**: Find the following polynomials $\mathrm{Irr}_{\mathbb{Q}}(a,x)$.

Hint: Set $x = a$, and work to eliminate the roots.

1 $\mathrm{Irr}_{\mathbb{Q}}(\sqrt{5},x)$

2 $\mathrm{Irr}_{\mathbb{Q}}(\sqrt[3]{5},x)$

3 $\mathrm{Irr}_{\mathbb{Q}}(\sqrt{2}+\sqrt{3},x)$

4 $\mathrm{Irr}_{\mathbb{Q}}(\sqrt[3]{2}+\sqrt{2},x)$

5 $\mathrm{Irr}_{\mathbb{Q}}(\sqrt{\sqrt{2}-1},x)$

6 $\mathrm{Irr}_{\mathbb{Q}}(\sqrt[3]{\sqrt{5}-1},x)$

7 $\mathrm{Irr}_{\mathbb{Q}}(\sqrt{\sqrt[4]{3}+1},x)$

8 $\mathrm{Irr}_{\mathbb{Q}}\left(\sqrt{\sqrt{\sqrt{2}-1}+1},x\right)$

For Problems **9** through **12**: Find all of the roots of the polynomial.

9 $\text{Irr}_{\mathbb{Q}}(\sqrt{2}+\sqrt{3}, x)$. (See Problem **3**.)
10 $\text{Irr}_{\mathbb{Q}}(\sqrt[3]{2}+\sqrt{2}, x)$. (See Problem **4**.)
11 $\text{Irr}_{\mathbb{Q}}(\sqrt{\sqrt{2}-1}, x)$. (See Problem **5**.)
12 $\text{Irr}_{\mathbb{Q}}(\sqrt[3]{\sqrt{5}-1}, x)$. (See Problem **6**.)
13 $\text{Irr}_{\mathbb{Q}}\left(\sqrt{\sqrt[4]{3}+1}, x\right)$. (See Problem **7**.)
14 $\text{Irr}_{\mathbb{Q}}\left(\sqrt{\sqrt{\sqrt{2}-1}+1}, x\right)$. (See Problem **8**.)

15 Let a be a root of the equation

$$x^5+\sqrt{2}x^3+\sqrt{3}x^2+\sqrt{5}x+\sqrt{7}.$$

Show that $\mathbb{Q}(a)$ is a finite extension of $\mathbb{Q}$ with dimension at most 80.

16 Let K be a finite extension of a field F. If u and v are in K, prove that $F(u)(v) = F(v)(u)$.

17 Prove that $\mathbb{Q}(\sqrt{2})$ is not isomorphic to $\mathbb{Q}(\sqrt{3})$.

18 Find all of the automorphisms of $\mathbb{Q}(\sqrt{2}, \sqrt{3})$.

Interactive Problems

19 Define the field $\mathbb{Q}(\sqrt{-3})$ in *Sage*, then find $1/(5+\sqrt{-3})$.

20 Define the field $\mathbb{Q}(\sqrt{5})$ in *Sage*. Does the polynomial x^2+4x-1 factor in this field?

14.3 Splitting Fields

We have already seen that given an irreducible polynomial $f(x)$ in $F[x]$, we can construct a field $F[x]/\langle f(x)\rangle$ for which $f(x)$ has a root in this new field. This raises an interesting question: Can we construct a field for which $f(x)$ factors *completely* in the new field? In this section, we will learn how this can be done. Let us demonstrate with some examples.

Motivational Example 14.8
Find a field for which $f(x) = x^3+x^2-2x-1$ factors completely.
SOLUTION: We begin by showing that this polynomial is irreducible over the rationals.

```
InitDomain(0,"x")
factor(x^3 + x^2 - 2*x - 1)
    x^3 + x^2 - 2*x - 1
```

The **InitDomain(0, "x")** command defines $\mathbb{Q}[x]$, so we see that the polynomial is irreducible over $\mathbb{Q}[x]$.

If a is defined to be one root of this polynomial, we can define $\mathbb{Q}(a)$ in *Sage* as follows, and find the new factorization over $\mathbb{Q}(a)$.

```
AddFieldVar("a")
Define(a^3, - a^2 + 2*a + 1)
factor(x^3 + x^2 - 2*x - 1)
    (x - a) * (x - a^2 + 2) * (x + a^2 + a - 1)
```

This shows that the polynomial $x^3 + x^2 - 2x - 1$ factors completely as

$$(x - a)(x - a^2 + 2)(x + a^2 + a - 1)$$

in the field $\mathbb{Q}(a)$. □

In this case, creating an extension field allowed the polynomial to factor completely in the new field. In fact, this is very similar to what we discovered for finite fields. However, this will not always be the case.

Motivational Example 14.9

Find a field for which $x^3 - 2$ factors completely.

The factorization of this polynomial in $\mathbb{Q}(\sqrt[3]{2})$ is

$$(x - \sqrt[3]{2})(x^2 + \sqrt[3]{2}x + \sqrt[3]{4}).$$

Since the other two roots are complex, the quadratic term must be irreducible over $\mathbb{Q}(\sqrt[3]{2})$, since it is irreducible over the real numbers.

```
InitDomain(0, "x")
AddFieldVar("a")
Define(a^3, 2)
factor(x^3 - 2)
    (x - a) * (x^2 + a*x + a^2)
```

How can we get the polynomial $x^3 - 2$ to factor completely into linear terms? We can define a new element, b, to be a root of the irreducible quadratic. That is, we use an "extension of an extension" $\mathbb{Q}(\sqrt[3]{2}, b)$, where b satisfies $a^2 + ab + b^2 = 0$, that is, $b^2 = -\sqrt[3]{4} - b\sqrt[3]{2}$.

```
Define(b^2, -a^2 - a*b)
factor(x^3 - 2)
    (x + b + a) * (x - b) * (x - a)
```

Notice that $\mathbb{Q}(\sqrt[3]{2})$ is a 3-dimensional extension of $\mathbb{Q}$, and $\mathbb{Q}(\sqrt[3]{2}, b)$ is a 2-dimensional extension of $\mathbb{Q}(\sqrt[3]{2})$. Thus, by Proposition 14.2, $\mathbb{Q}(\sqrt[3]{2}, b)$ is a 6-dimensional extension of $\mathbb{Q}$. ∎

Example 14.10

A longer example of this process is the polynomial $x^4 - x + 1$. Find a field extension for which this factors.

SOLUTION:

```
InitDomain(0, "x")
factor(x^4 - x + 1)
    x^4 - x + 1
```

We define a to be a root of the irreducible polynomial.

```
AddFieldVar("a")
Define(a^4, a - 1)
factor(x^4 - x + 1)
    (x - a) * (x^3 + a*x^2 + a^2*x + a^3 - 1)
```

This leaves a third-degree polynomial still unfactored, so we will define b to be a root of this polynomial.

```
AddFieldVar("b")
Define(b^3, 1 - a^3 - a^2*b - a*b^2)
factor(x^4 - x + 1)
    (x - a) * (x - b) * (x^2 + (b + a)*x + b^2 + a*b + a^2)
```

This still leaves an unfactored quadratic. Let c be a root of this quadratic

```
AddFieldVar("c")
Define(c^2, - a^2 - a*b - b^2 - a*c - b*c)
factor(x^4 - x + 1)
    (x + c + b + a) * (x - a) * (x - b) * (x - c)
```

We finally did it! Each time we create an extension in *Sage* that forces another root to the equation, the remaining polynomial refuses to factor in the new field extension. Thus, it requires three field extensions before it finally factors completely. By this time, the final extension is a $4 \cdot 3 \cdot 2 = 24$ dimensional over the rational numbers $\mathbb{Q}$. ∎

From this example it is easy to see that this procedure could be carried out over any polynomial.

LEMMA 14.4

Let F be a field, and let $f(x)$ be a polynomial in $F[x]$ of degree n whose leading coefficient is c_n. Then there is a finite extension K of F such that

$$f(x) = c_n \cdot (x - u_1) \cdot (x - u_2) \cdot (x - u_3) \cdots (x - u_n),$$

where $u_1, u_2, u_3, \ldots u_n$ are elements in K. Furthermore, the dimension of K over F is at most $n!$.

PROOF: The proof is by induction on n. If $n = 1$, then $f(x)$ is a linear function, so its only root is in F. Thus $K = F$, and the degree of K over F is $1 = 1!$.

Suppose that this is true for polynomials of degree less than n. Let $p(x)$ be an irreducible factor of $f(x)$, and consider the field $E = F[x]/\langle p(x)\rangle$. By Proposition 14.4, E is a finite extension of F whose dimension over F is the degree of $p(x)$, which is at most n. Then $u_n = x + \langle p(x)\rangle$ is a root of $p(x)$ in the field E, and since $p(x)$ is a factor of $f(x)$, $(x - u_n)$ is a factor of $f(x)$ in the field E. Thus, we can write $f(x) = g(x)\cdot(x - u_n)$ for some $g(x)$ in $E[x]$. Note that $g(x)$ has degree $(n-1)$, and has the same leading coefficient as $f(x)$. Thus, we can use the induction hypothesis to show that there is a field K that is a finite extension of E with dimension at most $(n-1)!$ such that $g(x)$ factors completely as

$$g(x) = c_n \cdot (x - u_1)\cdot(x - u_2)\cdot(x - u_3)\cdot\cdots\cdot(x - u_{n-1}).$$

Thus,

$$f(x) = c_n \cdot (x - u_1)\cdot(x - u_2)\cdot(x - u_3)\cdot\cdots\cdot(x - u_{n-1})\cdot(x - u_n).$$

By Proposition 14.2, the dimension of K over F is the product of the dimension of E over F times the dimension of K over E. Thus, the dimension of K over F is at most $n\cdot(n-1)! = n!$. ∎

DEFINITION 14.8 If K is a field for which the polynomial $f(x)$ in $F[x]$ factors as

$$f(x) = c_n \cdot (x - u_1)\cdot(x - u_2)\cdot(x - u_3)\cdots(x - u_n),$$

then the field $F(u_1, u_2, u_3, \ldots u_n)$ is called the *splitting field* for the polynomial $f(x)$.

For example, the splitting field of x^3+x^2-2x-1 was found to be $\mathbb{Q}(a)$, where a is one root of the polynomial. Thus, the splitting field is a 3-dimensional extension of $\mathbb{Q}$. The splitting field of $x^3 - 2$ turned out to be a 6-dimensional extension of $\mathbb{Q}$. The splitting field of $x^4 - x + 1$ turned out to be a 24-dimensional extension of $\mathbb{Q}$. Lemma 14.4 points out that this is the largest possible dimension of a fourth-degree polynomial.

Computational Example 14.11

Find the splitting field for the polynomial $x^5 - 5x + 12$.

SOLUTION: When we factor this over the field $\mathbb{Q}(a)$, where a is a root of the polynomial,

```
InitDomain(0, "x")
factor(x^5 - 5*x + 12)
     x^5 - 5*x + 12
AddFieldVar("a")
Define(a^5, 5*a - 12)
factor(x^5 - 5*x + 12)
     (x - a) * (x^2 + (-1/4*a^4 - 1/4*a^3 - 1/4*a^2 + 3/4*a + 1)*x
     - 1/4*a^4 - 1/4*a^3 - 1/4*a^2 - 5/4*a + 2)
     * (x^2 + (1/4*a^4 + 1/4*a^3 + 1/4*a^2 + 1/4*a - 1)*x
     - 1/2*a^3 - 1/2*a - 1)
```

we find it does not split completely. We can let b be a root to the last polynomial, and try again.

```
AddFieldVar("b")
Define(b^2, 1 + a/2 + a^3/2 + b - (a + a^2 + a^3 + a^4)*b/4)
factor(x^5 - 5*x + 12)
     (x - b) * (x - a) * (x + (1/2*a + 1/2)*b + 1/2*a - 1/2) *
     (x + b + 1/4*a^4 + 1/4*a^3 + 1/4*a^2 + 1/4*a - 1) *
     (x + (-1/2*a - 1/2)*b - 1/4*a^4 - 1/4*a^3 - 1/4*a^2 + 1/4*a +
     3/2)
```

This time, the polynomial factors completely in $\mathbb{Q}(a, b)$. Hence the splitting field is 10-dimensional over $\mathbb{Q}$. ∎

If we had let b be a root of the *other* quadratic, would we get the same splitting field? The answer is yes, since the splitting fields are uniquely determined up to isomorphism. In order to prove this by induction, we actually have to prove slightly more.

PROPOSITION 14.8

Let ϕ be an isomorphism from the field F to a field E. Let

$$f(x) = c_0 + c_1x + c_2x^2 + c_3x^3 + \cdots + c_nx^n$$

be a polynomial in $F[x]$. Then

$$g(x) = \phi(c_0) + \phi(c_1)x + \phi(c_2)x^2 + \phi(c_3)x^3 + \cdots + \phi(c_n)x^n$$

is a polynomial in $E[x]$. Suppose that K is a splitting field of $f(x)$ over F, and L is a splitting field of $g(x)$ over E. Then there is an isomorphism μ from K to L, such that $\mu(t) = \phi(t)$ for all t in F.

PROOF: If $f(x)$ has degree 1, then the roots of $f(x)$ are in F, and the roots of $g(x)$ are in E. Thus, $K = E$, and $L = F$, and so the function $\mu(t) = \phi(t)$ satisfies the necessary conditions.

Let us use induction on the degree of the polynomial $f(x)$. That is, we will assume that the proposition is true for all polynomials of degree $(n-1)$. By Lemma 13.3, the isomorphism ϕ extends to an isomorphism from $F[x]$ to $E[x]$ in such a way that $\phi(x) = x$. Thus, if $p(x)$ is an irreducible factor of the polynomial $f(x)$, then $\phi(p(x))$ is an irreducible factor of the polynomial $g(x) = \phi(f(x))$. Note that every root of $p(x)$ is also a root of $f(x)$, so that $p(x)$ factors completely in the field K. Likewise, $\phi(p(x))$ factors completely in the field L.

Let u be a root of $p(x)$ in K, and let v be a root of $\phi(p(x))$ in L. By Proposition 14.7, there is an isomorphism θ mapping $F(u)$ to $E(v)$, such that $\theta(u) = v$, and $\theta(t) = \phi(t)$ for all t in F.

Since u is a root of $f(x)$, we can write $f(x) = (x-u) \cdot h(x)$, with $h(x)$ in $F(u)[x]$. Then

$$g(x) = \phi(f(x)) = \theta(f(x)) = \theta(x-u) \cdot \theta(h(x)) = (x-v) \cdot \theta(h(x)).$$

Since $h(x)$ has degree $(n-1)$, we can use the induction hypothesis. Obviously K is the splitting field of $h(x)$ over $F(u)$, and L is the splitting field of $\theta(h(x))$ over $E(v)$. Thus, by the induction hypothesis, the proposition is true for the polynomial $h(x)$, so there is an isomorphism μ such that $\mu(t) = \theta(t)$ for all t in $F(u)$. Since $\theta(t) = \phi(t)$ for all t in F, we have found an isomorphism with the necessary properties. ∎

COROLLARY 14.3

If $f(x)$ is a polynomial in $F[x]$, then all splitting fields of $f(x)$ are isomorphic.

PROOF: Simply let $F = E$, and let $\phi(t) = t$ for all t in F. Then by Proposition 14.8, any two splitting fields of $f(x) = g(x)$ will be isomorphic. ∎

In §13.3, we studied the properties of cyclotomic polynomials. It will be important later on to determine the splitting fields of these polynomials. For example, the ninth cyclotomic polynomial is given as

```
Cyclotomic(9, "x")
    x^6 + x^3 + 1
```

The splitting field found by defining a^6 to be $-1-a^3$.

```
InitDomain(0, "x")
AddFieldVar("a")
Define(a^6, - 1 - a^3)
factor(x^6 + x^3 + 1)
    (x - a) * (x - a^2) * (x - a^4) * (x + a^4 + a) * (x - a^5) *
    (x + a^5 + a^2)
```

This shows us that the splitting field is simply $\mathbb{Q}(a)$, where a is one root of the polynomial.

We can quickly generalize this result to apply to all cyclotomic polynomials.

PROPOSITION 14.9
The splitting field of the nth cyclotomic polynomial has dimension $\phi(n)$ over $\mathbb{Q}$, where $\phi(n)$ is Euler's totient function. In fact, the splitting field is given as $\mathbb{Q}(\omega_n)$, where ω_n is a primitive n^{th} root of unity.

PROOF: The generator

$$\omega_n = e^{(2\pi i/n)} = \cos\left(\frac{2\pi}{n}\right) + i\sin\left(\frac{2\pi}{n}\right)$$

is a root of the nth cyclotomic polynomial

$$\Phi_n(x) = (x - (\omega_n)^{k_1}) \cdot (x - (\omega_n)^{k_2}) \cdot (x - (\omega_n)^{k_3}) \cdot \cdots \cdot (x - (\omega_n)^{k_i}),$$

where $k_1, k_2, k_3, \ldots k_i$ are the integers from 1 to n that are coprime to n. Thus, the splitting field contains $\mathbb{Q}(\omega_n)$. Note that all powers of ω_n are in this field, and so the nth cyclotomic polynomial factors completely in $\mathbb{Q}(\omega_n)$.

Since $\Phi_n(x)$ is irreducible by Gauss's theorem on cyclotomic polynomials (13.2), and the dimension of $\Phi_n(x)$ is $\phi(n)$, the dimension of the splitting field $\mathbb{Q}(\omega_n)$ is $\phi(n)$. ∎

We now will show that splitting fields have special properties that most field extensions do not have. For example, we can define the splitting field of $x^3 - 2$ as follows:

```
InitDomain(0, "x")
AddFieldVar("a")
Define(a^3, 2)
AddFieldVar("b")
Define(b^2, - a^2 - a*b)
```

Note that $x^2 + 3$ factors in the splitting field, as does $x^6 + 108$. In fact, both polynomials factor completely in this field $\mathbb{Q}(a, b)$.

```
factor(x^2 + 3)
    (x - a^2*b - 1) * (x + a^2*b + 1)
factor(x^6 + 108)
    (x - b + a) * (x + b - a) * (x + b + 2*a) * (x + 2*b + a) *
    (x - 2*b - a) * (x - b - 2*a)
```

This last example suggests a startling fact: Whenever an irreducible polynomial in $\mathbb{Q}[x]$ has just one root in a splitting field, then the polynomial factors completely in the splitting field. This property characterizes splitting fields from other extensions of $\mathbb{Q}$.

LEMMA 14.5
Let K be the splitting field of a polynomial $f(x)$ in $F[x]$. Then if $p(x)$ is an irreducible polynomial in $F[x]$ for which there is one root in K, then $p(x)$ factors completely in K.

PROOF: Let $u_1, u_2, u_3, \ldots, u_n$ be the roots of $f(x)$ in K. Then

$$K = F(u_1, u_2, u_3, \ldots, u_n).$$

Suppose that $p(x)$ has one root v in K. Consider $p(x)$ as a polynomial in K, and let L be the splitting field of $p(x)$ over K. Let w be any other root of $p(x)$ in L besides v. To show that $K = L$, we need to show that w is in K, which would show that all roots of $p(x)$ are in K.

By Proposition 14.7, there is an isomorphism ϕ from $F(v)$ to $F(w)$ such that $\phi(v) = w$, and $\phi(t) = t$ for all t in F. (We let $f(t) = t$, the identity map, and let E and K both be the field F.) By Lemma 13.3 we can extend ϕ to an isomorphism from $F(v)[x]$ to $F(w)[x]$, and $\phi(f(x)) = f(x)$.

We now want to consider the field $K(w)$. We have

$$K(w) = F(u_1, u_2, u_3, \ldots, u_n, w) = F(w, u_1, u_2, u_3, \ldots, u_n).$$

Thus, $K(w)$ is the splitting field of $f(x)$ over the field $F(w)$. Since v is in K,

$$K = K(v) = F(u_1, u_2, u_3, \ldots, u_n, v) = F(v, u_1, u_2, u_3, \ldots, u_n),$$

so K is the splitting field of $f(x)$ over the field $F(v)$.

Consequently, Proposition 14.8 shows us that the isomorphism ϕ from $F(v)$ to $F(w)$ extends to an isomorphism μ from K to $K(w)$, and $\mu(v) = w$. Also, $\mu(t) = t$ for all t in F. Thus, we can use Corollary 14.3 to show that K and $K(w)$ have the same dimension over F. By Proposition 14.2, the dimension of $K(w)$ over F equals the dimension of $K(w)$ over K times the dimension of K over F. Therefore, the dimension of $K(w)$ over K must be 1, so w is in K. Therefore, every root of $p(x)$ is in K, so $p(x)$ factors completely in K. □

The fact that the splitting field of $x^6 + 108$ is the same as the splitting field of $x^3 - 2$ reveals another curious property of splitting fields. Rather than having to make an "extension of an extension" to define the splitting field $\mathbb{Q}(a, b)$, we could have defined the same field using a single extension of the element $w = \sqrt[6]{-108}$.

DEFINITION 14.9 We say that a finite extension of a field K is called a *simple extension* if it can be expressed as $K(a)$ for some element a.

The splitting field of $x^3 - 2$, even though it was originally described as an extension of an extension, is in fact a simple extension of $\mathbb{Q}$ of dimension 6.

Let us show, using the splitting fields, that an extension of an extension will usually form a simple extension.

PROPOSITION 14.10

Let F be a field, and let K be a finite-dimensional extension of F. Suppose that $K = F(u, v)$ with u, v in K. Let L be the splitting field of the polynomial $g(x) = \mathrm{Irr}_F(v, x)$, and suppose that there are no multiple roots of $g(x)$ in the field L. Then there is an element w of K such that $K = F(w)$.

PROOF: If F is a finite field, then K will also be a finite field, and we can simply let w be a generator of the multiplicative group K^*, using Proposition 13.4. Thus, we will assume that F is an infinite field. Let $f(x) = \mathrm{Irr}_F(u, x)$ and $g(x) = \mathrm{Irr}_F(v, x)$. Let E be the splitting field of $g(x)$ over the field $F(u)$. Since $g(x)$ factors completely in L without double roots, $g(x)$ will also factor completely in E without double roots. Let $v = v_1, v_2, v_3, \ldots, v_n$ be the distinct roots of $g(x)$ in E.

Since u is in E, there is at least one root of $f(x)$ in the field E. Even though $f(x)$ may not factor completely in the field E we can let $u = u_1, u_2, u_3, \ldots, u_m$ be the roots of $f(x)$ over E.

Since F is an infinite field, we can pick some element y of F, such that

$$y \neq \frac{u_h - u}{v - v_k} \qquad \text{for all } 1 \leq h \leq m, \quad 1 < k \leq n.$$

Finally, we let $w = u + yv$. Let us show that $K = F(w)$. To show that v is in $F(w)$, let $p(x) = f(w - yx)$, and note that $p(v) = f(u + yv - yv) = f(u) = 0$ so v is a root of $p(x)$. If one of the other roots of $g(x)$ is a root of $h(x)$, then $w - yv_k = u_h$ for some h and k, so $u + yv - yv_j = u_h$, giving us

$$y = \frac{u_h - u}{v - v_k},$$

and we specifically chose y so that it would avoid these values. Thus, there is only one root in common between $g(x)$ and $p(x)$ in the field E.

Let $r(x) = \mathrm{Irr}_{F(w)}(v, x)$. Then $r(x)$ divides the polynomials $g(x)$ and $p(x)$, since both polynomials have v as a root. In fact, we have seen that $g(x)$ and $p(x)$ have no other roots in common, so $r(x)$ has only one root in the field E. But $g(x)$ splits completely in E, and has no multiple roots in E. Thus, $r(x)$ has degree 1, and in fact $r(x) = x - v$. This proves that v is in $F(w)$. To see that u is in $F(w)$, we note that $u = yv - w$. Thus, $F(u, v)$ is contained in $F(w)$ while $F(w)$ is obviously contained in $F(u, v)$. Therefore, $F(u, v) = F(w)$. □

COROLLARY 14.4

Let K be a finite-dimensional extension of F, with $K = F(u_1, u_2, u_3, \ldots u_n)$ and suppose that none of the polynomials $\mathrm{Irr}_F(u_i, x)$ have multiple roots in *each of their splitting fields. Then there exists an element w in K such that $K = F(w)$.*

PROOF: We will proceed by induction on n. If $n = 1$, we can let $w = u_1$, and there is nothing to prove. Suppose that the corollary is true for the previous case, so that we found a u in K such that $F(u) = F(u_1, u_2, u_3, \ldots, u_{n-1})$. Let $v = u_n$, and since $g(x) = \mathrm{Irr}_F(u_n, x)$ does not have a multiple root in its splitting field L, we can use Proposition 14.10 to find a w in K such that $F(w) = F(u, v)$. But then $F(w) = F(u_1, u_2, u_3, \ldots, u_{n-1}, u_n)$. Thus, the corollary is true for all positive values of n. ∎

Sage has a function **SimpleExtension** that finds one of the many elements w for which the field $\mathbb{Q}(a, b, \ldots) = \mathbb{Q}(w)$.

Computational Example 14.12

We have seen the splitting field of $x^3 - 2$ is $\mathbb{Q}(\sqrt[3]{2}, \omega_3\sqrt[3]{2})$. Find a single element w such that the splitting field is $\mathbb{Q}(w)$.

SOLUTION: First we set up the splitting field as $\mathbb{Q}(a, b)$, where $a^3 = 2$ and $b^2 = -a^2 - a \cdot b$.

```
InitDomain(0, "x")
AddFieldVar("a")
Define(a^3, 2)
AddFieldVar("b")
Define(b^2, - a^2 - a*b)
```

We then can find an element w by the command

```
SimpleExtension(a, b)
   2*b + a
```

Thus, $\mathbb{Q}(a, b) = \mathbb{Q}(a + 2b)$, which is a simple extension. This element turns out to be a sixth root of -108. ∎

How does this command work? The key is in the proof of Proposition 14.10. Within the proof, we found that $F(u, v) = F(u + yv)$, where y is any number such that

$$y \neq \frac{u_h - u}{v - v_k}$$

whenever u_h is a root of $\mathrm{Irr}_F(u, x)$, and v_k is a root of $\mathrm{Irr}_F(v, x)$.

Example 14.13

Consider $\mathbb{Q}(\sqrt[3]{2}, \sqrt{2})$. This is not a splitting field, but it is contained in the splitting field of $f(x) = (x^3 - 2)(x^2 - 2)$, which does not have multiple roots, so we can still apply Proposition 14.10 to show that $\mathbb{Q}(\sqrt[3]{2}, \sqrt{2}) = \mathbb{Q}(w)$ for some element w. But what is that element?

SOLUTION: Note that $\mathrm{Irr}_{\mathbb{Q}}(\sqrt[3]{2}, x) = x^3 - 2$, which has roots of $\sqrt[3]{2}$, $\omega_3\sqrt[3]{2}$, and $\omega_3^2\sqrt[3]{2}$. Likewise, $\mathrm{Irr}_{\mathbb{Q}}(\sqrt{2}, x) = x^2 - 2$, which has roots of $\pm\sqrt{2}$. Hence, we must pick a rational value of y that is not equal to

$$\frac{\omega_3^i\sqrt[3]{2} - \sqrt[3]{2}}{\sqrt{2} \pm \sqrt{2}}.$$

That is, y cannot equal 0, $(\omega_3 - 1)\sqrt[3]{2}/(2\sqrt{2})$, or $(\omega_3^2 - 1)\sqrt[3]{2}/(2\sqrt{2})$. Any other rational value of y will do, so for convenience we can take $y = 1$. Then $w = u + yv = \sqrt[3]{2} + \sqrt{2}$.

We can also have *Sage* find an element for us.

```
InitDomain(0)
AddFieldVar("a")
Define(a^3, 2)
AddFieldVar("b")
Define(b^2, 2)
SimpleExtension(a, b)
    b + a
```

□

There is in fact an easier way to find a simple extension in this case. Merely note that $\sqrt[6]{2} \in \mathbb{Q}(\sqrt[3]{2}, \sqrt{2})$, since $\sqrt[6]{2} = \sqrt{2}/\sqrt[3]{2}$. Yet $\sqrt{2} = \sqrt[6]{2}^3$, and $\sqrt[3]{2} = \sqrt[6]{2}^2$. So $\mathbb{Q}(\sqrt[3]{2}, \sqrt{2}) = \mathbb{Q}(\sqrt[6]{2})$.

The fact that we can convert an extension of an extension to a simple extension will simplify many of the proofs involving splitting fields. In particular, it will allow us to explore the automorphisms of the splitting fields. In the next chapter we will discover that the automorphisms of the splitting fields determine much of the information about the roots of the polynomial, and whether they can be expressed in terms of square roots and cube roots. This beautiful correlation is referred to as Galois theory.

Problems for §14.3

For Problems **1** through **6**: Find a single number w such that the following field can be written as $\mathbb{Q}(w)$.

1 $\mathbb{Q}(\sqrt{2}, \sqrt[5]{2})$ **3** $\mathbb{Q}(\sqrt{2}, \sqrt{3}, \sqrt{5})$ **5** $\mathbb{Q}(\sqrt[3]{2}, \omega_3)$
2 $\mathbb{Q}(\sqrt{2}, \sqrt{5})$ **4** $\mathbb{Q}(\sqrt[3]{2}, i)$ **6** $\mathbb{Q}(\omega_3, \omega_5)$

7 Show by direct computation that if a and b are two distinct roots of $x^3 - 2$, then $(a + 2b)^6 = -108$.

Hint: Use the fact that $b^2 = -ab - a^2$ to simplify as you go along.

8 Use either a calculator's *Solve* function or De Moivre's theorem (11.2) to find decimal approximations of the three roots of $x^3 - 2 = 0$. Verify that $a^2 + ab + b^2 = 0$ whenever a and b are two of the three roots.

9 The polynomial $x^3 + x - 1$ has one real root $a \approx 0.6823278038\ldots$. Show that the splitting field of this polynomial is 6-dimensional over $\mathbb{Q}$.
Hint: If $(x - a)$ is one factor, what is the other? Show that this other factor is irreducible in $\mathbb{R}$, and hence is irreducible in $\mathbb{Q}(a)$.

10 Find the splitting field of $x^4 - 6x^2 - 7$.

11 Find the splitting field of $x^4 + x^2 + 1 = (x^2 + x + 1)(x^2 - x + 1)$.

12 Find the splitting field of $x^4 - 2x^2 - 1$.

13 Find the splitting field of $x^4 - x^2 + 1$.

14 Let $F = Z_2(t)$ be the rational functions of t modulo 2. Let K be the splitting field of $x^2 - t$ (that is, $K = F(\sqrt{t})$). Show that K is isomorphic to F, even though K is an extension of F of order 2.
Hint: Let ϕ be a homomorphism that sends $\sqrt{t}$ to t.

15 Suppose $f(x)$ and $g(x)$ are two polynomials in $\mathbb{Q}[x]$. Suppose that the splitting field of $f(x)$ is of dimension n over $\mathbb{Q}$, and the splitting field of $g(x)$ is of dimension m over $\mathbb{Q}$. Prove that the splitting field of $f(x) \cdot g(x)$ has dimension no more than $n \cdot m$.

16 Let m and n be distinct integers. Show directly that $\mathbb{Q}(\sqrt{m}, \sqrt{n}) = \mathbb{Q}(\sqrt{m} + \sqrt{n})$.
Hint: $(\sqrt{m} + \sqrt{n})$, $(\sqrt{m} + \sqrt{n})^2$, and $(\sqrt{m} + \sqrt{n})^3$ are all in $\mathbb{Q}(\sqrt{m} + \sqrt{n})$. Find a way of obtaining $\sqrt{m}$ and $\sqrt{n}$ from these three expressions.

Interactive Problems

For Problems **17** through **20**: Define the splitting field of the polynomial in *Sage*. Determine the dimension of the splitting field over $\mathbb{Q}$.

17 $x^3 + x^2 - 4x + 1$

18 $x^5 + x^4 - 4x^3 - 3x^2 + 3x + 1$

19 $x^5 - 2$

20 $x^5 + 3x^3 + 5x + 10$

Chapter 15

Galois Theory

This chapter covers the beautiful topic of Galois theory, whose origins lie in the theory of equations. Although methods of solving the quadratic equation were known to the ancient mathematicians, interest in the theory of equations was sparked in the 1540s when the cubic equation was solved by Scipione del Ferro and Niccoló Tartaglia, and a method for solving the fourth-degree equation was discovered soon after by Lodovico Ferrari. This raised the question on whether fifth-degree equations, or even higher-order equations, might have similar solutions. Mathematicians tried for centuries, but were unable to find such a formula, raising some doubts about whether such a formula could exist. An important step was taken in 1770 by Lagrange, who pointed out that the formulas for the quadratic, cubic, and quartic equations hinged on finding a combination of the roots of the polynomial that would remain unchanged whenever the roots were permutated. Finally, in 1824, Niels Abel proved that no such formula could exist. (See the Historical Diversion on page 241.) Independently, Évariste Galois gave a sharper result of determining which fifth-degree polynomials, or even higher-order polynomials, could be solved in terms of square roots, cube roots, or higher-order roots. Unfortunately, Galois did not live to see himself get the credit for his work (see the Historical Diversion on page 525.)

15.1 The Galois Group of an Extension Field

In the last chapter, we explored the extensions of a field, and found that any finite extension could be entered into *Sage* fairly easily. In particular, we explored the splitting fields of several polynomials. In this section, we will explore the automorphisms on the field extensions, and discover that the group of automorphisms contains much information about the polynomial. In particular, it will tell us whether the roots of the polynomial can be expressed in terms of square roots and cube roots.

DEFINITION 15.1 Let K be a finite extension of the field F. An *F-automorphism* of K is a ring automorphism ϕ on the field K that fixes every element of F. That is, $\phi(x) = x$ whenever x is in F.

Note that there is at least one F-automorphism of K, the identity automorphism. Since we have seen that the set of group automorphisms of a group forms another group, it is not surprising that the same thing happens for F-automorphisms of a field.

PROPOSITION 15.1

If K is a finite extension of a field F, then the set of all F-automorphisms of K forms a group under the operation of composition of functions.

PROOF: By Lemma 11.5, the set of all ring automorphisms of a ring forms a group. So we only need to show that the set of F-automorphisms of K is a subgroup of the group of all automorphisms. If ϕ_1 and ϕ_2 are two F-automorphisms of K, then $\phi_1(x) = \phi_2(x) = x$ for all x in F. Thus, $(\phi_1 \cdot \phi_2)(x) = \phi_1(\phi_2(x)) = x$ for all x in F. Thus, $\phi_1 \cdot \phi_2$ is an F-automorphism of K. Note also that $\phi_1^{-1}(x) = x$ for all x in F, so ϕ_1^{-1} is also an F-automorphism of K. Since the set of all F-automorphisms of K is closed under multiplications and inverses, this set is a subgroup of the group of automorphisms of K. Thus, the set of F-automorphisms of K is a group. ∎

DEFINITION 15.2 The set of all F-automorphisms of K is denoted $\text{Gal}_F(K)$, and is called the *Galois group of K over F.*

Example 15.1

Find the Galois group of the set of complex numbers $\mathbb{C}$ over the real numbers.

SOLUTION: According to Proposition 11.4, $\mathbb{C}$ has two automorphisms that fix the real numbers: the identity automorphism, and the automorphism that sends each number to its complex conjugate. So there are exactly two elements of $\text{Gal}_{\mathbb{R}}(\mathbb{C})$. In other words, $\text{Gal}_{\mathbb{R}}(\mathbb{C})$ is isomorphic to Z_2. ∎

We want to find a way to compute the Galois group of any finite extension of a field F. Since we can define finite extensions in terms of polynomials, it is natural to ask what must happen to the roots of a polynomial.

LEMMA 15.1

Let K be a finite extension of F, and let $f(x)$ be a polynomial in $F[x]$. If u is a root of $f(x)$, and ϕ is in $\text{Gal}_F(K)$, then $\phi(u)$ is also a root of $f(x)$.

PROOF: Let $f(x) = c_0 + c_1x + c_2x^2 + c_3x^3 + \cdots + c_nx^n$. Since u is a root of $f(x)$ we have that

$$c_0 + c_1u + c_2u^2 + c_3u^3 + \cdots + c_nu^n = 0.$$

Historical Diversion

Évariste Galois (1811–1832)

Évariste Galois was born near Paris, and was home schooled until he was 12 years old, when he entered the Lycée Louis-le-Grand. Galois performed well for the first two years, obtaining the first-place prize in Latin, but soon became bored, and turned his attention to mathematics.

He obtained a copy of Legendre's *Éléments de Géométrie*, and mastered the material on the first reading. At 15 he was reading Legrange's original papers, such as the "Reflections on the algebraic solutions of equations," which probably motivated him into the theory of equations. However, his classwork lagged as a result of his mathematical ambition.

In 1828, he applied for the prestigious École Polytechnique, but failed due to lack of explanations in the oral examination. He had to settle for the inferior École Normale, where he found some professors that understood his situation. The following year, Galois published a paper on continued fractions, and began working with the theory of polynomial equations, submitting two papers to the Academy of Sciences. Augustin Cauchy, the referee, refused to accept these papers, but recognized their importance, suggesting that they be combined into one paper and entered into the Academy's Grand Prize in Mathematics. He submitted his combined paper to the Academy's secretary Joseph Fourier, but unfortunately Fourier died shortly thereafter, and his submission was lost. Galois did publish three papers that year, one that laid the foundations of Galois theory, and another that introduced the concept of a finite field. He was the first person to use the word group (groupe) to refer to a group of permutations.

In the July Revolution of 1830, the last Bourbon king, Charles X, was sent into exile. But his successor, Louis-Philippe, was also disappointing. Galois proposed a toast to king Louis-Philippe with a dagger above his cup, which was interpreted as a death threat. He was arrested, but soon acquitted. In 1832 his rebellious nature caused him to be challenged to a duel. Although the details of the duel are still unclear, what is known is that the night before the duel, anticipating his own death, he spent the night writing a letter to Auguste Chevalier, outlining the connection between group theory and the solutions of polynomial equations by radicals. The next morning he was shot in the abdomen, and died the following day, at the age of 20.

Image source: Wikimedia Commons

Since ϕ is a ring homomorphism, we have that

$$\begin{aligned} 0 = \phi(0) &= \phi(c_0 + c_1 u + c_2 u^2 + c_3 u^3 + \cdots + c_n u^n) \\ &= \phi(c_0) + \phi(c_1)\phi(u) + \phi(c_2)\phi(u^2) + \phi(c_3)\phi(u^3) + \cdots + \phi(c_n)\phi(u^n). \end{aligned}$$

Since $c_0, c_1, c_2, \ldots c_n$ are in F, we have

$$0 = c_0 + c_1\phi(u) + c_2\phi(u)^2 + c_3\phi(u)^3 + \cdots + c_n\phi(u)^n.$$

Therefore, $\phi(u)$ is also a root of $f(x)$. □

Computational Example 15.2

Let us use this lemma to find the Galois group of the splitting field of $x^3 - 2$.
SOLUTION: The splitting field is defined in *Sage* by letting $a^3 = 2$, and $b^2 = -a^2 - ab$.

```
InitDomain(0, "x")
AddFieldVar("a")
Define(a^3, 2)
AddFieldVar("b")
Define(b^2, - a^2 - a*b)
factor(x^3 - 2)
    (x + b + a) * (x - b) * (x - a)
```

The three roots of $x^3 - 2$ are a, b, and $-a - b$. Thus, Lemma 15.1 tells us that if $F(x)$ is an automorphism on $\mathbb{Q}(a, b)$, then $F(a)$ is either a, b, or $-a-b$, while $F(b)$ is either a, b, or $-a - b$. Let us try to find an automorphism such that $F(a) = b$ and $F(b) = a$.

```
F = FieldHomo()
HomoDef(F, a, b)
HomoDef(F, b, a)
CheckHomo(F)
    True
```

We have successfully defined one automorphism of the Galois group. (Any nonzero homomorphism on a field must be an automorphism in light of Proposition 10.5, and the fact that the kernel is always an ideal.) We can similarly define an automorphism $G(x)$ on $\mathbb{Q}(a, b)$ such that $G(a) = b$, and $G(b) = -a - b$.

```
G = FieldHomo()
HomoDef(G, a, b)
HomoDef(G, b, - a - b)
CheckHomo(G)
    True
```

With these two automorphisms we can actually produce three more:

$$G(G(x)), \qquad F(G(x)), \quad \text{and} \quad G(F(x)).$$

Sage can show us that all five of these automorphisms are different, and if we include the identity automorphism, we have found six automorphisms on $\mathbb{Q}(a,b)$. Note that the Galois group is not abelian, since $F(G(x)) \neq G(F(x))$.

```
F(G(a))
    a
G(F(a))
    -b - a
```

□

It seems as though we must have found all of the automorphisms at this point, but this still needs to be proved. We begin by showing that there will always be an automorphism that moves one root of an irreducible polynomial to another.

PROPOSITION 15.2

Let K be the splitting field of some polynomial $f(x)$ over F, and let u and v be two elements of K. Then there exists an F-automorphism ϕ such that $\phi(u) = v$ if, and only if, $\mathrm{Irr}_F(u,x) = \mathrm{Irr}_F(v,x)$.

PROOF: If there is some ϕ such that $\phi(u) = v$, we can let $g(x) = \mathrm{Irr}_F(u,x)$ and $h(x) = \mathrm{Irr}_F(v,x)$. Then u is a root of $g(x)$, and v is a root of $h(x)$. By Lemma 15.1, u is a root of $h(x)$ and v is a root of $g(x)$, since $u = \phi^{-1}(v)$. So $g(x)$ is a multiple of $h(x)$, and vice versa. Since both have a leading coefficient of 1, we have that $g(x) = h(x)$.

Now suppose that $\mathrm{Irr}_F(u,x) = \mathrm{Irr}_F(v,x)$. Then by Corollary 14.2 there is an isomorphism ϕ from $F(u)$ to $F(v)$ such that $\phi(u) = v$, and $\phi(x) = x$ for all x in F. Since K is a splitting field of $f(x)$ over F, it is a splitting field of $f(x)$ over both $F(u)$ and $F(v)$. Therefore ϕ extends to an F-automorphism of K (which we will also denote ϕ) by Proposition 14.8. Therefore, ϕ is in $\mathrm{Gal}_F(K)$, and $\phi(u) = v$. □

The next lemma will be important in determining the subgroups of the Galois group.

LEMMA 15.2

Let K be a finite extension of F, and let ϕ be an F-automorphism of K. Then the set of all elements x such that $\phi(x) = x$ forms a subfield of K containing F.

PROOF: Let E be the set of all elements x such that $\phi(x) = x$. Since ϕ is an F-automorphism, by definition E must contain the elements of F. If x and y are in E, note that

$$\phi(x + y) = \phi(x) + \phi(y) = x + y,$$
$$\phi(x \cdot y) = \phi(x) \cdot \phi(y) = x \cdot y,$$
$$\phi(-x) = -\phi(x) = -x,$$
$$\phi(x^{-1}) = \phi(x)^{-1} = x^{-1}, \qquad \text{if } x \neq 0.$$

Thus, $x + y$, $x \cdot y$, and $-x$ are in E whenever x and y are, and x^{-1} is in E whenever $x \neq 0$ is in E. Thus, E is a subfield of K. ▯

Next we want to work on finding an upper bound on the number of elements in $\text{Gal}_F(K)$.

PROPOSITION 15.3

Let $K = F(u_1, u_2, u_3, \ldots, u_n)$ be a finite extension field of F. If ϕ_1 and ϕ_2 are two F-automorphisms in $\text{Gal}_F(K)$, and

$$\phi_1(u_1) = \phi_2(u_1), \qquad \phi_1(u_2) = \phi_2(u_2), \qquad \ldots \qquad \phi_1(u_n) = \phi_2(u_n),$$

then $\phi_1(x) = \phi_2(x)$ for all x in K. In other words, an F-automorphism in $\text{Gal}_F(K)$ is completely determined by its action on $u_1, u_2, u_3, \ldots, u_n$.

PROOF: Consider the F-automorphism $\phi_2^{-1}(\phi_1(x))$. It is clear that this automorphism fixes $u_1, u_2, u_3, \ldots u_n$, as well as the elements of F. By Lemma 15.2, the set E of all elements x such that $\phi_2^{-1}(\phi_1(x)) = x$ forms a subfield of K. But K is by Lemma 14.3 the smallest field containing u_1, u_2,$u_3, \ldots, u_n$, and F. Thus, $K = E$, and so $\phi_1(x) = \phi_2(x)$ for all x in K. ▯

We can now apply this proposition to the field $\mathbb{Q}(a, b)$ of Example 15.2. Any $\mathbb{Q}$-automorphism is determined by where it sends the elements a and b. By Lemma 15.1, these elements can only be sent to a, b, or $-a - b$. Yet an automorphism cannot send two elements to the same element. Thus, there are at most six $\mathbb{Q}$-automorphisms on the field $\mathbb{Q}(a, b)$. Yet we have found precisely six $\mathbb{Q}$-automorphisms of $\mathbb{Q}(a, b)$. Thus, we have found all of the $\mathbb{Q}$-automorphisms, and the Galois group of $\mathbb{Q}(a, b)$ contains exactly six elements. Furthermore, we observed that $\text{Gal}_{\mathbb{Q}}(\mathbb{Q}(a, b))$ was non-commutative, so we find that $\text{Gal}_{\mathbb{Q}}(\mathbb{Q}(a, b))$ must be isomorphic to S_3.

We can find an upper bound for the number of F-automorphisms in any splitting field using a similar argument.

COROLLARY 15.1

If K is the splitting field of a polynomial $f(x)$ of degree n in $F[x]$, then $\text{Gal}_F(K)$ is isomorphic to a subgroup of S_n.

PROOF: Since $f(x)$ has degree n in $F[x]$, there are at most n roots of $f(x)$ in K. Call these roots $u_1, u_2, \ldots, u_m$. Since K is the splitting field of $f(x)$ over F, we can write $K = F(u_1, u_2, u_3, \ldots, u_m)$. If ϕ is in $\mathrm{Gal}_F(K)$, then $\phi(u_1), \phi(u_2), \phi(u_3), \ldots, \phi(u_m)$ will be distinct roots of $f(x)$ by Lemma 15.1. Hence, ϕ will act as a permutation on the roots of $f(x)$. By Proposition 15.3, ϕ is completely determined by this permutation on the roots of $f(x)$. Thus, $\mathrm{Gal}_F(K)$ is isomorphic to a subgroup of S_m, and since m is not larger than n, $\mathrm{Gal}_F(K)$ is isomorphic to a subgroup of S_n. ∎

We immediately see from this corollary that the Galois group of a finite extension must be a finite group.

Let us look at one more example of a Galois group of a field. Consider the field $\mathbb{Q}(\sqrt[3]{2})$, which is a subfield of the field $\mathbb{Q}(a, b)$. Note that in this subfield all of the elements are *real*. Thus, in this field $\mathbb{Q}(\sqrt[3]{2})$ there is only one root to the polynomial $x^3 - 2$. Hence, if $\phi(x)$ is a $\mathbb{Q}$-automorphism of $\mathbb{Q}(\sqrt[3]{2})$, then $\phi(\sqrt[3]{2})$ must be $\sqrt[3]{2}$. By Proposition 15.3, the $\mathbb{Q}$-automorphism is completely determined by where ϕ sends $\sqrt[3]{2}$. Thus, $\mathrm{Gal}_{\mathbb{Q}}(\mathbb{Q}(\sqrt[3]{2}))$ is merely the trivial group.

In order to find the Galois group of a field, it is very helpful to know ahead of time the exact size of the Galois group. The next proposition allows us to compute the size of the Galois group for an important class of field extensions.

PROPOSITION 15.4

Suppose K is the splitting field of a polynomial $f(x)$ in $F[x]$, and that K can be expressed as a simple extension $K = F(w)$. If $\mathrm{Irr}_F(w, x)$ has no double roots in K, then the number of F-automorphisms in $\mathrm{Gal}_F(K)$ is precisely the dimension of K over F.

PROOF: Let d be the dimension of K over F. Then if $g(x) = \mathrm{Irr}_F(w, x)$, $g(x)$ has degree d. Since K is a splitting field and contains one root of $g(x)$, by Lemma 14.5, $g(x)$ splits completely in K. Since there are no double roots of $g(x)$ in K, then there are d roots $w = w_1, w_2, w_3 \cdots w_d$. Since $g(x)$ is irreducible, $\mathrm{Irr}_F(w_i, x) = \mathrm{Irr}_F(w, x)$ so Proposition 15.2 states that there is an F-automorphism that sends w to w_i for $1 \le i \le d$. Hence, there are at least d F-automorphisms. But by Proposition 15.3, the F-automorphism of $F(w)$ is determined by where it sends w, which must be one of the d roots. So $|\mathrm{Gal}_F(K)| = d$. ∎

We are ready to try a more complicated example.

Computational Example 15.3

Find the Galois group for the splitting field of the polynomial $x^4 - 2x^3 + x^2 + 1$.

SOLUTION: First we verify that this polynomial is irreducible.

```
InitDomain(), "x")
factor(x^4 - 2*x^3 + x^2 + 1)
    x^4 - 2*x^3 + x^2 + 1
```

Sage shows this polynomial is irreducible over $\mathbb{Q}$. Let us define a to be one root of this polynomial, and see how this polynomial factors over $\mathbb{Q}(a)$.

```
AddFieldVar("a")
Define(a^4, 2*a^3 - a^2 - 1)
factor(x^4 - 2*x^3 + x^2 + 1)
    (x - a) * (x + a - 1) * (x^2 - x + a^2 - a)
```

This tells us that if a is a root, then $1 - a$ is another root. However, it did not factor completely, so we have to define b to be a root of the irreducible quadratic.

```
AddFieldVar("b")
Define(b^2, b + a - a^2)
factor(x^4 - 2*x^3 + x^2 + 1)
    (x - b) * (x - a) * (x + a - 1) * (x + b - 1)
```

So the four roots are a, $1 - a$, b, and $1 - b$. Any $\mathbb{Q}$-automorphism will map each of these roots to another root, and so the Galois group will be a subgroup of S_4. But which permutations will give rise to a $\mathbb{Q}$-automorphism? A little trial and error will help.

Proposition 15.2 says that there will be some $\mathbb{Q}$-automorphism that sends any one of these four roots to any other of the four roots. So there is a $\mathbb{Q}$-automorphism that sends a to $1-a$. But where would it send the other three roots? Note that if $f(a) = 1 - a$, then $f(1 - a) = f(1) - f(a) = a$. So we only have to determine if $f(b)$ is b or $1-b$. *Sage* can show that both of these work, and can draw a picture of how these two $\mathbb{Q}$-automorphisms act on the four roots of the polynomial.

```
F = FieldHomo()
HomoDef(F, a, 1 - a)
HomoDef(F, b, b)
CheckHomo(F)
    True
G = FieldHomo()
HomoDef(G, a, 1 - a)
HomoDef(G, b, 1 - b)
CheckHomo(G)
    True
CircleGraph([a, 1 - a, b, 1 - b], F)
CircleGraph([a, 1 - a, b, 1 - b], G)
```

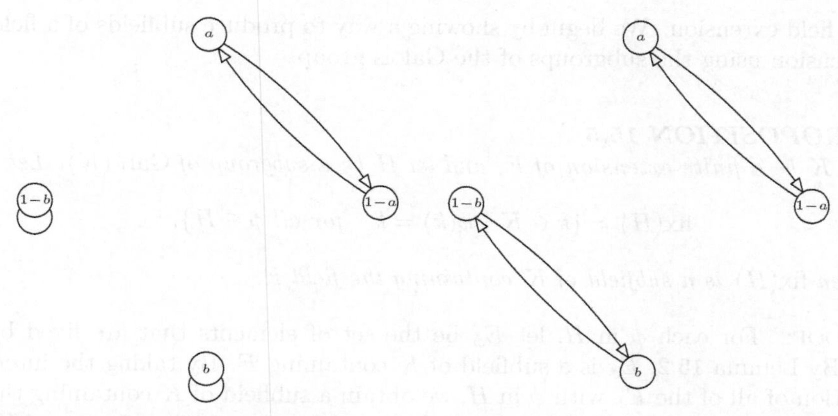

FIGURE 15.1: Automorphisms of splitting field for $x^4 - 2x^3 + x^2 + 1$

The circle graphs of these two automorphisms are depicted in Figure 15.1. If we number the four roots

1) a 2) $1 - a$ 3) b 4) $1 - b$

we can view these two $\mathbb{Q}$-automorphisms as $P(2,1)$ and $P(2,1,4,3)$. But Proposition 15.4 indicates that we must have eight $\mathbb{Q}$-automorphisms, so let us try mapping a to b. Then $1 - a$ would have to map to $1 - b$, but b could map to either a or $1 - a$. *Sage* shows that mapping b to a yields another $\mathbb{Q}$-automorphism, which would correspond to the permutation $P(3,4,1,2)$. If we find the subgroup generated by these three $\mathbb{Q}$-automorphisms

```
M = Group(P(2, 1), P(2, 1, 4, 3), P(3, 4, 1, 2)); M
    {P(), P(2, 1), P(1, 2, 4, 3), P(2, 1, 4, 3), P(3, 4, 1, 2),
    P(4, 3, 1, 2), P(3, 4, 2, 1), P(4, 3, 2, 1)}
```

we see that we have at least eight $\mathbb{Q}$-automorphisms. Since this is the number predicted by Proposition 15.4, we are done. Hence, we found the Galois group as a subgroup of S_4 of order 8. The command

```
RootCount(M, 2)
    6
```

shows that this group has six solutions to $x^2 = e$. Thus, the Galois group is isomorphic to D_4. ∎

This example shows the usefulness of Proposition 15.4 in finding the Galois group. In fact, sometimes the Galois group can be determined using only Corollary 15.1 and Proposition 15.4.

One of the tools we will use for finding the $\mathbb{Q}$-automorphisms is the close connection between the subgroups of the Galois group, and the subfields of the field extension. We begin by showing a way to produce subfields of a field extension using the subgroups of the Galois group.

PROPOSITION 15.5
Let K be a finite extension of F, and let H be a subgroup of $\mathrm{Gal}_F(K)$*. Let*

$$\mathrm{fix}(H) = \{k \in K \mid \phi(k) = k \quad \textit{for all } \phi \in H\}.$$

Then $\mathrm{fix}(H)$ *is a subfield of K containing the field F.*

PROOF: For each ϕ in H, let E_ϕ be the set of elements that are fixed by ϕ. By Lemma 15.2, E_ϕ is a subfield of K containing F. By taking the intersection of all of the E_ϕ with ϕ in H, we obtain a subfield of K containing the field F. ∎

DEFINITION 15.3 The field $\mathrm{fix}(H)$ is called the *fixed field* of the subgroup H.

Let us go back to Example 15.2, where we considered the Galois group of $\mathbb{Q}(a, b)$, where a and b were two roots of $x^3 - 2$.
The Galois group was described as

$$\{I(x), F(x), G(x), G(G(x)), F(G(x)), G(F(x))\},$$

where $I(x)$ represents the identity automorphism that sends every element to itself. The subgroups of $\mathrm{Gal}_{\mathbb{Q}}(\mathbb{Q}(a, b))$ are as follows:

$$H_1 = \{I(x)\}, \qquad H_2 = \{I(x), F(x)\}, \qquad H_3 = \{I(x), F(G(x))\},$$
$$H_4 = \{I(x), G(F(x))\}, \qquad H_5 = \{I(x), G(x), G(G(x))\},$$
$$H_6 = \{I(x), F(x), G(x), G(G(x)), F(G(x)), G(F(x))\}.$$

Example 15.4
Let us find the six fixed fields of $\mathbb{Q}(a, b) = \mathbb{Q}(\sqrt[3]{2}, \omega_3\sqrt[3]{2})$.
SOLUTION: The field $\mathrm{fix}(H_1)$ is the set of elements fixed by the identity mapping, which is of course all of $\mathbb{Q}(a, b)$. The field $\mathrm{fix}(H_2)$ contains the elements fixed by the mapping $F(x)$, which maps a to b, and b to a. Notice that the third root, $-a-b$, is fixed by the automorphism F. Thus, $\mathrm{fix}(H_2) = \mathbb{Q}(-a-b)$. By a similar argument, we see that $\mathrm{fix}(H_3) = \mathbb{Q}(a)$, and $\mathrm{fix}(H_4) = \mathbb{Q}(b)$. The field $\mathrm{fix}(H_5)$ is a little bit trickier, since $G(x)$ moves a, b, and $-a-b$. With a little bit of experimenting, we notice that

$$\begin{aligned} G(a^2b) &= b^2(-a-b) = (-a^2-ab)(-a-b) = a^3 + a^2b + a^2b + ab^2 \\ &= 2 + 2a^2b + a(-a^2-ab) = a^2b. \end{aligned}$$

If we substitute two of the roots of $x^3 - 2$ for a and b, that is, let $a = \sqrt[3]{2}$ and $b = \omega_3 \sqrt[3]{2}$, we find that $a^2 b$ is $2\omega_3 = -1 + \sqrt{-3}$. This agrees with our previous observation that $\sqrt{-3}$ is in the field $\mathbb{Q}(a, b)$. Since -1 is already rational, we can write the fixed field $\text{fix}(H_5)$ as $\mathbb{Q}(\sqrt{-3})$.

Finally, the only elements of $\mathbb{Q}(a, b)$ that are fixed by *all* $\mathbb{Q}$-automorphisms are the elements of $\mathbb{Q}$. Hence $\text{fix}(H_6) = \mathbb{Q}$. ∎

Notice that we have found six different subfields of $\mathbb{Q}(a, b)$ by using the six subgroups of the Galois group. We will discover in the next section that this is *all* of the subfields of $\mathbb{Q}(a, b)$. Thus, we have found a convenient way of finding all of the subfields of a given field.

Here is another example, although a bit easier.

Example 15.5

Find the fixed fields for the field $\mathbb{Q}(\sqrt[3]{2})$.

SOLUTION: Since the only $\mathbb{Q}$-automorphism is the identity automorphism, which fixes the whole group, the only fixed field of $\mathbb{Q}(\sqrt[3]{2})$ is $\mathbb{Q}(\sqrt[3]{2})$, even though there is the obvious subfield $\mathbb{Q}$ within this field. ∎

We were hoping to be able to find *all* subfields of a field by looking at the fixed fields, but in this example we failed. We will understand why the field $\mathbb{Q}(\sqrt[3]{2})$ is not as well behaved as $\mathbb{Q}(\sqrt[3]{2}, \omega_3 \sqrt[3]{2})$ in the next section.

Problems for §15.1

1 The Galois group $\text{Gal}_{\mathbb{Q}}(\mathbb{Q}(\sqrt{2}, \sqrt{3}))$ is given by $\{\phi_0, \phi_1, \phi_2, \phi_3\}$, where

$$\begin{array}{lcl} \phi_0(\sqrt{2}) = \sqrt{2} & \text{and} & \phi_0(\sqrt{3}) = \sqrt{3}, \\ \phi_1(\sqrt{2}) = \sqrt{2} & \text{and} & \phi_1(\sqrt{3}) = -\sqrt{3}, \\ \phi_2(\sqrt{2}) = -\sqrt{2} & \text{and} & \phi_2(\sqrt{3}) = \sqrt{3}, \\ \phi_3(\sqrt{2}) = -\sqrt{2} & \text{and} & \phi_3(\sqrt{3}) = -\sqrt{3}. \end{array}$$

Give the multiplication table for $\text{Gal}_{\mathbb{Q}}(\mathbb{Q}(\sqrt{2}, \sqrt{3}))$.

2 For each of the automorphisms in Problem 1, find the fixed field of the automorphism.

3 Find the automorphisms of the field $\mathbb{Q}(\sqrt[6]{2})$.

4 Find the fixed field of each of the automorphisms in Problem 3.

5 The four solutions of $x^4 - 2 = 0$ are $r_1 = \sqrt[4]{2}$, $r_2 = i\sqrt[4]{2}$, $r_3 = -\sqrt[4]{2}$, and $r_4 = -i\sqrt[4]{2}$. Thus, $K = \mathbb{Q}(\sqrt[4]{2}, i)$ is the splitting field of $x^4 - 2$. Determine the eight automorphisms of the field K, by finding where each automorphism maps the four roots.

Hint: If $\phi(r_1) = r_2$, then $\phi(-r_1) = -r_2$. It helps to use permutations to represent the automorphisms.

6 Determine what familiar group the automorphism group of Problem 5 is isomorphic to?

7 Label the three solutions of $x^3 - 3 = 0$ as $\sqrt[3]{3}$, r_2, and r_3. Determine the six automorphisms of the splitting field of $x^3 - 3$ by finding where each automorphism maps the three roots.

8 Find the Galois group of the field $\mathbb{Q}(\sqrt{2}, \sqrt{5})$ over $\mathbb{Q}$.
Hint: Use Problem 1 as a model.

9 If E is a finite extension of $\mathbb{Q}$, and ϕ is an automorphism on E, show that ϕ is a $\mathbb{Q}$-automorphism of E.
Hint: $\phi(1) = 1$ implies that $\phi(n) = n$ for all integers n.

10 Find the Galois Group for GF(4), over Z_2.

11 Find the Galois Group for GF(9), that is, the "complex numbers modulo 3," over Z_3.

12 Find the Galois Group for GF(81), over Z_3.

13 The irreducible polynomial $x^3 + x - 1$ has one real root and two complex roots. Using just this information, show that the Galois group of the splitting field is isomorphic to S_3.
Hint: The complex conjugate, which switches the two complex roots, is one of the $\mathbb{Q}$-automorphisms in the Galois group.

14 The irreducible polynomial $x^5 - 5x + 2$ has three real roots and two complex roots. Using just this information, show that the Galois group of the splitting field is isomorphic to S_5. (See the hint for Problem 13.)

15 Find, up to isomorphism, all possible Galois groups of the splitting field of a cubic polynomial $ax^3 + bx^2 + cx + d$.

16 Find, up to isomorphism, all possible Galois groups of the splitting field of a fourth degree polynomial $ax^4 + bx^3 + cx^2 + dx + e$.
Hint: By the second Sylow theorem (7.4), all subgroups of S_4 of order 8 are isomorphic to each other, hence isomorphic to D_4.

Interactive Problems

For Problems **17** through **22**: Find the set of $\mathbb{Q}$-automorphisms for the splitting fields of the following polynomials.

17 $x^3 - 3x - 1$
18 $x^3 - 3x + 3$
19 $x^4 - 5x^2 + 5$
20 $x^4 - x^2 + 1$
21 $x^4 + x^2 - 1$
22 $x^5 - x^4 - 4x^3 + 3x^2 + 3x - 1$

15.2 The Galois Group of a Polynomial in $\mathbb{Q}$

In this section, we will concentrate on polynomials with rational coefficients, which will give us some concrete examples of Galois groups. In fact, many of the applications of Galois theory involve extensions of the rational numbers, such as proving that a fifth-degree equation cannot be solved in terms of roots. By working with rational numbers, we will avoid the problem of a splitting field having multiple roots. (In fields of finite characteristic, this can cause a problem.) We want to show that this situation will never happen if we work with extension fields of the field of rational numbers.

One advantage of working with a familiar field is that we can borrow a tool from calculus, namely the derivative. It is not often that we will use a calculus result in algebra, but in this case it greatly simplifies the proof.

LEMMA 15.3
If $f(x)$ is an irreducible polynomial on $\mathbb{Q}[x]$, then $f(x)$ does not have multiple roots in the splitting field of $f(x)$.

PROOF: Since we are working in $\mathbb{Q}[x]$, we can use the familiar tools of calculus. Suppose that K is the splitting field of $f(x)$, and u is a multiple root of $f(x)$ in K. Then

$$f(x) = (x-u)^2 \cdot g(x).$$

Since we are working in a field extension of $\mathbb{Q}$, we can take the derivative of both sides to get

$$f'(x) = 2(x-u) \cdot g(x) + (x-u)^2 g'(x).$$

Thus, u is a root of $f'(x)$, which has lower degree than $f(x)$. Note that $f'(x)$ is not 0, since it has degree of at least one.

Since $f'(x)$ is also in $\mathbb{Q}[x]$, we see that $\mathrm{Irr}_{\mathbb{Q}}(u, x)$ has degree less than the degree of $f(x)$, and so $\mathrm{Irr}_{\mathbb{Q}}(u, x)$ is a divisor of $f(x)$. But this contradicts the fact that $f(x)$ is irreducible. Therefore, $f(x)$ cannot have multiple roots in its splitting field. □

Because of this lemma, we know from Proposition 14.10 that any splitting field can be expressed as a simple extension $\mathbb{Q}(w)$, and also we will be able to use Proposition 15.4 to predict the size of the Galois group of the splitting field. We can relate the Galois group of the splitting field directly to the polynomial.

DEFINITION 15.4 Let $f(x)$ be a polynomial in $\mathbb{Q}$. The *Galois group* of $f(x)$ is the Galois group of the splitting field of $f(x)$ over $\mathbb{Q}$.

We have already seen some examples of Galois groups of splitting fields. The Galois group of the splitting field of $x^3 - 2$ over $\mathbb{Q}$ was isomorphic to S_3. We also computed the Galois group of the splitting field of $x^4 - 2x^3 + x^2 + 1$, and found that the Galois group is isomorphic to D_4. Let us compute the Galois groups of some other polynomials.

Example 15.6
Find, up to isomorphism, the Galois group for the polynomial $x^3 + x^2 - 2x - 1$.
SOLUTION: This polynomial is irreducible, as *Sage* can verify:

```
InitDomain(0, "x")
factor(x^3 + x^2 - 2*x - 1)
    x^3 + x^2 - 2*x - 1
```

Thus, we can let a denote one of the roots, and try to factor this in $\mathbb{Q}(a)$.

```
AddFieldVar("x")
Define(a^3, -a^2 + 2*a + 1)
factor(x^3 + x^2 - 2*x - 1)
    (x - a) * (x - a^2 + 2) * (x + a^2 + a - 1)
```

Since this factors completely, we see that the splitting field of $x^3 + x^2 - 2x - 1$ is $\mathbb{Q}(a)$. This is a 3-dimensional extension of $\mathbb{Q}$, so by Proposition 15.4, the Galois group has three elements, hence is isomorphic to Z_3. □

Computational Example 15.7
Find, up to isomorphism, the Galois group of the polynomial $x^5 - 5x + 12$.
SOLUTION: In the last chapter, we were able to find a splitting field by making two extensions, one of dimension 5, and one of dimension 2.

```
InitDomain(0, "x")
factor(x^5 - 5*x + 12)
    x^5 - 5*x + 12
AddFieldVar("a")
Define(a^5, 5*a - 12)
factor(x^5 - 5*x + 12)
    (x - a) * (x^2 + (-1/4*a^4 - 1/4*a^3 - 1/4*a^2 + 3/4*a + 1)*x
    - 1/4*a^4 - 1/4*a^3 - 1/4*a^2 - 5/4*a + 2)
    * (x^2 + (1/4*a^4 + 1/4*a^3 + 1/4*a^2 + 1/4*a - 1)*x
    - 1/2*a^3 - 1/2*a - 1)
AddFieldVar("b")
Define(b^2, 1 + a/2 + a^3/2 + b - (a + a^2 + a^3 + a^4)*b/4)
factor(x^5 - 5*x + 12)
    (x - b) * (x - a) * (x + (1/2*a + 1/2)*b + 1/2*a - 1/2) *
    (x + b + 1/4*a^4 + 1/4*a^3 + 1/4*a^2 + 1/4*a - 1) *
    (x + (-1/2*a - 1/2)*b - 1/4*a^4 - 1/4*a^3 - 1/4*a^2 + 1/4*a +
    3/2)
```

If we define

```
c = (a^4 + a^3 + a^2 - 3*a - 4*b - 4)/4
d = (a - 4 - a^2 + a^3 - a^4 - 4*b - a*b + a^2*b - a^3*b + a^4*b)/8
e = (12 - 3*a - a^2 - 3*a^3 - a^4 + 4*b + a*b
- a^2*b + a^3*b - a^4*b)/8
```

we see that the following product simplifies to the original polynomial.

```
(x - a)*(x - b)*(x - c)*(x - d)*(x - e)
    x^5 - 5*x + 12
```

Thus, the five roots are a, b, c, d, and e. Any $\mathbb{Q}$-automorphism on the splitting field must send a and b to one of these five roots. Let us try to define a homomorphism f that sends $f(a) = b$, and $f(b) = a$.

```
F = FieldHomo()
HomoDef(F, a, b)
HomoDef(F, b, a)
CheckHomo(F)
    True
```

Not only does *Sage* verify that this is a homomorphism, but it can also draw a circle graph describing how this homomorphism acts on the five roots. The left side of Figure 15.2 is produced by the command

```
CircleGraph([a, b, "c", "d", "e"], F)
```

Note that we put the elements **c**, **d**, and **e** in quotes. This way, the single letter notations for these elements appear in the circle graph. Not every

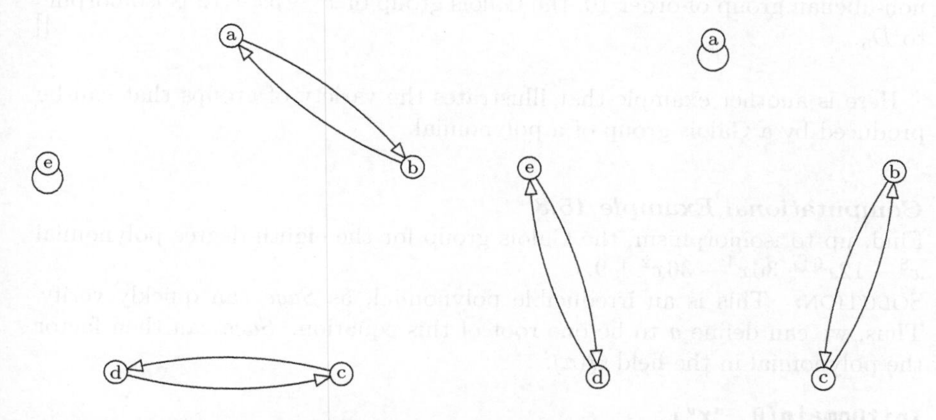

FIGURE 15.2: Automorphisms of splitting field for $x^5 - 5x + 12$

possible way of mapping a and b to the roots a, b, c, d, and e will produce a homomorphism. However, there is a homomorphism that maps $f(a) = a$ and $f(b) = c$. The commands

```
G = FieldHomo()
HomoDef(G, a, a)
HomoDef(G, b, c)
CheckHomo(G)
    True
CircleGraph([a, b, "c", "d", "e"], G)
```

produce the right side of Figure 15.2.

Once we have found two $\mathbb{Q}$-automorphisms, we can find more by considering the group generated by these two elements. By Corollary 15.1, the Galois group is a subgroup of S_5. We already have a natural ordering of the five roots, so the first permutation can be written **P(2,1,4,3)**, or $(1\,2)(3\,4)$, while the above permutation can be described as **P(1,3,2,5,4)**, or $(2\,3)(4\,5)$. Since the Galois group is a subgroup of S_5, we can ask *Sage* to find the subgroup generated by these two permutations.

```
G = Group(P(2,1,4,3), P(1,3,2,5,4)); G
    {P(), P(2, 1, 4, 3), P(1, 3, 2, 5, 4), P(3, 1, 5, 2, 4),
    P(2, 4, 1, 5, 3), P(4, 2, 5, 1, 3), P(3, 5, 1, 4, 2),
    P(5, 3, 4, 1, 2), P(4, 5, 2, 3, 1), P(5, 4, 3, 2, 1)}
```

This produces exactly 10 permutations. Proposition 15.4 states that the size of the Galois group is equal to the dimension of the splitting field. Since the splitting field is a 2-dimensional extension of a 5-dimensional extension, the Galois group contains exactly 10 elements. Thus, we have found all of the $\mathbb{Q}$-automorphisms of the splitting field. The multiplication table of the Galois group reveals that the group is non-abelian. Since there is only one non-abelian group of order 10, the Galois group of $x^5 - 5x + 12$ is isomorphic to D_5. ∎

Here is another example that illustrates the variety of groups that can be produced by a Galois group of a polynomial.

Computational Example 15.8

Find, up to isomorphism, the Galois group for the eighth-degree polynomial $x^8 - 12x^6 + 36x^4 - 36x^2 + 9$.

SOLUTION: This is an irreducible polynomial, as *Sage* can quickly verify. Thus, we can define a to be one root of this equation. *Sage* can then factor the polynomial in the field $\mathbb{Q}(a)$.

```
InitDomain(0, "x")
AddFieldVar("a")
```

```
Define(a^8, 12*a^6 - 36*a^4 + 36*a^2 - 9)
factor(x^8 - 12*x^6 + 36*x^4 - 36*x^2 + 9)
    (x - a) * (x + a) * (x - 1/3*a^5 + 3*a^3 - 3*a) *
    (x + 1/3*a^5 - 3*a^3 + 3*a) *
    (x - 2/3*a^7 + 22/3*a^5 - 17*a^3 + 10*a) *
    (x - 1/3*a^7 + 10/3*a^5 - 5*a^3 - a) *
    (x + 1/3*a^7 - 10/3*a^5 + 5*a^3 + a) *
    (x + 2/3*a^7 - 22/3*a^5 + 17*a^3 - 10*a)
```

The factorization can also be found by evaluating the following:

```
b = a^5/3 - 3*a^3 + 3*a
c = a^7/3 - 10*a^5/3 + 5*a^3 + a
d = 2*a^7/3 - 22*a^5/3 + 17*a^3 - 10*a
(x - a)*(x + a)*(x - b)*(x + b)*(x - c)*(x + c)*(x - d)*(x + d)
    x^8 - 12*x^6 + 36*x^4 - 36*x^2 + 9
```

This shows that the roots are $\pm a$, $\pm b$, $\pm c$, and $\pm d$, which are all expressed in terms of a. Hence, the splitting field for this polynomial is simply $\mathbb{Q}(a)$. Since this is an eighth-dimensional extension of $\mathbb{Q}$, the Galois group will have eight elements. But which group is this isomorphic to? Let us find a couple of $\mathbb{Q}$-automorphisms to find out.

By Proposition 15.2, there is a $\mathbb{Q}$-automorphism f for which $f(a) = b$. Let us find this $\mathbb{Q}$-automorphism.

```
F = FieldHomo()
HomoDef(F, a, b)
CheckHomo(F)
    True
```

We can have *Sage* draw a circle graph to find where the other seven roots are mapped to,

```
CircleGraph([a, "b", "c", "d", -a, "-b", "-c", "-d"], F)
```

producing the left-hand side of Figure 15.3.

We can express this element of the Galois group as **P(2,5,8,3,6,1,4,7)**, or $(1\,2\,5\,6)(3\,8\,7\,4)$.

By Proposition 15.2, we can also find a $\mathbb{Q}$-automorphism that sends a to c.

```
G = FieldHomo()
HomoDef(G, a, c)
CheckHomo(G)
    True
CircleGraph([a, "b", "c", "d", -a, "-b", "-c", "-d"], G)
```

This produces the circle graph on the right side of Figure 15.3. This element of the Galois group acts like the permutation **P(3,4,5,6,7,8,1,2)**, or $(1\,3\,5\,7)(2\,4\,6\,8)$. With these two permutations, we can see if we can generate the whole Galois group.

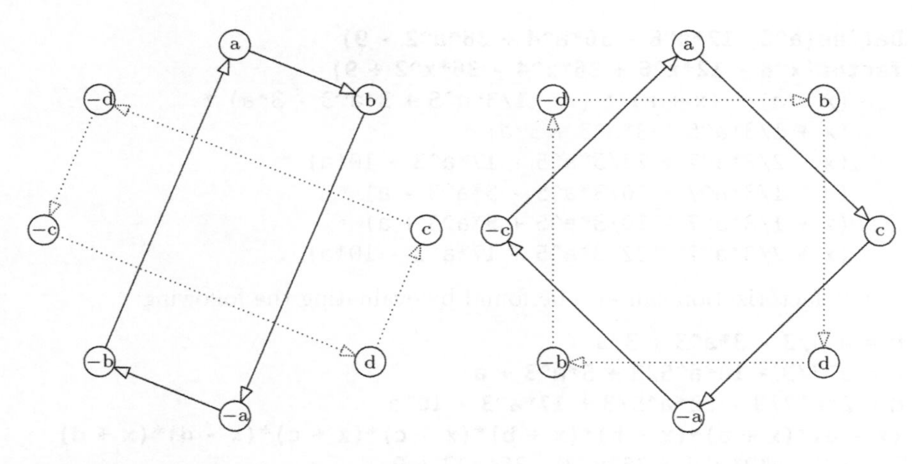

FIGURE 15.3: Two automorphisms for $x^8 - 12x^6 + 36x^4 - 36x^2 + 9$

```
G = Group( P(2,5,8,3,6,1,4,7), P(3,4,5,6,7,8,1,2) ); G
    {P(), P(2, 5, 8, 3, 6, 1, 4, 7), P(7, 8, 1, 2, 3, 4, 5, 6),
    P(8, 3, 6, 1, 4, 7, 2, 5), P(5, 6, 7, 8, 1, 2, 3, 4),
    P(6, 1, 4, 7, 2, 5, 8, 3), P(3, 4, 5, 6, 7, 8, 1, 2),
    P(4, 7, 2, 5, 8, 3, 6, 1)}
RootCount(G, 2)
    2
```

Sage produces eight elements, so this is the entire Galois group. Since there is only one element of order 2, this must be the quaternion group Q. ∎

Here is one more example that at first seems difficult because the splitting field is so large, but it is in fact easy to find the Galois group.

Example 15.9

Find, up to isomorphism, the Galois group of the polynomial $x^4 - x + 1$.

SOLUTION: In the last chapter we saw that the splitting field was 24 dimensional over $\mathbb{Q}$. We know from Corollary 15.1 that the Galois group is a subgroup of S_4. But S_4 has 24 elements, so the Galois group of $x^4 - x + 1$ must be isomorphic to S_4. ∎

Sage has a way of determining the Galois group, up to isomorphism, for irreducible polynomials up to degree around 15 (although some polynomials of degree 14 or 15 take an unreasonable amount of time). Applying **GaloisType** to a polynomial gives the name of the Galois group.

```
InitDomain(0, "x")
GaloisType(x^8 - 12*x^6 + 36*x^4 - 36*x^2 + 9)
  Q8
GaloisType(x^5 - 5*x + 12)
  D5
GaloisType(x^5 - x + 1)
  S5
```

In this way, we quickly redid the last two examples. However, this only gives an isomorphic group to the Galois group, instead of explicitly showing the elements of the group. The last example shows that the Galois group for the polynomial $x^5 - x + 1$ is S_5.

Finally, we wish to explore a whole class of polynomials at one time. In the last chapter, we computed the splitting field of the cyclotomic polynomials, and determined that $K = \mathbb{Q}(\omega_n)$, where

$$\omega_n = e^{(2\pi i/n)} = \cos\left(\frac{2\pi}{n}\right) + i\sin\left(\frac{2\pi}{n}\right).$$

We can use Proposition 14.9, along with some of the facts observed from §13.3, to find the Galois group of the nth cyclotomic polynomial $\Phi_n(x)$.

PROPOSITION 15.6

Let ω_n be the nth root of unity, and let $K = \mathbb{Q}(\omega_n)$. Then

$$\mathrm{Gal}_{\mathbb{Q}}(K) \approx Z_n^*.$$

PROOF: Let $g(x) = \Phi_n(x)$, which by Gauss' theorem on cyclotomic polynomials (13.2), is irreducible. The roots of $g(x)$ are of the form $(\omega_n)^k$, where k is coprime to n. Hence, K is the splitting field of $g(x)$. Since the degree of $\Phi_n(x)$ is $\phi(n)$, we know from Proposition 15.4 that the size of $\mathrm{Gal}_{\mathbb{Q}}(K)$ is $\phi(n)$.

To show that $\mathrm{Gal}_{\mathbb{Q}}(K)$ is isomorphic to Z_n^*, note that every ϕ in $\mathrm{Gal}_{\mathbb{Q}}(K)$ is determined by where it sends ω_n, and that it must send it to one of the roots $(\omega_n)^k$ for some k coprime to n. Thus, there is a natural mapping

$$f : \mathrm{Gal}_{\mathbb{Q}}(K) \to Z_n^*$$

defined by $f(\phi) =$ (the value k for which $\phi(\omega_n) = (\omega_n)^k$). This mapping is well defined since $(\omega_n)^n = 1$. This mapping is a homomorphism, for if $f(\phi) = k$ and $f(\mu) = m$, then

$$(\phi \cdot \mu)(\omega_n) = \phi(\mu(\omega_n)) = \phi((\omega_n)^m) = (\phi(\omega_n))^m = (\omega_n)^{k \cdot m},$$

so

$$f(\phi \cdot \mu) = k \cdot m = f(\phi) \cdot f(\mu).$$

Finally, an element in the kernel of this homomorphism sends ω_n to ω_n, so $\text{Ker}(f)$ is just the identity element of $\text{Gal}_{\mathbb{Q}}(K)$. Thus, f is an isomorphism from $\text{Gal}_{\mathbb{Q}}(K)$ to Z_n^*. Since we know that both $\text{Gal}_{\mathbb{Q}}(K)$ and Z_n^* have $\phi(n)$ elements, the mapping is onto, so $\text{Gal}_{\mathbb{Q}}(K) \approx Z_n^*$. ∎

From all of these examples, we have seen a host of different groups produced as Galois groups of polynomials: S_3, Z_3, D_5, Z_5, Q, D_4, S_4, and all groups of the form Z_n^*. It is natural to ask whether *all* finite groups can be expressed as a Galois group of some polynomial in $\mathbb{Q}[x]$. This is still an open problem, known as the *inverse Galois problem.* There has been much progress made on this problem, and it is very likely to be solved soon.

While we are working with cyclotomic polynomials and n^{th} roots of unity, let us prove one more proposition that will be useful later on.

PROPOSITION 15.7

Let F be a finite extension of $\mathbb{Q}$ that contains the nth roots of unity. Then if u is a root of the polynomial $f(x) = x^n - c$ for some $c \neq 0$ in F, then $K = F(u)$ is the splitting field of $f(x)$, and $\text{Gal}_F(K)$ is abelian.

PROOF: Since u is a root of $x^n - c$, we have that $u^n = c$. But $(\omega_n)^k \cdot u$ is also a root of this polynomial for all integers $k = 0, 1, 2, \ldots, n-1$, since

$$\left((\omega_n)^k \cdot u\right)^n = (\omega_n)^{k \cdot n} \cdot u^n = 1 \cdot c = c.$$

Since there are n distinct roots of the polynomial $x^n - c$ in K, the polynomial factors completely in $K[x]$. Thus, K is the splitting field of $f(x)$.

To show that $\text{Gal}_F(K)$ is abelian, note that any F-automorphism is determined by where u is sent, which must be of the form $(\omega_n)^k \cdot u$. Thus, if ϕ_1 and ϕ_2 are two F-automorphisms of K, where $\phi_1(u) = (\omega_n)^k \cdot u$ and $\phi_2(u) = (\omega_n)^m \cdot u$, then

$$(\phi_1 \cdot \phi_2)(u) = \phi_1(\phi_2(u)) = \phi_1((\omega_n)^m \cdot u) = (\phi_1(\omega_n))^m \phi_1(u) = (\omega_n)^m \cdot (\omega_n)^k \cdot u.$$

while

$$(\phi_2 \cdot \phi_1)(u) = \phi_2(\phi_1(u)) = \phi_2((\omega_n)^k \cdot u) = (\phi_2(\omega_n))^k \phi_2(u) = (\omega_n)^k \cdot (\omega_n)^m \cdot u.$$

Thus, $\phi_1 \cdot \phi_2 = \phi_2 \cdot \phi_1$, and so the Galois group is abelian. ∎

To introduce the problem of whether a fifth-degree polynomial can, in general, be solved in terms of square roots, cube roots, or fifth roots, we will have *Sage* try to solve some polynomial equations for us. *Sage* can solve polynomials with the command

```
var("x")
Solve(x^2 - x + 2, x)
    [x == -1/2*I*sqrt(7) + 1/2, x == 1/2*I*sqrt(7) + 1/2]
```

which obviously uses the quadratic equation. Note that the "double equals" == is *Sage*'s way of expressing an equation. Let's try changing the x^2 to an x^3:

```
Solve(x^3 - x + 2, x)
    [x == -1/2*(1/9*sqrt(26)*sqrt(3) - 1)^(1/3)*(I*sqrt(3) + 1)
    - 1/6*(-I*sqrt(3) + 1)/(1/9*sqrt(26)*sqrt(3) - 1)^(1/3),
    x == 1/2*(1/9*sqrt(26)*sqrt(3) - 1)^(1/3)*(-I*sqrt(3) + 1)
    - 1/6*(I*sqrt(3) + 1)/(1/9*sqrt(26)*sqrt(3) - 1)^(1/3),
    x == (1/9*sqrt(26)*sqrt(3) - 1)^(1/3) +
    1/3/(1/9*sqrt(26)*sqrt(3) - 1)^(1/3)]
```

Sage was still able to solve this, but what a mess! The answer involves $\sqrt{26} \cdot \sqrt{3} = \sqrt{78}$. Apparently *Sage* is using a formula that finds the roots of any cubic equation.

Let us try a forth-degree equation:

```
Solve(x^4 - x + 2, x)
```

Sage's answer is particularly long, but it can be expressed as

$$\left[x == -\frac{1}{2}\sqrt{-A - \frac{2}{\sqrt{A}}} - \frac{\sqrt{A}}{2}, \qquad x == \frac{1}{2}\sqrt{-A - \frac{2}{\sqrt{A}}} - \frac{\sqrt{A}}{2},\right.$$

$$\left. x == \frac{\sqrt{A}}{2} - \frac{1}{2}\sqrt{\frac{2}{\sqrt{A}} - A}, \qquad x == \frac{1}{2}\sqrt{\frac{2}{\sqrt{A}} - A} + \frac{\sqrt{A}}{2}\right]$$

where

$$A = \frac{\sqrt[3]{\frac{1}{2}(9 + i\sqrt{6063})}}{3^{2/3}} + \frac{8}{\sqrt[3]{\frac{3}{2}(9 + i\sqrt{6063})}}.$$

Once again, *Sage* was able to express the answer in terms of square roots and cube roots, yet this seems even more of a mess.

The equations for the cubic equation and the fourth-degree equation were discovered in 1539 and 1545 [4, p. 2]. The natural question is whether there is a similar formula for fifth-degree polynomials. Let us try to solve a fifth-degree polynomial in *Sage.*

```
Solve(x^5 - x + 2, x)
    [0 == x^5 - x + 2]
```

We see that *Sage* does not know of any formula for the fifth-degree polynomial, so *Sage* keeps the equation intact. Here is the way we can get a list of *approximate* roots:

```
(x^5 - x + 2 == 0).roots(x, ring = CC, multiplicities = false)
    [-1.26716830454212,
    -0.260963880386455 - 1.17722615339419*I,
    -0.260963880386455 + 1.17722615339419*I,
    0.894548032657517 - 0.534148546174733*I,
    0.894548032657517 + 0.534148546174733*I]
```

Here, **CC** is *Sage*'s way of expressing the complex numbers $\mathbb{C}$. Is *Sage* not smart enough to solve the equation exactly? No, because it is *impossible* to find a formula for the roots of a fifth-degree polynomial in terms of square roots, cube roots, or any other roots. The reason why is based on the properties of the Galois groups. The next section will reveal how the Galois groups are related to the splitting field.

Problems for §15.2

1 Find a polynomial whose Galois group is Z_6.
Hint: See Proposition 15.6.

2 Prove that if a fourth-degree polynomial in $\mathbb{Q}[x]$ has a Galois group isomorphic to Z_4, then the roots of the polynomial can be rearranged as r_1, r_2, r_3, and r_4 such that

$$r_1^2 r_2 + r_2^2 r_3 + r_3^2 r_4 + r_4^2 r_1$$

yields a real rational number.

Hint: There is a $\mathbb{Q}$-automorphism such that the roots map in a four-cycle: $r_1 \to r_2 \to r_3 \to r_4 \to r_1$. Note that the $\mathbb{Q}$-automorphisms fix the above expression, so the result must be in the fixed field of the Galois group.

3 Prove that if a fifth-degree polynomial in $\mathbb{Q}[x]$ has a Galois group isomorphic to D_5, then the roots of the polynomial can be rearranged as r_1, r_2, r_3, r_4, and r_5 such that

$$r_1 r_2 + r_2 r_3 + r_3 r_4 + r_4 r_5 + r_5 r_1$$

yields a real rational number.

Hint: See the hint for Problem 2. Note that here we must also consider a "flip" that exchanges $r_1 \leftrightarrow r_4$ and $r_2 \leftrightarrow r_3$.

4 Find a way similar to Problem 2 to test whether a Galois group of a fifth-degree polynomial is isomorphic to Z_5.

5 Find a way similar to Problem 3 to test whether a Galois group of a fourth-degree polynomial is D_4.

6 The roots of $x^4 - x^3 - 4x^2 + 4x + 1$ are approximately 1.827090915, 1.338261213, −1.956295201, and −0.209056927. Use trial and error to find an arrangement of these four roots such that

$$r_1^2 r_2 + r_2^2 r_3 + r_3^2 r_4 + r_4^2 r_1$$

yields an integer. (See Problem 2.)

7 The roots of the equation $x^5 - 5x - 12$ are approximately 1.842085966, $0.351854083 \pm 1.709561043i$, and $-1.272897224 \pm 0.7197986815i$. Use trial and error to find an arrangement of these five roots such that

$$r_1 r_2 + r_2 r_3 + r_3 r_4 + r_4 r_5 + r_5 r_1$$

yields a real integer. (See Problem 3.)

8 The roots of $x^4 - x^3 - 4x^2 + 4x + 1$ are approximately 1.827090915, 1.338261213, −1.956295201, and −0.209056927. Show that whenever a is a root, then $a^2 - 2$ is also a root. Show that, in fact, the operation $a \mapsto a^2 - 2$ permutes the four roots in a 4-cycle. Using this, prove that the Galois group must be isomorphic to Z_4.

Hint: If a is one of the roots, the splitting field is $\mathbb{Q}(a)$.

For Problems **9** through **14**: Find a group isomorphic to the Galois group of the polynomial

9 $x^2 - 3$

10 $x^3 - 3$

11 $x^2 - 4$

12 $x^3 - 8$

13 $(x^2 - 2)(x^2 - 3)$

14 $(x - 1)^2(x - 3)^3(x^2 - 5)$

15 Prove that if G is a group of order n that is isomorphic to a Galois group of some polynomial in $\mathbb{Q}[x]$, then G is isomorphic to a Galois group of an nth-degree polynomial in $\mathbb{Q}[x]$.

Hint: Use Corollary 14.4.

Interactive Problems

For Problems **16** through **21**: Use *Sage* to find the Galois group of the polynomial. Determine the number of elements in the Galois group, and display a multiplication table of the subgroup of S_n isomorphic to the Galois group.

16 $x^4 - 2$

17 $x^5 - 2$

18 $x^5 + 15x + 12$

19 $x^5 + x^4 - 4x^3 - 3x^2 + 3x + 1$

20 $x^4 - 10x^2 + 1$

21 $x^8 - 108x^6 + 1548x^4 - 3888x^2 + 1296$

22 Use *Sage* to find the Galois group of $x^5 + 20x + 16$. How many elements are in the Galois group? (This may take longer than the above problems.)

15.3 The Fundamental Theorem of Galois Theory

In this section we will clarify the relationship between subgroups of the Galois group, and the subfields of the extension field. Often there is a beautiful correspondence between these two, which is at the heart of Galois theory. The natural correlation is to map to each subgroup of $\mathrm{Gal}_F(K)$ the fixed field of the subgroup. However, we ended §15.1 with what seemed to be a bad example—$\mathbb{Q}(\sqrt[3]{2})$. The only fixed field was $\mathbb{Q}(\sqrt[3]{2})$, even though there was the obvious subfield. The way we will deal with exceptions like this one is to consider only field extensions for which the original field appears as one of the fixed fields.

DEFINITION 15.5 Let K be a finite extension of F. We say that K is a *Galois extension* if the fixed field of $\mathrm{Gal}_F(K)$ is the field F.

Although this definition successfully rules out $\mathbb{Q}(\sqrt[3]{2})$ from being a Galois extension, we need to find a simple test for determining whether a finite extension is a Galois extension. The following proposition takes us one step in that direction.

PROPOSITION 15.8
Let F be a field, and K a Galois extension of F. If $f(x)$ is an irreducible polynomial in $F[x]$ that has at least one root in K, then $f(x)$ factors completely in K. Furthermore, $f(x)$ has no multiple roots in the field K.

PROOF: Since $f(x)$ has at least one root in the field K, let $u_1, u_2, u_3, \ldots, u_n$ be the set of all roots of $f(x)$ in K. Consider the polynomial

$$g(x) = (x - u_1) \cdot (x - u_2) \cdot (x - u_3) \cdots (x - u_n).$$

By Lemma 13.3, any automorphism in $\mathrm{Gal}_F(K)$ extends to an automorphism on $K[x]$ with $\phi(x) = x$. Thus,

$$\phi(g(x)) = (x - \phi(u_1)) \cdot (x - \phi(u_2)) \cdot (x - \phi(u_3)) \cdots (x - \phi(u_n)).$$

By Lemma 15.1, $\phi(u_1), \phi(u_2), \phi(u_3), \ldots, \phi(u_n)$ will all be roots of $f(x)$ and so this list is a permutation of the list $u_1, u_2, u_3, \ldots, u_n$. Therefore, $\phi(g(x)) = g(x)$ for all ϕ in $\mathrm{Gal}_F(K)$.

Now, since K is a Galois extension of F, the fixed field of $\mathrm{Gal}_F(K)$ is the field F. Thus, $g(x)$ is a polynomial in $F[x]$. Since $g(x)$ certainly divides the polynomial $f(x)$, and $f(x)$ is irreducible in $F[x]$, we have that $f(x)$ and $g(x)$ have the same degree. Thus, n is the degree of $f(x)$, and so $f(x)$ factors completely in the field K. Furthermore, $f(x)$ has no multiple roots in the field K. □

This proposition allows us to immediately rule out certain field extensions from being a Galois extension. Clearly $\mathbb{Q}(\sqrt[3]{2})$ is ruled out because $\mathbb{Q}(\sqrt[3]{2})$ is not a splitting field. But there are even some splitting fields that are not Galois extensions, according to this proposition.

Motivational Example 15.10

Let $Z_2(t)$ be the field of rational functions in t, with coefficients in Z_2. This field can be defined in *Sage* by the command

```
InitDomain(2)
RationalFunctions("t")
```

Note this is different that the **AddFieldVar** command, which only defined the polynomial ring $Z_2[t]$. Here are some examples of elements in $Z_2(t)$:

```
(t^3 + t + 1)/t
    (t^3 + t + 1)/t
1/(t^2 + 1) + (t^3 + t + 1)/t
    (t^5 + t^2 + 1)/(t^3 + t)
```

We can now consider polynomials over this field.

```
AddFieldVar("x")
x^2 + t
    x^2 + t
```

Note that there is no element whose square is equal to $-t$, since this polynomial is irreducible.

```
factor(x^2 + t)
    x^2 + t
```

Suppose we define a new element a that solves this equation.

```
AddFieldVar("a")
Define(a^2 + t, 0)
a^2
    t
(x + a)*(x + a)
    x^2 + t
```

Now $x^2 + t$ factors in $Z_2(t)(a)$ as $(x + a)(x + a)$. Note, however, that there is a *double root* in this factorization! Thus, by Proposition 15.8, $Z_2(t)(a)$ is *not* a Galois extension of $Z_2(t)$. ∎

One immediate consequence from Proposition 15.8 is that a Galois extension can be written as a simple extension.

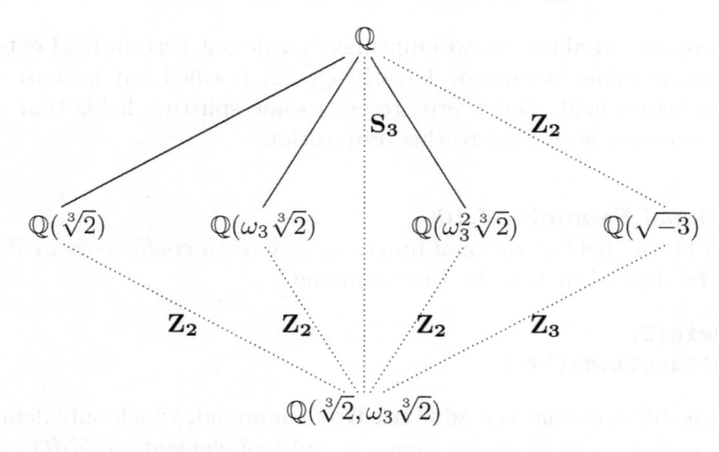

FIGURE 15.4: Subfield diagram for splitting field of $x^3 - 2$

COROLLARY 15.2

Let F be a field, and let K be a Galois extension of F. Then there exists an element w of K such that $K = F(w)$.

PROOF: Since K is a Galois extension of F, K is finite dimensional over F. Thus, $K = F(u_1, u_2, u_3, \ldots, u_n)$ for elements $u_1, u_2, u_3, \ldots, u_n$ in K. But the polynomials $\text{Irr}_F(u_i, x)$ all have a root in K, and so factor completely in the field K without multiple roots. Then we can use Corollary 14.4 to show that there is an element w in K such that $F(w) = K$. ☐

In order to introduce the correlation between the subgroups of the Galois group and the subfields of the Galois extension, let us consider the familiar splitting field of $x^3 - 2$. Since $\sqrt[3]{2}$ and $\omega_3\sqrt[3]{2}$ are two roots, we can express the splitting field as $\mathbb{Q}(\sqrt[3]{2}, \omega_3\sqrt[3]{2})$. The subfields of this Galois extension are $\mathbb{Q}$, $\mathbb{Q}(\sqrt[3]{2})$, $\mathbb{Q}(\omega_3\sqrt[3]{2})$, $\mathbb{Q}(\omega_3^2\sqrt[3]{2})$, $\mathbb{Q}(\sqrt{-3})$, and the whole field $\mathbb{Q}(\sqrt[3]{2}, \omega_3\sqrt[3]{2})$. We can draw a diagram of these subfields, showing which subfields are subfields of other subfields. This is shown in Figure 15.4.

The dotted lines in this diagram indicate which subfields are Galois extensions of the subfield above it. Also, whenever we have a Galois extension, the corresponding Galois group is shown in boldface. For example, this diagram indicates that the splitting field of $x^3 - 2$ is a Galois extension of $\mathbb{Q}(\sqrt{-3})$. This is true by Proposition 15.7, since $\mathbb{Q}(\sqrt{-3})$ contains the cube roots of unity.

Now let us compare this figure with the subgroups of the Galois group S_3, shown in Figure 15.5. Once again, we draw lines connecting two subgroups if one subgroup is contained in the other subgroup. We draw a dotted line to indicate that the smaller subgroup is a normal subgroup of the larger. Whenever the subgroup is a normal subgroup, the quotient group is indicated

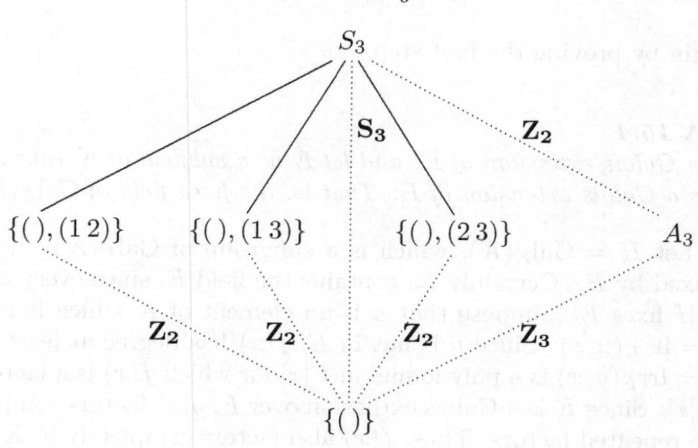

FIGURE 15.5: Subgroup diagram for Galois group of $x^3 - 2$

in boldface.

The pattern is now obvious. The two pictures are the same, except that the subfields are replaced by a subgroup of S_3. This feature of Galois extensions is the heart of Galois theory. In fact, there is a natural way that the subfields of K and the subgroups of $\text{Gal}_F(K)$ are related: For each subfield E of K, we can consider $\text{Gal}_E(K)$, the set of automorphisms of K that fix E. This is a subgroup of $\text{Gal}_F(K)$. On the other hand, given a subgroup H of $\text{Gal}_F(K)$, we can consider the fixed field $\text{fix}(H)$, which is a subfield of K. To show that, indeed, the two pictures will be essentially the same, we need four steps.

1. Show that if we start with a subfield E, then form the Galois group $\text{Gal}_E(K)$, and find the fixed field of this subgroup, we get back E.
2. Show that if we start with a subgroup H of $\text{Gal}_F(K)$, find the fixed field, then find the Galois group of the fixed field, we get back H. These first two steps establish a one-to-one correspondence between the subfields and the subgroups of the Galois group.
3. Show that if a subgroup N is a normal subgroup of another subgroup H, then the corresponding subfields form a Galois extension. Thus, a dotted line on the second picture corresponds to a dotted line on the first.
4. Show that if one subfield E is a Galois extension of another, L, then the corresponding Galois groups will have a normal subgroup relation. Furthermore, the quotient group of the Galois groups will be isomorphic to the Galois group of the Galois extension. Thus, a dotted line on the first picture corresponds to a dotted line on the second, and the boldface groups in the pictures will be isomorphic.

Let us begin by proving the first step.

LEMMA 15.4

Let K be a Galois extension of F, and let E be a subfield of K containing F. Then K is a Galois extension of E. That is, the fixed field of $\mathrm{Gal}_E(K)$ is E.

PROOF: Let $H = \mathrm{Gal}_E(K)$, which is a subgroup of $\mathrm{Gal}_F(K)$. Let E_0 be the field fixed by H. Certainly E_0 contains the field E, since every automorphism in H fixes E. Suppose that u is an element of K which is not in E. Let $f(x) = \mathrm{Irr}_E(u, x)$. Since u is not in E, $f(x)$ has degree at least 2. Note that $g(x) = \mathrm{Irr}_F(u, x)$ is a polynomial in $F[x]$ for which $f(x)$ is a factor in the domain $E[x]$. Since K is a Galois extension over F, $g(x)$ factors completely in K with no repeated factors. Thus, $f(x)$ also factors completely in K with no repeated factors, so there are at least two solutions to the equation $f(x) = 0$ in K. One solution is of course u, so let v be another solution. By Proposition 15.2, there is an E-automorphism in H such that $\phi(u) = v$. Thus, u is not in E_0. Therefore, $E_0 = E$, and so K is a Galois extension of E. □

We are now ready to proceed to the second step.

LEMMA 15.5

Let K be a Galois extension of F. If H is a subgroup of the Galois group $\mathrm{Gal}_F(K)$, and E is the fixed field of H, then $H = \mathrm{Gal}_E(K)$.

PROOF: Let n be the dimension of the field K over E. By Lemma 15.4, K is a Galois extension of E. Thus, by Corollary 15.2, there exists an element w in K such that $K = E(w)$. If $f(x) = \mathrm{Irr}_E(w, x)$, then the degree of $f(x)$ is n by Corollary 14.1. Since K is a Galois extension of E, by Proposition 15.8, the polynomial $f(x)$ factors completely in the field K, and there are no multiple roots. Thus, by Proposition 15.4, the number of E-automorphisms of K is the dimension of K over E, which is n.

Suppose that H contains m E-automorphisms. Let $v_1, v_2, v_3, \ldots, v_m$ be the images of w under the automorphisms in the subgroup H. That is, for each v_i there is an h in H such that $v_i = h(w)$.

Consider the polynomial

$$g(x) = (x - v_1) \cdot (x - v_2) \cdot (x - v_3) \cdots (x - v_m).$$

If ϕ is an automorphism in H, then $\phi(v_i) = \phi(h(w)) = v_j$ for some j. Also, since ϕ is one-to-one, the images of $\phi(v_1), \phi(v_2), \phi(v_3), \ldots, \phi(v_m)$ must all be distinct. Thus, each ϕ in H is a permutation on the elements $v_1, v_2, \cdots v_m$. Hence, $\phi(g(x)) = g(x)$. Since E is the fixed field $\mathrm{fix}(H)$ of the subgroup H, we see that $g(x)$ is in $E[x]$. Thus, $f(x) = \mathrm{Irr}_E(w, x)$ divides $g(x)$ so m is at least n. Thus,

$$|H| \leq |\mathrm{Gal}_E(K)| = n \leq m = |H|.$$

Therefore, $H = \mathrm{Gal}_E(K)$. ∎

Lemmas 15.4 and 15.5 show that there is a one-to-one correspondence between the subgroups of $\mathrm{Gal}_F(K)$ and the subfields of K containing F. We now consider the special significance of the normal subgroups of $\mathrm{Gal}_F(K)$.

LEMMA 15.6

Let K be a Galois extension of F, and let E be a subfield of K containing another subfield L. Suppose that $\mathrm{Gal}_E(K)$ is a normal subgroup of $\mathrm{Gal}_L(K)$. Then every L-automorphism of K maps elements of E to elements of E. Furthermore, E is a Galois extension of L.

PROOF: First, we want to show that if u is in E, and ϕ is in $\mathrm{Gal}_L(K)$, then $v = \phi(u)$ is in E. Since $\mathrm{Gal}_E(K)$ is a normal subgroup of $\mathrm{Gal}_L(K)$, for any f in $\mathrm{Gal}_E(K)$ we have that $\psi = \phi^{-1} \cdot f \cdot \phi$ is in $\mathrm{Gal}_E(K)$. Then $f \cdot \phi = \phi \cdot \psi$, or $f(\phi(u)) = \phi(\psi(u))$.

Since u is in E, $\psi(u) = u$, so

$$f(v) = f(\phi(u)) = \phi(\psi(u)) = \phi(u) = v.$$

Thus, v is fixed by every automorphism f in $\mathrm{Gal}_E(K)$. By Lemma 15.4, K is a Galois extension of E, so the fixed field of $\mathrm{Gal}_E(K)$ is E. Thus, v is in E.

To show that the fixed field of $\mathrm{Gal}_L(E)$ is L, consider an element u in E that is not in L. By Lemma 15.4, K is a Galois extension of L. Since u is not in the fixed field of $\mathrm{Gal}_L(K)$, there is an L-automorphism ϕ that moves u to another element, v. But ϕ moves all elements of E to elements of E, so we can consider the restriction of ϕ on the field E, denoted ϕ'. This is an automorphism of E, since the inverse is $(\phi^{-1})'$. Thus, there is an L-automorphism of E that moves the element u, so the fixed field of $\mathrm{Gal}_L(E)$ is only L. Therefore, E is a Galois extension of L. ∎

There is only one step left to show why Figures 15.4 and 15.5 are so similar.

LEMMA 15.7

Suppose that K is a Galois extension of F, and let E be a subfield of K that is also a Galois extension of a smaller subfield L, which contains F. Then there exists a surjective homomorphism f from $\mathrm{Gal}_L(K)$ to $\mathrm{Gal}_L(E)$ whose kernel is $\mathrm{Gal}_E(K)$.

PROOF: By Lemma 15.4, K is a Galois extension of L. We begin by showing that if ϕ is an L-automorphism of K, and u is in E, then $\phi(u)$ is in E. Let $g(x) = \mathrm{Irr}_L(u, x)$. Since E is a Galois extension of L, by Proposition 15.8, $g(x)$ factors completely in $E[x]$, which is of course the same factorization in $K[x]$. By Lemma 15.1, $\phi(u)$ is a root of $g(x)$ in K, but all of the roots are also in E. Thus, $\phi(u)$ is in E.

Next, we define the mapping f that sends an L-automorphism of K to its restriction on the field E. We denote the restriction of ϕ on the field E by ϕ'. Since ϕ maps elements of E to elements of E, we see that ϕ' is an L-automorphism of E. However, $(\phi^{-1})'$ is also an L-automorphism of E, and $(\phi^{-1})' \cdot \phi'$ is clearly the identity mapping on E. Thus, ϕ' is an element of $\text{Gal}_L(E)$.

To show that f is a homomorphism, note that

$$f(\phi_1 \cdot \phi_2) = (\phi_1 \cdot \phi_2)' = \phi_1' \cdot \phi_2' = f(\phi_1) \cdot f(\phi_2).$$

The kernel of this homomorphism is simply the L-automorphisms of K that fix the elements of E, which is of course $\text{Gal}_E(K)$.

Finally, to show that this homomorphism is surjective, let ψ be an L-automorphism of E. Since K is a splitting field of E, we can use Proposition 14.8 to extend ψ to an L-automorphism of K, which we will call ϕ. Then $f(\phi) = \psi$, and we have shown that f is surjective. ∎

Lemmas 15.4 through 15.7 explain the amazing similarity in the diagrams of the subfields, and the subgroups of the Galois group. By putting these four pieces together, we get the fundamental theorem of Galois theory.

THEOREM 15.1: The Fundamental Theorem of Galois Theory

Let K be a Galois extension of the field F. Then there is a one-to-one correspondence between the subfields of K containing F and the subgroups of $\text{Gal}_F(K)$*, given by mapping E to the subgroup* $\text{Gal}_E(K)$*. The dimension of K over the subfield E is* $|\text{Gal}_E(K)|$*. Furthermore, a subfield E is a Galois extension of L if, and only if,* $\text{Gal}_E(K)$ *is a normal subgroup of* $\text{Gal}_L(K)$*, in which case* $\text{Gal}_L(E)$ *is isomorphic to* $\text{Gal}_L(K)/\text{Gal}_E(K)$*.*

PROOF: If $\text{Gal}_E(K) = \text{Gal}_L(K)$ for two subfields E and L of K, then by Lemma 15.4, both E and L are the fixed field of the subgroup $\text{Gal}_E(K) = \text{Gal}_L(K)$, so $E = L$. Thus, the mapping $E \to \text{Gal}_E(K)$ is one-to-one. But if H is any subgroup of $\text{Gal}_F(K)$, then we can consider E to be the fixed field $\text{fix}(H)$, and by Lemma 15.5 $\text{Gal}_E(K) = H$. Thus, the correspondence is also onto. Also by Proposition 15.4, the dimension of K over E is $|\text{Gal}_E(K)|$, since K is a Galois extension of E.

If E is also a Galois extension of another subfield L, then by Lemma 15.7 there is a surjective homomorphism from $\text{Gal}_L(K)$ to $\text{Gal}_L(E)$, whose kernel is $\text{Gal}_E(K)$. Thus, $\text{Gal}_E(K)$ is a normal subgroup of $\text{Gal}_L(K)$, and by the first isomorphism theorem (4.1), $\text{Gal}_L(E)$ is isomorphic to $\text{Gal}_L(K)/\text{Gal}_E(K)$.

Finally, suppose that $\text{Gal}_E(K)$ is a normal subgroup of $\text{Gal}_L(K)$. By Lemma 15.6, E is a Galois extension of L. ∎

The fundamental theorem of Galois theory has many applications. With this theorem one can prove that it is impossible to trisect an angle with only a straightedge and a compass, and also that it is impossible to construct a line $\sqrt[3]{2}$ times the length of a given line [6, p. 433]. This finally puts to rest two of the three famous unsolved problems introduced by the ancient Greeks [12, p. 109]. (The last problem involves showing that π is not in an algebraic extension of $\mathbb{Q}$.) Both of these problems require a field extension of order 3, while any straightedge and compass construction involves a series of field extensions of order 2. Of course 3 does not divide any power of 2, so a field extension of dimension 3 cannot be a subfield of a field created by a sequence of extensions of order 2. Another important application shows that a fifth-degree equation cannot be solved in terms of radicals. We will explore both of these applications in the next section.

Problems for §15.3

1 The Galois group $\mathrm{Gal}_{\mathbb{Q}}(\mathbb{Q}(\sqrt{2}, \sqrt{3}))$ is given in Problem 1 of §15.1. Find the five subgroups of the Galois group, and for each subgroup H find the fixed field $\mathrm{fix}(H)$ of that subgroup.

2 Find all of the subfields of the field $\mathbb{Q}(\sqrt{2}, \sqrt{5})$.

Hint: First do Problem 8 of §15.1, and use the fundamental theorem of Galois theory, as was done in Problem 1.

3 There are 10 subfields of the field $K = \mathbb{Q}(\sqrt[4]{2}, i)$: $\mathbb{Q}$, $\mathbb{Q}(\sqrt[4]{2}, i)$, $\mathbb{Q}(\sqrt[4]{2})$, $\mathbb{Q}(i)$, $\mathbb{Q}(i\sqrt[4]{2})$, $\mathbb{Q}(\sqrt{2})$, $\mathbb{Q}(i\sqrt{2})$, $\mathbb{Q}(\sqrt{2}, i)$, $\mathbb{Q}((1+i)\sqrt[4]{2})$, and $\mathbb{Q}((1-i)\sqrt[4]{2})$. Match each of the 10 subfields with the 10 subgroups of $\mathrm{Gal}_{\mathbb{Q}}(K)$ so that each subfield is the fixed field $\mathrm{fix}(H)$ of the corresponding subgroup of $\mathrm{Gal}_{\mathbb{Q}}(K)$.

Hint: See Problem 5 of §15.1 to find $\mathrm{Gal}_{\mathbb{Q}}(K)$, which is isomorphic to D_4. Next find the 10 subgroups of this group. Finding the fixed field for some of the subgroups is obvious. Can the fundamental theorem of Galois theory help with the remaining subgroups?

4 Let F be the splitting field of $\Phi_5(x) = x^4 + x^3 + x^2 + x + 1$ over $\mathbb{Q}$. Show that there is only one nontrivial subfield besides $\mathbb{Q}$ of F, and find this subfield.

Hint: Use Proposition 15.6 to find $\mathrm{Gal}_{\mathbb{Q}}(F)$, and find that there is only one nontrivial subgroup of this group.

5 Let F be the splitting field of $\Phi_6(x)$ over $\mathbb{Q}$. Show that the only nontrivial subfield of F is $\mathbb{Q}$.

6 Let F be the splitting field of $\Phi_{15}(x)$ over $\mathbb{Q}$. Find 3 elements of $\mathrm{Gal}_{\mathbb{Q}}(F)$ that have order 2.

7 Let F be the splitting field of $\Phi_7(x)$ over $\mathbb{Q}$. Find an element of $\mathrm{Gal}_{\mathbb{Q}}(F)$ that has order 2, and another element of order 3.

8 Show that $\omega_7 + \omega_7^6$ is in the fixed field of the automorphism of order 2 from Problem 7, and that $\omega_7^3 + \omega_7^5 + \omega_7^6$ is in the fixed field of the automorphism of order 3.

9 Let E be a finite extension of a field F with dimension n. Show that $|\mathrm{Gal}_F(E)| = n$ if, and only if, E is a Galois extension of F.

10 Let E be a finite extension of a field F, and let $\phi(x)$ be an F-automorphism in $\mathrm{Gal}_F(E)$. Suppose that $\phi(u) = u$ for some element u in E. Show that ϕ is in $\mathrm{Gal}_{F(u)}(E)$.

11 If E is a Galois extension of F, show that there can only be a finite number of subfields of E that contain F.

12 Show that if E is a Galois extension of F with dimension p, where p is a prime, prove that $\mathrm{Gal}_F(E)$ is isomorphic to Z_p.

13 Suppose E is a Galois extension of F such that $\mathrm{Gal}_F(E) \approx D_5$. Determine the number of subfields of E that contains F.

14 Give an example for which $F \subseteq K \subseteq E$ are three different fields, and E is a Galois extension of F, and E is a Galois extension of K, but K is not a Galois extension of F.

15 Give an example for which $F \subseteq K \subseteq E$ are three different fields, and K is a Galois extension of F, and E is a Galois extension of K, but E is not a Galois extension of F.

Interactive Problems

16 Let $f(x) = \Phi_8(x) = x^4 + 1$, and let F be the splitting field of $f(x)$ over $\mathbb{Q}$. Use *Sage* to find the fixed fields for the 3 elements of $\mathrm{Gal}_{\mathbb{Q}}(F)$ of order 2. Express these fields in the form $\mathbb{Q}(\sqrt{a})$, where a is rational.

17 Let $f(x) = \Phi_{12}(x) = x^4 - x^2 + 1$, and let F be the splitting field of $f(x)$ over $\mathbb{Q}$. Use *Sage* to find the fixed fields for the 3 elements of $\mathrm{Gal}_{\mathbb{Q}}(F)$ of order 2. Express these fields in the form $\mathbb{Q}(\sqrt{a})$, where a is rational.

18 Let $f(x) = \Phi_{16}(x) = x^8 + 1$, and let F be the splitting field of $f(x)$ over $\mathbb{Q}$. Use *Sage* to find the fixed fields for the 4 elements of $\mathrm{Gal}_{\mathbb{Q}}(F)$ of order 4. Express these fields in the form $\mathbb{Q}(\sqrt{a})$, where a is rational.

19 Let $f(x) = \Phi_{15}(x) = x^8 - x^7 + x^5 - x^4 + x^3 - x + 1$, and let F be the splitting field of $f(x)$ over $\mathbb{Q}$. Use *Sage* to find the fixed fields for the 4 elements of $\mathrm{Gal}_{\mathbb{Q}}(F)$ of order 4. Express these fields in the form $\mathbb{Q}(\sqrt{a})$, where a is rational.

20 Let $f(x) = \Phi_{20}(x) = x^8 - x^6 + x^4 - x^2 + 1$, and let F be the splitting field of $f(x)$ over $\mathbb{Q}$. Use *Sage* to find the fixed fields for the 4 elements of $\mathrm{Gal}_{\mathbb{Q}}(F)$ of order 4. Express these fields in the form $\mathbb{Q}(\sqrt{a})$, where a is rational.

15.4 Applications of Galois Theory

There are two main results of Galois theory. One result is that one can demonstrate that it is impossible to find a formula for the solutions to a fifth-degree polynomial in terms of square roots, cube roots, or fifth roots. This finally closed the door on a centuries-old problem, ever since the solutions for a third- and fourth-degree equations were discovered in the 16th century. But the other application of Galois theory applies to problems thousands of years old. The three great construction problems of antiquity are trisecting an angle, duplicating the cube, and squaring the circle. (See Historical Diversion on page 5.) Galois theory proved once and for all that these problems are impossible to construct with only a straight edge and compass.

We will begin with the problem of determining whether or not a polynomial can be solved in terms of radicals. The first step is to show that, in $\mathbb{Q}$, a Galois extension is the same thing as a splitting field.

PROPOSITION 15.9

Let E be a finite extension of $\mathbb{Q}$. If $f(x)$ is a polynomial in $E[x]$, then the splitting field of $f(x)$ is a Galois extension of E.

PROOF: Let K be the splitting field of $f(x)$ in $E[x]$. If u is an element of K not in E, then $g(x) = \mathrm{Irr}_E(u, x)$ has degree > 1. By Lemma 14.5, $g(x)$ factors completely in the field K. Thus, the splitting field of $g(x)$ is contained in the field K. However, $g(x)$ is a factor of $\mathrm{Irr}_{\mathbb{Q}}(u, x)$, which by Lemma 15.3 does not have multiple roots in K. Therefore, $g(x)$ cannot have multiple roots in K, so there exist at least two roots of $g(x)$ in K. Let v be a root of $g(x)$ different from u. Then $g(x) = \mathrm{Irr}_E(v, x)$, and so by Proposition 15.2 there exists a ϕ in $\mathrm{Gal}_E(K)$ such that $\phi(u) = v$. Thus, u is not in the fixed field of $\mathrm{Gal}_E(K)$. Since E is obviously contained in the fixed field of $\mathrm{Gal}_E(K)$, we find that the fixed field is E, so K is a Galois extension of E. □

The next step is to give a clear definition of what it means for a polynomial to be solvable by radicals.

DEFINITION 15.6 A field K is called a *radical extension* of F if $K = F(u_1, u_2, \ldots, u_n)$, where a power of each u_i is contained in $F(u_1, u_2, \ldots, u_{i-1})$.

Example 15.11

Express the splitting field of $x^4 - 8x^2 - 8x - 2$ as a radical extension.

SOLUTION: We can have *Sage* solve for the roots explicitly.

```
var("x")
solve(x^4 - 8*x^2 - 8*x - 2, x)
    [x == -sqrt(2) - 1/2*sqrt(-4*sqrt(2) + 8),
    x == -sqrt(2) + 1/2*sqrt(-4*sqrt(2) + 8),
    x == sqrt(2) - 1/2*sqrt(4*sqrt(2) + 8),
    x == sqrt(2) + 1/2*sqrt(4*sqrt(2) + 8)]
```

We see that the solutions are

$$x = -\sqrt{2} \pm \sqrt{2 - \sqrt{2}}, \qquad \text{or} \qquad x = \sqrt{2} \pm \sqrt{2 + \sqrt{2}}.$$

How would we express the splitting field as a radical extension? It is apparent that we first must include $\sqrt{2}$ in this field. But then it seems we need to include $\sqrt{2+\sqrt{2}}$ and $\sqrt{2-\sqrt{2}}$ in our field. Note, however, that the product of these two numbers is $\sqrt{2}$. Thus, all four roots are in the field $\mathbb{Q}(\sqrt{2}, \sqrt{2+\sqrt{2}})$. This is a radical extension of $\mathbb{Q}$ of dimension 4, and the splitting field of $x^4 - 8x^2 - 8x - 2$ must be at least 4. Hence, we have found that the splitting field is a radical extension of $\mathbb{Q}$. □

DEFINITION 15.7 The polynomial equation $f(x) = 0$ is said to be *solvable by radicals* if there is a radical extension of $\mathbb{Q}$ that contains the splitting field of $f(x)$.

This definition agrees with our intuitive understanding of what it means for a polynomial to be solved in terms of radicals.

Computational Example 15.12

Use *Sage* to find a radical extension that contains the splitting field of the polynomial $x^3 - x + 3$.

SOLUTION: We can have *Sage* find the exact roots of the equation.

```
var("x")
Solve(x^3 - x + 2, x)
    [x == -1/2*(1/9*sqrt(26)*sqrt(3) - 1)^(1/3)*(I*sqrt(3) + 1)
    - 1/6*(-I*sqrt(3) + 1)/(1/9*sqrt(26)*sqrt(3) - 1)^(1/3),
    x == 1/2*(1/9*sqrt(26)*sqrt(3) - 1)^(1/3)*(-I*sqrt(3) + 1)
    - 1/6*(I*sqrt(3) + 1)/(1/9*sqrt(26)*sqrt(3) - 1)^(1/3),
    x == (1/9*sqrt(26)*sqrt(3) - 1)^(1/3) +
    1/3/(1/9*sqrt(26)*sqrt(3) - 1)^(1/3)]
```

This result reveals that the splitting field is contained in radical extension

$$\mathbb{Q}\left(\sqrt{26 \cdot 3}, \sqrt[3]{1 - \sqrt{26 \cdot 3}/9}, \sqrt{-3}\right).$$

This is in fact overkill, since the splitting field is at most a 6-dimensional extension of $\mathbb{Q}$, while the above radical extension may be up to a 12-dimensional extension of $\mathbb{Q}$. Yet the point is that there is *some* radical extension of $\mathbb{Q}$ that contains the roots of $x^3 - x + 2$, because the roots can be solved in terms of square roots and cube roots. ∎

Not all radical extensions of $\mathbb{Q}$ are Galois extensions. For example, $\mathbb{Q}(\sqrt[3]{2})$ is not a Galois extension, since this extension is not the splitting field of a polynomial. In order to utilize Galois theory, we need to show that a radical extension is contained in some extension that is both a radical extension and a Galois extension.

LEMMA 15.8

Let E be a radical extension of $\mathbb{Q}$. Then E is contained in a radical extension K of $\mathbb{Q}$ such that K is a Galois extension of $\mathbb{Q}$.

PROOF: Let $E = \mathbb{Q}(u_1, u_2, u_3, \ldots, u_n)$ be a radical extension of $\mathbb{Q}$. Then for every $i = 1, 2, 3, \ldots, n$, there is a k_i for which

$$(u_i)^{k_i} = v, \text{ for which } v \in \mathbb{Q}(u_1, u_2, u_3, \ldots, u_{i-1}).$$

Note that if $n = 0$, then $E = \mathbb{Q}$, and the lemma is obviously true. We will prove this by induction on n. That is, we will assume that the lemma is true for the field

$$\mathbb{Q}(u_1, u_2, u_3, \ldots, u_{n-1}).$$

That is, this field is contained in a radical extension L of $\mathbb{Q}$ that is also a Galois extension of $\mathbb{Q}$.

By Corollary 15.2, there exists an element w of L such that $L = \mathbb{Q}(w)$.

Let $g(x) = \text{Irr}_{\mathbb{Q}}(w, x)$ and $p(x) = \text{Irr}_{\mathbb{Q}}(u_n, x)$. Let K be the splitting field of $g(x) \cdot p(x)$ over $\mathbb{Q}$. By Proposition 15.9, K is a Galois extension of $\mathbb{Q}$. Since w is in K, L is a subfield of K. The only thing left to show is that K is a radical extension of L.

Let $v_1, v_2, v_3, \ldots, v_m$ be all of the roots of $p(x)$ in K. Since $p(x)$ is irreducible, by Proposition 15.2 there is a $\mathbb{Q}$-automorphism ϕ_i that sends u_n to v_i. Since $(u_n)^{k_n} = b$ is in L, we have

$$(v_i)^{k_n} = (\phi_i(u_n))^{k_n} = \phi_i((u_n)^{k_n}) = \phi_i(b).$$

Now, L is a Galois extension of $\mathbb{Q}$, so by the fundamental theorem of Galois theory (15.1), $\text{Gal}_L(K)$ is a normal subgroup of $\text{Gal}_{\mathbb{Q}}(K)$. So by Lemma 15.6 $\mathbb{Q}$-automorphisms of K map elements of L to elements of L. Thus, $\phi_i(b)$ is in L, and so $K = L(v_1, v_2, v_3, \ldots v_m)$ is a radical extension of L. ∎

Lemma 15.8, when combined with the definition of a polynomial solvable by radicals, tells us that if a polynomial is solvable by radicals, then the splitting field of the polynomial is contained in a field extension of $\mathbb{Q}$ that is both a radical extension and a Galois extension. What can we say about such an extension? Surprisingly, the answer has a connection with the Jordan-Hölder theorem (8.2).

LEMMA 15.9

Let K be a Galois extension of $\mathbb{Q}$ that is also a radical extension, and let E be a subfield of K. If E is a Galois extension of $\mathbb{Q}$, then $\mathrm{Gal}_{\mathbb{Q}}(E)$ is a solvable group.

PROOF: Since K is a radical extension of $\mathbb{Q}$, we can write

$$K = \mathbb{Q}(u_1, u_2, u_3, \ldots, u_n)$$

where some power of each u_i, $(u_i)^{k_i}$, is in $\mathbb{Q}(u_1, u_2, u_3, \ldots, u_{i-1})$.

Let m be the least common multiple of all of the k_i, and let $u_0 = \omega_m$, the mth root of unity. We would like to add u_0 in the front of the sequence of u's to get a larger field

$$M = \mathbb{Q}(u_0, u_1, u_2, u_3, \ldots, u_n).$$

Since $(u_0)^m = 1$, we see that M is still a radical extension of $\mathbb{Q}$. To show that $M = K(u_0)$ is a Galois extension of $\mathbb{Q}$, note that by Corollary 15.2, $K = \mathbb{Q}(w)$ for some element w in K. If $f(x) = \mathrm{Irr}_{\mathbb{Q}}(w, x)$, then M is the splitting field of the polynomial $f(x) \cdot (x^m - 1)$. Thus, by Proposition 15.9, M is a Galois extension of $\mathbb{Q}$.

Consider the sequence of subfields

$$\begin{aligned}
E_0 &= \mathbb{Q}(u_0), \\
E_1 &= \mathbb{Q}(u_0, u_1), \\
E_2 &= \mathbb{Q}(u_0, u_1, u_2), \\
E_3 &= \mathbb{Q}(u_0, u_1, u_2, u_3), \\
&\cdots\cdots\cdots \\
E_n &= \mathbb{Q}(u_0, u_1, u_2, u_3, \ldots, u_n) = M.
\end{aligned}$$

By Proposition 15.7, each of these fields is a Galois extension of the previous field, since the mth roots of unity were designed to be in all of these fields. Also, by Proposition 15.6, E_0 is a Galois extension of $\mathbb{Q}$.

We can now apply the fundamental theorem of Galois theory (15.1). We find that $\mathrm{Gal}_{E_k}(M)$ is a normal subgroup of $\mathrm{Gal}_{E_{k-1}}(M)$, and the quotient group

$$\mathrm{Gal}_{E_{k-1}}(M)/\mathrm{Gal}_{E_k}(M)$$

is isomorphic to $\mathrm{Gal}_{E_{k-1}}(E_k)$.

By Proposition 15.7, each of these quotient groups are abelian. Also, by Proposition 15.6, $\mathrm{Gal}_{\mathbb{Q}}(E_0)$ is isomorphic to Z_m^*, which is abelian. Thus, the sequence of subgroups

$$\mathrm{Gal}_{\mathbb{Q}}(M) \supseteq \mathrm{Gal}_{E_0}(M) \supseteq \mathrm{Gal}_{E_1}(M) \supseteq \cdots \supseteq \mathrm{Gal}_{E_n}(M) = \{e\}$$

is a subnormal series for which all of the quotient groups are abelian. Therefore, the composition series of $\mathrm{Gal}_{\mathbb{Q}}(M)$ will consists of only prime, cyclic factors. By the solvability theorem (8.3), $\mathrm{Gal}_{\mathbb{Q}}(M)$ is a solvable group.

To finish the theorem, we note that E is a Galois extension of $\mathbb{Q}$, so by the fundamental theorem of Galois theory (15.1), $\mathrm{Gal}_E(M)$ is a normal subgroup of $\mathrm{Gal}_{\mathbb{Q}}(M)$, and $\mathrm{Gal}_{\mathbb{Q}}(E)$ is isomorphic to $\mathrm{Gal}_{\mathbb{Q}}(M)/\mathrm{Gal}_E(M)$. Using Proposition 8.3 we see that $\mathrm{Gal}_{\mathbb{Q}}(E)$ is solvable. □

The light is beginning to appear at the end of the tunnel. We know that any subgroup of a solvable group must be solvable. Thus, we can immediately tell whether a polynomial is solvable by radicals from its Galois group.

THEOREM 15.2: Galois' Criterion Theorem

Let $f(x)$ be a polynomial with rational coefficients. Then the equation $f(x) = 0$ is solvable by radicals only if the Galois group of $f(x)$ is a solvable group.

PROOF: Suppose that $f(x)$ is a polynomial that is solvable by radicals. Let E be the splitting field of $f(x)$. By Lemma 15.8, there is a field K containing E which is a Galois extension of $\mathbb{Q}$, and also is a radical extension of $\mathbb{Q}$. By Proposition 15.9, E is a Galois extension of $\mathbb{Q}$. Thus, we can use Lemma 15.9 to show that the Galois group of $f(x)$, $\mathrm{Gal}_{\mathbb{Q}}(E)$, is a solvable group. □

Galois' criterion theorem is able to show us that there are some polynomials whose roots cannot be expressed in terms of square roots, cube roots, and other roots. In fact we found one of them using *Sage*, namely $x^5 - x + 1$.

```
InitDomain(0, "x")
GaloisType(x^5 - x + 1)
    S5
```

COROLLARY 15.3

There is no formula, using only the field operations and extraction of roots, for the zeros of all fifth-degree polynomial equations.

PROOF: We have already shown that the Galois group of $x^5 - x + 1$ is isomorphic to S_5. But S_5 is not solvable, since it contains the non-cyclic simple subgroup A_5. Thus, by Galois' criterion theorem (15.2) this particular equation cannot be solved with a formula involving only field operations and extraction of roots, so certainly there can be no general formula. □

This corollary was actually first proven by Abel, but Galois was unaware of Abel's proof when he proved the Galois' criterion theorem. Abel and Galois's theorems ended the long search for a formula that finds the roots of a fifth-degree polynomial. In fact, Galois' criterion theorem works in the other direction as well—if the Galois group is solvable, then the polynomial *is* solvable by radicals [2, p. 558]. Since a fourth-degree equation is a subgroup of S_4, which is solvable, there must be a formula for the roots of a fourth-degree polynomial. The change of the structure between S_4 and S_5 is what changes the behavior of fifth-degree polynomials from fourth-degree polynomials.

Galois theory also can be used to analyze *construction problems.* The ancient Greeks had a very rigorous definition of what it meant for a certain geometrical object to be constructed using an unmarked straightedge and compass.

- Given two points that have already been constructed, we can construct the straight line going through the points.
- Given two points that have already been constructed, we can construct a circle with a center at one point and passing through the other.
- Given two lines that have been constructed, the point of intersection becomes a constructed point.
- Given a circle and a line, or two circles, that have been constructed, the points of intersection, if they exist, become constructed points.

Starting with a unit segment, what other lengths can be constructed?

DEFINITION 15.8 *We say that a real number x is* constructible *if we can create a line segment whose length is $|x|$ times the length of the unit segment in a finite number of steps, using only the allowed procedures using the straightedge and compass.*

Using plane geometry, it is not hard to see that if x and $y \neq 0$ are constructible numbers, then $x + y$, $x - y$, $x \cdot y$, and x/y are constructible. (See Problems 9 through 11.) Thus, the set of constructible numbers forms a subfield F of $\mathbb{R}$. We can characterize the field F by the following theorem.

THEOREM 15.3: The Constructible Criterion Theorem
If x is a constructible number, then x is contained in an extension of $\mathbb{Q}$ of dimension 2^n for some n.

PROOF: Since it is possible to construct a line perpendicular to a given line and going through a particular point, we can use Cartesian coordinates to describe the constructible numbers. By first constructing two perpendicular lines, the x- and y-axes, given two constructible numbers a and b, we can construct the point (a, b) using two perpendicular lines. Likewise, if some point is constructible, we can drop perpendiculars to construct the numbers of the coordinates. Thus, we see that the point (a, b) is constructible if, and only if, a and b are constructible numbers.

We can likewise characterize the lines and circles that are constructible. A line is clearly constructible if and only if the x and y intercepts are constructible numbers. Thus, the line $ax + by + c = 0$ is constructible if, and only if, a, b, and c can be made to be constructible numbers. Likewise, the circle $x^2 + y^2 + ax + by + c = 0$ is constructible if and only if a, b, and c are constructible numbers.

Now, the intersection of two lines can be found using only field operations. But the intersection of a line and a circle, or two circles, will involve a quadratic equation. The solution of a quadratic equation involves taking a square root of an element in F. This means that a number x is a constructible number if, and only if, there is a sequence of fields $\mathbb{Q} = F_1, F_2, F_3, \ldots F_n$ with $x \in F_n$ such that each $F_{k+1} = F_k(\sqrt{a_k})$, where $a_k > 0$ is in F_k.

Thus, we see that whenever x is a constructible number, it is contained in a radical extension of $\mathbb{Q}$. Furthermore, since only square roots are involved in the field extensions, the dimension of the radical extension will be 2^n for some n. ∎

We can now show that two of the three great construction problems of antiquity are impossible.

COROLLARY 15.4

It is impossible, using only an unmarked straightedge and compass, to trisect an angle or duplicate the cube.

PROOF: Suppose it were possible to trisect any angle. Then in particular it would be possible to trisect a $60°$ angle, hence we would be able to construct a $20°$ angle, so $x = \cos(20°)$ would be a constructible number. But we saw from Example 11.6 that x is a root of the irreducible polynomial $8x^3 - 6x - 1$. Hence, x cannot be in an field extension of $\mathbb{Q}$ of dimension 2^n. So we cannot trisect a $60°$ angle with a straightedge and compass, let alone the general angle.

Likewise, we can show that duplicating the cube is impossible. This would involve constructing a line of length $\sqrt[3]{2}$, which is a root of $x^3 - 2$. Again, since $\mathbb{Q}(\sqrt[3]{2})$ is a three-dimensional extension of $\mathbb{Q}$, this cannot be contained in a field of dimension 2^n, so $\sqrt[3]{2}$ is not a constructible number. ∎

The last construction problem involves knowledge that π is transcendental, so it requires a more difficult proof to show that π is not a constructible number.

There are many other applications of Galois theory. For example, one can

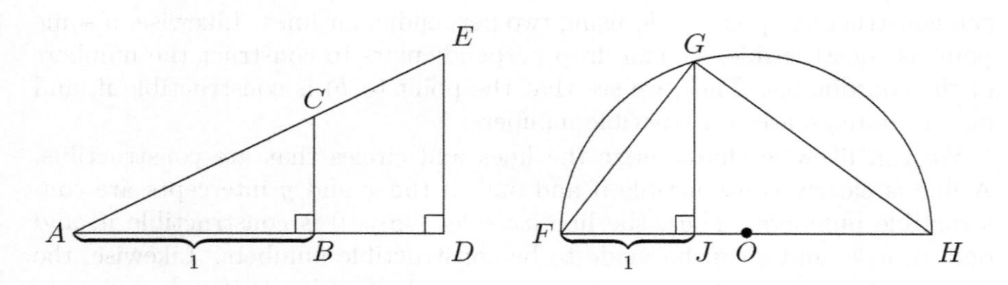

FIGURE 15.6: Geometric constructions used for Problems 10 through 13

use the fundamental theorem of Galois theory to prove that any polynomial in the complex numbers has a root in the complex numbers. This is known as the *fundamental theorem of algebra.*

Problems for §15.4

For Problems **1** through **8**: Express the splitting field of the following polynomials as a radical extension of $\mathbb{Q}$.

1 $x^3 - 2$
2 $x^4 - 3$
3 $x^4 + 5$
4 $x^5 + 6$
5 $x^4 - 5x^2 + 6$
6 $x^4 + 2x^2 - 2$
7 $x^6 + x^3 - 6$
8 $x^6 + 3x^3 - 5$

9 Show that if a and b are positive constructible numbers, and $a > b$, then $a + b$ and $a - b$ are constructible.

10 Show that if a and b are positive constructible numbers, and $b > 1$, then $a \cdot b$ is constructible.
Hint: In Figure 15.6, let $\overline{BC} = a$, and $\overline{AD} = b$. What is $\overline{DE}$?

11 Show that if a and b are positive constructible numbers, and $b > 1$, then a/b is constructible.
Hint: In Figure 15.6, let $\overline{DE} = a$, and $\overline{AD} = b$. What is $\overline{BC}$?

12 Explain how we could redraw Figure 15.6 so that we can construct $a \cdot b$ and a/b for the case where $b < 1$ in Problems 10 and 11.

13 Show that if a is a positive constructible number, then $\sqrt{a}$ is constructible.
Hint: In Figure 15.6, let $\overline{HJ} = a$. Note that since a triangle inscribed in a semi-circle is always a right triangle, $\angle FGH$ is a right angle.

14 Although constructible numbers are defined to be real numbers, we can consider extensions involving square roots of complex numbers as well. That is, we can consider the complex number x to be constructible if there is a sequence of fields $\mathbb{Q} = F_1, F_2, F_3, \ldots F_n$ with $x \in F_n$ such that each $F_{i+1} = F_i(\sqrt{a_i})$, where a_i is in F_i. Show that $a + bi$ is constructible in the new sense if, and only if, a and b are real constructible numbers by Definition 15.8.

Hint: First show that

$$\sqrt{a+bi} = \frac{\sqrt{\sqrt{a^2+b^2}+a}}{\sqrt{2}} + \frac{bi}{\sqrt{2}\sqrt{\sqrt{a^2+b^2}+a}}.$$

15 The converse of the Constructible Criterion Theorem (15.3) is not always true. That is, just because a complex number z is contained in a field extension F of $\mathbb{Q}$ of dimension 2^n does not mean that z is constructible in the sense of Problem 14. However, by adding the condition that F is a Galois extension of $\mathbb{Q}$, show that z is constructible.

Hint: Use Proposition 7.8 to show $\mathrm{Gal}_{\mathbb{Q}}(F)$ is solvable, with all composition factors isomorphic to Z_2.

16 Use Problem 15 to show that a pentagon is constructible.

17 Use Problem 15 to show that a 17-gon is constructible. This was first discovered by Gauss.

18 Use Problem 15 to show that a 257-gon is constructible. Note that 257 is prime.

Interactive Problems

19 Express $\mathbb{Q}(\omega_5)$ as a radical extension using only square roots. This, along with Problem 14, can be used to show how to construct a regular pentagon.

Hint: Find a subnormal sequence for the Galois group, and then find generators for the fixed fields for each subgroup.

20 Express $\mathbb{Q}(\omega_{17})$ as a radical extension using only square roots. This, along with Problem 14, can be used to show how to construct a regular 17-gon. See the hint for Problem 19.

Appendix: Sage vs. Mathematica®

This textbook incorporates either *Sage* or *Mathematica* to help students visualize the important concepts of abstract algebra. It is recommended that one of the two programs be used with the book, but it is not necessary to have both. This section compares the two programs, and gives instructions for how to use these programs with the files on the included CD.

Mathematica is a symbolic manipulator package published by Wolfram Research, Inc. That is, it is a general purpose mathematical program used by scientists, engineers, and analysts. Its main feature that sets it apart from other symbolic manipulators is the graphics capabilities. In *Mathematica* 10.0, one can plot a 3-dimensional object, then use the mouse to rotate the object in three dimensions to see it from all possible angles.

Sage is also a symbolic manipulator, but has the advantage of being open source. This means that it is totally free. It has slightly less graphic capabilities than *Mathematica*, but it can still graph three dimensional objects, and rotate them. *Sage* is also capable of interfacing with GAP, which stands for "Groups, Algorithms, and Programming." Hence *Sage* is particularly suited for abstract algebra. *Mathematica*, however, was never designed to work problems involving abstract algebra. The reason why *Mathematica* is able to do the abstract algebra calculations is because of the supporting software provided with the textbook.

IMPORTANT: In order to use either *Sage* or *Mathematica* for this textbook, you will also need to install the supporting files into your computer. Simply put the CD provided into the computer, and the installation program should start running. If this program does not start automatically in any of the Windows versions, click on the "Start" icon, and select "Run." At this menu, select "Browse." Next, find the drive for the CD, and select the file "AbstractAlgebraSetup.exe." Hit "OK" to start the setup program running. Follow the instructions to install either the *Mathematica* or *Sage* supporting files, or both, onto the computer. Another option would be to copy the **`math`** and/or **`sage`** folders directly from the CD to the computer. The latter method will work for any operating system. Note that this only loads the supporting files, so you will also have to install *Mathematica* or *Sage* programs as well.

Included in the supporting files are two *Mathematica* packages, "group.m" and "ring.m." The first of these is used for chapters 0-8 of the text, while the other is used in the remaining chapters. Both files are in the `math` folder on the CD provided with this book. These two files allow *Mathematica* to work with groups as fluently as *Sage*. There are, however, a few things that *Sage* can do that *Mathematica* cannot, due to the algorithms.

Sage also has a package "absalgtext.sage" in the `sage` directory, which causes the interface with GAP to be seamless. Because of this file, the commands for *Sage* are nearly the same as the commands for *Mathematica*.

Also in the supporting files are *Mathematica* and *Sage* files for each chapter of the textbook. *Mathematica* calls these files *notebooks*, and the file names are "group00.nb" through "group08.nb," and "ring09.nb" through "ring15.nb" in the `math` directory. *Sage* calls these files *worksheets*, but they do the same thing. The files "group00.sws" through "group08.sws," and "ring09.sws" through "ring15.sws" are the corresponding *Sage* files in the `sage` folder. These notebooks/worksheets allow a student to walk through the examples in the book, along with other similar examples. Included in these notebooks/worksheets are all the theorems and proofs in the textbook.

Once the supporting files have been installed, then one of the packages can be loaded into *Mathematica* with either of the two commands:

<< c:\math\group.m

<< c:\math\ring.m

This will only have to be done once in each *Mathematica* session. Likewise, each *Sage* session must begin with the command that loads the file **`absalgtext.sage`**, but unlike *Mathematica*, this file will automatically be loaded when any worksheet is loaded. *Sage* will even search two possible locations for this file.

Because of the similarities of the two systems, this book only shows the input and output for *Sage*. The main reason for this is that switching back and forth between two systems proved to be distracting, as seen in the first edition of this book. Those using *Mathematica* can open the notebooks to see the corresponding *Mathematica* commands, and still follow along closely with the book.

Mathematica is not free, but price information can be obtained from

`http://www.wolfram.com`

However, one can obtain a 30-day *Mathematica* product trial.

There is also a free *Mathematica Player* available from Wolfram, which will be able to open the notebooks provided with this textbook. However, one cannot execute any of the *Mathematica* commands with *Mathematica Player*.

To load one of the supporting files in *Mathematica*, click on "File" and then slide down to "Open." One can locate one of the 16 notebooks with the .nb extension in the `C:\math` directory.

Although *Sage* is a totally free program, it takes some effort to install. This is because it runs on Linux, not Windows. As a result, one has three options.

1. Run *Sage* in VirtualBox, which is a free Windows program. This will make *Sage* particularly slow, especially if you have less than 6 GB of RAM on your computer. (4 GB is minimum.)

2. Create a hard drive partition that can boot to a Linux operating system, such as Ubuntu. The computer will then be able to boot to either Windows or Linux. This is actually the preferred method, and is easier than it might first appear.
3. Run *Sage* on a cloud, using a web browser. This would be ideal in a university setting, since the professor can upload the software to the cloud, as explained below, and give students access to it.

Each of the three options requires some instructions to set up.

Using VirtualBox

After installing the software from the CD, the VirtualBox installer will be in the `C:\InterAbstAlg` directory. One can also download the most recent version from `www.virtualbox.org`. Double click on the file to start installing VirtualBox, and follow the standard installation instructions.

Start VirtualBox, select the `File` menu, and select the `Import Appliance` option. At this point you can either click on the right file icon, and browse for the *Sage* package, or type in `C:\InterAbstAlg\sage-5.12.ova`. This is the version supplied with this book, but you may want to first replace this file with the most recent version at

`http://www.sagemath.org/download-windows.html`.

Finally, hit `Next`, and then `Import`.

Before *Sage* will be able to read the worksheet files supplied with the textbook, we will need to set up a shared directory between Windows and Linux. This is accomplished by selecting the `Settings` menu, and selecting the `Shared Folders` option. This brings up a short form to fill out. Under "Folder path", type `C:\sage`, and click on the "Auto-mount" box. Finally hit "OK" twice to set up the shared directory. Now every time that VirtualBox is run, the directory `C:\sage` will appear on the Linux side as the directory `/media/sf_sage`.

While we are in the settings menu, click on the "General" menu, and select the "Advanced" tab. Change the "Shared Clipboard" option to "Bidirectional." This will allow you to cut and paste text to and from *Sage*.

We are now ready to click on the big blue *Sage* button. It will take some time to load, but will finally display the *Sage* banner. Click on "Upload," and on the menu select "Browse." Navigate through "Filesystem," then the "media" directory, and finally the "sf_sage" subdirectory. Select one of the worksheets, such as "group00.sws". Finally, click on "Open" and then "Upload Worksheet."

Using Linux

This assumes that you have installed a Linux system on your computer. For Ubuntu-based Linux distributions (Linux Mint, *buntu, etc.), we can use a PPA. To install *Sage*, open a Terminal window, and type

```
sudo apt-add-repository -y ppa:aims/sagemath
sudo apt-get update
sudo apt-get install sagemath-upstream-binary
```

For other Linux distributions, a tarball can be downloaded from the main website `www.sagemath.org`.

There is also a database file to be installed that allows GAP to perform advanced group operations, which are only needed for the **StructureDescription** and **GaloisType** commands.

```
sudo sage -i database_gap
```

In order to view animations in *Sage*, you will also have to install either *ImageMagick* or *FFmpeg*. The following installs *ImageMagick*, along with other recommended programs.

```
sudo apt-get install gfortran
sudo apt-get install imagemagick texlive dvipng
```

You will have to manually copy the files in the `sage` directory on the CD to a directory named `sage` in your home directory. However, the worksheets are expecting these files to be in the `media/sf_sage` directory because of VirtualBox. Rather than constantly change the *Sage* commands, we can add a symbolic link to make these directories the same.

```
cd /media
sudo ln -s  /sage sf_sage
```

You can now exit the terminal window with **exit**. It is important at this point to run the Software Update on the computer. This final step links the *Sage* and *ImageMagick* programs, so they can interact.

At this point the big blue *Sagemath* button should appear in the Applications menu. Clicking on this will cause *Sage* to appear in a web browser. Click on "Upload," and on the menu select "Browse." One can either navigate through "Filesystem" then the "media" directory, and finally the "sf_sage" subdirectory, or find the "sage" directory inside your home directory. Select one of the worksheets, such as "group00.sws." Finally, click on "Open" and then "Upload Worksheet."

Using SageMathCloud

At the time this book was published, *SageMathCloud* had some trouble displaying some of the more complicated math formulas, so it was harder to read. Nonetheless, it gives another option for running *Sage* under Windows or Mac OS.

To begin, sign up at

`https://cloud.sagemath.com`

Once signed up, click on "Projects," and then "Create New Project." After creating and opening the new project, you can download the .sws files that are in the `C:\sage` directory one at a time, along with the "absalgtext.sage" file. Select one of the worksheets, such as "group00.sws".

Once the Notebook/Worksheet is Loaded

The first cell of every *Mathematica* notebook or *Sage* worksheet is the Initialization cell. This must be executed first before any other commands in the notebook will work, since it loads either the "group.m," "ring.m," or "absalgtext.sage" file. Note that in *Sage*, the initialization cell is executed automatically when the worksheet is loaded. The very first time that the initialization is done in *Sage*, there may also be additional databases that are automatically downloaded from the Internet.

For *Mathematica*, click on this cell, and hit Shift and Enter at the same time. When done, the message

```
Initialization Done
```

will appear. It will always prompt you if you want to run the initialization cell first.

Both of the programs are interactive systems. Every expression that one types into the computer is immediately evaluated, and the result is shown. This is known as a read-evaluate-print loop. To create a new cell in *Sage*, move the cursor to a point between two cells, and a long blue strip will appear. Clicking on this strip inserts a new input cell. In *Mathematica*, click between two cells and start typing, and a new cell will be created.

In either system, try computing 3^{90}, using the Shift-Enter combination.

```
3^90
    8727963568087712425891397479476727340041449
```

If *Sage* is not running under SageMathCloud, there is an "Evaluate" button that appears when the cell is selected. Clicking on this button is an alternative to the Shift-Enter combination.

Mathematica adds `In[]` and `Out[]` numbers.

In[2] := **3^90**
Out[2]:= 8 727 963 568 087 712 425 891 397 479 476 727 340 041 449

Mathematica will number all of the input and output statements, but the prompt does not appear until *after* some expression is entered. Note that the numbers correspond to the cells evaluated in the current *session*, not the current notebook. So when the notebooks are first opened, none of the "In[n] :=" or "Out[n] :=" will be present. Likewise, if a second notebook is opened and a cell is evaluated, it might start with a value other than "In[1]". It is suggested that the cells be evaluated in the order that they appear, but there is nothing to prevent executing the statements in any order, or executing a statement more than once. The “In[n] :=” and ”Out[n] :=” will show which commands have been run and in what order. Any cell that does not have a “In[n] :=” has not been evaluated yet, even through it appears to have a corresponding output.

Had we put a semi-colon before pressing the Shift-Enter, we would get a different effect. It computes the expression, but does not display the answer. For example, entering

```
a = 3^300;
```

will assign the variable a a 144-digit number, but will not display this number. Actually, *Sage* would not display the number even without the semi-colon, because the value is assigned to the variable. To see the value of a, one can enter

```
a
    136891479058588375991326027382088315966463695625337436\
    471480190078368997177499076593800206155688941388250484\
    4059799404281351273276569577456 6001
```

Note that both *Sage* and *Mathematica* use the backslash to show that the number is continued on the next line.

In both programs, a variable is a sequence of letters and/or digits, but must begin with a letter. Variables are case sensitive, so a is a different variable than A. Keywords, such as **if** or **quit**, are not allowed as variables, but the list of keywords is too long to give here. None of the lower-case letters are keywords, so we can safely use the 26 variables a through z.

Mathematica does not automatically expand an expression, although it might rearrange the factors and terms.

(y^2 + 3y − 1)(y^2 − 2y + 4)
$(4 - 2y + y^2)(-1 + 3y + y^2)$

Because we have not yet assigned a value to y, *Mathematica* assumes that it is an indeterminate, so that it expresses the answer in terms of y. Also note that *Mathematica* assumes that a number and letter next to each other are to be multiplied together. In *Sage*, we must explicitly use the * for every multiplication.

```
(y^2 + 3*y - 1)*(y^2 - 2*y + 4)
    Traceback (click to the left of this block for traceback)
    ...
    NameError: name 'y' is not defined
```

This time, we get an error message, since *Sage* has not been told what y is. Unlike with *Mathematica*, we must declare y to be a variable in *Sage* before we can use it as a variable. The simplest way to do this is with the command:

```
var("y")
    y
```

We can try the expression again.

```
(y^2 + 3*y - 1)*(y^2 - 2*y + 4)
    (y^2 + 3*y - 1)*(y^2 - 2*y + 4)
```

If we want to expand this, we can use the **expand** function.

```
expand(_)
    y^4 + y^3 - 3*y^2 + 14*y - 4
factor(_)
    (y^2 + 3*y - 1)*(y^2 - 2*y + 4)
```

Note that the underscore (_) is a *Sage* abbreviation for the last output. The corresponding symbol in *Mathematica* is the percent sign (%).

Expand[%]
$-4 + 14y - 3y^2 + y^3 + y^4$
Factor[%]
$(4 - 2y + y^2)(-1 + 3y + y^2)$

There are other syntax differences between *Sage* and *Mathematica*: *Sage* uses parentheses for functions, as the standard notation, but *Mathematica* uses square brackets for functions. Also, every function name in *Mathematica* is capitalized.

Most calculations in *Mathematica* and *Sage* are also exact, but you do have the option of finding a decimal approximation using the **N** function. For example, the first 50 digits of $\sqrt{2}$ are computed in *Mathematica* as

N[Sqrt[2], 50]
1.4142135623730950488016887242096980785696718753769

The same command in *Sage* requires a bit more syntax.

```
N(sqrt(2), digits=50)
    1.4142135623730950488016887242096980785696718753769
```

Both *Sage* and *Mathematica* will point out any mistakes in the input line. For example, if one types

```
(4 = 3)*2
    Traceback (click to the left of this block for traceback)
    ...
    SyntaxError: invalid syntax
```

To find out more information, click on the left side of the error message, and it will expand. The last few lines are as follows:

```
(4 = 3)*2
      (_sage_const_4  = _sage_const_3 )*_sage_const_2
                      ^

   SyntaxError: invalid syntax
```

Sage points to the error with a caret (ˆ). The same typo also produces an error in *Mathematica*, but for a different reason.

(4 = 3)*2
6

Mathematica returns an answer, but also displays a strange message,

"Set::setraw : Cannot assign to raw object 4. ≫"

in a separate Messages window. Because the equal sign in *Mathematica* is used to assign a value to a variable, *Mathematica* thinks we are trying to assign the value 3 to the number 4, which of course cannot be done. But besides this, this value of 3 is multiplied by 2 to get the answer displayed.

Ironically, had we used a double equal sign, neither *Mathematica* nor *Sage* command would have produced an error.

(4 == 3)*2
2 False

The double equal sign is used to test if two expressions are equal. *Mathematica* sees no problem in symbolically multiplying **False** with an integer. *Sage* produces a different answer.

```
(4 == 3)*2
   0
```

Sage converts `False` to 0, and `True` to 1 if needed. Other features of *Sage* will be introduced in the textbook as the need arises. With a little practice, you will find both programs are relatively easy to use.

Answers to Odd-Numbered Problems

Section 0.1

1) $q = 25$, $r = 15$
3) $q = -19$, $r = 22$
5) $q = 166$, $r = 13$
7) $q = 0$, $r = 35$
9) $q = 0$, $r = 0$
11) $2^n = 2 \cdot 2^{n-1} < 2(n-1)! < n(n-1)! = n!$
13) If $(n-1)^3 + 2(n-1) = 3k$, then $n^2 + 2n = 3(k + n^2 + n + 1)$.
15) If $6^{n-1} + 4 = 20k$, then $6^n + 4 = 20(6k - 1)$.
17) $(n-1)((n-1)+1)/2 + n = n(n+1)/2$.
19) $(n-1)((n-1)+1)(2(n-1)+1)/6 + n^2 = n(n+1)(2n+1)/6$.
21) $(n-1)((n-1)+1)((n-1)+2)/3 + n(n+1) = n(n+1)(n+2)/3$.
23) $2 \cdot 24 + (-1) \cdot 42 = 6$.
25) $2 \cdot 102 + (-3) \cdot 66 = 6$.
27) $14 \cdot 1999 + (-965) \cdot 29 = 1$.
29) $5 \cdot (-602) + 12 \cdot 252 = 14$.
31) $0 \cdot 0 + 1 \cdot 7 = 7$.
33) Since xy is a common multiple, by the well-ordering axiom there is a least common multiple, say $z = ax = by$. Note that $\gcd(a, b) = 1$, else we can divide by $\gcd(a, b)$ to produce an even smaller common multiple. Then there is a u and v such that $ua + vb = 1$, so $uaxy + vbxy = xy$, hence $z(uy + vx) = xy$.
35) $2^8 \cdot 5^3$.
37) $2^2 \cdot 3 \cdot 5^2 \cdot 19$.
39) $7^4 \cdot 11$.
41) $u = -13717445541839$, $v = 97393865569283$.
43) $3^2 \cdot 17^2 \cdot 379721$.
45) $449 \cdot 494927 \cdot 444444443$.

Section 0.2

1) $\{e, n, o, r, t, x, y\}$.
3) a) Not one-to-one, $f(-1) = f(1) = 1$. b) Not onto, $f(x) \neq -1$.
5) a) One-to-one, $x^3 = y^3 \Rightarrow x = y$. b) Onto, $f(\sqrt[3]{y}) = y$.
7) a) Not one-to-one, $f(0) = f(4) = 0$. b) Not onto, $f(x) \neq -5$, since $x^2 - 4x + 5$ has complex roots.
9) a) One-to-one, if x even, y odd, then $y = x + 1/2$. b) Not onto, $f(x) \neq 3$.
11) a) One-to-one, if x even, y odd, then $x = 2y - 1$ is odd. b) Not onto, $f(x) \neq 4$.
13) a) Not one-to-one $f(0) = f(3) = 1$. b) Onto, either $f(2y - 2) = y$ or $f(2y + 1) = y$.

15) If $2x^2 + x = 2y^2 + y = c$, then x and $y = (-1 \pm \sqrt{1+8c})/4$. If $x \neq y$, then $|x - y| = \sqrt{1+8c}/2$, which is never an integer when c is an integer.
17) Suppose f were onto. Although f might not be one-to-one, if $f(a_1) = f(a_2)$, we could remove a_1 from the set A and still have the function be onto. In this way we can create the set $\tilde{A}$ such that $\tilde{f} : \tilde{A} \to B$ is a bijection. By lemma 0.5, $|\tilde{A}| = |B|$, but $|B| > |A| \geq |\tilde{A}|$.
19) There are more than a million people in London, so if we let A be the set of people in London, and B be the set of numbers from 0 to 1,000,000, we have a mapping $f : A \to B$, with $|A| > |B|$, so this cannot be onto. Thus, two people in London have the same number of hairs. (Alternative solution: find two bald men in London.)
21) $2x(x - |x|)$.
23) $2x + 3$.
25) $f(x) = \begin{cases} x - 1 & \text{if } x \text{ is even,} \\ (x+3)/2 & \text{if } x \text{ is odd.} \end{cases}$
27) Given $c \in C$, there is a $b \in B$ such that $f(b) = c$. Then there is an $a \in A$ such that $g(a) = b$. Then $f(g(a)) = c$.
29) There are two different elements x and y in A such that $g(x) = g(y) = z$. Then $f(g(x)) = f(z) = f(g(y))$.
31) If $x \geq 0$ and $y < 0$, $f(x) = f(y)$ means $y = x^2 \geq 0$. Onto is proven by finding the inverse: $f^{-1}(x) = \begin{cases} \sqrt{x} & \text{if } x \geq 0, \\ x & \text{if } x < 0. \end{cases}$
33) Not associative, $(x * y) * z = 4x + 2y + z$, $x * (y * z) = 2x + 2y + z$.
35) Associative, $(x * y) * z = x * (y * z) = xyz - xy - xz - yz + x + y + z$.
37) No. $1 - 1 = 0 \notin S$.
39) Yes.
41) No. $1/2 = .5 \notin S$.
43) **`g(x) = - x - 2*floor(-x)`**
Since $g(x) = f^{-1}(x)$, both $f(g(x))$ and $g(f(x))$ is x.

Section 0.3

1) 18
3) 37
5) 2
7) 13
9) 7
11) 86
13) Since $10^n \textbf{ mod } 9 = 1^n \textbf{ mod } 9 = 1$, we find that

$$a_0 + 10a_1 + 10^2a_2 + 10^3a_3 + \cdots + 10^m a_m \textbf{ mod } 9 = a_0 + a_1 + a_2 + \cdots + a_m \textbf{ mod } 9.$$

15) 140
17) 235
19) 523
21) 3270
23) 9554

25) 3776
27) 5
29) 17
31) 28
33) 2
35) 27
37) 28
39) 358981174162646211769
41) 8570104178046812602269593145496421011
43) 862056187088917362

Section 0.4

1) $\{0, 1, \frac{1}{2}, 2, \frac{1}{3}, \frac{3}{2}, \frac{2}{3}, 3, \frac{1}{4}, \frac{4}{3}, \frac{3}{5}, \frac{5}{2}, \frac{2}{5}, \frac{5}{3}, \frac{3}{4}, 4\}$.
3) If $a_n = b_n/b_{n+1}$, then $\lfloor a_n \rfloor = (b_n - (b_n \textbf{ mod } b_{n+1}))/b_{n+1}$. Then $1/a_{n+1} = (b_{n+1} + 2(b_n - (b_n \textbf{ mod } b_{n+1})) - b_n)/b_{n+1}$. This simplifies to give $a_{n+1} = b_{n+1}/(b_n + b_{n+1} - 2(b_n \textbf{ mod } b_{n+1}))$. Hence, $b_{n+2} = b_n + b_{n+1} - 2(b_n \textbf{ mod } b_{n+1})$.
5) $a_{2n} = b_{2n}/b_{2n+1} = b_n/(b_n + b_{n+1}) = (b_n/b_{n+1})/((b_n/b_{n+1}) + 1) = a_n/(a_n + 1)$.
7) Let $x = p/q$ be a rational number, and assume the statement is true for smaller $p + q$. If $x \geq 1$, then $a_m = x - 1$ for some m, and $a_{2m+1} = x$. If $x < 1$, then $a_m = x/(1 - x)$ for some m, and $a_{2m} = x$.
9) Because a_i can only be one of q possible integers for $i > 0$, at some point we must have $a_i = a_j$. Because a_{n+1} is determined solely on a_n, $a_{2i-j} = a_i$, and the sequence will repeat from this point on.
11) $x = n.d_1d_2 \ldots d_1 + 10^{-i} \cdot 0.d_{i+1}d_{i+2} \cdots d_{i+j} + 10^{-i-j} \cdot 0.d_{i+1}d_{i+2} \ldots d_{i+j} + 10^{-i-2j} \cdot 0.d_{i+1}d_{i+2} \cdots d_{i+j} + \cdots$. The series is geometric after the first term, so the sum is $n.d_1d_2 \ldots d_1 + 10^{-i} \cdot 0.d_{i+1}d_{i+2} \ldots d_{i+j}/(1 - 10^{-j})$, which is rational.
13) If $p^2/q^2 = 3$ with p and q coprime, then $3|p$, but replacing $p = 3r$ shows $3|q$ too.
15) If $p^2/q^2 = 6$ with p and q coprime, then $2|p$, but replacing $p = 2r$ shows $2|q$ too.
17) If $p^3/q^3 = 4$ with p and q coprime, then $2|p$, but replacing $p = 2r$ shows $2|q$ too.
19) If $a + b$ were rational, and a was rational, then $b = (a + b) - a$ would be rational.
21) If $a \cdot b$ were rational, and a was rational and nonzero, then $b = (a \cdot b)/a$ would be rational.
23) If $e = p/q$, then $q!e$ will be an integer. But the series for $q!e$ is

$$\frac{q!}{0!} + \frac{q!}{1!} + \frac{q!}{2!} + \cdots \frac{q!}{q!} + \frac{q!}{(q+1)!} + \frac{q!}{(q+2)!} + \cdots.$$

The terms up to $q!/q!$ will be integers, but the remaining terms simplify to

$$\frac{1}{q+1} + \frac{1}{(q+1)(q+2)} + \frac{1}{(q+1)(q+2)(q+3)} + \cdots.$$

This sum is clearly less than $1/2 + 1/6 + 1/24 + \cdots$, which sums to $2 - e < 1$. Thus, the sum is not an integer.
25) $y = (4/3)^{10}(3/4)^{(1/x)}$.

Section 1.1

1) 8 steps: Stay, RotLft, RotRt, Rot180, Flip (along horizontal axis), Spin (along vertical axis), FlipLft (exchanges NE and SW corners), and FlipRt.

	Stay	RotLft	Rot180	RotRt	Flip	Spin	FlipLft	FlipRt
Stay	Stay	RotLft	Rot180	RotRt	Flip	Spin	FlipLft	FlipRt
RotLft	RotLft	Rot180	RotRt	Stay	FlipLft	FlipRt	Spin	Flip
Rot180	Rot180	RotRt	Stay	RotLft	Spin	Flip	FlipRt	FlipLft
RotRt	RotRt	Stay	RotLft	Rot180	FlipRt	FlipLft	Flip	Spin
Flip	Flip	FlipRt	Spin	FlipLft	Stay	Rot180	RotRt	RotLft
Spin	Spin	FlipLft	Flip	FlipRt	Rot180	Stay	RotLft	RotRt
FlipLft	FlipLft	Flip	FlipRt	Spin	RotLft	RotRt	Stay	Rot180
FlipRt	FlipRt	Spin	FlipLft	Flip	RotRt	RotLft	Rot180	Stay

3) $e = e \cdot e' = e'$, so $e = e'$.
5) If $a \cdot b = a \cdot c$, then $a^{-1} \cdot (a \cdot b) = a^{-1} \cdot (a \cdot c)$, so $b = c$.
7) 50% (18 of 36).
9) After a flip and a rotation, Terry will be facing the opposite direction, so it would be a flip.
11) $(\texttt{FlipRt}\cdot\texttt{Spin})^{-1} \neq \texttt{FlipRt}\cdot\texttt{Spin}$. Other answers are possible.
13) Stay = FlipRt·FlipRt, RotRt = FlipRt·FlipLft, RotLft = FlipLft·FlipRt, Spin = FlipRt·FlipLft·FlipRt. Other answers are possible.
15) Such a routine is impossible, since it involves three flips. See Problem 8.

Section 1.2

1)

	0	1	2	3	4
0	0	1	2	3	4
1	1	2	3	4	0
2	2	3	4	0	1
3	3	4	0	1	2
4	4	0	1	2	3

3)

	0	1	2	3	4	5	6	7
0	0	1	2	3	4	5	6	7
1	1	2	3	4	5	6	7	0
2	2	3	4	5	6	7	0	1
3	3	4	5	6	7	0	1	2
4	4	5	6	7	0	1	2	3
5	5	6	7	0	1	2	3	4
6	6	7	0	1	2	3	4	5
7	7	0	1	2	3	4	5	6

5)

	0	2	4	6	8	10
0	0	2	4	6	8	10
2	2	4	6	8	10	0
4	4	6	8	10	0	2
6	6	8	10	0	2	4
8	8	10	0	2	4	6
10	10	0	2	4	6	8

7)

	1	3	5	7
1	1	3	5	7
3	3	1	7	5
5	5	7	1	3
7	7	5	3	1

9)

	1	5	7	11
1	1	5	7	11
5	5	1	11	7
7	7	11	1	5
11	11	7	5	1

11)

	1	5	7	11	13	17
1	1	5	7	11	13	17
5	5	7	17	1	11	13
7	7	17	13	5	1	11
11	11	1	5	13	17	7
13	13	11	1	17	7	5
17	17	13	11	7	5	1

13) Simply define $x \sim y$ if x and y belong to the same subset.
15) 7.
17) 19.
19) 3.
21) 67.
23) 103.
25) 277.
27) $n = 5$, 8, or 12.

Section 1.3

1) $(a \cdot a) \cdot b \neq a \cdot (a \cdot b)$.
3) Yes, this is a group.
5) 0 has no inverse.
7) Not closed, no identity, hence no inverses.
9) Yes, this is a group.
11) 3 has no inverse.
13) Yes, this is a group.
15) Note that y has an inverse, z, so that $y \cdot z = e$. But then $x = x \cdot (y \cdot z) = (x \cdot y) \cdot z = z$, so $y \cdot x = e$.
17) If both $x \cdot y_1$ and $x \cdot y_2 = e$, then by Problem 15, $y_2 \cdot x = e$, so $y_2 = y_2 \cdot (x \cdot y_1) = (y_2 \cdot x) \cdot y_1 = y_1$.
19) $a^{-1} \cdot (a \cdot x) = a^{-1} \cdot (a \cdot y)$, so $x = y$.
21) If $(a \cdot b)^2 = a^2 \cdot b^2$, then $a \cdot b \cdot a \cdot b = a \cdot a \cdot b \cdot b$.
23) $a^2 \neq e$ if, and only if, $a^{-1} \neq a$, so these elements pair off, leaving an even number of elements. Since the identity is one of the remaining elements, there must be another.
25)

·	a	b	c	d
a	b	d	a	c
b	d	c	b	a
c	a	b	c	d
d	c	a	d	b

27)

	a	b	c	d	e	f	g	h
b	b	g	d	f	a	h	e	c
g	g	e	f	h	b	c	a	d
h	h	f	b	a	c	e	d	g
c	c	h	g	b	d	a	f	e
a	a	b	c	d	e	f	g	h
d	d	c	e	g	f	b	h	a
e	e	a	h	c	g	d	b	f
f	f	d	a	e	h	g	c	b

29) $18 \to 6$, $54 \to 18$, $162 \to 54$, $486 \to 162$, $50 \to 20$, $250 \to 100$, $98 \to 42$, $686 \to 294$. Conjecture $(p-1)n/(2p)$.

Section 2.1

1) 1, 5, 7, and 11.
3) 1, 3, 5, 7, 9, 11, 13, and 15.
5) 2 and 5.
7) No generators
9) No generators
11) 5 and 11.
13) 40.
15) 168.
17) 288.
19) 480.
21) If n has an odd prime factor p, then $p-1$ will be even. If n is 2^q for some $q > 1$, then 2^{q-1} is even. In all cases, there is an even factor in the formula for $\phi(n)$.
23) 1, 2, 4, 5, 8, 10, 11, 13, 16, 17, 19, 20.
25) Z_n^* is cyclic if n is a power of an odd prime.
27) The only other cases where Z_n^* is cyclic is $n = 2$ and $n = 4$.

Section 2.2

1) If $b \cdot a = a \cdot b$, then $e = b^2 \cdot a^2 = b \cdot (b \cdot a) \cdot a = b \cdot a \cdot b \cdot a$. If $b \cdot a \cdot b \cdot a = e$, then $b \cdot a = b \cdot (b \cdot a \cdot b \cdot a) \cdot a = b^2 \cdot (a \cdot b) \cdot a^2 = a \cdot b$.
3) $b^3 \cdot a = b^2 \cdot (a^2 \cdot b) = b \cdot (a^2 \cdot b) \cdot a \cdot b = (a^2 \cdot b) \cdot a \cdot (a^2 \cdot b) \cdot b = a^2 \cdot (a^2 \cdot b) \cdot a^2 \cdot b^2 = a^4 \cdot (a^2 \cdot b) \cdot a \cdot b^2 = a^6 \cdot (a^2 \cdot b) \cdot b^2 = a^5 \cdot a^3 \cdot b^3 = a^3 \cdot b^3$.
5) $a \cdot b$.
7) $b^2 \cdot c^3$.
9) $b \cdot c^3$.
11) $a \cdot b^2 \cdot c^2$.
13) $b^2 \cdot c^2$.
15) c^3.
17) There are 24 ways of rearranging four books.

19)

```
InitGroup("e")
AddGroupVar("b", "c")
Define(b^3, e)
Define(c^4, e)
Define((b^2*c)^2, e)
Group(b, c)
    {e, b^2*c*b^2, c, c^2, b^2*c, b, b*(b*c)^2, c*b,
    b^2, b^2*c*b, c*b^2, b*(b*c)^2*b, b^2*c^2, b*c,
    b*(b*c)^2*c, c*b*c, b*c*b^2, (b*c)^2*b, b*c^2, c*b*c^2,
    (b*c)^2, b*c*b, (b*c)^2*c, (c*b)^2}
```

Section 2.3

1) $\{0\}$, $\{0, 2, 4, 6, 8, 10\}$, $\{0, 3, 6, 9\}$, $\{0, 4, 8\}$, $\{0, 6\}$, and the whole group.
3) $\{0\}$, $\{0, 3, 6, 9, 12, 15, 18\}$, $\{0, 7, 14\}$, and the whole group.
5) $\{1\}$, $\{1, 3\}$, $\{1, 5\}$, $\{1, 7\}$, and the whole group.
7) Six elements of order 4, eight elements of order 3, and nine elements of order 2.
9) Because the corners can only rotate, every third repetition will bring the corners back to the initial state. If all 6 of the edges move, then after 6 repetitions the edges will be back in the right place, but possibly flipped. But then after 12 repetitions the edges will also be back to normal, making the order at most 12. If 5 of the edges move, then it will take 5 repetitions to get the edges into place, possibly flipped, so 10 repetitions to get the edge pieces into the right position, but then the corners may be twisted, so the order could be at most 30.
11) $R_6(G) = 6$, $R_3(G) = 3$, $R_2(G) = 2$, so two elements of order 6. ($6 - 3 - 2$ subtracts the identity element twice.)
13) Every element of Z_n will have an order that divides n. Counting the number of elements of order k for each divisor and summing over all divisors will give the number of elements of Z_n.
15) If G were cyclic, Corollary 2.1 shows that there are at most p solutions to $x^p = e$.
17) $(a \cdot b)^n = a \cdot b \cdot (a \cdot b)^{n-1} = a \cdot b \cdot a \cdot (b \cdot a)^{n-2} \cdot b = a \cdot (b \cdot a)^{n-1} \cdot b$; $(a \cdot b)^n = e \Leftrightarrow a \cdot (b \cdot a)^{n-1} \cdot b = e \Leftrightarrow (b \cdot a)^{n-1} = a^{-1}b^{-1} \Leftrightarrow (b \cdot a)^n = e$.
19) If $x, y \in H$, then $x = a^2$, $y = b^2$, so $(a \cdot b^{-1})^2 = x \cdot y^{-1} \in H$. If $a^2 = 1$, then $a^2 - 1 = (a + 1)(a - 1)$ is a multiple of p, so $a = 1$ or $a = p - 1$. If $x = a^2 = b^2$, then $(a \cdot b^{-1})^2 = 1$ so $b = a$ or $b = a(p - 1)$. Since $x \mapsto x^2$ is two to one, H contains half the elements of Z_p^*.
21) The subgroup has 10 elements: $\{e, a, a^2, a^3, a^4, b^2, a \cdot b^2, a^2 \cdot b^2, a^3 \cdot b^2, a^4 \cdot b^2\}$.
23) Even for a non-abelian group, the order of the inverse element is the same as the order of the element.
25) $f \cdot r \cdot b$ flips the top edge efficiently, and in the process cycles the remaining 5 edges, so this has order 30.

Section 3.1

1) $\{\{0,5\},\{1,6\},\{2,7\},\{3,8\},\{4,9\}\}$.
3) $\{\{0,5,10\},\{1,6,11\},\{2,7,12\},\{3,8,13\},\{4,9,14\}\}$.
5) $\{\{1,14\},\{2,13\},\{4,11\},\{7,8\}\}$.
7) $\{\{1,9\},\{3,11\},\{5,13\},\{7,15\}\}$.
9) Left cosets: {Stay, Spin}, {FlipRt, RotLft}, {RotRt, FlipLft}. Right cosets: {Stay, Spin}, {FlipRt, RotRt}, {RotLft, FlipLft}.
11) 5.
13) 7.
15) 5.
17) 8.
19) 4.
21) 36.
23) Since $(n-1)^2 = 1$ in Z_n^*, $\{1, n-1\}$ is a subgroup of order 2, so $|Z_n^*|$ is even for $n > 3$.
25) Since $y \in Hx$, $y = hx$ for some $h \in H$, so $Hy = H \cdot (hx) = (H \cdot h)x = Hx$.
27) Possible orders are 1, p, q, and pq, so a non-trivial subgroup either has order p or q. But any group of prime order is cyclic.
29) $\{1\}$, $\{1,3,9,11\}$, $\{1,5,9,13\}$, $\{1,7\}$, $\{1,9\}$, $\{1,15\}$, $\{1,7,9,15\}$, and the whole group.
31) Left cosets: $\{\{e,c^2,c,c^3\},\{a,a\cdot c^2,a\cdot c,a\cdot c^3\},\{b,b\cdot c^2,b\cdot c,b\cdot c^3\},\{a\cdot b\cdot c,a\cdot b,a\cdot b\cdot c^3,a\cdot b\cdot c^2\},\{b^2,b^2\cdot c,b^2\cdot c^2,b^2\cdot c^3\},\{a\cdot b^2,a\cdot b^2\cdot c,a\cdot b^2\cdot c^2,a\cdot b^2\cdot c^3\}\}$. Right cosets: $\{\{e,c^2,c,c^3\},\{a,a\cdot b\cdot c,b\cdot c^2,b^2\cdot c^3\},\{b,b^2\cdot c,a\cdot c^2,a\cdot b\cdot c^3\},\{a\cdot b,b^2\cdot c^2,b\cdot c,a\cdot c^3\},\{b^2,a\cdot c,a\cdot b\cdot c^2,b\cdot c^3\},\{a\cdot b^2,a\cdot b^2\cdot c,a\cdot b^2\cdot c^2,a\cdot b^2\cdot c^3\}\}$.

Section 3.2

1) 24, 28, 1, 0, 23, 9, 24, 11, 28
3) 5, 9, 0, 4, 24, 9, 8, 12, 26, 19
5) THIS IS EASY
7) MAKE IT SO
9) If $n = pqr$, $\phi(n) = (p-1)(q-1)(r-1)$. If x is coprime to n, use Proposition 3.1, otherwise suppose x is a multiple of p, but not a multiple of qr. Then $x^{rs} \equiv x \pmod{p}$, and since $rs \equiv 1 \pmod{(q-1)(r-1)}$, Proposition 3.2 shows that $x^{rs} \equiv x \pmod{qr}$ as well. Finish with the Chinese remainder theorem (0.7).
11) $f^{-1}(x) = x^{11}$ **mod** 51.
13) $f^{-1}(x) = x^{29}$ **mod** 91.
15) $f^{-1}(x) = x^{35}$ **mod** 221.
17) $f^{-1}(x) = x^{103}$ **mod** 1001.
19) 1835, 1628, 1084. Inverse $= x^{157}$ **mod** 2773.
21) Answers will vary.
23) Answers will vary.

Section 3.3

1) Since $e \in H$, $H = e \cdot H \subseteq H \cdot H$. But H is closed with respect to multiplication, so $H \cdot H \subseteq H$.

3) Since $e \in H$, $a \in a \cdot H$, so $a \in H \cdot b$. But $a \in H \cdot a$ as well, so $H \cdot b = H \cdot a$, hence $a \cdot H = H \cdot a$.

5) Any element of $h \in H$ is also in G, so $h \cdot n \cdot h^{-1} \in N$.

7) Three possible answers: $\{e, c^2\}$, $\{e, a \cdot b^2 \cdot c\}$, or $\{e, a \cdot b^2 \cdot c^3\}$.

9) If $g \in G$ and $h \in Z$, then $g \cdot h \cdot g^{-1} = h \cdot g \cdot g^{-1} = h \in Z$.

11) Let a be a generator of H, and let m be the smallest positive integer for which $a^m \in K$. For a given $g \in G$, $g \cdot a \cdot g^{-1} \in H$, so $g \cdot a \cdot g^{-1} = a^n$ for some n. Then for $a^{sm} \in K$, $g \cdot a^{sm} \cdot g^{-1} = (g \cdot a \cdot g^{-1})^{sm} = (a^n)^{sm} = (a^m)^{sn} \in K$.

13) Let $f(x) = mx + b \in G$, and $t(x) = qx \in T$, so $f^{-1}(x) = (x - b)/m$. Then $(f \cdot t \cdot f^{-1})(x) = f(t(f^{-1}(x))) = qx - bq + b \notin T$. If $f(x) = 2x + 3$, then Tf is the set of functions $k(2x + 3)$, whereas fT is the set of functions $kx + 3$.

15) If $g_1 = h_1 \cdot k_1$ and $g_2 = h_2 \cdot k_2$, then $g_1 \cdot g_2^{-1} = h_1 \cdot k_1 \cdot k_2^{-1} \cdot h_2^{-1} = (h_1 \cdot h_2^{-1}) \cdot (h_2 \cdot k_1 \cdot k_2^{-1} \cdot h_2^{-1}) \in H \cdot K$, since K is normal.

17) $g \cdot H \cdot K \cdot g^{-1} = (g \cdot H \cdot g^{-1}) \cdot (g \cdot K \cdot g^{-1}) = H \cdot K$.

19) Subgroups are $\{e\}$, with cosets $\{e\}$, $\{a\}$, $\{a^2\}$, $\{a^3\}$, $\{b\}$, $\{a \cdot b\}$, $\{a^2 \cdot b\}$, and $\{a^3 \cdot b\}$; $\{e, a^2\}$, with cosets $\{e, a^2\}$, $\{a, a^3\}$, $\{b, a^2 \cdot b\}$, and $\{a \cdot b, a^3 \cdot b\}$; $\{e, a, a^2, a^3\}$, with cosets $\{e, a, a^2, a^3\}$ and $\{b, a \cdot b, a^2 \cdot b, a^3 \cdot b\}$; $\{e, b, a^2, a^2 \cdot b\}$, with cosets $\{e, b, a^2, a^2 \cdot b\}$ and $\{a, a \cdot b, a^3, a^3 \cdot b\}$; $\{e, a \cdot b, a^2, a^3 \cdot b\}$, with cosets $\{e, a \cdot b, a^2, a^3 \cdot b\}$ and $\{a, b, a^2 \cdot b, a^3\}$; and the whole group, with one coset containing the whole group.

Section 3.4

1)

	{0,5}	{1,6}	{2,7}	{3,8}	{4,9}
{0,5}	{0,5}	{1,6}	{2,7}	{3,8}	{4,9}
{1,6}	{1,6}	{2,7}	{3,8}	{4,9}	{0,5}
{2,7}	{2,7}	{3,8}	{4,9}	{0,5}	{1,6}
{3,8}	{3,8}	{4,9}	{0,5}	{1,6}	{2,7}
{4,9}	{4,9}	{0,5}	{1,6}	{2,7}	{3,8}

3)

	{0,5,10}	{1,6,11}	{2,7,12}	{3,8,13}	{4,9,14}
{0,5,10}	{0,5,10}	{1,6,11}	{2,7,12}	{3,8,13}	{4,9,14}
{1,6,11}	{1,6,11}	{2,7,12}	{3,8,13}	{4,9,14}	{0,5,10}
{2,7,12}	{2,7,12}	{3,8,13}	{4,9,14}	{0,5,10}	{1,6,11}
{3,8,13}	{3,8,13}	{4,9,14}	{0,5,10}	{1,6,11}	{2,7,12}
{4,9,14}	{4,9,14}	{0,5,10}	{1,6,11}	{2,7,12}	{3,8,13}

5)

	{1,4}	{2,8}	{7,13}	{11,14}
{1,4}	{1,4}	{2,8}	{7,13}	{11,14}
{2,8}	{2,8}	{1,4}	{11,14}	{7,13}
{7,13}	{7,13}	{11,14}	{1,4}	{2,8}
{11,14}	{11,14}	{7,13}	{2,8}	{1,4}

7)

	{1,7}	{3,5}	{9,15}	{11,13}
{1,7}	{1,7}	{3,5}	{9,15}	{11,13}
{3,5}	{3,5}	{9,15}	{11,13}	{1,7}
{9,15}	{9,15}	{11,13}	{1,7}	{3,5}
{11,13}	{11,13}	{1,7}	{3,5}	{9,15}

9)

	{1,5}	{7,11}	{13,17}	{19,23}
{1,5}	{1,5}	{7,11}	{13,17}	{19,23}
{7,11}	{7,11}	{1,5}	{19,23}	{13,17}
{13,17}	{13,17}	{19,23}	{1,5}	{7,11}
{19,23}	{19,23}	{13,17}	{7,11}	{1,5}

11)

	$\{e, b, b^2\}$	$\{a, a\cdot b, a\cdot b^2\}$
$\{e, b, b^2\}$	$\{e, b, b^2\}$	$\{a, a\cdot b, a\cdot b^2\}$
$\{a, a\cdot b, a\cdot b^2\}$	$\{a, a\cdot b, a\cdot b^2\}$	$\{e, b, b^2\}$

13) Each element of G/N is a set of functions $f(x) = px + k$ for which the p is the same for all functions in the coset.
15) If xN and yN are two elements in G/N, then $(xN)\cdot(yN) = x\cdot y\cdot N = y\cdot x\cdot N = (yN)\cdot(xN)$.
17) If h_1N and h_2N are two elements of H/N, then h_1 and h_2 are in H, and $(h_1N)\cdot(h_2N)^{-1} = (h_1\cdot h_2^{-1})\cdot N \in H/N$. So H/N is a subgroup of G/N.
19) $|Z_{105}^*| = 48$, $H = \{1, 11, 16, 46, 71, 86\}$, coset $\{2, 22, 36, 37, 67, 92\}$ has order 4.

Section 4.1

1) If $f(x) = a$ and $f(y) = b$, then $f^{-1}(a\cdot b) = x\cdot y = f^{-1}(a)\cdot f^{-1}(b)$.
3) **Stay** $\to e$, **RotRt** $\to b$, **RotLft** $\to b^2$, **Spin** $\to a$, **FlipRt** $\to a\cdot b$, **FlipLft** $\to a\cdot b^2$.
5) $Z_6 = \{0, 1, 2, 3, 4, 5\} \approx Z_7^*$ with order $\{1, 3, 2, 6, 4, 5\}$.
7) $Z_6 = \{0, 1, 2, 3, 4, 5\} \approx Z_{14}^*$ with order $\{1, 3, 9, 13, 11, 5\}$.
9) $Z_{10} = \{0, 1, 2, 3, \ldots, 9\} \approx Z_{11}^*$ with order $\{1, 2, 4, 8, 5, 10, 9, 7, 3, 6\}$.
11) $Z_{12} = \{0, 1, 2, \ldots, 11\} \approx Z_{13}^*$ with order $\{1, 2, 4, 8, 3, 6, 12, 11, 9, 5, 10, 7\}$.
13) $Z_{12}^* = \{1, 5, 7, 11\} \approx Z_8^*$ with order $\{1, 3, 5, 7\}$.
15) Let g be a generator, and consider the function $f(x) : \mathbb{Z} \to G$ defined by $f(x) = g^x$.
17) $a^m = e_1$ if and only if $\phi(a^m) = \phi(e_1) = e_2$ if and only if $\phi(a)^m = e_2$.
19) $Z_{16}^* = \{1, 3, 5, 7, 9, 11, 13, 15\} \approx Z_{15}^*$ with order $\{1, 2, 7, 11, 4, 8, 13, 14\}$.
21) $Z_{30}^* = \{1, 7, 11, 13, 17, 19, 23, 29\} \approx Z_{15}^*$ with order $\{1, 7, 11, 13, 2, 4, 8, 14\}$.

Section 4.2

1) If $a, b \in \text{Im}(\phi)$, then $a = \phi(x)$, $b = \phi(y)$ for some $x, y \in G$. Then $a\cdot b = \phi(x\cdot y) = \phi(y\cdot x) = b\cdot a$.
3) $\phi(x\cdot y) = \phi(x + y) = 2(x + y) = 2x + 2y = \phi(x) + \phi(y) = \phi(x)\cdot\phi(y)$, since $\cdot$ is addition in this group.

5) $\phi(x \cdot y) = \phi(x + y) = x + y + 3$, but $\phi(x) \cdot \phi(y) = \phi(x) + \phi(y) = (x + 3) + (y + 3) = x + y + 6$.
7) $\phi(x \cdot y) = 2(x \cdot y) = 2xy$, but $\phi(x) \cdot \phi(y) = (2x) \cdot (2y) = 4xy$.
9) $\phi(x \cdot y) = \phi(x + y) = e^{x+y} = e^x \times e^x = \phi(x) \cdot \phi(y)$. Image is the positive real numbers.
11) $\phi(f \cdot g) = \phi(f(t) + g(t)) = f(3) + g(3) = \phi(f) + \phi(g) = \phi(f) \cdot \phi(g)$. The kernel is the set of polynomials with 3 as a root, hence $t - 3$ is a factor.
13) $\phi(1) = 1$, $\phi(7) = 13$, $\phi(11) = 1$, $\phi(13) = 7$, $\phi(17) = 13$, $\phi(19) = 19$, $\phi(23) = 7$, $\phi(29) = 19$.
15) $\phi(\pm 1) = 1$, $\phi(\pm i) = 3$, $\phi(\pm j) = 5$, $\phi(\pm k) = 7$. The 3, 5, and 7 can be permuted.
17) For each element $h \in H$, $f^{-1}(h)$ is a coset of K, where $K = \text{Ker} f$. Hence $|f^{-1}(h)| = |K|$. Since each element in H produces a different coset of K, the size of $f^{-1}(H)$ is $|H| \cdot |K|$.
19) Many solutions, since b can map to either **RotLft** or **RotRt**, and a can map to **FlipLft**, **FlipRt**, or **Spin**. Any of these combinations will work.

Section 4.3

1) Z_{10}, Z_5, Z_2, and the trivial group.
3) Z_{15}^*, Z_4, Z_8^*, Z_2, and the trivial group.
5) Q, Z_8^*, Z_2, and the trivial group.
7) Z_{24}^*, Z_8^*, Z_2, and the trivial group.
9) If K is the kernel, it is sufficient to show that G/K is cyclic. If g is a generator of G, then gK is a generator of G/K, since every element can be expressed as $g^m \cdot K = (gK)^m$.
11) Ten homomorphisms, one sending all elements to e, three sending $\{1, 3\}$ to e, $\{5, 7\}$ to a, $a \cdot b$, or $a \cdot b^2$, respectively, three sending $\{1, 5\}$ to e, $\{3, 7\}$ to a, $a \cdot b$, or $a \cdot b^2$, respectively, and three sending $\{1, 7\}$ to e, $\{3, 5\}$ to a, $a \cdot b$, or $a \cdot b^2$, respectively.
13) Since $\{0, 2, 4\}$ and $\{0, 3\}$ are normal subgroups of Z_6, $\phi^{-1}(\{0, 2, 4\})$ and $\phi^{-1}(\{0, 3\})$ are normal subgroups of G.
15) H and K must be normal, since they have index 2. Then $H \cdot K$ is a subgroup with more than half the elements, so $H \cdot K = G$. By the second isomorphism theorem, $G/K \approx K/(H \cap K) \approx Z_2$. So $H \cap K$ contains half the elements of K, hence a fourth of the elements of G, so $G/(H \cap K)$ contains 4 elements. For every element $a \in G$, a^2 is in both H and K, so every element in the quotient group is of order 1 or 2. Thus, $G/(H \cap K) \approx Z_8^*$.
17) $\{\{\{1, 4\}, \{2, 8\}\}, \{\{7, 13\}, \{14, 11\}\}\} \approx \{\{1, 2, 4, 8\}, \{7, 11, 13, 14\}\}$.
19) Let $\phi(a) = \phi(c) = -1$, and $\phi(b) = 1$.

Section 5.1

1) $\begin{pmatrix} 1\,2\,3\,4\,5 \\ 5\,1\,2\,4\,3 \end{pmatrix}$.

3) $\begin{pmatrix} 1\,2\,3\,4\,5\,6 \\ 3\,4\,5\,1\,2\,6 \end{pmatrix}$.

5) $\begin{pmatrix} 1\,2\,3\,4\,5\,6\,7 \\ 4\,2\,5\,6\,7\,3\,1 \end{pmatrix}$.

7) $\binom{12345678}{85132746}$.

9)

	$\binom{123}{123}$	$\binom{123}{132}$	$\binom{123}{213}$	$\binom{123}{231}$	$\binom{123}{312}$	$\binom{123}{321}$
$\binom{123}{123}$	$\binom{123}{123}$	$\binom{123}{132}$	$\binom{123}{213}$	$\binom{123}{231}$	$\binom{123}{312}$	$\binom{123}{321}$
$\binom{123}{132}$	$\binom{123}{132}$	$\binom{123}{123}$	$\binom{123}{312}$	$\binom{123}{321}$	$\binom{123}{213}$	$\binom{123}{231}$
$\binom{123}{213}$	$\binom{123}{213}$	$\binom{123}{231}$	$\binom{123}{123}$	$\binom{123}{132}$	$\binom{123}{321}$	$\binom{123}{312}$
$\binom{123}{231}$	$\binom{123}{231}$	$\binom{123}{213}$	$\binom{123}{321}$	$\binom{123}{312}$	$\binom{123}{123}$	$\binom{123}{132}$
$\binom{123}{312}$	$\binom{123}{312}$	$\binom{123}{321}$	$\binom{123}{132}$	$\binom{123}{123}$	$\binom{123}{231}$	$\binom{123}{213}$
$\binom{123}{321}$	$\binom{123}{321}$	$\binom{123}{312}$	$\binom{123}{231}$	$\binom{123}{213}$	$\binom{123}{132}$	$\binom{123}{123}$

11) $\binom{1234}{1342}, \binom{1234}{1423}, \binom{1234}{3241}, \binom{1234}{4213}, \binom{1234}{2431}, \binom{1234}{4132}, \binom{1234}{2314}, \binom{1234}{3124}$.

13) $\binom{12345}{13245}$.

15) $x = \binom{12345}{24315}, \binom{12345}{54123}, \binom{12345}{14532}$, or $\binom{12345}{34251}$.

17) **Right·Last·Left**.

19) **Right·First·Right**.

Section 5.2

1) $(1\,3\,2\,4\,6)$.

3) $(1\,8)(2\,6\,7)(4\,5)$.

5) Product is $(1\,2)(n+1\ \ n+2)$. When $n = 2$, we easily get $(1\,2)(3\,4)$, so assume that product is correct for $n-1$. Then by induction, the product is $(1\,2)(n\ \ n+1)(n\ \ n+1\ \ n+2) = (1\,2)(n+1\ \ n+2)$.

7) If $f = \phi_1 \cdot \phi_2$, where ϕ_1 and ϕ_2 are disjoint, then $f^n = e$ if and only if $\phi_1^n = e$ and $\phi_2^n = e$.

9) Consider $\sigma_H : H \mapsto \mathbb{R}$ as the signature function restricted to the elements of H. The kernel is $H \cap A_n$. If the image is $\{1\}$, then all permutations in H are even, otherwise the image is $\{1, -1\}$, and the first isomorphism theorem shows $|H/(H \cap A_n)| = 2$.

11) If $\phi_1, \phi_2 \in S_\infty$, then ϕ_2 is one-to-one and onto, so ϕ_2^{-1} is too. Thus, $\phi_1 \cdot \phi_2^{-1} \in S_\infty$. Example: $(1\,2)(3\,4)(5\,6)\ldots(2n-1\ \ 2n)\ldots \in S_\infty$, but $\notin S_\Omega$.

13) 144.

15) Let H be the subgroup generated by the n-cycle $\phi = (1\,2\,3\,\ldots\,n)$. Then ϕ^{j-i} will map i to j.

17) If $\phi = (i_1\ i_2\ i_3\ \ldots i_r)$ and $f = (j_1\ j_2\ j_3\ \ldots j_s)$, then $x \cdot \phi \cdot x^{-1} = (x(i_1)\ x(i_2)\ x(i_3)\ \ldots x(i_r))$, and $x \cdot f \cdot x^{-1} = (x(j_1)\ x(j_2)\ x(j_3)\ \ldots x(j_s))$.

19) a^2 is a 3-cycle, $a^3 = ()$, b^2 is a product of two 3-cycles, b^3 is a product of three 2-cycles, $b^6 = ()$.

21) $a \cdot b \cdot a^{-1} = (2\,4\,3\,5\,6\,7)$. In general, $a \cdot b \cdot a^{-1}$ will have the same cycle structure as b.

Section 5.3

1) $\{\binom{1234}{1234}, \binom{1234}{2143}, \binom{1234}{3412}, \binom{1234}{4321}\}$.

3) (), (1 2 3 4)(5 6 7 8), (1 3)(2 4)(5 7)(6 8), (1 4 3 2)(5 8 7 6), (1 5)(2 8)(3 7)(4 6), (1 6)(2 5)(3 8)(4 7), (1 7)(2 6)(3 5)(4 8), (1 8)(2 7)(3 6)(4 5).

5) (), (1 2)(3 4)(5 6)(7 8), (1 3)(2 4)(5 7)(6 8), (1 4)(2 3)(5 8)(6 7), (1 5)(2 6)(3 7)(4 8),(1 6)(2 5)(3 8)(4 7), (1 7)(2 8)(3 5)(4 6), (1 8)(2 7)(3 6)(4 5).

7) S_6 contains a subgroup generated by (12), (34), and (56).

9) Applying Corollary 5.2: 35 divides $5! \cdot |N|$, so 7 divides $|N|$, hence $H = N$, and H is normal.

11) Applying Corollary 5.2: 200 divides $8! \cdot |N|$, so 5 divides $|N|$, hence either $H = N$, or $|N| = 5$.

13) Applying Corollary 5.2: 189 divides $7! \cdot |N|$, so 3 divides $|N|$, hence either $H = N$, $|N| = 3$, or $|N| = 9$.

15) Applying Corollary 5.2: $3|H|$ divides $3! \cdot |N|$, so $H = N$ or $|N| = |H|/2$.

17) Any non-trivial subgroup would have order p. Applying Corollary 5.2 gives p^2 dividing $p! \cdot |N|$, so N must be a multiple of p, giving $H = N$.

19) Since the set is finite, for a given element a, the set $\{a, a^2, a^3, \cdots\}$ must repeat, so $a^m = a^n$ for some $m < n$. Then by the cancellation laws, $a^{n-m} = 1$, so $a^{n-m-1} \cdot a = 1$. Thus, a has an inverse, a^{n-m-1}.

21) (), (1 2 3)(4 5 6)(7 8 9)(10 11 12), (1 3 2)(4 6 5)(7 9 8)(10 12 11), (1 4 7 10)(2 6 8 12)(3 5 9 11), (1 5 7 11)(2 4 8 10)(3 6 9 12), (1 6 7 12)(2 5 8 11)(3 4 9 10), (1 7)(2 8)(3 9)(4 10)(5 11)(6 12), (1 8 3 7 2 9)(4 11 6 10 5 12), (1 9 2 7 3 8)(4 12 5 10 6 11), (1 10 7 4)(2 12 8 6)(3 11 9 5), (1 11 7 5)(2 10 8 4)(3 12 9 6), (1 12 7 6)(2 11 8 5)(3 10 9 4)

Section 5.4

1) 532.

3) 2195.

5) 3928.

7) 37387.

9) 29035.

11) $P(3, 4, 2, 6, 5, 1)$.

13) $P(1, 5, 6, 7, 2, 3, 4)$.

15) $P(3, 4, 6, 5, 7, 1, 2)$.

17) $P(6, 4, 8, 5, 1, 2, 3, 7)$.

19) $A_4 = \{1, 4, 5, 8, 9, 12, 13, 16, 17, 20, 21, 24\}$, the numbers congruent to 0 or 1 (mod 4). But **NthPerm(25)** is not in A_5.

21) $P[4, 5, 1, 6, 2, 3] = (1463)(25)$ is the only solution.

Section 6.1

1) $\{(0,0), (0,1), (1,0), (1,1), (2,0), (2,1), (3,0), (3,1)\}$ corresponds to the order $\{1, 11, 2, 7, 4, 14, 8, 13\}$.

3)

	(0,1)	(0,3)	(0,5)	(0,7)	(1,1)	(1,3)	(1,5)	(1,7)
(0,1)	(0,1)	(0,3)	(0,5)	(0,7)	(1,1)	(1,3)	(1,5)	(1,7)
(0,3)	(0,3)	(0,1)	(0,7)	(0,5)	(1,3)	(1,1)	(1,7)	(1,5)
(0,5)	(0,5)	(0,7)	(0,1)	(0,3)	(1,5)	(1,7)	(1,1)	(1,3)
(0,7)	(0,7)	(0,5)	(0,3)	(0,1)	(1,7)	(1,5)	(1,3)	(1,1)
(1,1)	(1,1)	(1,3)	(1,5)	(1,7)	(0,1)	(0,3)	(0,5)	(0,7)
(1,3)	(1,3)	(1,1)	(1,7)	(1,5)	(0,3)	(0,1)	(0,7)	(0,5)
(1,5)	(1,5)	(1,7)	(1,1)	(1,3)	(0,5)	(0,7)	(0,1)	(0,3)
(1,7)	(1,7)	(1,5)	(1,3)	(1,1)	(0,7)	(0,5)	(0,3)	(0,1)

5) Consider the natural homomorphism $\phi : G \to K$ defined by $\phi(h,k) = k$. The kernel is $\overline{H}$, so by the 1st isomorphism theorem, $G/\overline{H} \approx K$. Similarly, $G/\overline{K} \approx H$.
7) 1 element of order 2, 2 elements of order 3, 2 elements of order 4.
9) 3 elements of order 2, 8 elements of order 3, no elements of order 4.
11) 7 elements of order 2, 8 elements of order 3, no elements of order 4.
13) 7 elements of order 2, 8 elements of order 3, 8 elements of order 4.
15) $R_2(Z_2 \times Z_6) = 2 \cdot 2 = 4$, whereas $R_2(Z_{12}) = 2$.
17) Suppose $R_2(A \times B) = R_2(A) \cdot R_2(B) = 10$, with $R_2(A) \geq R_2(B)$. If $R_2(A) = 5$, A would have an even number of elements, but by Problem 20 $R_2(A)$ would be even. Thus, $R_2(A) = 10$, meaning that A has at least 10 elements, so B would have at most 2. Then $B \approx Z_2$, and $R_2(B) \neq 1$.
19) Put elements of Z_{21}^* in the order $\{1, 2, 4, 8, 16, 11, 13, 5, 10, 20, 19, 17\}$.

Section 6.2

1) Since $x^n = e$ for all $x \in Z_n \times Z_n$, we see that $Z_n \times Z_n$ is not cyclic.
3) Z_{32}, $Z_{16} \times Z_2$, $Z_8 \times Z_4$, $Z_8 \times Z_2 \times Z_2$, $Z_4 \times Z_4 \times Z_2$, $Z_4 \times Z_2 \times Z_2 \times Z_2$, and $Z_2 \times Z_2 \times Z_2 \times Z_2 \times Z_2$.
5) Only Z_{210}.
7) $Z_{450} \approx Z_2 \times Z_9 \times Z_{25}$, $Z_2 \times Z_9 \times Z_5 \times Z_5$, $Z_2 \times Z_3 \times Z_3 \times Z_{25}$, $Z_2 \times Z_3 \times Z_3 \times Z_5 \times Z_5$.
9) $Z_{600} \approx Z_8 \times Z_3 \times Z_{25}$, $Z_2 \times Z_4 \times Z_3 \times Z_{25}$, $Z_2 \times Z_2 \times Z_2 \times Z_3 \times Z_{25}$, $Z_8 \times Z_3 \times Z_5 \times Z_5$, $Z_2 \times Z_4 \times Z_3 \times Z_5 \times Z_5$, $Z_2 \times Z_2 \times Z_2 \times Z_3 \times Z_5 \times Z_5$.
11) $Z_{900} \approx Z_4 \times Z_9 \times Z_{25}$, $Z_2 \times Z_2 \times Z_9 \times Z_{25}$, $Z_4 \times Z_3 \times Z_3 \times Z_{25}$, $Z_2 \times Z_2 \times Z_3 \times Z_3 \times Z_{25}$, $Z_4 \times Z_9 \times Z_5 \times Z_5$, $Z_2 \times Z_2 \times Z_9 \times Z_5 \times Z_5$, $Z_4 \times Z_3 \times Z_3 \times Z_5 \times Z_5$, $Z_2 \times Z_2 \times Z_3 \times Z_3 \times Z_5 \times Z_5$.
13) Two for Z_{16}, four for $Z_8 \times Z_2$, 12 for $Z_4 \times Z_4$, and eight for $Z_4 \times Z_2 \times Z_2$.
15) $Z_{16} \times Z_8 \times Z_2$.
17) $Z_4 \times Z_2 \times Z_5$.
19) For each permutation written in terms of disjoint cycles, we can add "1-cycles" so that every number from 1 to n is mentioned. Then the sum of the sizes of the sycles will add to n. Thus, there is a one-to-one correspondence between cycle structures and partitions of n.
21) The exact value is $c = \pi\sqrt{2/3} \approx 2.5651$.

Section 6.3

1) 6: $\phi(b) = b$ or b^2 (order 3), $\phi(a) = a$, $a \cdot b$, or $a \cdot b^2$ (order 2).
3) 8: $\phi(2) = 2$, 7, 8, or 13 (order 4), forcing $\phi(4) = 4$. $\phi(11) = 11$ or 14 (order 2).
5) 48: $\phi(a) =$ one of the 8 elements of order 3, which determines $\phi(a^2)$. $\phi(b) =$ one of the six remaining elements of order 3.
7) Note that any automorphism must fix the identity element, leaving $n - 1$ elements.
9) $\phi(x) = x^{-1}$ is clearly one-to-one and onto, and $\phi(x \cdot y) = y^{-1} \cdot x^{-1} = x^{-1} \cdot y^{-1} = \phi(x) \cdot \phi(y)$ since the group is abelian. If a has order greater than 2, $\phi(a) \neq a$, so this is non-trivial.
11) If Aut(G) is cyclic, then so is Inn(G) with a generator $x \mapsto g^{-1}xg$. For each $y \in G$, $y^{-1}xy = g^{-n}xg^n$ for some n, plugging in $x = g$ yields $y^{-1}gy = g$, or $gy = yg$. Since $gy = gy$ for all y, Inn(G) $\approx \{e\}$, and G is abelian.
13) (), $(b, a^2 \cdot b)(a \cdot b, a^3 \cdot b)$, $(a, a^3)(a \cdot b, a^3 \cdot b)$, $(a, a^3)(b, a^2 \cdot b)$.
15) All automorphism are inner: (), $(b, b^2)(a \cdot b, a \cdot b^2)$, $(a, a \cdot b, a \cdot b^2)$, $(a, a \cdot b^2)(b, b^2)$, $(a, a \cdot b^2, a \cdot b)$, $(a, a \cdot b)(b, b^2)$.
17) Aut($\mathbb{Z}$) $\approx Z_2$, with $\phi_0(x) = x$, $\phi_1(x) = -x$.
19) Eight automorphisms: (), (2, 7)(8, 13), (2, 8)(7, 13), (2, 13)(7, 8), (2, 8)(11, 14), (2, 13, 8, 7)(11, 14), (7, 13)(11, 14), (2, 7, 8, 13)(11, 14).
21) There are 20 automorphisms, generated by $f(a) = a$, $f(b) = b^2$, and $g(a) = a \cdot b$, $g(b) = b$.

Section 6.4

1) (7, 7).
3) (7, 5).
5) (1, 1).
7) A nontrivial homomorphism from Z_2 to Aut(Z_8^*) $\approx S_3$ must send 1 to a 2-cycle. But Proposition 6.7 shows such homomorphisms are equivalent, so we may assume $\phi_1 = (3\,5)$. $Z_8^* \rtimes Z_2 \approx D_4$.

	(1,0)	(1,1)	(3,0)	(3,1)	(5,0)	(5,1)	(7,0)	(7,1)
(1,0)	(1,0)	(1,1)	(3,0)	(3,1)	(5,0)	(5,1)	(7,0)	(7,1)
(1,1)	(1,1)	(1,0)	(5,1)	(5,0)	(3,1)	(3,0)	(7,1)	(7,0)
(3,0)	(3,0)	(3,1)	(1,0)	(1,1)	(7,0)	(7,1)	(5,0)	(5,1)
(3,1)	(3,1)	(3,0)	(7,1)	(7,0)	(1,1)	(1,0)	(5,1)	(5,0)
(5,0)	(5,0)	(5,1)	(7,0)	(7,1)	(1,0)	(1,1)	(3,0)	(3,1)
(5,1)	(5,1)	(5,0)	(1,1)	(1,0)	(7,1)	(7,0)	(3,1)	(3,0)
(7,0)	(7,0)	(7,1)	(5,0)	(5,1)	(3,0)	(3,1)	(1,0)	(1,1)
(7,1)	(7,1)	(7,0)	(3,1)	(3,0)	(5,1)	(5,0)	(1,1)	(1,0)

9) A nontrivial homomorphism from Z_4 to Aut(Z_3) $\approx Z_2$ must send 1 and 3 to the 2-cycle (1 2). There will only be one element of order 2.
11) Since Aut($\mathbb{Z}$) $\approx Z_2$, we see that $\phi_1(x) = -x$. So $(x, a) \cdot (y, b) = (x + y, a + b)$ when a is even, but $(x, a) \cdot (y, b) = (x - y, a + b)$ when a is odd.

13) $\psi_\sigma((g_1, g_2, \ldots g_n) \cdot (h_1, h_2, \ldots h_n)) = \psi_\sigma(g_1 \cdot h_1, g_2 \cdot h_2, \ldots g_n \cdot h_n) = (g_{\sigma^{-1}(1)} \cdot h_{\sigma^{-1}(1)}, g_{\sigma^{-1}(2)} \cdot h_{\sigma^{-1}(2)}, \ldots g_{\sigma^{-1}(n)} \cdot h_{\sigma^{-1}(n)}) = \psi_\sigma(g_1, g_2, \ldots g_n) \cdot \psi_\sigma(h_1, h_2, \ldots h_n)$. Since $\phi_{\sigma^{-1}}$ is an inverse function, we see it is an automorphism.
15) By Problems 13 and 14, ψ is a homomorphism from H to $\mathrm{Aut}(G^n)$. Thus, the semi-direct product would have size $|G^n| \cdot |H| = |G|^n \cdot |H|$.
17) A nontrivial homomorphism from Z_8^* to $\mathrm{Aut}(Z_8^*) \approx S_3$ must be two-to-one, and send two of the elements to a 2-cycle. Proposition 6.7 shows that it does not matter which 2-cycle, and since the non-identity elements of Z_8^* are essentially equivalent, there is isomorphically only one $Z_8^* \rtimes Z_8^* \approx Z_2 \times D_4$.
19) Z_3 Wr $S_2 \approx Z_3 \times S_3$.
21) Z_2 Wr $S_3 \approx Z_2 \times S_4$.

Section 7.1

1) $\{1, -1\}$.
3) $\{e, b^2\}$.
5) Yes, if x and y are in the center, then $x \cdot y = y \cdot x$.
7) Clearly if $a \in Z(a)$ and $b \in Z(b)$, then (a, b) will commute with all elements in $A \times B$. But if either a or b are not in the center, then there is an element of $A \times B$ that would not commute with (a, b). Thus,

$$Z(A \times B) = \{(a, b) \mid a \in Z(a) \text{ and } b \in Z(B)\}.$$

9) Let $H = \{e, a\}$. Since H is normal, $g \cdot a \cdot g^{-1}$ is in H for all g. But $g \cdot a \cdot g^{-1} \neq e$ since $a \neq e$. So $g \cdot a \cdot g^{-1} = a$, so $g \cdot a = a \cdot g$.
11) For any $g \in G$, let $b = \phi^{-1}(g)$. Then $\phi(z) \cdot g = \phi(z) \cdot \phi(b) = \phi(z \cdot b) = \phi(b \cdot z) = \phi(b) \cdot \phi(z) = g \cdot \phi(z)$.
13) Clearly $\phi(H)$ will have the same size as H, since ϕ is one-to-one. But $\phi(H)$ must also be a subgroup of G. Since there is only one subgroup of size $|H|$, we have $\phi(H) = H$.
15) For a cyclic group of order n, for each d that divides n there is only one subgroup of order d, which by Problem 13 is characteristic. For infinite cyclic groups, which would be isomorphic to $\mathbb{Z}$, there is only one non-trivial automorphism, $\phi(x) = -x$, and clearly $\phi(h) \in H$ for any subgroup.
17) Let $\phi(x) = g \cdot x \cdot g^{-1}$ be an inner automorphism of G. Since N is normal, $\phi(n) \in N$ for all $n \in N$, so ϕ can be restricted to form an automorphism on N. Then $\phi(h) \in H$ for all $h \in H$, since H is a characteristic subgroup of N. Hence, H is a normal subgroup of G.
19) The non-trivial element in the center is $\phi(x) = x^2$. The group $G/Z(G) \approx S_4$.

Section 7.2

1) $N_{D_4}(\{e\}) = N_{D_4}(\{a^2\}) = D_4$, $N_{D_4}(\{a\}) = N_{D_4}(\{a^3\}) = \{e, a, a^2, a^3\}$, $N_{D_4}(\{b\}) = N_{D_4}(\{a^2 \cdot b\}) = \{e, a^2, b, a^2 \cdot b\}$, $N_{D_4}(\{a \cdot b\}) = N_{D_4}(\{a^3 \cdot b\}) = \{e, a^2, a \cdot b, a^3 \cdot b\}$.
3) $N_{D_4}(\{e, a^2\}) = D_4$, $N_{D_4}(\{e, b\}) = N_{D_4}(\{e, a^2 \cdot b\}) = \{e, a^2, b, a^2 \cdot b\}$, $N_{D_4}(\{e, a \cdot b\}) = N_{D_4}(\{e, a^3 \cdot b\}) = \{e, a^2, a \cdot b, a^3 \cdot b\}$.

5) No, since $N_G(\{e\}) = G$ for all groups.
7) $x \in N_G(\{g\}) \Leftrightarrow x \cdot g = g \cdot x \Leftrightarrow x \cdot g^{-1} = g^{-1} \cdot x \Leftrightarrow x \in N_G(\{g^{-1}\})$.
9) If $z \in Z(G)$ and $g \in S$, then $z \cdot g \cdot z^{-1} = g \cdot z \cdot z^{-1} = g \in S$.
11) $\{e, a, a^2, a^3\}$.
13) $\{e, a^2, b, a^2 \cdot b\}$.
15) $\{e, a^2, b, a^2 \cdot b\}$.
17) $\{e, a, a^2, a^3, a^4\}$.
19) D_5.
21) $N_{D_6}(\{e\}) = N_{D_6}(\{a^3\}) = D_6$, $N_{D_6}(\{a\}) = N_{D_6}(\{a^2\}) = N_{D_6}(\{a^4\}) = N_{D_6}(\{a^5\}) = \{e, a, a^2, a^3, a^4, a^5\}$, $N_{D_6}(\{b\}) = N_{D_6}(\{a^3 \cdot b\}) = \{e, a^3, b, a^3 \cdot b\}$, $N_{D_6}(\{a \cdot b\}) = N_{D_6}(\{a^4 \cdot b\}) = \{e, a \cdot b, a^3, a^4 \cdot b\}$, $N_{D_6}(\{a^2 \cdot b\}) = N_{D_6}(\{a^5 \cdot b\}) = \{e, a^2 \cdot b, a^3, a^5 \cdot b\}$.

Section 7.3

1) $\{e\}$, $\{b^2\}$, $\{b, b^3\}$, $\{a, a \cdot b^2\}$, and $\{a \cdot b, a \cdot b^3\}$.
3) $\{e\}$, $\{a, a \cdot b, a \cdot b^2, a \cdot b^3, a \cdot b^4\}$, $\{b, b^4\}$, and $\{b^2, b^3\}$.
5) $\{(1,1)\}$, $\{(1,3)\}$, $\{(7,1)\}$, $\{(7,3)\}$, $\{(3,3),(5,3)\}$, $\{(3,1),(5,1)\}$, $\{(1,5),(7,5)\}$, $\{(3,5),(5,5)\}$, $\{(1,7),(7,7)\}$, $\{(3,7),(5,7)\}$.
7) If $g \cdot x \cdot g^{-1} = x^{-1}$ for some g, then $g^2 \cdot x = x \cdot g^2$, and since g has odd order, $(g^2)^k = g$ for some k. Thus, $g \cdot x = x \cdot g$, and so $g \cdot x \cdot g^{-1} = x$.
9) If N is a nontrivial normal subgroup, $|N| \geq 13$, so $|N| = 30$, 20, or 15 (divisors of 60). $|N| \neq 15$, so $|N|$ is even, hence classes of size 1 and 15 are in N. Since $|N| \geq 28$, $|N| = 30$, but there is no class of size 14.
11) $|N| \geq 57$, so $|N| = 252$, 168, 126, 84, 72, or 63 (divisors of 504). $|N| \neq 63$, so $|N|$ is even, hence classes of size 1 and 63 are in N, making $|N| \geq 120$. Seven divides $|N|$, so all classes of order 72 are in N, making $|N| \geq 280$.
13) $|N| \geq 85$, so $|N| = 546$, 364, 273, 182, 156, or 91 (divisors of 1092). 13 divides $|N|$, hence both classes of size 84 are in N, making $|N| \geq 260$. Seven divides $|N|$, so all three classes of order 156 are in N, making $|N| \geq 728$.
15) The next largest group would be A_7, with 2520 elements. (Only 72 more elements then $L_2(17)$.) The next largest group $L_2(19)$ has 3420 elements.
17) $|N| \geq 316$, so $|N| = 10080$, 6720, 5040, 4032, 3360, 2880, 2520, 2240, 2016, 1680, 1440, 1344, 1260, 1120, 1008, 960, 840, 720, 672, 630, 576, 560, 504, 480, 448, 420, 360, 336, or 320 (divisors of 20160). $|N|$ is even, so classes of size 1 and 315 are in N, making $|N| \geq 1576$. $|N| \neq 2240$, so $|N|$ is a multiple of 3, so the class of size 2240 is in N, making $|N| \geq 3816$. Seven divides $|N|$, so both classes of size 2880 are in N, making $|N| \geq 9576$. Five divides $|N|$, so both classes of size 4032 are in N, making $|N| \geq 16380$. A_8 has a conjugacy class of size 112 (all 3-cycles).
19) 20160 elements, same as A_8 and $L_3(4)$ from Problem 17. This group is in fact isomorphic to A_8.
21) Nontrivial normal subgroups are $\{1, 13016\}$ and $\{1, 6212, 13016, 19853, 24132, 25315, 33108, 38807\}$.

Section 7.4

1) Not possible, for the identity will be in a conjugacy class of size 1.

3) This is possible. In fact, it is the sizes of the conjugacy classes of D_5.
5) Not possible, for the size of the group, 25, is the square of a prime, and so by Corollary 7.3 the group would be abelian.
7) Applying Corollary 5.2 yields p^n divides $p!\,|N|$, but since p^2 does not divide $p!$, $H = N$.
9) Let K be any p-Sylow subgroup of size p^n, and divide G into families, where u and v are related if $u = h \cdot v \cdot k$ for $h \in H$ and $k \in K$. Then $|G| = p^n \cdot m = \sum p^i \cdot p^n/|H \cap (u_j \cdot K \cdot u_j^{-1})|$, so $|H \cap (u_j \cdot K \cdot u_j^{-1})| = p^i$ for some j, meaning that H is completely contained in a p-Sylow subgroup.
11) There are either one or eight 7-Sylow subgroups. If not unique, there are 48 elements of order 7, leaving 8 elements for a unique 2-Sylow subgroup.
13) There is only one 3-Sylow subgroup H, and only one 11-Sylow subgroup N, so both are normal, and $G \approx H \times N$. Thus, $G \approx Z_{99}$ or $Z_3 \times Z_3 \times Z_{11}$.
15) There is only one 17-Sylow subgroup N, 1 or 51 5-Sylow subgroups, and 1 or 85 3-Sylow subgroups. Either a 3-Sylow subgroup H or 5-Sylow subgroup K is normal, so $H \cdot K$ is a subgroup of order 15 $\approx Z_{15}$. Then $G \approx Z_{15} \times Z_{17} \approx Z_{255}$, or $G \approx Z_{17} \rtimes_\phi Z_{15}$. But there is no nontrivial homomorphism between Z_{15} and Z_{17}^*.
17) Factors of $|G|$ are 1, p, p^2, q, pq, p^2q. There are either 1 or q p-Sylow subgroups, and either 1, p, or p^2 q-Sylow subgroups. If neither are unique, $q \equiv 1 \pmod{p}$, implying $p < q$, so $p^2 \equiv 1 \pmod{q}$. Then we have $p^2(q-1)$ elements of order q, leaving only p^2 elements for a normal p-Sylow subgroup.
19) Only cases not covered by Problems 16 through 18 or Proposition 7.8 are 30, 36, 42, and 48. If $G = 30$, there aren't enough elements for both 10 3-Sylow subgroups and 6 5-Sylow subgroups. If $G = 36$, there is a 3-Sylow subgroup of order 9, and applying Corollary 5.2 gives a normal subgroup of size 3 or 9. If $G = 42$, there is only one 7-Sylow subgroup. If $G = 48$, there is a 2-Sylow subgroup of order 16, and applying Corollary 5.2 gives a normal subgroup of size 8 or 16.
21) There must be one 7-Sylow subgroup N, which must be normal, and at least one 3-Sylow subgroup H, and since $N \cdot H$ must be the whole group, we see that it is either the direct product, giving Z_{21}, or a semi-direct product $Z_7 \rtimes_\phi Z_3$. Since $\mathrm{Aut}(Z_7) \approx Z_7^* \approx Z_6$, we have two elements of order 3 that Z_3 can map to. But these two choices are equivalent through an automorphism of Z_3. So there is only one possible semi-direct product $Z_7 \rtimes Z_3$, which can be defined by $a^7 = e$, $b^3 = e$, $b \cdot a = a^2 \cdot b$.

Section 8.1

1) $A_{1,1} = A_{1,2} = B_{1,1} = Z_{12}$, $A_{2,1} = \{0, 6\}$, $A_{2,2} = B_{1,3} = \{0\}$, $B_{1,2} = \{0, 2, 4, 6, 8, 10\}$. The arrows show the isomorphisms $Z_{12}/Z_{12} \approx Z_{12}/Z_{12}$, $Z_{12}/Z_{12} \approx \{0, 2, 4, 6, 8, 10\}/\{0, 2, 4, 6, 8, 10\}$, $Z_{12}/\{0, 3, 6, 9\} \approx \{0, 4, 8\}/\{0\}$, $\{0, 3, 6, 9\}/\{0, 6\} \approx Z_{12}/\{0, 2, 4, 6, 8, 10\}$, $\{0, 6\}/\{0\} \approx \{0, 2, 4, 6, 8\}/\{0, 4, 8\}$, $\{0\}/\{0\} \approx \{0\}/\{0\}$.
3) $Z_{24}^* \supseteq \{1, 5, 7, 11\} \supseteq \{1, 5\} \supseteq \{1\}$.

5) $Z_{12} \times Z_{18} \supseteq \{0,3,6,9\} \times Z_{18} \supseteq \{0,6\} \times Z_{18} \subseteq \{0\} \times Z_{18} \supseteq \{0\} \times \{0,3,6,9,12,15\} \supseteq \{0\} \times \{0,9\} \supseteq \{0\} \times \{0\}$.
7) $D_4 \subseteq \{e, b, b^2, b^3\} \subseteq \{e, b^2\} \subseteq \{e\}$.
9) $D_6 \subseteq \{e, b, b^2, b^3, b^4, b^5\} \subseteq \{e, b^3\} \subseteq \{e\}$.
11) A_4 and $\{(), (12)(34), (13)(24), (14)(23)\}$ must be in the series, and then we have three choices, $\{(), (12)(34)\}$, $\{(), (13)(24)\}$, or $\{(), (14)(23)\}$ for the next term in the series.
13) S_5 and Z_{120}.
15) Pick a cyclic group of prime order.
17) Since all of the A_i and B_j are normal subgroups of G, then $A_{i,j} = (A_{i-1} \cap B_j) \cdot A_i$ and $B_{j,i} = (B_{j-1} \cap A_i) \cdot B_j$ are normal subgroups of G using Problem 17 from §4.3.
19) $M \supseteq \{e, a, a^2, a^3, a^4, b^2, a \cdot b^2, a^2 \cdot b^2, a^3 \cdot b^2, a^4 \cdot b^2\} \supseteq \{e, a, a^2, a^3, a^4\} \supseteq \{e\}$.

Section 8.2

1) Use induction on n. If G is abelian, it is obviously soluble. Otherwise, $Z(G)$ is nontrivial by Corollary 7.2, and so by induction both $Z(G)$ and $G/Z(G)$ are soluble.
3) $[z \cdot x \cdot z^{-1}, z \cdot y \cdot z^{-1}] = (z \cdot x \cdot z^{-1})^{-1} \cdot (z \cdot y \cdot z^{-1})^{-1} \cdot (z \cdot x \cdot z^{-1}) \cdot (z \cdot y \cdot z^{-1}) = z \cdot x^{-1} \cdot y^{-1} \cdot x \cdot y \cdot z^{-1} = z \cdot [x, y] \cdot z^{-1}$.
5) Since G' is a normal subgroup of G, either $G' = G$ or $G' = \{e\}$. But $G' \neq \{e\}$ since G is not abelian. Note that if $G = A_5 \times A_5$, then $G' = G$.
7) $(S_3)' = A_3$, $(S_3)'' = \{()\}$.
9) $(D_5)' = \{e, b, b^2, b^3, b^4\}$, $(D_5)'' = \{e\}$.
11) $G' = \{e, a, a^2\}$, $G'' = \{e\}$.
13) By Problem 12, G' is a characteristic subgroup of G, and G'' is a characteristic subgroup of G', and so on. By Problem 16 of §7.1, the characteristic subgroup of a characteristic subgroup is characteristic, so all of G', G'', etc. will be characteristic subgroups of G. Finally, by Problem 12 of §7.1, all of the subgroups in the series will be normal subgroups of G.
15) If $G = S_4$, then $G_1 = [S_4, S_4] = A_4$, but $G_2 = [S_4, A_4] = A_4$, so G_n will never go to $\{e\}$.
17) If $G = N_0 \supseteq N_1 \supseteq \cdots \supseteq N_k = \{e\}$ is a chief series, then $G' \subseteq N_1$ by lemma 8.3. Define $G_1 = G'$, $G_2 = [G', G_1]$, $G_3 = [G', G_2]$, …, and suppose by induction that $G_i \subseteq N_i$. We must show that $[G', N_i] \subseteq N_{i+1}$, since this would indicate that $G_k = \{e\}$. Since N_i/N_{i+1} is cyclic, there is a generator nN_{i+1}. For $x, y \in G$, we have $x \cdot n \cdot x^{-1} N_{i+1} = n^q N_{i+1}$ for some q, and $y \cdot n \cdot y^{-1} N_{i+1} = n^r N_{i+1}$ for some r. Then $y^{-1} \cdot x^{-1} \cdot y \cdot x \cdot n^{-1} \cdot x^{-1} \cdot y^{-1} \cdot x \cdot y \cdot nN_{i+1} = N_{i+1}$, so $[x^{-1} \cdot y^{-1} \cdot x \cdot y, n] \in N_{i+1}$. Thus, $[G', N_i] \subseteq N_{i+1}$.
19) Since $A' = A$, the derived series never goes to $\{e\}$.

Section 8.3

1) $D_4 \supseteq \{e, a, a^2, a^3\} \subseteq \{e\}$.
3) $Z_{15}^* \supseteq \{1, 2, 4, 8\} \supseteq \{1\}$.
5) $Z_{26}^* \supseteq \{1\}$. (Group is cyclic.)
7) $Z_2 \times Z_3 \times Z_4 \supseteq Z_2 \times Z_3 \times \{0\} \supseteq \{e\}$.

9) $Z_2 \times Z_2 \times Z_2 \times Z_3 \times Z_3 \supseteq Z_2 \times Z_2 \times \{0\} \times Z_3 \times \{0\} \supseteq Z_2 \times \{0\} \times \{0\} \times \{0\} \times \{0\} \supseteq \{e\}$.
11) No. $Z_2 \times Z_{12} \approx Z_4 \times Z_6$, so one polycyclic series of length 2 has quotient groups isomorphic to Z_2 and Z_{12}, while another polycyclic series has quotient groups isomorphic to Z_4 and Z_6.
13) Z_{16}, $Z_8 \times Z_2$, $Z_4 \times Z_4$, $Z_4 \times Z_2 \times Z_2$, and $Z_2 \times Z_2 \times Z_2 \times Z_2$ are the only groups that are abelian, and by the fundamental theorem of finite abelian groups (6.2) these are all non-isomorphic. $R_2(Z_2 \times D_8) = 12$, $R_2(D_{16}) = 10$, $R_2(G) = 6$ from section 6.4, and $R_2(D) = 2$. $R_2(B) = R_2(C) = 8$, but B has only 2 elements along the diagonal, whereas C has 3. Finally, $R_2(M) = R_2(Z_2 \times Q) = R_2(Z_4 \rtimes Z_4) = 4$, but $Z_2 \times Q$ has only 2 elements along the diagonal, M has 4 elements along the diagonal, and $Z_4 \rtimes Z_4$ has 3 elements along the diagonal.
15) By Problem 22, G is abelian, hence solvable. But for G/N to be cyclic, then G/N would be of order 2, and N would have the same properties. Thus, a polycyclic series would not reach $\{e\}$ in a finite number of steps.
17) If $a = (1\,2\,3)$, $b = (1\,2)(3\,4)$, and $c = (1\,3)(2\,4)$, then $a^3 = b^2 = c^2 = e$, $a^{-1} \cdot b \cdot a = c$, $a^{-1} \cdot c \cdot a = b \cdot c$, $b^{-1} \cdot c \cdot b = c$.
19) $C \supseteq \{1, F, G, H\} \supseteq \{1, F\} \supseteq \{1\}$; if $a = I$, $b = G$, and $c = F$, then $a^4 = b^2 = c^2 = e$, $a^{-1} \cdot b \cdot a = c$, $a^{-1} \cdot c \cdot a = b$, $b^{-1} \cdot c \cdot b = c$.

Section 8.4

1) $(0, 1, 1, 0, 0, 1, P(2, 4, 3, 1, 5, 6))$.
3) $(1, 1, 1, 1, 1, 1, P(2, 5, 3, 6, 4, 1))$.
5) $(0, 1, 0, 1, 1, 1, (1\,3\,2\,4)(5\,6))$.
7) $(1, 1, 1, 1, 1, 1, (1\,2\,6\,4)(3\,5))$.
9) 1 element of order 1, 391 of order 2, 64880 of order 3, 2520 of order 4, 2304 of order 5, 173840 of order 6, 1440 of order 8, 2304 of order 10, 201600 of order 12, 184320 of order 15, 115200 of order 24, and 184320 elements of order 30.
11) Since each 5-Sylow subgroup contains 4 different elements of order 5, there are $2304/4 = 576$ 5-Sylow subgroups.
13) Each 3-Sylow subgoup has 8 elements besides the identity, so there must be at least 100 3-Sylow subgroups. The divisors of 11520 that are congruent to 1 (mod 3) are 640, 256, and 160. But by Lemma 7.3 with $H = G$, the number of 3-Sylow subgroups is a divisor of 2880. Thus, there are 160 3-Sylow subgroups.
15) The size of group is $8! \cdot (12!/2) \cdot 3^7 \cdot 2^{11} = 43252003274489856000$. The only nontrivial element in the center flips all 12 edges. (Rotating all 8 corners clockwise can't be done, since 8 is not a multiple of 3.)
17) $a^{-1} \cdot b \cdot a \cdot b \cdot a$.
19) $b^{-2} \cdot a \cdot b^{-1} \cdot a^{-1} \cdot b^{-1} \cdot a^2 \cdot b \cdot a^{-1} \cdot b \cdot a^{-1} \cdot b^{-1}$

Section 9.1

1) $(-x) \cdot y = (-x) \cdot y + [x \cdot y + -(x \cdot y)] = [(-x) \cdot y + x \cdot y] + -(x \cdot y) = [(-x) + x] \cdot y + -(x \cdot y) = 0 \cdot y + -(x \cdot y) = -(x \cdot y)$.

3) $(-x)\cdot(-y) = -((-x)\cdot y) = -(-(x\cdot y)) = x\cdot y$.
5) $(x-y)\cdot z = (x+(-y))\cdot z = x\cdot z + (-y)\cdot z = x\cdot z + -(y\cdot z) = x\cdot z - y\cdot z$.
7) Either $(a\cdot b)\cdot x = 0$ or $x\cdot(a\cdot b) = 0$ for some non-zero x. In the first case, $a\cdot(b\cdot x) = 0$, so either a is a zero divisor, or $b\cdot x = 0$, making b a zero divisor. The second case is similar.
9) $\overline{i}\cdot\overline{j} = (-i)\cdot(-j) = k$, yet $\overline{i\cdot j} = \overline{k} = -k$. What is true is that $\overline{x_1}\cdot\overline{x_2} = \overline{x_2\cdot x_1}$.
11) $|x_1\cdot x_2| = \sqrt{x_1\cdot x_2\cdot\overline{x_1\cdot x_2}} = \sqrt{x_1\cdot x_2\cdot\overline{x_2}\cdot\overline{x_1}} = \sqrt{x_1\cdot\overline{x_1}\cdot x_2\cdot\overline{x_2}} = \sqrt{x_1\cdot\overline{x_1}}\sqrt{\cdot x_2\cdot\overline{x_2}} = |x_1||x_2|$.
13) $(x+i)\cdot(x-i) = x^2 + i\cdot x - x\cdot i + 1 \neq x^2+1$. (For example, if $x = j$.)
15) This set is not closed under multiplication. For example, $\sqrt[3]{2}\cdot\sqrt[3]{2} = \sqrt[3]{4}$.
17) Since G is an abelian group, we only need to check the associate law and the two distributive laws. But these are both trivial, since both sides would evaluate to 0.
19)

+	0	a	b	c
0	0	a	b	c
a	a	c	0	b
b	b	0	c	a
c	c	b	a	0

·	0	a	b	c
0	0	0	0	0
a	0	c	c	0
b	0	c	c	0
c	0	0	0	0

21)

```
u = x*y
v = y*z
u*z - x*v
    0
```

Section 9.2

1) $[[a,b],[a,b]]$.
3) $[[b,a],[a,b]]$.
5) $[[0,a],[a,a+b]]$.
7) $[[a,a],[a,a]]$.
9) $[[a,b],[3\cdot a, 3\cdot b]]$.
11) $b^2 = c^2 = a+b+c$, $b\cdot a = c$, $c\cdot b = a$.
13) By induction in m: $m(x+y) = (m-1)(x+y) + (x+y) = (m-1)x + (m-1)y + x + y = mx + my$.
15) By induction in m: $(mn)x = ((m-1)n+n)x = ((m-1)n)x + nx = (m-1)(nx) + nx = m(nx)$.
17) Since the additive group is abelian, it can be written as $Z_{n_1}\times Z_{n_2}\times\cdots\times Z_{n_r}$. Then the r elements $(1,0,\ldots 0)$, $(0,1,\ldots,0),\ldots,(0,0,\ldots,1)$ form a basis.
19)

```
InitRing("a", "b", "c")
DefineRing([2, 2, 2],[[a, a, c] [b, b, a +  b  +  c], [a, a, c]])
```

21) Here is one way:
```
InitRing("a", "b")
DefineRing([2, 2],[[a, a],[b, b]])
```

Section 9.3

1) Both $x \cdot a = x$ and $x \cdot b = x$ for all x in the ring, but there is no r for which $r \cdot c = c$, since $r \cdot c = 0$.
3) $(x + y) \cdot (x^{-1} - x^{-2} \cdot y + x^{-3} \cdot y^2) = e + x^{-1} \cdot y - x^{-1} \cdot y - x^{-2} \cdot y^2 + x^{-2} \cdot y^2 + x^{-3} \cdot y^3 = e$.
5) This is actually a field, with 6 as the unity.
7) Since $(x+y)^2 = x^2 + x \cdot y + y \cdot x + y^2 = x + y$, we have that $x \cdot y + y \cdot x = 0$. By Problem 6, $x \cdot y = -x \cdot y$, and so $x \cdot y = y \cdot x$.
9) Obviously 0 and e satisfy $a^2 = a$. If $a \neq 0$, then a^{-1} exists, and $a = a^2 \cdot a^{-1} = a \cdot a^{-1} = e$.
11) $(x + y)^2 = x^2 + x \cdot y + y \cdot x + y^2 = x^2 + 2xy + y^2$. $(x + y)^3 = (x + y)(x^2 + 2xy + y^2) = x^3 + y \cdot x^2 + 2x^2 \cdot y + 2y \cdot x \cdot y + x \cdot y^2 + y^3 = x^3 + 3x^2y + 3xy^2 + y^3$.
13) e and g.
15) $2a + b$ is the only irreducible element.
17) a, b, d, and f are prime.
19)
```
InitRing("a", "c")
T4 = DefineRing([2,2], [[a,c],[0,0]]
```

Section 10.1

1) Subring. $(x_1 - x_2) + (y_1 - y_2)\sqrt{5}$ and $(x_1x_2 + 5y_1y_2) + (x_1y_2 + x_2y_1)\sqrt{5}$ are in the set.
3) Not a subring, since not closed under subtraction.
5) Subring. $(x_1 2^{y_1} - x_2 2^{y_1})/2^{(y_1+y_2)}$ and $(x_1x_2)/2^{(y_1+y_2)}$ are in the set.
7) Subring. $(x_1 - x_2) + (y_1 - y_2)\sqrt[3]{2} + (z_1 - z_2)\sqrt[3]{4}$ and $(x_1x_2 + 2y_1z_2 + 2y_2z_1) + (x_1y_2 + x_2y_1 + 2z_1z_2)\sqrt[3]{2} + (x_1z_2 + y_1y_2 + z_1x_2)\sqrt[3]{4}$ are in the set.
9) Not a subring, since not closed under multiplication. $(1+\sqrt{2})(1-\sqrt{2}) = -1$.
11) If $a, b \in A$, then $a \cdot y = b \cdot y = 0$, so $(a - b) \cdot y = 0$ and $(a \cdot b) \cdot y = 0$, so $a - b$ and $a \cdot b$ are in A.
13) If $a, b \in Z$, and $x \in R$, then $(a - b) \cdot x = a \cdot x - b \cdot x = x \cdot a - x \cdot b = x \cdot (a - b)$ and $(a \cdot b) \cdot x = a \cdot (x \cdot b) = x \cdot (a \cdot b)$, so $a - b$ and $a \cdot b$ are in Z.
15) 2 and 3 are in $2\mathbb{Z} \cup 3\mathbb{Z}$, but $2 + 3 = 5 \notin 2\mathbb{Z} \cup 3\mathbb{Z}$.
17) $\{0\}$, $\{0, a\}$, $\{0, b\}$, $\{0, c\}$, and the whole ring.
19) $\{0\}$, $\{0, a, 2a, -a\}$, $\{0, b\}$, $\{0, 2a\}$, $\{0, b, 2a, 2a + b\}$, $\{0, 2a + b\}$, $\{0, a + b, 2a, -a + b\}$, and the whole ring.

Section 10.2

1) If $a \in X + Y$ and $z \in R$, then $a = x + y$ for some $x \in X$ and $y \in Y$. Then $a \cdot z = (x \cdot z) + (y \cdot z) \in X + Y$. Likewise, $z \cdot a \in X + Y$.
3) If $a \in X \cdot Y$, and $z \in R$, then $a = x_1 \cdot y_1 + x_2 \cdot y_2 + \cdots + x_n \cdot y_n$, so $a \cdot z = x_1 \cdot (y_1 \cdot z) + x_2 \cdot (y_2 \cdot z) + \cdots x_n \cdot (y_n \cdot z) \in X \cdot Y$. Likewise, $z \cdot a \in X \cdot Y$.
5) If $a \in X \cdot Y$, then $a = x_1 \cdot y_1 + x_2 \cdot y_2 + \cdots + x_n \cdot y_n \in X$. Likewise, $a \in Y$, so $a \in X \cap Y$.

7) $\{0\}$, $\{0, a, 2a, 3a\}$, $\{0, 2a\}$, $\{0, b\}$, $\{0, 2a + b, b, 2a\}$, and the whole ring.
9) $\{0\}$, $\{0, c\}$, $\{0, a, b, c\}$, $\{0, c, d, f\}$, and the whole ring.
11)

$+$	$\{0,2a\}$	$\{a,3a\}$	$\{b,2a+b\}$	$\{a+b,3a+b\}$
$\{0,2a\}$	$\{0,2a\}$	$\{a,3a\}$	$\{b,2a+b\}$	$\{a+b,3a+b\}$
$\{a,3a\}$	$\{a,3a\}$	$\{0,2a\}$	$\{a+b,3a+b\}$	$\{b,2a+b\}$
$\{b,2a+b\}$	$\{b,2a+b\}$	$\{a+b,3a+b\}$	$\{0,2a\}$	$\{a,3a\}$
$\{a+b,3a+b\}$	$\{a+b,3a+b\}$	$\{b,2a+b\}$	$\{a,3a\}$	$\{0,2a\}$

$\cdot$	$\{0,2a\}$	$\{a,3a\}$	$\{b,2a+b\}$	$\{a+b,3a+b\}$
$\{0,2a\}$	$\{0,2a\}$	$\{0,2a\}$	$\{0,2a\}$	$\{0,2a\}$
$\{a,3a\}$	$\{0,2a\}$	$\{a,3a\}$	$\{0,2a\}$	$\{a,3a\}$
$\{b,2a+b\}$	$\{0,2a\}$	$\{0,2a\}$	$\{b,2a+b\}$	$\{b,2a+b\}$
$\{a+b,3a+b\}$	$\{0,2a\}$	$\{a,3a\}$	$\{b,2a+b\}$	$\{a+b,3a+b\}$

13) $\{0, d\}$ and $\{0, f\}$ are left ideals.
15)

$+$	A	$1+A$	$2+A$	$3+A$	$4+A$	$5+A$
A	A	$1+A$	$2+A$	$3+A$	$4+A$	$5+A$
$1+A$	$1+A$	$2+A$	$3+A$	$4+A$	$5+A$	A
$2+A$	$2+A$	$3+A$	$4+A$	$5+A$	A	$1+A$
$3+A$	$3+A$	$4+A$	$5+A$	A	$1+A$	$2+A$
$4+A$	$4+A$	$5+A$	A	$1+A$	$2+A$	$3+A$
$5+A$	$5+A$	A	$1+A$	$2+A$	$3+A$	$4+A$

$\cdot$	A	$1+A$	$2+A$	$3+A$	$4+A$	$5+A$
A	A	A	A	A	A	A
$1+A$	A	$1+A$	$2+A$	$3+A$	$4+A$	$5+A$
$2+A$	A	$2+A$	$4+A$	A	$2+A$	$4+A$
$3+A$	A	$3+A$	A	$3+A$	A	$3+A$
$4+A$	A	$4+A$	$2+A$	A	$4+A$	$2+A$
$5+A$	A	$5+A$	$4+A$	$3+A$	$2+A$	$1+A$

17) If $a, b \in A$, then $a \cdot y = b \cdot y = 0$, so $(a - b) \cdot y = 0$, hence $a - b \in A$. If $z \in R$, then $(a \cdot z) \cdot y = z \cdot (a \cdot y) = 0$, so A is an ideal.
19) Problem 14 of §10.1 shows it is a subring, so suppose a is nilpotient, so that $a^m = 0$. If $x \in R$, $(a \cdot x)^m = a^m \cdot x^m = 0$, so $a \cdot x$ is nilpotient.
21) Nontrivial ideals: $\{0, b\}$, $\{0, 2a\}$, and $\{0, b, 2a, 2a + b\}$.

Section 10.3

1) $\phi(x \cdot y) = \phi(x) \cdot \phi(y) = \phi(y) \cdot \phi(x) = \phi(y \cdot x)$. Since ϕ is one-to-one, $x \cdot y = y \cdot x$.
3) If $x \cdot y = 0$ with non-zero x and y, then $0 = \phi(0) = \phi(x \cdot y) = \phi(x) \cdot \phi(y)$. Since ϕ is one-to-one, $\phi(x)$ and $\phi(y)$ are non-zero.
5) $\{0, a, b, c\}$ gives a copy of T_4 inside of T_8.
7) T_4^{op} has an element c for which $c \cdot x = 0$ for all x, T_4 has no such element.
9) $\{0, e, a, b, c, d, f, g\} \mapsto \{0, e, d, f, c, a, b, g\}$ or $\{0, e, f, d, c, b, a, g\}$.
11) No, $2\mathbb{Z}$ has an element x for which $x + x = x^2$, $3\mathbb{Z}$ has no such element.

13) No, $\mathbb{R}$ has no element for which $x^2 + e = 0$.
15) Z_{21}, $3Z_{63}$, $7Z_{147}$, and $21Z_{441}$.
17) Z_{210}, $2Z_{420}$, $3Z_{630}$, $5Z_{1050}$, $6Z_{1260}$, $7Z_{1470}$, $10Z_{2100}$, $14Z_{2940}$, $15Z_{3150}$, $21Z_{4410}$, $30Z_{6300}$, $35Z_{7350}$, $42Z_{8820}$ $70Z_{14700}$ $105Z_{22050}$ and $210Z_{44100}$.
19) $\{\{0,6a\},\{a,7a\},\{2a,8a\},\{3a,9a\},\{4a,10a\},\{5a,11a\}\} \leftrightarrow$
$\{0,b,2b,3b,4b,5b\}$.
21) 4 rings: Z_6, $2Z_{12}$, $3Z_{18}$ and $6Z_{36}$.

Section 10.4

1) $\{0,1,2,3,4,5\} \mapsto \{0,0,0,0,0,0\}$, $\{0,1,2,3,4,5\}$, $\{0,3,0,3,0,3\}$, or $\{0,4,2,0,4,2\}$.
3) $2 = \phi(1 \cdot 1) \neq \phi(1) \cdot \phi(1) = 4$.
5) No. $4 = \phi(1 \cdot 1) \neq \phi(1) \cdot \phi(1) = 16$.
7) Yes, since clearly $\phi(x+y) = (x+y) \textbf{ mod } 5 = \phi(x) + \phi(y)$, and $\phi(x \cdot y) = (x \cdot y) \textbf{ mod } 5 = \phi(x) \cdot \phi(y)$.
9) No. $0 = \phi(0) = \phi(1+1) \neq \phi(1) + \phi(1) = 2$.
11) $\phi(x) + \phi(y) = a + c - (b+d)i = \phi(x+y)$, $\phi(x) \cdot \phi(y) = (a - bi)(c - di) = ac - bd - (bc + ad)i = \phi(x \cdot y)$.
13) If $x^n = 0$, then $\phi(x^n) = \phi(0) = 0$, so $\phi(x)$ is nilpotent.
15) The homomorphism $\phi : R \mapsto R/I$, given by $\phi(x) = x + I$, restricted to the ideal K, produces $\phi' : K \mapsto (K+I)/I$. The kernel of ϕ' is $K \cap I$, and so by the first isomorphism theorem for rings (10.2), $K/(K \cap I) \approx (K+I)/I$.
17) The kernel cannot be $\{0,c\}$, otherwise the image of 4 elements would have a unity. So the two possible kernels are $\{0,a,b,c\}$ and $\{0,c,d,f\}$. The image would be isomorphic to Z_2, so it is either $\{0,a\}$ or $\{0,b\}$. So
$\{0,e,a,b,c,d,f,g\} \mapsto \{0,a,0,0,0,a,a,a\}$, $\{0,b,0,0,0,b,b,b\}$,
$\{0,a,a,a,0,0,0,a\}$, or $\{0,b,b,b,0,0,0,b\}$.
19)

```
I3 = Ideal(R, b)
Q = Coset(R, I3)
i = RingHomo(R, Q)
HomoDef(i, a, a + I3)
HomoDef(i, b, I3)
FinishHomo(i)
    'Homomorphism defined'
```

Section 11.1

1) b.
3) $2ax^2 + ax + b$.
5) $2ax^3 + 2ax^2 + bx$.
7) 4.
9) Since the additive group is of order n, $nx = 0$ for all x. But $m \cdot 1 \neq 0$ for all $m < n$.
11) $(-x)^2 = -x$, but also $(-x)^2 = x^2 = x$. So $-x = x$ for all x, hence $2x = 0$.

13) D would have no zero divisors, so we can use Proposition 11.2, and the characteristic is a prime number p. Then the additive order of all non-zero elements is p.
15)

$+$	0	e	a	b
0	0	e	a	b
e	e	0	b	a
a	a	b	0	e
b	b	a	e	0

$\cdot$	0	e	a	b
0	0	0	0	0
e	0	e	a	b
a	0	a	b	e
b	0	b	e	a

17) $(x+y)^4 = ((x+y)^2)^2 = (x^2+y^2)^2 = (x^2)^2 + (y^2)^2 = x^4 + y^4$.
19) For the ring of Example 10.6, $x = a$, $y = b$.
21) x^2+1, x^2+x+2, x^2+2x+2, $2x^2+2$, $2x^2+x+1$, $2x^2+2x+1$.
23) x^3+x+1, x^3+x^2+1.
25) $(i+2)(i+3) = 0$ in this ring, so it is not a field. Primes that are one more than a multiple of 4 will fail to form a field, but primes that are one less than a multiple of 4 will form a field.

Section 11.2

1) $\left(\frac{-a}{b}\right) + \left(\frac{a}{b}\right) = \left(\frac{-a\cdot b + a\cdot b}{b^2}\right) = \left(\frac{0}{b^2}\right) = \left(\frac{0}{z}\right)$.
3) $\left(\frac{u}{v}\right)\cdot\left(\left(\frac{x}{y}\right)\cdot\left(\frac{z}{w}\right)\right) = \left(\frac{u}{v}\right)\cdot\left(\frac{xz}{yw}\right) = \left(\frac{uxz}{vyw}\right) = \left(\frac{ux}{vy}\right)\cdot\left(\frac{z}{w}\right) = \left(\left(\frac{u}{v}\right)\cdot\left(\frac{x}{y}\right)\right)\cdot\left(\frac{z}{w}\right)$.
5) Isomorphism given by $0 \mapsto \{(0,1),(0,2),(0,3),(0,4)\}$,
$1 \mapsto \{(1,1),(2,2),(3,3),(4,4)\}$, $2 \mapsto \{(2,1),(4,2),(1,3),(3,4)\}$,
$3 \mapsto \{(3,1),(1,2),(4,3),(2,4)\}$, $4 \mapsto \{(4,1),(3,2),(2,3),(1,4)\}$.
7) Every rational number p/q can be put in the form $(2p)/(2q)$, so there is a natural mapping from $\mathbb{Q}$ to the quotient field.
9) $((1+i)x+i+2)((1+2i)x+i) = (x+1+2i)(2x+2i) = 2x^2+2x+2+2i$.
11) $(x^3+x^2+1)/(x^4+x)$.
13) $x^2/(x^3+1)$.
15) $(x^3+x^2)/(x^2+x+1)$.
17) $(x^3+x^2)/(x^4+x^2+1)$.
19) The square of every element is the same as replacing every x with x^2. Reason: because of Problem 16 of §11.1, $\phi(x) = x^2$ is a ring homomorphism.
21) Cross multiplying, $(3x+3a)(2x) = ((a+1)x)((1-a)x+5+a) = 6x^2+6ax$.

Section 11.3

1)

$$e^i = 1 + \frac{i}{1!} + \frac{-1}{2!} + \frac{-i}{3!} + \frac{1}{4!} + \frac{i}{5!} + \cdots$$
$$= \left(1 - \frac{1}{2!} + \frac{1}{4!} - \cdots\right) + i\left(\frac{1}{1!} - \frac{1}{3!} + \frac{1}{5!} - \cdots\right) = \cos 1 + i\sin 1.$$

3)

$$1 + \frac{i}{n} = \sqrt{1 + \frac{1}{n^2}}\left(\cos(\tan^{-1}(1/n)) + i\sin(\tan^{-1}(1/n))\right),$$

so

$$\left(1+\frac{i}{n}\right)^n = \left(1+\frac{1}{n^2}\right)^{n/2} \left(\cos(n\tan^{-1}(1/n)) + i\sin(n\tan^{-1}(1/n))\right).$$

But

$$\lim_{n\to\infty}\left(1+\frac{1}{n^2}\right)^{n/2} = 1 \quad \text{and} \quad \lim_{n\to\infty} n\tan^{-1}(1/n) = 1$$

by L'Hôpital's rule.
5) $\ln 2 - \pi/6 + 2k\pi i$, where $k \in \mathbb{Z}$.
7) $\sqrt{2}/2 \pm i\sqrt{2}/2$, $-\sqrt{2}/2 \pm i\sqrt{2}/2$.
9) $-2i$, $\pm\sqrt{3} + i$.
11) $\ldots, e^{-7\pi/4}, e^{-3\pi/4}, e^{\pi/4}, e^{5\pi/4}, e^{9\pi/4}, \ldots$.
13) $(1)^{i\ln 2/(2\pi)}$.
15) From DeMoivre's theorem, all solutions $z^n = 1$ are of the form $z = \cos(2k\pi/n) + i\sin(2k\pi/n) = (\cos(2\pi/n) + i\sin(2\pi/n))^k$. Thus, ω_n generates the group. A generator of this group would be ω_n^k, where k is coprime to n, hence a primitive n-th root of unity.
17) False: $(2^2)^{1/2} = 4^{1/2} = \pm 2$, yet $2^{(2\cdot 1/2)} = 2^1 = 2$.
19)

```
H = [I, 1/2 + I*sqrt(3)/2, I/2 + sqrt(3)/2,
     1, -I/2 + sqrt(3)/2, 1/2 - I*sqrt(3)/2,
     -I, -1/2 - I*sqrt(3)/2, -I/2 - sqrt(3)/2,
     -1, I/2 - sqrt(3)/2, -1/2 + I*sqrt(3)/2]
CircleGraph(H, Mult(sqrt(3) +  I/2))
```

Section 11.4

1) Since $x^2 \geq 0$ and $e > 0$, then $x^2 + e > 0$.
3) Since $x < y$, $y - x > 0$, and $x + y > 0$, so $y^2 - x^2 = (y - x)\cdot(x + y) > 0$.
5) Note that $a^{-1} > 0$, since $a \cdot a^{-1} = e > 0$. If $a^{-1} < b^{-1}$, then by Problem 4 we have $e < e$, a contradiction. Since $a^{-1} \neq b^{-1}$, $a^{-1} > b^{-1}$.
7) Since $(x+y)^2 \geq 0$, $x^2 + 2xy + y^2 \geq 0$, so $x^2 + y^2 \geq -2xy$.
9) If $\phi(f(x)) = f(-x)$, then $\phi(f(x) + g(x)) = f(-x) + g(-x) = \phi(f(x)) + \phi(g(x))$, and $\phi(f(x)\cdot g(x)) = f(-x)\cdot g(-x) = \phi(f(x)\cdot\phi(g(x))$. Clearly, this is one-to-one and onto, since $\phi^{-1} = \phi$. Hence $P' = \phi(P)$ gives another ordering of $\mathbb{Z}[x]$. The positive elements would be polynomials of even degree with a positive leading coefficient, and polynomials of odd degree with a negative leading coefficient.
11) It is clear that addition forms an abelian group under addition, since the derivatives only act as a place holder. A linear operator of a linear operator is linear, so multiplication is closed. Since linear differential operators are functions from $\mathbb{Z}[x]$ to $\mathbb{Z}[x]$, multiplication is associative. The property of linear operators is that $T(y_1 + y_2) = T(y_1) + T(y_2)$, so the distributive law holds.

13) By the product rule, $(p_n(x)y^{(n)}(x))' = p_n'(x)y^{(n)}(x) + p_n(x)y^{(n+1)}(x)$. By keeping only the highest derivative of y, we see that the coefficient polynomial is unaffected. By induction this is true for higher derivatives as well, so $\phi(T \cdot S) = \phi(T) \cdot \phi(S)$.
15) If $a^3 + 2b^3 + 4c^4 - 6abc = 0$, we can assume a, b and c are integers with no prime factors in common. Then a^3 is even, so we can replace $a = 2x$, and find that b is even. Replacing $b = 2y$ shows c is even, a contradiction.
17) Since f is an automorphism, $f(y) = f(x \cdot f(x) \cdot f(f(x))) = f(x) \cdot f(f(x)) \cdot f(f(f(x))) = f(x) \cdot f(f(x)) \cdot x = y$. Since f sends y to itself, y must be rational.
19)

```
InitDomain(0)
FieldHomo(F)
HomoDef(F, sqrt(5), - sqrt(5))
CheckHomo(F)
    True
```

Section 12.1

1) $q(x) = x + 2$, $r(x) = -4x + 2$.
3) $q(x) = x^2 + x$, $r(x) = 0$.
5) $q(x) = x^2 + 2x + 1$, $r(x) = 2x + 2$.
7) $f(x) = 3x^2 - 2x + 1$.
9) If $x^2 + 5$ has a root a in $\mathbb{R}$, then $a^2 + 5 = 0$. But $a^2 \geq 0$, so $a^2 + 5 \geq 5$. Finally, apply Proposition 12.3.
11) Irreducible.
13) Since $f(x)$ and $g(x)$ are coprime in $F[x]$, by Theorem 12.2 there are $s(x)$ and $t(x)$ in $F[x]$ such that $f(x) \cdot s(x) + g(x) \cdot t(x) = 1$. Then $s(x)$ and $t(x)$ are in the larger ring $K[x]$, hence $f(x)$ and $g(x)$ are coprime in $K[x]$, too.
15) $x^2 - 2x + 5$.
17) $x + 3$.
19) 1.
21) $f(x) = 11x^3/6 - 19x^2/2 + 50x/3 - 8$, $f(5) = 67$.

Section 12.2

1) If $x = p/q$ is a root in lowest terms, $p^3 - 3pq^2 + 3q^3 = 0$, so $q = 3r$ for some r. Then $9r^3 - 3rq^2 + q^3 = 0$, so q is a multiple of 3, a contradiction.
3) $(x + 4)(x^2 + 3x + 3)$.
5) $(x + 1)(x + 4)(x^2 + 3)$.
7) $(x^2 + x + 1)(x^3 + x^2 + 1)$.
9) 2 and 10 are irreducible.
11) $\{1, 3, 5, 7\}$ are units, and all other elements are multiples of 2. This forces 2 to be irreducible, with 2 and 6 associates. The only factorizations of 4 are $2 \cdot 2$, $2 \cdot 6$, $6 \cdot 2$, $6 \cdot 6$, all equivalent to $2 \cdot 2$.
13) If $f(x) = g(x) \cdot h(x)$, then $f(x) \textbf{ mod } p = (g(x) \textbf{ mod } p) \cdot (h(x) \textbf{ mod } p)$. Since p does not divide the leading coefficient, $g(x) \textbf{ mod } p$ and $h(x) \textbf{ mod } p$ will have

the same degree as $g(x)$ and $h(x)$. But we know that $f(x)$ **mod** p is irreducible, so we have a contradiction.
15) If $f(x) = g(x) \cdot h(x)$ in $\mathbb{Q}[x]$, let a be the common denominator of $g(x)$, and b be the common denominator of $h(x)$, so that $ag(x) = G(x)$ and $bh(x) = H(x)$ are in $\mathbb{Z}[x]$. Then $abf(x) = G(x) \cdot H(x)$, and by Gauss' Lemma, each prime factor of ab divides either $G(x)$ or $H(x)$, so we can eliminate prime factors of ab one at a time, to produce a factorization of $f(x)$ in $\mathbb{Z}[x]$.
17) $f(0) = 41$, $f(\pm 1) = 43$, $f(\pm 2) = 61$, $f(\pm 3) = 131$, $f(\pm 4) = 313$. This gives 9 values for which $f(a)$ is prime.
19) In order for $f(x)$ to be a unit, it must be a constant, but since fractional constants are not allowed, the only units are ± 1. Likewise, for 2 to factor, one of the factors would be ± 1, so 2 is irreducible. But x factors as $2 \cdot x/2 = 2 \cdot 2 \cdot x/4 = \cdots$ so 2 is a factor of x an unlimited number of times.
21) 7, 13, 19, and 37 are prime in $\mathbb{Z}[\sqrt[3]{2}]$. Since random elements have a factorization, this ring is apparently a UFD.

Section 12.3

1) $\langle\{2, 1+\sqrt{-5}\}\rangle = \{a + b\sqrt{-5} \mid (a+b)$ **mod** $2 = 0\}$, so this is not all of $\mathbb{Z}[\sqrt{-5}]$. If $\langle\{2, 1+\sqrt{-5}\}\rangle = \langle c\rangle$ for some c, then c can't be a unit, but both 2 and $1+\sqrt{-5}$ must be multiples of c. This is impossible, since both 2 and $1+\sqrt{-5}$ are irreducible.
3) 2, 3, 9, and 10 are prime.
5) 2, 3, 4, 8, 10, 14, 15, 16 are prime.
7) A prime ideal P is $\langle p\rangle$ for some prime p. If there were an ideal B containing P, and $b \in B$ but $b \notin P$, then $\langle\{p, b\}\rangle$ would be in B. But $\gcd(p, b) = 1$, so $\langle\{p, b\}\rangle$ would be all of $\mathbb{Z}$.
9) Let B be an ideal that properly contains I, and let $b \in B$, $b \notin I$. Since R/I is a field, $b + I$ has an inverse $c + I$, and $bc + I = e + I$, so $e - bc \in I \in B$. Since $bc \in B$, this means $e \in B$, hence $B = R$.
11) The quotient ring $\mathbb{Z}[x]/I$ would have only 2 elements, I, and $1 + I$. Thus, the quotient ring is isomorphic to Z_2, which is a field.
13) If I is a prime ideal, by Proposition 12.7, R/I has no zero divisors. But R/I will be finite, so R/I will be a field. Then by Problem 9, I is a maximal ideal.
15) Z_6 has 2 non-trivial ideals, $\langle 2\rangle$ and $\langle 3\rangle$, so both are maximal.
17) Let $x \in P$, and $y \in R$. Since P is not all of I, there is a $z \in I$ with $z \notin P$. Then $x \cdot (y \cdot z) \in P$, since $y \cdot z \in I$. So $(x \cdot y) \cdot z \in P$, forcing $x \cdot y \in P$ since $z \notin P$. Hence, P is an ideal of R.
19)

```
InitDomain(11)
AddFieldVar("a")
Define(a^2, 6)
R = Ring(a)
```

Since $\mathbb{Z}[\sqrt{6}]/\langle 11\rangle$ is a field, $\langle 11\rangle$ is a prime ideal, hence 11 is prime.

Section 12.4

1) By letting $\mu(x)$ be the smallest n for which $x \in S_n$, then $\mu(x) \geq 0$ for all x. If $\mu(x \cdot y) = n$, then $\langle x \cdot y \rangle + S_{n-1} = R$, so $\langle x \rangle + S_{n-1} = R$, hence $\mu(x) \leq n = \mu(x \cdot y)$. If y is a unit, pick $q = x \cdot y^{-1}$ and $r = 0$. Otherwise, let $n = \mu(y)$, so that $x \in \langle y \rangle + S_{n-1}$, that is, there is an $r \in S_{n-1}$ for which $x = y \cdot q + r$. Then $\mu(r) < n = \mu(y)$, so μ is a Euclidean valuation on R.

Now suppose R is a Euclidean domain with a valuation $\mu(x)$, and we want to show that S_n contains all nonzero elements for which $\mu(x) \leq n$. Clearly if $\mu(y) = 0$, then y is a unit, so $y \in S_0$. Suppose that it is true for all smaller values of n. If $\mu(y) = n$, then every x can be written as $y \cdot q + r$, with $\mu(r) < n$, so $r \in S_{n-1}$. Thus $R = \langle y \rangle + S_{n-1}$, so $y \in S_n$. Since S_n contains all nonzero elements for which $\mu(x) \leq n$, then every element of R is in some S_n.

3) $\mu(u) = \mu(u \cdot 1) \geq \mu(1)$. But if u is a unit, $\mu(1) = \mu(u \cdot u^{-1}) \geq \mu(u)$, too.

5) $10 = 2 \cdot 5 = (2 + \sqrt{-6}) \cdot (2 - \sqrt{-6})$. Note that 2 is irreducible, since $N(2) = 4$, but $N(x) \neq 2$ for any x. Yet neither $2 \pm \sqrt{-6}$ are multiples of 2.

7) $(2n)^2 = 4n^2$, and $(2n+1)^2 = 4n^2 + 4n + 1$, so a square is either 0 or 1 (mod 4). So the sum of two squares cannot be 3 (mod 4).

9) Since $a^2 + b^2 = (a + bi) \cdot (a - bi)$, which are both prime by Problem 8, this is the prime factorization of p.

11) Let $\overline{q} = (1 - \sqrt{4n+1})/2$, and $\overline{x + yq} = x + y\overline{q}$. If $x^2 + xy - ny^2 = 0$ for integers x and $y \neq 0$, then $x/y = (-1 \pm \sqrt{1+4n})/2$, which is irrational. Thus $y = 0$, and this forces $x = 0$. $N((x_1 + y_1q)(x_2 + y_2q)) = (x_1 + y_1q)(x_2 + y_2q)\overline{(x_1 + y_1q)(x_2 + y_2q)} = (x_1 + y_1q)(x_1 + y_1\overline{q})(x_2 + y_2q)(x_2 + y_2\overline{q}) = N(x_1 + y_1q) \cdot N(x_2 + y_1\overline{q})$.

13) Let $t = x \cdot y^{-1} = u + vq \in \mathbb{Q}(\sqrt{-3})$, and round u and v to the nearest integers i and j. If $p = i + jq$, then $N(p - t) = a^2 + ab + b^2$, where a and b are both less than $1/2$, so $N(p - t) \leq 3/4$. Hence $\mu(r) = |N(r)| = |N(p \cdot y - x)| = |N(p - t) \cdot N(y)| < |N(y)| = \mu(y)$.

15) $x + yq$ is a unit if and only if $x^2 + xy + 5y^2 = 1$. If $y \neq 0$, then the left side is at least 5, while $y = 0$ gives $x^2 = 1$.

17) By Problem 15, the only units are ± 1, so $S_0 = \{-1, 0, 1\}$. If $m \in S_1$ but $m \notin S_0$, then $2 \in \langle m \rangle + S_0$, so either 1, 2, or 3 is in $\langle m \rangle$. By Problem 16 both 2 and 3 are prime, so m would have to be ± 2 or ± 3. But then $(1 + \sqrt{-19})/2 \notin \langle m \rangle + S_0$, so $S_i = S_0$ for all i.

19) We can pick x by Problem 18, so that $z - x$ is in the rectangle, hence is in one of the 5 circles in the diagram below. If $z - x$ is within one unit of 0, q, or $-\overline{q}$, we can let $y = x$, $x + q$ or $x - \overline{q}$ respectively, and $|z - y| < 1$. If $z - x$ is within a half unit of $q/2$ or $-\overline{q}/2$, then letting $y = 2x + q$ or $2x - \overline{q}$ makes $|2z - y| < 1$. This argument extends to $\mathbb{Z}[(1 + \sqrt{-43})/2]$, $\mathbb{Z}[(1 + \sqrt{-67})/2]$,

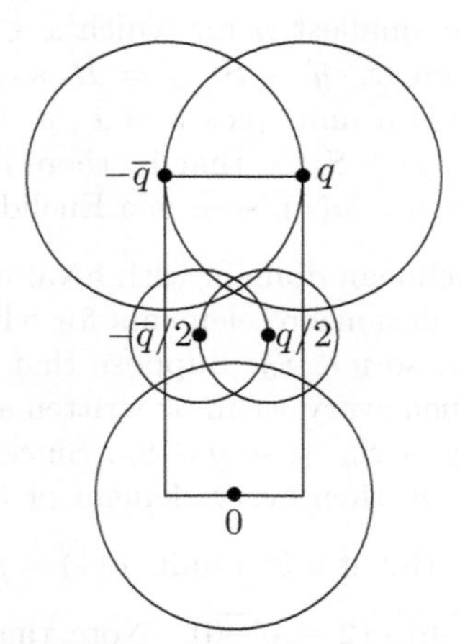

and $\mathbb{Z}[(1+\sqrt{-163})/2]$, except that one must include more circles of radius 1/3, 1/4, etc. to completely cover the rectangle, so that either $|z-y|$, $|2z-y|$, $|3z-y|$, $|4z-y|$, $|5z-y|$, $|6z-y|$, or $|7z-y|$ is less than 1.
21) From Problem 20, we can find y such that $2x = my$. Since 2 is prime in R, either m or y is even. But if y is a multiple of 2, then x is a multiple of m, and we are assuming $x \notin \langle m \rangle$. Hence, m is a multiple of 2, and y is not, so $x = (m/2)\cdot y$. This result can be extended to $\mathbb{Z}[(1+\sqrt{-43})/2]$, $\mathbb{Z}[(1+\sqrt{-67})/2]$, and $\mathbb{Z}[(1+\sqrt{-163})/2]$, except that one must also use that 3 is prime, and prove that either m is a multiple of 2 or a multiple of 3. In the last ring, one even has to include the primes 5 and 7.
23) 3, 7, 11, 19, 23, and 31 are prime in $\mathbb{Z}[i]$. In general, primes for which $p \equiv 3 \pmod 4$.

Section 13.1

1) $4w$. 3) 3. 5) $w+1$. 7) $4w+1$. 9) $3w+2$.

11) Each element can be written as ay^2+by+c, where y is the root of x^3+x+1 in the extension field. To save space, every such element is written as abc.

+	000	001	010	011	100	101	110	111
000	000	001	010	011	100	101	110	111
001	001	000	011	010	101	100	111	110
010	010	011	000	001	110	111	100	101
011	011	010	001	000	111	110	101	100
100	100	101	110	111	000	001	010	011
101	101	100	111	110	001	000	011	010
110	110	111	100	101	010	011	000	001
111	111	110	101	100	011	010	001	000

·	000	001	010	011	100	101	110	111
000	000	000	000	000	000	000	000	000
001	000	001	010	011	100	101	110	111
010	000	010	100	110	011	001	111	101
011	000	011	110	101	111	100	001	010
100	000	100	011	111	110	010	101	001
101	000	101	001	100	010	111	011	110
110	000	110	111	001	101	011	010	100
111	000	111	101	010	001	110	100	011

13) The simplest irreducible polynomial of degree 4 in $Z_2[x]$ is $x^4 + x + 1$, which also happens to be the Conway polynomial. Every element of $Z_2[x]/\langle x^4 + x + 1\rangle$ is of the form $ay^3 + by^2 + cy + d$, where y is the root of $x^4 + x + 1$ in the extension field. To save space, denote each element by the integer obtained by replacing y with 2, for example $y^3 + y + 1$ is denoted by 11.

+	0	1	2	3	4	5	6	7	8	9	10	11	12	13	14	15
0	0	1	2	3	4	5	6	7	8	9	10	11	12	13	14	15
1	1	0	3	2	5	4	7	6	9	8	11	10	13	12	15	14
2	2	3	0	1	6	7	4	5	10	11	9	8	14	15	12	13
3	3	2	1	0	7	6	5	4	11	10	9	8	15	14	13	12
4	4	5	6	7	0	1	2	3	12	13	14	15	8	9	10	11
5	5	4	7	6	1	0	3	2	13	12	15	14	9	8	11	10
6	6	7	4	5	2	3	0	1	14	15	12	13	10	11	8	9
7	7	6	5	4	3	2	1	0	15	14	13	12	11	10	9	8
8	8	9	10	11	12	13	14	15	0	1	2	3	4	5	6	7
9	9	8	11	10	13	12	15	14	1	0	3	2	5	4	7	6
10	10	11	8	9	14	15	12	13	2	3	0	1	6	7	4	5
11	11	10	9	8	15	14	13	12	3	2	1	0	7	6	5	4
12	12	13	14	15	8	9	10	11	4	5	6	7	0	1	2	3
13	13	12	15	14	9	8	11	10	5	4	7	6	1	0	3	2
14	14	15	12	13	10	11	8	9	6	7	4	5	2	3	0	1
15	15	14	13	12	11	10	9	8	7	6	5	4	3	2	1	0

·	0	1	2	3	4	5	6	7	8	9	10	11	12	13	14	15
0	0	0	0	0	0	0	0	0	0	0	0	0	0	0	0	0
1	0	1	2	3	4	5	6	7	8	9	10	11	12	13	14	15
2	0	2	4	6	8	10	12	14	3	1	7	5	11	9	15	13
3	0	3	6	5	12	15	10	9	11	8	13	14	7	4	1	2
4	0	4	8	12	3	7	11	15	6	2	14	10	5	1	13	9
5	0	5	10	15	7	2	13	8	14	11	4	1	9	12	3	6
6	0	6	12	10	11	13	7	1	5	3	9	15	14	8	2	4
7	0	7	14	9	15	8	1	6	13	10	3	4	2	5	12	11
8	0	8	3	11	6	14	5	13	12	4	15	7	10	2	9	1
9	0	9	1	8	2	11	3	10	4	13	5	12	6	15	7	14
10	0	10	7	13	14	4	9	3	15	5	8	2	1	11	6	12
11	0	11	5	14	10	1	15	4	7	12	2	9	13	6	8	3
12	0	12	11	7	5	9	14	2	10	6	1	13	15	3	4	8
13	0	13	9	4	1	12	8	5	2	15	11	6	3	14	10	7
14	0	14	15	1	13	3	2	12	9	7	6	8	4	10	11	5
15	0	15	13	2	9	6	4	11	1	14	12	3	8	7	5	10

15) By Proposition 13.1, $\mathbb{R}[x]/\langle x^2 + 1\rangle$ is a field with a root to the equation $x^2 = -1$. Thus, there is an isomorphism sending $\mathbb{R}[x]/\langle x^2 + 1\rangle$ to $\mathbb{C}$, which sends this root to i.

17) For a given a, there are at most two solutions to $x^2 = a$, and when $a = 0$

there is only one solution. Hence, more than half of the elements are squares. For a fixed k, the sets $\{x^2 \mid x \in F\}$ and $\{k - y^2 \mid y \in F\}$ both contain more than half the elements, so there must be an element in the intersection, yielding a solution to $x^2 + y^2 = k$.
19) $\{0, 1, y^5 + y^4 + y^3 + y, y^5 + y^4 + y^3 + y + 1\}$ is a subfield of order 4, where y is the root of $x^6 + x + 1$ in the field extension. $\{0, 1, y^3 + y^2 + y, y^3 + y^2 + y + 1, y^4 + y^2 + y, y^4 + y^2 + y + 1, y^4 + y^3, y^4 + y^3 + 1\}$ is a subfield of order 8.

Section 13.2

1) The generators are $1 + i$, $1 + 2i$, $2 + i$, $2 + 2i$.
3) {5, 7, 10, 11, 14, 15, 17, 19, 20, 21}.
5) {2, 5, 13, 15, 17, 18, 19, 20, 22, 24, 32, 35}.
7) The Frobenius automorphism $f : x \to x^p$ must send a generator to a generator.
9) We can let x be the solution given from Problem 8. Then $(x + i)(x - i) = x^2 + 1$ would be a multiple of p, and clearly neither $x + i$ nor $x - i$ is a multiple of p. Therefore, p is not prime in $Z[i]$.
11) All subfields contain the multiplicative identity, and this element generates a subfield of order p. So this subfield is in all of the subfields of F, and since it is one of the subfields, there are no other elements in the intersection.
13) The subfields would have to have order p^n for some prime p. Consider the polynomial $x^{(p^n)} - x$. There are at most p^n roots, but all elements from both subfields would be roots.
15) If n is a multiple of d, then by Corollary 13.5, $p^n - 1$ is a multiple of $p^d - 1$, and so $x^{(p^n-1)} - 1$ is divisible by $x^{(p^d-1)} - 1$, and so $x^{(p^n)} - x$ is divisible by $x^{(p^d)} - x$ in $\mathbb{Z}[x]$. Since $x^{(p^n)} - x$ factors completely in F with no double roots, so does $x^{(p^d)} - x$, and these p^d elements will form a subfield since these elements are fixed by the automorphism $x \to x^{p^d}$. Problem 13 gives uniqueness.
17) A field of order p^n can be described by $Z_p[x]/\langle f(x)\rangle$, where $f(x)$ is an irreducible polynomial in $Z_p[x]$ of degree n. An automorphism would be determined by where it sends one of the roots of $f(x)$, and there are n possible roots. Thus, there are at most n automorphisms, and we found n Frobenius automorphisms.
19) $2 \in Z_p^*$, and $2^n = -1$ in Z_p, and $2^{2n} = 1$. So the order of the element 2 is $2n$. Since Z_p^* has order $p - 1 = 2^n$, by Lagrange's theorem $2n$ divides 2^n, so n is a power of 2.
21) $x^{32} - x = x(x + 1)(x^5 + x^2 + 1)(x^5 + x^3 + 1)(x^5 + x^3 + x^2 + x + 1)(x^5 + x^4 + x^2 + x + 1)(x^5 + x^4 + x^3 + x + 1)(x^5 + x^4 + x^3 + x^2 + 1)$, giving the 6 irreducible polynomials of degree 5.

Section 13.3

1) $x^2 - x + 1$.
3) $x^4 - x^3 + x^2 - x + 1$.
5) ω_n is a root of $x^n - 1 = (x - 1) \cdot (1 + x + x^2 + x^3 + \cdots x^{n-1})$, so ω_n is a root of the latter factor, which produces the sum of the nth roots of unity.

7) GF(1024).
9) $\Phi_1(x) = x - 1$ and $\Phi_2(x) = x + 1$, so assume that it is true for previous n. Plugging in $x = 0$ into Proposition 13.7 gives $0^n - 1 = -1 \cdot 1 \cdots \Phi_n(0)$, so $\Phi_n(0) = 1$.
11) If ω_n^k is a root of $\phi_n(-x)$, then $-\omega_n^k$ has order n. Since n is odd, this means that ω_n^k has order $2n$, so it is a root of $\Phi_{2n}(x)$. Since every root of $\Phi_n(-x)$ is a root of $\Phi_{2n}(x)$, we only need to show that these polynomials have the same degree, which is easy since $\phi(2n) = \phi(n)$.
13) Since $\Phi_3(x) = x^2 + x + 1$, $\Phi_{54}(x) = x^{18} - x^9 + 1$.
15) Let $f(x)$ be an irreducible polynomial of degree n over Z_p, and let r be a root of $f(x)$ in $GF(p^n)$. If $r^m = 1$ for some $m < p^n - 1$, then $f(x)$ cannot be a factor of $\Phi_{(p^n-1)}(x)$, lest r be a double root of $x^{(p^n-1)} - 1$, and then it would contradict Lemma 13.5. However, if $r^m \neq 1$ for any $m < p^n - 1$, then $f(x)$ is a factor of $x^{(p^n-1)} - 1$, yet not a factor of any $x^m - 1$ for $m < p^n - 1$, so $f(x)$ must be a factor of $\Phi_{(p^n-1)}(x)$.
17) $\Phi_{15}(x) = (x^4 + x + 1)(x^4 + x^3 + 1)$. But $\Phi_5(x) = x^4 + x^3 + x^2 + x + 1$ is also irreducible.
19) $\Phi_8(x) = (x^2 + x + 2)(x^2 + 2x + 2)$. But $x^2 + 1$ is also irreducible.

Section 13.4

1) If there were a p that did not have such values of a, b, c, d, then in the ring of integer quaternions modulo p, every non-zero element $a+bi+cj+dk$ would have an inverse, $(a - bi - cj - dk)/(a^2 + b^2 + c^2 + d^2)$. But this would make the ring a finite skew field, which is impossible.
3) Pick $x_1 = a_1 + a_2 i + a_3 j + a_4 k$ and $x_2 = b_1 + b_2 i + b_3 j + b_4 k$. The identity follows from expanding out $|x_1|^2|x_2|^2 = |x_1 x_2|^2$.
5) If $x = a_1 + a_2 i + a_3 j + a_4 k$, pick b_1 so that $b_1 \equiv a_1 (\text{mod } m)$ and $-m/2 < b_1 < m/2$. Likewise pick b_2, b_3, and b_4. Then $y = b_1 + b_2 i + b_3 j + b_4 k$ will differ from x by a multiple of m, and $|y|^2 < 4(m/2)^2 = m^2$.
7) Since $1 = 0^2 + 0^2 + 0^2 + 1^2$, the theorem is true for $n = 1$. Problem 3 shows the set of integers for which the theorem is true is closed under multiplication, and Problem 6 proves the theorem for prime numbers. Since all integers > 1 can be factored into a product of primes, the theorem is true for all positive integers.
9) If x is an integer quaternion, then clearly $N(x)$ will be a non-negative integer. Also, if $x = [(2a + 1) + (2b + 1)i + (2c + 1)j + (2d + 1)k]/2$, then $|x|^2 = a^2 + a + b^2 + b + c^2 + c + d^2 + d + 1$.
11) Clearly $N(1) = 1$. If u is a unit, then there is an inverse u^{-1}, so $N(u \cdot u^{-1}) = 1$, so $N(u)$ must be 1. If $N(u) = 1$, then there are q and r such that $1 = uq + r$, with $N(r) < N(u) = 1$, which forces $r = 0$, so u is a unit. There are 24 units: ± 1, $\pm i$, $\pm j$, $\pm k$, and $(\pm 1 \pm i \pm j \pm k)/2$, where in the last case each $\pm$ is independent.
13) Since $p = a^2 + b^2 + c^2 + d^2$, $p = (a + bi + cj + dk) \cdot (a - bi - cj - dk)$. Note that both factors have $N(x) = p$, so by Problem 12 both factors are irreducible.

15) Given a right ideal I, pick an element with the smallest positive norm, y. For any other element x, $x = y \cdot q + r$ for some q and r, with $N(r) < N(y)$. But $r = x - y \cdot q$ would be in the right ideal, so $r = 0$. Then $x \in y \cdot R$, so $I \subseteq y \cdot R \subseteq I$. Thus, $I = y \cdot R$.
17) For w to be in the center, $c_2 = c_3 = c_4 = c_5 = c_6 = c_7 = c_8 = c_9 = 0$, so we get a field isomorphic to $\mathbb{Q}$.
19) The center is $\{1, -1\}$, and the group $U/\{1, -1\} \approx A_4$.

Section 14.1

1) $\{1, \sqrt{2}\}$.
3) $\{1, \sqrt{2}, \sqrt{3}, \sqrt{6}\}$.
5) $\{1, \sqrt[3]{2}, \sqrt[3]{4}\}$.
7) Since ω_8 satisfies a polynomial equation of degree 4, $\mathbb{Q}(\omega_8)$ is a 4-dimensional extension of $\mathbb{Q}$. Hence, $\{1, \omega_8, \omega_8^2 = i, \omega_8^3\}$ is a basis.
9) $\{1, \sqrt[3]{2}, \sqrt[3]{4}, \omega_3, \omega_3\sqrt[3]{2}, \omega_3\sqrt[3]{4}\}$.
11) If the set is linearly independent, it is a basis. Otherwise, one vector will be a linear combination of the others, so we can delete that vector. Repeat the process until we have a linearly independent set.
13) If $M = R$, then M is an abelian group under addition, and $r \cdot m \in M$, since $m \in R$. The distributive and associate laws come from the ring R.
15) If N is an R-submodule, then N is a subgroup of R under addition, and $rn \in N$ whenever $r \in R$ and $n \in N$. Since $N \subseteq R$, we see that N is closed under multiplication, so N is a subring, and hence a left ideal of R, Clearly a left ideal of R would be an R-submodule.
17) Since $(a_1 + b_1) - (a_2 + b_2) = (a_1 - a_2) + (b_1 - b_2) \in A + B$, we see $A + B$ is a subgroup under addition. Also, $r(a + b) = r \cdot a + r \cdot b \in A + B$ whenever $r \in R$, so $A + B$ is an R-submodule.
19) Given two countable sets $A = \{a_1, a_2, a_3, \ldots\}$ and $B = \{b_1, b_2, b_3, \ldots\}$, we can show $A \times B$ is countable by forming a list containing every (a_i, b_j) in increasing order of $i + j$:

$$\{(a_1, b_1), (a_1, b_2), (a_2, b_1), (a_1, b_3), (a_2, b_2), (a_3, b_1), (a_1, b_4), \cdots\}.$$

To show that an n-dimensional vector space V of $\mathbb{Q}$ is countable, assume by induction that an $(n-1)$-dimensional vector space U is. There is the obvious one-to-one correspondence between V and $U \times \mathbb{Q}$.
21) $\langle 2, 1, 2\rangle$.

Section 14.2

1) $x^2 - 5$.
3) $x^4 - 10x^2 + 1$.
5) $x^4 + 2x^2 - 1$.
7) $x^8 - 4x^6 + 6x^4 - 4x^2 - 2$.
9) $\sqrt{2} + \sqrt{3}, -\sqrt{2} - \sqrt{3}, -\sqrt{2} + \sqrt{3}, \sqrt{2} - \sqrt{3}$.
11) $\sqrt{\sqrt{2} - 1}, -\sqrt{\sqrt{2} - 1}, \sqrt{-\sqrt{2} - 1}, -\sqrt{-\sqrt{2} - 1}$.
13) Eight roots: $\pm\sqrt{1 + \sqrt[4]{3}}, \pm\sqrt{1 - \sqrt[4]{3}}, \pm\sqrt{1 + i\sqrt[4]{3}}, \pm\sqrt{1 - i\sqrt[4]{3}}$.

15) $\mathbb{Q}(a)$ is a subfield of a 5-dimensional extension of $\mathbb{Q}(\sqrt{2}, \sqrt{3}, \sqrt{5}, \sqrt{7})$, which is at most a 16-dimensional extension of $\mathbb{Q}$.
17) $\mathbb{Q}(\sqrt{2})$ contains a root to the polynomial $x^2 - 2$, whereas $\mathbb{Q}(\sqrt{3})$ does not.
19) $1/(5 + \sqrt{-3}) = 5/28 - \sqrt{-3}/28$.

Section 14.3

1) $\sqrt[10]{2}$, or $\sqrt{2} + \sqrt[5]{2}$.
3) $\sqrt{2} + \sqrt{3} + \sqrt{5}$.
5) Since $\mathbb{Q}(\sqrt[3]{2}, \omega_3) = \mathbb{Q}(\sqrt[3]{2}, \omega_3\sqrt[3]{2})$, we have already seen that this can be expressed as $\mathbb{Q}(\sqrt[3]{2} + 2\omega_3\sqrt[3]{2})$.
7) $(a+2b)^3 = a^3 + 6a^2b + 12ab^2 + 8b^3 = 18 + 6a^2b + 12a(-ab - a^2) = -6 - 6a^2b$. $(a+2b)^6 = (-6 - 6a^2b)^2 = 36(a^4b^2 + 2a^2b + 1) = 36(2a(-ab - a^2) + 2a^2b + 1) = 36(-4 + 1) = -108$.
9) $x^3 + x - 1 = (x - a)(x^2 + ax + (a^2 + 1))$. Using quadratic equation on second factor produces $\sqrt{-3a^2 - 4}$, so other 2 roots are complex. Thus, $\mathbb{Q}(a)$ is not the splitting field, hence a 2-dimensional extension of the 3-dimensional extension is needed.
11) Both quadratics factor in $\mathbb{Q}(\sqrt{-3})$.
13) This happens to be $\Phi_{12}(x)$, so the splitting field is $\mathbb{Q}(\omega_{12}) = \mathbb{Q}((\sqrt{3}+i)/2)$.
15) The splitting field of $f(x)$ is a simple extension $\mathbb{Q}(w)$, and the splitting field of $g(x)$ is a simple extension $\mathbb{Q}(y)$. Then $f(x) \cdot g(x)$ splits in $\mathbb{Q}(w, y)$, which has dimension at most $n \cdot m$.
17) Splitting field $= \mathbb{Q}(a)$, where $a^3 = -a^2 + 4a - 1$; 3-dimensional extension.
19) Splitting field $= \mathbb{Q}(a, b)$, where $a^5 = 2$ and $b^4 = -ab^3 - a^2b^2 - a^3b - a^4$; 20-dimensional extension.

Section 15.1

1)

$\cdot$	ϕ_0	ϕ_1	ϕ_2	ϕ_3
ϕ_0	ϕ_0	ϕ_1	ϕ_2	ϕ_3
ϕ_1	ϕ_1	ϕ_0	ϕ_3	ϕ_2
ϕ_2	ϕ_2	ϕ_3	ϕ_0	ϕ_1
ϕ_3	ϕ_3	ϕ_2	ϕ_1	ϕ_0

3) Besides the identity automorphism, there is the automorphism sending $\sqrt[6]{2} \mapsto -\sqrt[6]{2}$.
5) $\phi_0(x) = x$ for all x; ϕ_1 fixes $\pm\sqrt[4]{2}$, $i\sqrt[4]{2} \leftrightarrow -i\sqrt[4]{2}$; ϕ_2 fixes $\pm i\sqrt[4]{2}$, $\sqrt[4]{2} \leftrightarrow -\sqrt[4]{2}$; ϕ_3: $\sqrt[4]{2} \leftrightarrow -\sqrt[4]{2}$, $i\sqrt[4]{2} \leftrightarrow -i\sqrt[4]{2}$; ϕ_4: $\sqrt[4]{2} \to i\sqrt[4]{2} \to -\sqrt[4]{2} \to -i\sqrt[4]{2} \to \sqrt[4]{2}$; ϕ_5: $\sqrt[4]{2} \to -i\sqrt[4]{2} \to -\sqrt[4]{2} \to i\sqrt[4]{2} \to \sqrt[4]{2}$; ϕ_6: $\sqrt[4]{2} \leftrightarrow i\sqrt[4]{2}$, $-\sqrt[4]{2} \leftrightarrow -i\sqrt[4]{2}$; ϕ_7: $\sqrt[4]{2} \leftrightarrow -i\sqrt[4]{2}$, $-\sqrt[4]{2} \leftrightarrow i\sqrt[4]{2}$.
7) $\phi_0(x) = x$ for all x; ϕ_1 fixes $\sqrt[3]{3}$, $r_2 \leftrightarrow r_3$; ϕ_2 fixes r_2, $\sqrt[3]{3} \leftrightarrow r_3$; ϕ_3 fixes r_3, $\sqrt[3]{3} \leftrightarrow r_2$; ϕ_4: $\sqrt[3]{3} \to r_2 \to r_3 \to \sqrt[3]{3}$; ϕ_5: $\sqrt[3]{3} \to r_3 \to r_2 \to \sqrt[3]{3}$.
9) $\phi(1) = 1$, since 1 is the multiplicative identity. $\phi(2) = \phi(1+1) = \phi(1) + \phi(1) = 2$, and likewise $\phi(n) = n$ for all integers n. Then $\phi(p/q) = \phi(p)/\phi(q) = p/q$, so ϕ fixes the rationals, hence $\phi \in \mathrm{Gal}_{\mathbb{Q}}(E)$.
11) The only non-trivial automorphism is the Frobenius automorphism $f(x) = x^3$.

13) Since $x^3 + x + 1$ is irreducible, $\mathbb{Q}(a)$ has dimension 3, where a is the real root. But the complex roots are not in $\mathbb{Q}(a)$, since they are complex, so there must be another extension of dimension 2 to produce the splitting field, forcing the Galois group to have order 6, hence isomorphic to S_3.
15) Z_1, Z_2, Z_3, or S_3. (Possible subgroups of S_3.)
17) 3 automorphisms, sending a to a, $2 - a^2$, or $a^2 - a - 2$, where a is a root.
19) 4 automorphisms, sending a to a, $-a$, $3a - a^3$, or $a^3 - 3a$, where a is a root. Group is isomorphic to Z_4.
21) 8 automorphisms, sending $\{a, b\}$ to $\{a, b\}$, $\{-a, b\}$, $\{a, -b\}$, $\{-a, -b\}$, $\{b, a\}$, $\{-b, a\}$, $\{b, -a\}$, or $\{-b, -a\}$ where a is a root, and b is a root of $x^2 + a^2 + 1$. Group is isomorphic to D_4.

Section 15.2

1) Since $Z_7^* \approx Z_6$, we can consider $\Phi_7(x) = x^6 + x^5 + x^4 + x^3 + x^2 + x + 1$.
3) Since D_5 has an element of order 5, there must be an element in the Galois group with order 5, that is, a 5-cycle of the roots $\phi_1 : r_1 \to r_2 \to r_3 \to r_4 \to r_5 \to r_1$. But D_5 also contains an element of order 2, which must fix one of the roots, say r_5, which forces $\phi_2 : r_1 \leftrightarrow r_4$, $r_2 \leftrightarrow r_3$. (Any other element of order 2 generates with ϕ_1 more than 10 elements.) Letting $k = r_1r_2 + r_2r_3 + r_3r_4 + r_4r_5 + r_5r_1$, we see that both $\phi_1(k) = k$, and $\phi_2(k) = k$, and since ϕ_1 and ϕ_2 generate the Galois group, k is in the fixed field, so $k \in \mathbb{Q}$.
5) If the Galois group is D_4, the roots of the polynomial can be rearranged such that $r_1r_2 + r_2r_3 + r_3r_4 + r_4r_1$ is rational.
7) One solution: $r_1 = 1.842085966$, $r_2 = 0.351854083 - 1.709561043i$, $r_3 = -1.272897224 + 0.7197986815i$, $r_4 = -1.272897224 - 0.7197986815i$, $r_5 = 0.351854083 + 1.709561043i$, $r_1r_2 + r_2r_3 + r_3r_4 + r_4r_5 + r_5r_1 = 5$.
9) Z_2.
11) Trivial group (polynomial factors).
13) $Z_2 \times Z_2$.
15) If some polynomial $f(x)$ in $\mathbb{Q}[x]$ has Galois group G, then the splitting field of $f(x)$ can be written as $\mathbb{Q}(w)$ for some w (Corollary 14.4). Then $g(x) = \mathrm{Irr}_{\mathbb{Q}}(w, x)$ will have the degree n, and will have the same splitting field. Thus, the Galois group of $g(x)$ will also be G.
17) $\mathrm{Gal}_{\mathbb{Q}}(K) \approx Z_5 \rtimes Z_4$, with 20 elements.
19) $\mathrm{Gal}_{\mathbb{Q}}(K) \approx Z_5$, with 5 elements.
21) $\mathrm{Gal}_{\mathbb{Q}}(K) \approx Z_2 \times Z_2 \times Z_2$, with 8 elements.

Section 15.3

1) $\mathrm{fix}(\{\phi_0\}) = \mathbb{Q}(\sqrt{2}, \sqrt{3})$, $\mathrm{fix}(\{\phi_0, \phi_1\}) = \mathbb{Q}(\sqrt{2})$, $\mathrm{fix}(\{\phi_0, \phi_2\}) = \mathbb{Q}(\sqrt{3})$, $\mathrm{fix}(\{\phi_0, \phi_3\}) = \mathbb{Q}(\sqrt{6})$, $\mathrm{fix}(\{\phi_0, \phi_1, \phi_2, \phi_3\}) = \mathbb{Q}$.
3) Using Problem 5 of §15.1 notation, $\{\phi_0\} \leftrightarrow \mathbb{Q}(\sqrt[4]{2}, i)$, $\{\phi_0, \phi_1\} \leftrightarrow \mathbb{Q}(\sqrt[4]{2})$, $\{\phi_0, \phi_2\} \leftrightarrow \mathbb{Q}(i\sqrt[4]{2})$, $\{\phi_0, \phi_3\} \leftrightarrow \mathbb{Q}(\sqrt{2}, i)$, $\{\phi_0, \phi_1, \phi_2, \phi_3\} \leftrightarrow \mathbb{Q}(\sqrt{2})$, $\{\phi_0, \phi_4, \phi_3, \phi_5\} \leftrightarrow \mathbb{Q}(i)$, $\{\phi_0, \phi_6\} \leftrightarrow \mathbb{Q}((1+i)\sqrt[4]{2})$, $\{\phi_0, \phi_7\} \leftrightarrow \mathbb{Q}((1-i)\sqrt[4]{2})$, $\{\phi_0, \phi_3, \phi_6, \phi_7\} \leftrightarrow \mathbb{Q}(1\sqrt{2})$, $\{\phi_0, \phi_1, \phi_2, \phi_3, \phi_4, \phi_5, \phi_6, \phi_7\} \leftrightarrow \mathbb{Q}$.
5) Since $Z_6^* \approx Z_2$, there is no non-trivial subgroups of $\mathrm{Gal}_{\mathbb{Q}}(F)$.

7) Since $Z_7^* \approx Z_6$, there is an element of order 2, namely 6, and two elements of order 3, namely 2 and 4. So the automorphism sending ω_7 to ω_7^6 has order 2, and the automorphisms sending ω_7 to ω_7^2 or ω_7^4 have order 3.
9) If E is a Galois extension of F, then by Corollary 15.2 $E = F(w)$ for some w, hence by Proposition 15.4 $|\mathrm{Gal}_F(E)| = n$. If E is not a Galois extension, then the fixed field of $\mathrm{Gal}_F(E)$ is K, strictly larger than F. Then E is a Galois extension of K, so $|\mathrm{Gal}_K(E)|$ is the dimension of E over K, which is $< n$.
11) $\mathrm{Gal}_F(E)$ is a finite group, so it can only have a finite number of subgroups. Since the fundamental theorem of Galois theory shows a one-to-one correspondence between the subgroups of $\mathrm{Gal}_F(E)$ and the subfields of E containing F, there are only a finite number of such subfields.
13) Since D_5 has 8 subgroups. there are 8 subfields of E containing F.
15) $F = \mathbb{Q}$, $K = \mathbb{Q}(\sqrt{2})$, $E = \mathbb{Q}(\sqrt[4]{2})$.
17) For $\phi(\omega_{12}) = \omega_{12}^5$, $\mathrm{fix}(\phi) = \mathbb{Q}(\omega_{12}^3) = \mathbb{Q}(\sqrt{-1})$. For $\phi(\omega_{12}) = \omega_{12}^7$, $\mathrm{fix}(\phi) = \mathbb{Q}(\omega_{12}^2) = \mathbb{Q}(\sqrt{-3})$. For $\phi(\omega_{12}) = \omega_{12}^{11}$, $\mathrm{fix}(\phi) = \mathbb{Q}(\omega_{12} + \omega_{12}^{11}) = \mathbb{Q}(\sqrt{3})$.
19) For $\phi(\omega_{15}) = \omega_{15}^2$ or ω_{15}^8, $\mathrm{fix}(\phi) = \mathbb{Q}(\omega_{15} + \omega_{15}^2 + \omega_{15}^4 + \omega_{15}^8) = \mathbb{Q}(\sqrt{-15})$. For $\phi(\omega_{15}) = \omega_{15}^7$ or ω_{15}^{13}, $\mathrm{fix}(\phi) = \mathbb{Q}(\omega_{15} + \omega_{15}^4 + \omega_{15}^7 + \omega_{15}^{13}) = \mathbb{Q}(\sqrt{-3})$.

Section 15.4

1) $\mathbb{Q}(\omega_3, \sqrt[3]{2})$.
3) $\mathbb{Q}(\sqrt{-1}, \sqrt[4]{-5})$.
5) $\mathbb{Q}(\sqrt{2}, \sqrt{3})$.
7) $\mathbb{Q}(\omega_3, \sqrt[3]{2}, \sqrt[3]{3})$.
9) Draw a circle of radius b at the end of a segment of length a. The points where the circle intersects the line extending from the segment mark off $a - b$ and $a + b$.
11) Since $\triangle ABC$ is similar to $\triangle ADE$, $\overline{BC}/\overline{AB} = \overline{DE}/\overline{AD}$ Thus, $\overline{BC} = a/b$.
13) $\triangle FJG$ is similar to $\triangle FGH$, which is in turn similar to $\triangle GJH$. Thus, $\overline{GJ}/\overline{FJ} = \overline{HJ}/\overline{GJ}$. Hence, $\overline{GJ}^2 = a$, so $\overline{GJ} = \sqrt{a}$.
15) Since $\mathrm{Gal}_{\mathbb{Q}}(F)$ is a group of order 2^n, by Proposition 7.8 we can form a composition series with each subgroup half the size of the previous. By finding the fixed fields of the subgroups, we can find fields $\mathbb{Q} = F_0 \subseteq F_1 \subseteq F_2 \subseteq \cdots \subseteq F_n = F$, each being a second-degree extension of the previous. By the quadratic equation, any second-degree extension can be expressed in terms of a (possibly complex) square root. So every element of F will be constructible.
17) The 17-gon can be constructed if ω_{17} is a constructible complex number. But $\mathbb{Q}(\omega_{17})$ is a Galois extension of $\mathbb{Q}$ of dimension $16 = 2^4$, so by Problem 15, ω_{17} is constructible.
19) $\mathbb{Q}(\omega_5 + \omega_5^4, \omega_5) = \mathbb{Q}(a = (-1 + \sqrt{5})/2, (a + \sqrt{a^2 - 4})/2)$.

Bibliography

The following list not only gives the books and articles mentioned in the text, but also additional references that may help students explore related topics.

Undergraduate textbooks on abstract algebra

1. J. B. Fraleigh, *A First Course in Abstract Algebra,* 8th ed., Addison Wesley, Boston (2009).

2. J. A. Gallian, *Contemporary Abstract Algebra,* 8th ed., Houghton Mifflin, Boston (2013).

3. J. Gilbert and L. Gilbert, *Elements of Modern Algebra,* 8th ed., PWS Publishing Co., Boston (2014).

4. L. J. Goldstein, *Abstract Algebra, A First Course,* Prentice-Hall, Englewood Cliffs, New Jersey (1973).

5. I. N. Herstein, *Abstract Algebra,* Macmillan Publishing Company, New York (1986).

6. T. W. Hungerford, *Abstract Algebra, An Introduction,* Saunders College Publishing, Philadelphia (1990).

7. J. J. Rotman, *A First Course in Abstract Algebra,* Prentice-Hall, Upper Saddle River, New Jersey (1996).

Graduate textbooks on abstract algebra

8. I. N. Herstein, *Topics in Algebra,* 2nd ed., Wiley, New York (1975).

9. J. F. Humphrey, *A Course in Group Theory,* Oxford University Press, Oxford (1996).

10. D. S. Malik, J. N. Mordeson, and M. K. Sen, *Fundamentals of Abstract Algebra,* McGraw-Hill, New York (1997).

Sources for historical information

11. D. M. Burton, *The History of Mathematics, An Introduction,* 6th ed., McGraw-Hill, Boston (2007).

12. J. H. Eves, *An Introduction to the History of Mathematics,* 6th ed., Saunders College Publishing, Fort Worth (1990).

Other sources

13. J. H. Conway, R. T. Curtis, S. P. Norton, R. A. Parker, and R. A. Wilson, *Atlas of Finite Groups,* Clarendon Press, Oxford (1985).

14. The GAP Group, *GAP Reference Manual,* Release 4.4.12, http://www.gap-system.org.

15. I. S. Reed and G. Solomon, "Polynomial Codes over Certain Finite Fields," *SIAM Journal of Applied Math.*, vol. 8, 1960, pp. 300–304.

16. "Reed-Solomon error correction," Wikipedia, the free encyclopedia, http://en.wikipedia.org.

Index

Page numbers are underlined in the index when they represent the definition or the main source of information about the subject indexed. Boldface page numbers refer to sections for which the entire section pertains to the topic. References to problems are in italics. Occasionally, both underlining and italics are appropriate, should a homework problem introduce a new concept. Note that in these cases, only the homework problems are indexed, and not the answers in the back, even though the answers often shed more light on the topic.